Pharmaceutical Biotechnology

Concepts and Applications

Gary Walsh

University of Limerick, Republic of Ireland

John Wiley & Sons, Ltd

D1354578

Other Wiley Editorial Offices

John Wiley & Sons Inc., 111 River Street, Hoboken, NJ 07030, USA

Jossey-Bass, 989 Market Street, San Francisco, CA 94103-1741, USA

Wiley-VCH Verlag GmbH, Boschstr. 12, D-69469 Weinheim, Germany

John Wiley & Sons Australia Ltd, 42 McDougall Street, Milton, Queensland 4064, Australia

John Wiley & Sons (Asia) Pte Ltd, 2 Clementi Loop #02-01, Jin Xing Distripark, Singapore 129809

John Wiley & Sons Canada Ltd, 6045 Freemont Blvd, Mississauga, Ontario, L5R 4J3

Wiley also publishes its books in a variety of electronic formats. Some content that appears in print may not be available in
electronic books.

Anniversary Logo Design: Richard J. Pacifico

Library of Congress Cataloging-in-Publication Data

Walsh, Gary, Dr.
    Pharmaceutical biotechnology : concepts and applications / Gary Walsh.
        p. ; cm.
    Includes bibliographical references.
    ISBN 978-0-470-01244-4 (cloth)
 1. Pharmaceutical biotechnology.   I. Title.
    [DNLM: 1. Technology, Pharmaceutical.   2. Biotechnology.   3.
Pharmaceutical Preparations. QV 778 W224p 2007]
    RS380.W35 2007
    615′.19–dc22   2007017884

British Library Cataloguing in Publication Data

A catalogue record for this book is available from the British Library

ISBN 978-0-470-01244-4 (HB)
ISBN 978-0-470-01245-1 (PB)

Typeset in 10.5/12.5 pt Times by Thomson Digital
Printed and bound in Great Britain by Antony Rowe Ltd., Chippenham, Wilts

Pharmaceutical Biotechnology

I dedicate this book to my beautiful daughter Alice.
To borrow a phrase:

'without her help, it would have been written in half the time'!

Contents

Preface

This book has been written as a sister publication to *Biopharmaceuticals: Biochemistry and Biotechnology*, a second edition of which was published by John Wiley and Sons in 2003. The latter textbook caters mainly for advanced undergraduate/postgraduate students undertaking degree programmes in biochemistry, biotechnology and related disciplines. Such students have invariably pursued courses/modules in basic protein science and molecular biology in the earlier parts of their degree programmes; hence, the basic principles of protein structure and molecular biology were not considered as part of that publication. This current publication is specifically tailored to meet the needs of a broader audience, particularly to include students undertaking programmes in pharmacy/pharmaceutical science, medicine and other branches of biomedical/clinical sciences. Although evolving from *Biopharmaceuticals: Biochemistry and Biotechnology*, its focus is somewhat different, reflecting its broader intended readership. This text, therefore, includes chapters detailing the basic principles of protein structure and molecular biology. It also increases/extends the focus upon topics such as formulation and delivery of biopharmaceuticals, and it contains numerous case studies in which both biotech and clinical aspects of a particular approved product of pharmaceutical biotechnology are overviewed. The book, of course, should also meet the needs of students undertaking programmes in core biochemistry, biotechnology or related scientific areas and be of use as a broad reference source to those already working within the pharmaceutical biotechnology sector.

As always, I owe a debt of gratitude to the various people who assisted in the completion of this textbook. Thanks to Sandy for her help in preparing various figures, usually at ridiculously short notice. To Gerard Wall, for all the laughs and for several useful discussions relating to molecular biology. Thank you to Nancy, my beautiful wife, for accepting my urge to write (rather than to change baby's nappies) with good humour – most of the time anyway! I am also grateful to the staff of John Wiley and Sons for their continued professionalism and patience with me when I keep overrunning submission deadlines. Finally, I have a general word of appreciation to all my colleagues at the University of Limerick for making this such an enjoyable place to work.

Gary Walsh
November 2006

Acronyms

ADCC	antibody-dependent cell cytoxicity
BAC	bacterial artificial chromosome
BHK	baby hamster kidney
cDNA	complementary DNA
CHO	Chinese hamster ovary
CNTF	ciliary neurotrophic factor
CSF	colony-stimulating factor
dsRNA	double-stranded RNA
EDTA	ethylenediaminetetraacetic acid
ELISA	enzyme-linked immunosorbent assay
EPO	erythropoietin
FGF	fibroblast growth factor
FSH	follicle-stimulating hormone
GDNF	glial cell-derived neurotrophic factor
GH	growth hormone
hCG	human chorionic gonadotrophin
HIV	human immunodeficiency virus
HPLC	high-performance liquid chromatography
IGF	insulin-like growth factor
ISRE	interferon-stimulated response element
JAK	Janus kinase
LAF	lymphocyte activating factor
LIF	leukaemia inhibitory factor
LPS	lipopolysaccharide
MHC	major histocompatibility complex
MPS	mucopolysaccharidosis
mRNA	messenger RNA
PDGF	platelet-derived growth factor
PEG	polyethylene glycol

PTK	protein tyrosine kinase
PTM	post-translational modification
rDNA	recombinant DNA
RNAi	RNA interference
rRNA	ribosomal RNA
SDS	sodium dodecyl sulfate
ssRNA	single-stranded RNA
STATs	signal transducers and activators of transcription
TNF	tumour necrosis factor
tPA	tissue plasminogen activator
tRNA	transfer RNA
WAP	whey acid protein
WFI	water for injections

1
Pharmaceuticals, biologics and biopharmaceuticals

1.1 Introduction to pharmaceutical products

Pharmaceutical substances form the backbone of modern medicinal therapy. Most traditional pharmaceuticals are low molecular weight organic chemicals (Table 1.1). Although some (e.g. aspirin) were originally isolated from biological sources, most are now manufactured by direct chemical synthesis. Two types of manufacturing company thus comprise the 'traditional' pharmaceutical sector: the chemical synthesis plants, which manufacture the raw chemical ingredients in bulk quantities, and the finished product pharmaceutical facilities, which purchase these raw bulk ingredients, formulate them into final pharmaceutical products, and supply these products to the end user.

In addition to chemical-based drugs, a range of pharmaceutical substances (e.g. hormones and blood products) are produced by/extracted from biological sources. Such products, some major examples of which are listed in Table 1.2, may thus be described as products of biotechnology. In some instances, categorizing pharmaceuticals as products of biotechnology or chemical synthesis becomes somewhat artificial. For example, certain semi-synthetic antibiotics are produced by chemical modification of natural antibiotics produced by fermentation technology.

1.2 Biopharmaceuticals and pharmaceutical biotechnology

Terms such as 'biologic', 'biopharmaceutical' and 'products of pharmaceutical biotechnology' or 'biotechnology medicines' have now become an accepted part of the pharmaceutical literature. However, these terms are sometimes used interchangeably and can mean different things to different people.

Although it might be assumed that 'biologic' refers to any pharmaceutical product produced by biotechnological endeavour, its definition is more limited. In pharmaceutical circles, 'biologic' generally refers to medicinal products derived from blood, as well as vaccines, toxins and allergen products. 'Biotechnology' has a much broader and long-established meaning. Essentially, it refers

Pharmaceutical biotechnology: concepts and applications   Gary Walsh
© 2007 John Wiley & Sons, Ltd   ISBN 978 0 470 01244 4 (HB)   978 0 470 01245 1 (PB)

Table 1.1   Some traditional pharmaceutical substances that are generally produced by direct chemical synthesis

Drug	Molecular formula	Molecular mass	Therapeutic indication
Acetaminophen (paracetamol)	$C_8H_9NO_2$	151.16	Analgesic
Ketamine	$C_{13}H_{16}C/NO$	237.74	Anaesthetic
Levamisole	$C_{11}H_{12}N_2S$	204.31	Anthelmintic
Diazoxide	C_8H_7C/N_2O_2S	230.7	Antihypertensive
Acyclovir	$C_8H_{11}N_5O_3$	225.2	Antiviral agent
Zidovudine	$C_{10}H_{13}N_5O_4$	267.2	Antiviral agent
Dexamethasone	$C_{22}H_{29}FO_5$	392.5	Anti-inflammatory and immunosuppressive agent
Misoprostol	$C_{22}H_{38}O_5$	382.5	Anti-ulcer agent
Cimetidine	$C_{10}H_{16}N_6$	252.3	Anti-ulcer agent

to the use of biological systems (e.g. cells or tissues) or biological molecules (e.g. enzymes or antibodies) for/in the manufacture of commercial products.

The term 'biopharmaceutical' was first used in the 1980s and came to describe a class of therapeutic proteins produced by modern biotechnological techniques, specifically via genetic engineering (Chapter 3) or, in the case of monoclonal antibodies, by hybridoma technology (Chapter 13). Although the majority of biopharmaceuticals or biotechnology products now approved or in development are proteins produced via genetic engineering, these terms now also encompass nucleic-acid-based, i.e. deoxyribonucleic acid (DNA)- or ribonucleic acid (RNA)-based products, and whole-cell-based products.

1.3   History of the pharmaceutical industry

The pharmaceutical industry, as we now know it, is barely 60 years old. From very modest beginnings, it has grown rapidly, reaching an estimated value of US$100 billion by the mid 1980s. Its current value is likely double or more this figure. There are well in excess of 10 000 pharmaceutical companies in existence, although only about 100 of these can claim to be of true international significance. These companies manufacture in excess of 5000 individual pharmaceutical substances used routinely in medicine.

Table 1.2   Some pharmaceuticals that were traditionally obtained by direct extraction from biological source material. Many of the protein-based pharmaceuticals mentioned are now also produced by genetic engineering

Substance	Medical application
Blood products (e.g. coagulation factors)	Treatment of blood disorders such as haemophilia A or B
Vaccines	Vaccination against various diseases
Antibodies	Passive immunization against various diseases
Insulin	Treatment of diabetes mellitus
Enzymes	Thrombolytic agents, digestive aids, debriding agents (i.e. cleansing of wounds)
Antibiotics	Treatment against various infections agents
Plant extracts (e.g. alkaloids)	Various, including pain relief

The first stages of development of the modern pharmaceutical industry can be traced back to the turn of the twentieth century. At that time (apart from folk cures), the medical community had at their disposal only four drugs that were effective in treating specific diseases:

- Digitalis (extracted from foxglove) was known to stimulate heart muscle and, hence, was used to treat various heart conditions.

- Quinine, obtained from the barks/roots of a plant (*Cinchona* genus), was used to treat malaria.

- Pecacuanha (active ingredient is a mixture of alkaloids), used for treating dysentery, was obtained from the bark/roots of the plant genus *Cephaelis*.

- Mercury, for the treatment of syphilis.

This lack of appropriate, safe and effective medicines contributed in no small way to the low life expectancy characteristic of those times.

Developments in biology (particularly the growing realization of the microbiological basis of many diseases), as well as a developing appreciation of the principles of organic chemistry, helped underpin future innovation in the fledgling pharmaceutical industry. The successful synthesis of various artificial dyes, which proved to be therapeutically useful, led to the formation of pharmaceutical/chemical companies such as Bayer and Hoechst in the late 1800s. Scientists at Bayer, for example, succeeded in synthesizing aspirin in 1895.

Despite these early advances, it was not until the 1930s that the pharmaceutical industry began to develop in earnest. The initial landmark discovery of this era was probably the discovery, and chemical synthesis, of the sulfa drugs. These are a group of related molecules derived from the red dye *prontosil rubrum*. These drugs proved effective in the treatment of a wide variety of bacterial infections (Figure 1.1). Although it was first used therapeutically in the early 1920s, large-scale industrial production of insulin also commenced in the 1930s.

The medical success of these drugs gave new emphasis to the pharmaceutical industry, which was boosted further by the commencement of industrial-scale penicillin manufacture in the early 1940s. Around this time, many of the current leading pharmaceutical companies (or their forerunners) were founded. Examples include Ciba Geigy, Eli Lilly, Wellcome, Glaxo and Roche. Over the next two to three decades, these companies developed drugs such as tetracyclines, corticosteroids, oral contraceptives, antidepressants and many more. Most of these pharmaceutical substances are manufactured by direct chemical synthesis.

1.4   The age of biopharmaceuticals

Biomedical research continues to broaden our understanding of the molecular mechanisms underlining both health and disease. Research undertaken since the 1950s has pinpointed a host of proteins produced naturally in the body that have obvious therapeutic applications. Examples include the interferons and interleukins (which regulate the immune response), growth factors, such as erythropoietin (EPO; which stimulates red blood cell production), and neurotrophic factors (which regulate the development and maintenance of neural tissue).

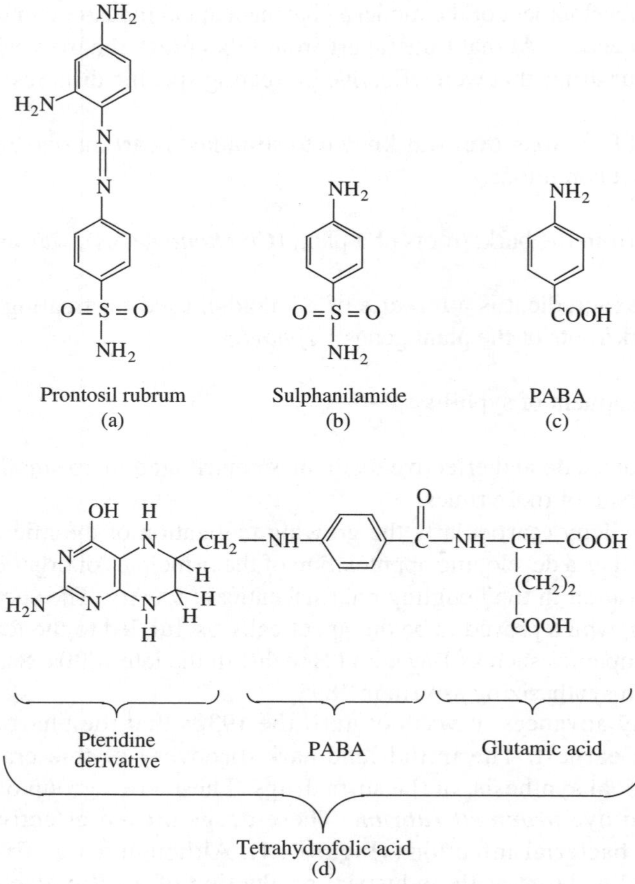

Figure 1.1 Sulfa drugs and their mode of action. The first sulfa drug to be used medically was the red dye prontosil rubrum (a). In the early 1930s, experiments illustrated that the administration of this dye to mice infected with haemolytic streptococci prevented the death of the mice. This drug, although effective *in vivo*, was devoid of *in vitro* antibacterial activity. It was first used clinically in 1935 under the name Streptozon. It was subsequently shown that prontosil rubrum was enzymatically reduced by the liver, forming sulfanilamide, the actual active antimicrobial agent (b). Sulfanilamide induces its effect by acting as an anti-metabolite with respect to *para*-aminobenzoic acid (PABA) (c). PABA is an essential component of tetrahydrofolic acid (THF) (d). THF serves as an essential cofactor for several cellular enzymes. Sulfanilamide (at sufficiently high concentrations) inhibits manufacture of THF by competing with PABA. This effectively inhibits essential THF-dependent enzyme reactions within the cell. Unlike humans, who can derive folates from their diets, most bacteria must synthesize it *de novo*, as they cannot absorb it intact from their surroundings

Although the pharmaceutical potential of these regulatory molecules was generally appreciated, their widespread medical application was in most cases rendered impractical due to the tiny quantities in which they were naturally produced. The advent of recombinant DNA technology (genetic engineering) and monoclonal antibody technology (hybridoma technology) overcame many such difficulties, and marked the beginning of a new era of the pharmaceutical sciences.

Recombinant DNA technology has had a fourfold positive impact upon the production of pharmaceutically important proteins:

- *It overcomes the problem of source availability.* Many proteins of therapeutic potential are produced naturally in the body in minute quantities. Examples include interferons (Chapter 8), interleukins (Chapter 9) and colony-stimulating factors (CSFs; Chapter 10). This rendered impractical their direct extraction from native source material in quantities sufficient to meet likely clinical demand. Recombinant production (Chapters 3 and 5) allows the manufacture of any protein in whatever quantity it is required.

- *It overcomes problems of product safety.* Direct extraction of product from some native biological sources has, in the past, led to the unwitting transmission of disease. Examples include the transmission of blood-borne pathogens such as hepatitis B and C and human immunodeficiency virus (HIV) via infected blood products and the transmission of Creutzfeldt–Jakob disease to persons receiving human growth hormone (GH) preparations derived from human pituitaries.

- *It provides an alternative to direct extraction from inappropriate/dangerous source material.* A number of therapeutic proteins have traditionally been extracted from human urine. Follicle-stimulating hormone (FSH), the fertility hormone, for example, is obtained from the urine of post-menopausal women, and a related hormone, human chorionic gonadotrophin (hCG), is extracted from the urine of pregnant women (Chapter 11). Urine is not considered a particularly desirable source of pharmaceutical products. Although several products obtained from this source remain on the market, recombinant forms have now also been approved. Other potential biopharmaceuticals are produced naturally in downright dangerous sources. Ancrod, for example, is a protein displaying anti-coagulant activity (Chapter 12) and, hence, is of potential clinical use. It is, however, produced naturally by the Malaysian pit viper. Although retrieval by milking snake venom is possible, and indeed may be quite an exciting procedure, recombinant production in less dangerous organisms, such as *Escherichia coli* or *Saccharomycese cerevisiae*, would be considered preferable by most.

- *It facilitates the generation of engineered therapeutic proteins displaying some clinical advantage over the native protein product.* Techniques such as site-directed mutagenesis facilitate the logical introduction of predefined changes in a protein's amino acid sequence. Such changes can be as minimal as the insertion, deletion or alteration of a single amino acid residue, or can be more substantial (e.g. the alteration/deletion of an entire domain, or the generation of a novel hybrid protein). Such changes can be made for a number of reasons, and several engineered products have now gained marketing approval. An overview summary of some engineered product types now on the market is provided in Table 1.3. These and other examples will be discussed in subsequent chapters.

Despite the undoubted advantages of recombinant production, it remains the case that many protein-based products extracted directly from native source material remain on the market. In certain circumstances, direct extraction of native source material can prove equally/more attractive than recombinant production. This may be for an economic reason if, for example, the protein is produced in very large quantities by the native source and is easy to extract/purify, e.g. human serum albumin (HSA; Chapter 12). Also, some blood factor preparations purified from donor blood actually contain several different blood factors and, hence, can be used to treat several haemophilia patient types. Recombinant blood factor preparations, on the other hand, contain but a single blood factor and, hence, can be used to treat only one haemophilia type (Chapter 12).

The advent of genetic engineering and monoclonal antibody technology underpinned the establishment of literally hundreds of start-up biopharmaceutical (biotechnology) companies in

Table 1.3 Selected engineered biopharmaceutical types/products that have now gained marketing approval. These and additional such products will be discussed in detail in subsequent chapters

Product description/type	Alteration introduced	Rationale
Faster acting insulins (Chapter 11)	Modified amino acid sequence	Generation of faster acting insulin
Slow acting insulins (Chapter 11)	Modified amino acid sequence	Generation of slow acting insulin
Modified tissue plasminogen activator (tPA; Chapter 12)	Removal of three of the five native domains of tPA	Generation of a faster acting thrombolytic (clot degrading) agent
Modified blood factor VIII (Chapter 12)	Deletion of 1 domain of native factor VIII	Production of a lower molecular mass product
Chimaeric/humanized antibodies (Chapter 13)	Replacement of most/virtually all of the murine amino acid sequences with sequences found in human antibodies	Greatly reduced/eliminated immunogenicity. Ability to activate human effector functions
'Ontak', a fusion protein (Chapter 9)	Fusion protein consisting of the diphtheria toxin linked to interleukin-2 (IL-2)	Targets toxin selectively to cells expressing an IL-2 receptor

the late 1970s and early 1980s. The bulk of these companies were founded in the USA, with smaller numbers of start-ups emanating from Europe and other world regions.

Many of these fledgling companies were founded by academics/technical experts who sought to take commercial advantage of developments in the biotechnological arena. These companies were largely financed by speculative monies attracted by the hype associated with the establishment of the modern biotech era. Although most of these early companies displayed significant technical expertise, the vast majority lacked experience in the practicalities of the drug development process (Chapter 4). Most of the well-established large pharmaceutical companies, on the other hand, were slow to invest heavily in biotech research and development. However, as the actual and potential therapeutic significance of biopharmaceuticals became evident, many of these companies did diversify into this area. Most either purchased small, established biopharmaceutical concerns or formed strategic alliances with them. An example was the long-term alliance formed by Genentech (see later) and the well-

Table 1.4 Pharmaceutical companies who manufacture and/or market biopharmaceutical products approved for general medical use in the USA and EU

Sanofi-Aventis	Hoechst AG
Bayer	Wyeth
Novo Nordisk	Genzyme
Isis Pharmaceuticals	Abbott
Genentech	Roche
Centocor	Novartis
Boehringer Manheim	Serono
Galenus Manheim	Organon
Eli Lilly	Amgen
Ortho Biotech	GlaxoSmithKline
Schering Plough	Cytogen
Hoffman-la-Roche	Immunomedics
Chiron	Biogen

established pharmaceutical company Eli Lilly. Genentech developed recombinant human insulin, which was then marketed by Eli Lilly under the trade name Humulin. The merger of biotech capability with pharmaceutical experience helped accelerate development of the biopharmaceutical sector.

Many of the earlier biopharmaceutical companies no longer exist. The overall level of speculative finance available was not sufficient to sustain them all long term (it can take 6–10 years and US$800 million to develop a single drug; Chapter 4). Furthermore, the promise and hype of biotechnology sometimes exceeded its ability actually to deliver a final product. Some biopharmaceutical substances showed little efficacy in treating their target condition, and/or exhibited unacceptable side effects. Mergers and acquisitions also led to the disappearance of several biopharmaceutical concerns. Table 1.4 lists many of the major pharmaceutical concerns which now manufacture/market biopharmaceuticals approved for general medical use. Box 1.1 provides a profile of three well-established dedicated biopharmaceutical companies.

Box 1.1

Amgen, Biogen and Genentech

Amgen, Biogen and Genentech represent three pioneering biopharmaceutical companies that still remain in business.

Founded in the 1980s as AMGen (Applied Molecular Genetics), Amgen now employs over 9000 people worldwide, making it one of the largest dedicated biotechnology companies in existence. Its headquarters are situated in Thousand Oaks, California, although it has research, manufacturing, distribution and sales facilities worldwide. Company activities focus upon developing novel (mainly protein) therapeutics for application in oncology, inflammation, bone disease, neurology, metabolism and nephrology. By mid 2006, seven of its recombinant products had been approved for general medical use (the EPO-based products 'Aranesp' and 'Epogen' (Chapter 10), the CSF-based products 'Neupogen' and 'Neulasta' (Chapter 10), as well as the interleukin-1 (IL-1) receptor antagonist 'Kineret', the anti-rheumatoid arthritis fusion protein Enbrel (Chapter 9) and the keratinocyte growth factor 'Kepivance', indicated for the treatment of severe oral mucositis. Total product sales for 2004 reached US$9.9 billion. In July 2002, Amgen acquired Immunex Corporation, another dedicated biopharmaceutical company founded in Seattle in the early 1980s.

Biogen was founded in Geneva, Switzerland, in 1978 by a group of leading molecular biologists. Currently, its global headquarters are located in Cambridge, MA, and it employs in excess of 2000 people worldwide. The company developed and directly markets the interferon-based product 'Avonex' (Chapter 8), but also generates revenues from sales of other Biogen-discovered products that are licensed to various other pharmaceutical companies. These include Schering Plough's 'Intron A' (Chapter 8) and a number of hepatitis B-based vaccines sold by SmithKline Beecham (SKB) and Merck (Chapter 13).

Genentech was founded in 1976 by scientist Herbert Boyer and the venture capitalist Robert Swanson. Headquartered in San Francisco, it employs almost 5000 staff worldwide and has 10 protein-based products on the market. These include hGHs (Nutropin, Chapter 11), the antibody-based products 'Herceptin' and 'Rituxan' (Chapter 13) and the thrombolytic agents 'Activase' and 'TNKase' (Chapter 12). The company also has 20 or so products in clinical trials. In 2004, it generated some US$4.6 billion in revenues.

1.5   Biopharmaceuticals: current status and future prospects

Approximately one in every four new drugs now coming on the market is a biopharmaceutical. By mid 2006, some 160 biopharmaceutical products had gained marketing approval in the USA and/or EU. Collectively, these represent a global biopharmaceutical market in the region of US$35 billion (Table 1.5), and the market value is estimated to surpass US$50 billion by 2010. The products include a range of hormones, blood factors and thrombolytic agents, as well as vaccines and monoclonal antibodies (Table 1.6). All but two are protein-based therapeutic agents. The exceptions are two nucleic-acid-based products: 'Vitravene', an antisense oligonucleotide, and 'Macugen', an aptamer (Chapter 14). Many additional nucleic-acid-based products for use in gene therapy or antisense technology are in clinical trials, although the range of technical difficulties that still beset this class of therapeutics will ensure that protein-based products will overwhelmingly predominate for the foreseeable future (Chapter 14).

Many of the initial biopharmaceuticals approved were simple replacement proteins (e.g. blood factors and human insulin). The ability to alter the amino acid sequence of a protein logically coupled to an increased understanding of the relationship between protein structure and function (Chapters 2 and 3) has facilitated the more recent introduction of several engineered therapeutic proteins (Table 1.3). Thus far, the vast majority of approved recombinant proteins have been produced in the bacterium *E. coli*, the yeast *S. cerevisiae* or in animal cell lines (most notably Chinese hamster ovary (CHO) cells or baby hamster kidney (BHK) cells. These production systems are discussed in Chapter 5.

Although most biopharmaceuticals approved to date are intended for human use, a number of products destined for veterinary application have also come on the market. One early such example is that of recombinant bovine GH (Somatotrophin), which was approved in the USA in the early 1990s and used to increase milk yields from dairy cattle. Additional examples of approved veterinary biopharmaceuticals include a range of recombinant vaccines and an interferon-based product (Table 1.7).

Table 1.5 Approximate annual market values of some leading approved biopharmaceutical products. Data gathered from various sources, including company home pages, annual reports and industry reports

Product (Company)	Product description (use)	Annual sales value (US$, billions)
Procrit (Amgen/Johnson & Johnson)	EPO (treatment of anaemia)	4.0
Epogen & Aranesp combined (Amgen)	EPO (treatment of anaemia)	4.0
Intron A (Schering Plough)	IFN-α (treatment of leukaemia)	0.3
Remicade (Johnson & Johnson)	Monoclonal antibody based (treatment of Crohn's disease)	1.7
Avonex (Biogen)	Interferon-β (IFN-β; treatment of multiple sclerosis)	1.2
Embrel (Wyeth)	Monoclonal antibody based (treatment of rheumatoid arthritis)	1.3
Rituxan (Genentech)	Monoclonal antibody based (non-Hodgkin's lymphoma)	1.5
Humulin (Eli Lilly)	Insulin (diabetes)	1.0

Table 1.6 Summary categorization of biopharmaceuticals approved for general medical use in the EU and/or USA by 2006

Product type	Examples	No. approved	Refer to
Blood factors	Factors VIII and IX	8	Chapter 12
Thrombolytic agents	tPA	6	Chapter 12
Hormones	Insulin, GH, gonadotrophins	33	Chapter 11
Haematopoietic growth factors	EPO, CSFs	8	Chapter 10
Interferons	IFN-α, -β, -γ	16	Chapter 8
Interleukin-based products	IL-2	3	Chapter 9
Vaccines	Hepatitis B-surface antigen	20	Chapter 13
Monoclonal antibodies	Various	30	Chapter 13
Nucleic acid based	Antisense and aptamer	2	Chapter 14
Additional products	Tumour necrosis factor (TNF), therapeutic enzymes	18	Various chapters

At least 1000 potential biopharmaceuticals are currently being evaluated in clinical trials, although the majority of these are in early stage trials. Vaccines and monoclonal antibody-based products represent the two biggest product categories. Regulatory factors (e.g. hormones and

Table 1.7 Some recombinant (r) biopharmaceuticals recently approved for veterinary application in the EU

Product	Company	Indication
Vibragen Omega (r-feline interferon omega; IFN-ω)	Virbac	Reduction of mortality/clinical symptoms associated with canine parvovirus
Fevaxyl Pentafel (combination vaccine containing r-feline leukaemia viral antigen as one component)	Fort Dodge Laboratories	Immunization of cats against various feline pathogens
Porcilis porcoli (combination vaccine containing r-*E. coli* adhesins)	Intervet	Active immunization of sows
Porcilis AR-T DF (combination vaccine containing a recombinant modified toxin from *Pasteurella multocida*)	Intervet	Reduction in clinical signs of progressive atrophic rhinitis in piglets
Porcilis pesti (combination vaccine containing r-classical swine fever virus E_2 subunit antigen)	Intervet	Immunization of pigs against classical swine fever
Bayovac CSF E2 (combination vaccine containing r-classical swine fever virus E_2 subunit antigen)	Intervet	Immunization of pigs against classical swine fever

cytokines) and gene therapy and antisense-based products also represent significant groupings. Although most protein-based products likely to gain marketing approval over the next 2–3 years will be produced in engineered *E. coli*, *S. cerevisiae* or animal cell lines, some products now in clinical trials are being produced in the milk of transgenic animals (Chapter 5). Additionally, plant-based transgenic expression systems may potentially come to the fore, particularly for the production of oral vaccines (Chapter 5).

Interestingly, the first generic biopharmaceuticals are already entering the market. Patent protection for many first-generation biopharmaceuticals (including recombinant human GH (rhGH), insulin, EPO, interferon-α (IFN-α) and granulocyte-CSF (G-CSF)) has now/is now coming to an end. Most of these drugs command an overall annual market value in excess of US$1 billion, rendering them attractive potential products for many biotechnology/pharmaceutical companies. Companies already/soon producing generic biopharmaceuticals include Biopartners (Switzerland), Genemedix (UK), Sicor and Ivax (USA), Congene and Microbix (Canada) and BioGenerix (Germany). Genemedix, for example, secured approval for sale of a recombinant CSF in China in 2001 and is also commencing the manufacture of recombinant EPO. Sicor currently markets hGH and IFN-α in eastern Europe and various developing nations. A generic hGH also gained approval in both Europe and the USA in 2006.

To date (mid 2006), no gene-therapy-based product has thus far been approved for general medical use in the EU or USA, although one such product ('Gendicine'; Chapter 14) has been approved in China. Although gene therapy trials were initiated as far back as 1989, the results have been disappointing. Many technical difficulties remain in relation to, for example, gene delivery and regulation of expression. Product effectiveness was not apparent in the majority of trials undertaken and safety concerns have been raised in several trials.

Only one antisense-based product has been approved to date (in 1998) and, although several such antisense agents continue to be clinically evaluated, it is unlikely that a large number of such products will be approved over the next 3–4 years. Aptamers represent an additional emerging class of nucleic-acid-based therapeutic. These are short DNA- or RNA-based sequences that adopt a specific three-dimensional structure, enabling them to bind (and thereby inhibit) specific target molecules. One such product (Macugen) has been approved to date. RNA interference (RNAi) represents a yet additional mechanism of achieving downregulation of gene expression (Chapter 14). It shares many characteristics with antisense technology and, like antisense, provides a potential means of treating medical conditions triggered or exacerbated by the inappropriate overexpression of specific gene products. Despite the disappointing results thus far generated by nucleic-acid-based products, future technical advances will almost certainly ensure the approval of gene therapy and antisense-based products in the intermediate to longer term future.

Technological developments in areas such as genomics, proteomics and high-throughput screening are also beginning to impact significantly upon the early stages of drug development (Chapter 4). By linking changes in gene/protein expression to various disease states, for example, these technologies will identify new drug targets for such diseases. Many/most such targets will themselves be proteins, and drugs will be designed/developed specifically to interact with. They may be protein based or (more often) low molecular mass ligands.

Additional future innovations likely to impact upon pharmaceutical biotechnology include the development of alternative product production systems, alternative methods of delivery and the development of engineered cell-based therapies, particularly stem cell therapy. As mentioned previously, protein-based biotechnology products produced to date are produced in either microbial or in animal cell lines. Work continues on the production of such products in transgenic-based production systems, specifically either transgenic plants or animals (Chapter 5).

Virtually all therapeutic proteins must enter the blood in order to promote a therapeutic effect. Such products must usually be administered parenterally. However, research continues on the development of non-parenteral routes which may prove more convenient, less costly and obtain improved patient compliance. Alternative potential delivery routes include transdermal, nasal, oral and bucal approaches, although most progress to date has been recorded with pulmonary-based delivery systems (Chapter 4). An inhaled insulin product ('Exubera', Chapters 4 and 11) was approved in 2006 for the treatment of type I and II diabetes.

A small number of whole-cell-based therapeutic products have also been approved to date (Chapter 14). All contain mature, fully differentiated cells extracted from a native biological source. Improved techniques now allow the harvest of embryonic and, indeed, adult stem cells, bringing the development of stem-cell-based drugs one step closer. However, the use of stem cells to replace human cells or even entire tissues/organs remains a long term goal (Chapter 14). Overall, therefore, products of pharmaceutical biotechnology play an important role in the clinic and are likely to assume an even greater relative importance in the future.

Further reading

Books

Crommelin, D. and Sindelar, R. 2002. *Pharmaceutical Biotechnology*, second edition. Taylor and Francis, London, UK.

Goldberg, R. 2001. *Pharmaceutical Medicine, Biotechnology and European Law*. Cambridge University Press.

Grindley, J. and Ogden, J. 2000. *Understanding Biopharmaceuticals. Manufacturing and Regulatory Issues*. Interpharm Press.

Kayser, O. and Muller, RH. 2004. *Pharmaceutical Biotechnology*. Wiley VCH, Weinheim, Germany.

Oxender, D. and Post, L. 1999. *Novel Therapeutics from Modern Biotechnology*. Springer Verlag.

Spada, S. and Walsh, G. 2005. *Directory of Approved Biopharmaceutical Products*. CRC Press, Florida, USA.

Articles

Mayhall, E., Paffett-Lugassy N., and Zon L.I. 2004. The clinical potential of stem cells. *Current Opinion in Cell Biology* **16**, 713–720.

Reichert, J. and Paquette, C. 2003. Therapeutic recombinant proteins: trends in US approvals 1982-2002. *Current Opinion in Molecular Therapy* **5**, 139–147.

Reichert, J. and Pavlov, A. 2004. Recombinant therapeutics – success rates, market trends and values to 2010. *Nature Biotechnology* **22**, 1513–1519.

Walsh, G. 2005. Biopharmaceuticals: recent approvals and likely directions. *Trends in Biotechnology* **23**, 553–558.

Walsh, G. 2006. Biopharmaceutical benchmarks 2006. *Nature Biotechnology* **24**, 769–776.

Weng, Z. and DeLisi, C. 2000. Protein therapeutics: promises and challenges of the 21st century. *Trends in Biotechnology* **20**, 29–36.

2
Protein structure

2.1 Introduction

Almost all products of modern pharmaceutical biotechnology, be they on the market or likely to gain approval in the short to intermediate term, are protein based. As such, an understanding of protein structure is central to this topic. A comprehensive treatment of the subject would easily constitute a book on its own, and many such publications are available. The aim of this chapter is to provide a basic overview of the subject in order to equip the reader with a knowledge of protein science sufficient to understand relevant concepts outlined in the remaining chapters of this book. The interested reader is also referred to the 'Further reading' section, which lists several excellent specialist publications in the field. Much additional information may also be sourced via the web sites mentioned within the chapter.

2.2 Overview of protein structure

Proteins are macromolecules consisting of one or more polypeptides (Table 2.1). Each polypeptide consists of a chain of amino acids linked together by peptide (amide) bonds. The exact amino acid sequence is determined by the gene coding for that specific polypeptide. When synthesized, a polypeptide chain folds up, assuming a specific three-dimensional shape (i.e. a specific conformation) that is unique to it. The conformation adopted is dependent upon the polypeptide's amino acid sequence, and this conformation is largely stabilized by multiple, weak non-covalent interactions. Any influence (e.g. certain chemicals and heat) that disrupts such weak interactions results in disruption of the polypeptide's native conformation, a process termed denaturation. Denaturation usually results in loss of functional activity, clearly demonstrating the dependence of protein function upon protein structure. A protein's structure currently cannot be predicted solely from its amino acid sequence. Its conformation can, however, be determined by techniques such as X-ray diffraction and nuclear magnetic resonance (NMR) spectroscopy.

Proteins are sometimes classified as 'simple' or 'conjugated'. Simple proteins consist exclusively of polypeptide chain(s) with no additional chemical components present or being required for biological activity. Conjugated proteins, in addition to their polypeptide components(s),

Pharmaceutical biotechnology: concepts and applications   Gary Walsh
© 2007 John Wiley & Sons, Ltd   ISBN 978 0 470 01244 4 (HB)   978 0 470 01245 1 (PB)

Table 2.1   Selected examples of proteins. The number of polypeptide chains and amino acid residues constituting the protein are listed, along with its molecular mass and biological function

Protein	No. polypeptide chains	Total no. amino acids	Molecular mass (Da)	Biological function
Insulin (human)	2	51	5 800	Complex, includes regulation of blood glucose levels
Lysozyme (egg)	1	129	13 900	Enzyme capable of degrading peptidoglycan in bacterial cell walls
IL-2 (human)	1	133	15 400	T-lymphocyte-derived polypeptide that regulates many aspects of immunity
EPO (human)	1	165	36 000	Hormone that stimulates red blood cell production
Chymotrypsin (bovine)	3	241	21 600	Digestive proteolytic enzyme
Subtilisin (*Bacillus amyloliquefaciens*)	1	274	27 500	Bacterial proteolytic enzyme
Tumour necrosis factor (human TNF-α)	3	471	52 000	Mediator of inflammation and immunity
Haemoglobin (human)	4	574	64 500	Gas transport
Hexokinase (yeast)	2	800	102 000	Enzyme capable of phosphorylating selected monosaccharides
Glutamate dehydrogenase (bovine)	~40	~8 300	~1 000 000	Enzyme interconverts glutamate and α-ketoglutarate and NH_4^+

contain one or more non-polypeptide constituents known as prosthetic group(s). The most common prosthetic groups found in association with proteins include carbohydrates (glycoproteins), phosphate groups (phosphoproteins), vitamin derivatives (e.g. flavoproteins) and metal ions (metalloproteins).

Table 2.2 The 20 commonly occurring amino acids. They may be subdivided into five groups on the basis of side-chain structure. Their three- and one-letter abbreviations are also listed (one-letter abbreviations are generally used only when compiling extended sequence data, mainly to minimize writing space and effort). In addition to their individual molecular masses, the percentage occurrence of each amino acid in an 'average' protein is also presented. These data were generated from sequence analysis of over 1000 different proteins

R group classification	Amino acid	Abbreviation		Molecular mass	Occurrence in 'average' protein (%)
		3 letters	1 letter		
Nonpolar, aliphatic	Glycine	Gly	G	75	7.2
	Alanine	Ala	A	89	8.3
	Valine	Val	V	117	6.6
	Leucine	Leu	L	131	9
	Isoleucine	Ile	I	131	5.2
	Proline	Pro	P	115	5.1
Aromatic	Tyrosine	Tyr	Y	181	3.2
	Phenylalanine	Phe	F	165	3.9
	Tryptophan	Trp	W	204	1.3
Polar but uncharged	Cysteine	Cys	C	121	1.7
	Serine	Ser	S	105	6
	Methionine	Met	M	149	2.4
	Threonine	Thr	T	119	5.8
	Asparagine	Asn	N	132	4.4
	Glutamine	Gln	Q	146	4
Positively charged	Arginine	Arg	R	174	5.7
	Lysine	Lys	K	146	5.7
	Histidine	His	H	155	2.2
Negatively charged	Aspartic acid	Asp	D	133	5.3
	Glutamic acid	Glu	E	147	6.2

2.2.1 Primary structure

Polypeptides are linear, unbranched polymers, potentially containing up to 20 different monomer types (i.e. the 20 commonly occurring amino acids) linked together in a precise predefined sequence. The primary structure of a polypeptide refers to its exact amino acid sequence, along with the exact positioning of any disulfide bonds present (described later). The 20 commonly occurring amino acids are listed in Table 2.2, along with their abbreviated and one-letter designations. The structures of these amino acids are presented in Figure 2.1. Nineteen of these amino acids contain a central (α) carbon atom, to which is attached a hydrogen atom (H), an amino group (NH_2) a carboxyl group (COOH), and an additional side chain (R) group – which differs from amino acid to amino acid. The amino acid proline is unusual in that its R group forms a direct covalent bond with the nitrogen atom of what is the free amino group in other amino acids (Figure 2.1).

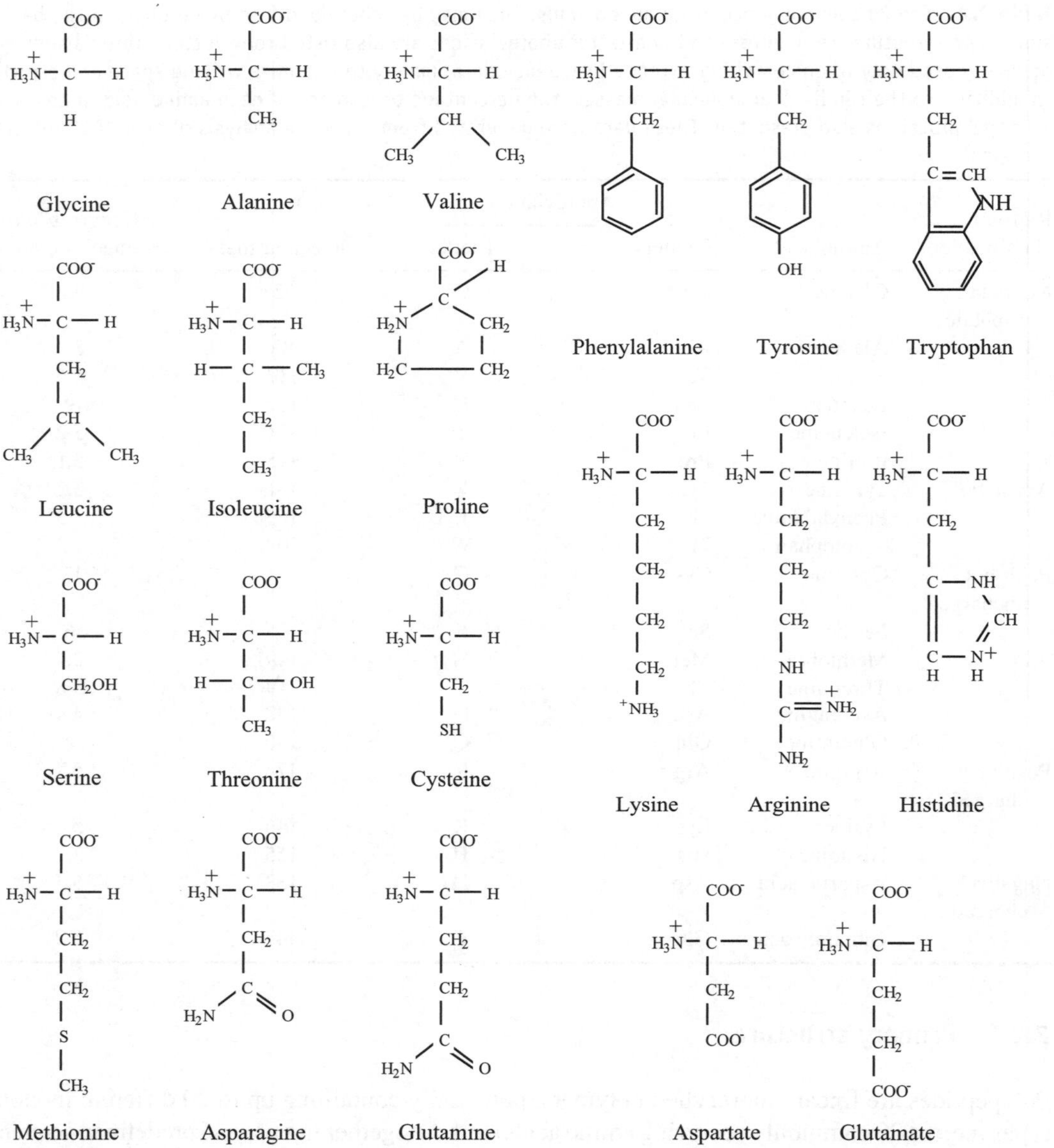

Figure 2.1 The chemical structure of the 20 amino acids commonly found in proteins

As will be evident from Section 2.2.2, peptide bond formation between adjacent amino acid residues entails the establishment of covalent linkages between the amino and carboxyl groups attached to their respective central (α) carbon atoms. Hence, the free functional (i.e. chemically reactive) groups in polypeptides are almost entirely present as part of the constituent amino acids' side chains (R groups). In addition to determining the chemical reactivity of a polypeptide, these R groups also very largely dictate the final conformation adopted by a polypeptide. Stabilizing/repulsive forces between different R groups (as well as between R groups and the surrounding aqueous media) largely dictate what final shape the polypeptide adopts, as will be described later.

The R groups of the non-polar, alipathic amino acids (Gly, Ala, Val, Leu, Ile and Pro) are devoid of chemically reactive functional groups. These R groups are noteworthy in that, when present in a polypeptide's backbone, they tend to interact with each other non-covalently (via hydrophobic interactions). These interactions have a significant stabilizing influence on protein conformation.

Glycine is noteworthy in that its R group is a hydrogen atom. This means that the α-carbon of glycine is not asymmetric, i.e. is not a chiral centre. (To be a chiral centre the carbon would have to have four different chemical groups attached to it; in this case, two of its four attached groups are identical.) As a consequence, glycine does not occur in multiple stereo-isomeric forms, unlike the remaining amino acids, which occur as either D or L isomers. Only L-amino acids are naturally found in polypeptides.

The side chains of the aromatic amino acids (Phe, Tyr and Trp) are not particularly reactive chemically, but they all absorb ultraviolet (UV) light. Tyr and Trp in particular absorb strongly at 280 nm, allowing detection and quantification of proteins in solution by measuring the absorbance at this wavelength.

Of the six polar but uncharged amino acids, two (cysteine and methionine) are unusual in that they contain a sulfur atom. The side chain of methionine is non-polar and relatively unreactive, although the sulfur atom is susceptible to oxidation. In contrast, the thiol ($-C-SH$) portion of cysteine's R group is the most reactive functional group of any amino acid side chain. *In vivo*, this group can form complexes with various metal ions and is readily oxidized, forming 'disulfide linkages' (covalent linkages between two cysteine residues within the same or even different polypeptide backbones). These help stabilize the three-dimensional structure of such polypeptides. Interchain disulfide linkages can also form, in which cysteines from two different polypeptides participate. This is a very effective way of covalently linking adjacent polypeptides.

Of the four remaining polar but uncharged amino acids, the R groups of serine and threonine contain hydroxyl (OH) groups and the R groups of asparagine and glutamine contain amide ($CONH_2$) groups. None are particularly reactive chemically; however, upon exposure to high temperatures or extremes of pH, the latter two can deamidate, yielding aspartic acid and glutamic acid respectively.

Aspartic and glutamic acids are themselves negatively charged under physiological conditions. This allows them to chelate certain metal ions, and also to markedly influence the conformation adopted by polypeptide chains in which they are found.

Lysine, arganine and histidine are positively charged amino acids. The arganine R group consists of a hydrophobic chain of four $-CH_2$ groups (Figure 2.1), capped with an amino (NH_2) group, which is ionized (NH_3^+) under most physiological conditions. However, within most polypeptides there is normally a fraction of un-ionized lysines, and these (unlike their ionized counterparts) are quite chemically reactive. Such lysine side chains can be chemically converted into various analogues. The arganine side chain is also quite bulky, consisting of three CH_2 groups, an amino group ($-NH_2$) and an ionized guanido group ($=NH_2^+$). The 'imidazole' side chain of histidine can be described chemically as a tertiary amine (R_3-N), and thus it can act as a strong nucleophilic catalyst (the nitrogen atom houses a lone pair of electrons, making it a 'nucleus lover' or nucleophile; it can donate its electron pair to an 'electron lover' or electrophile). As such, the histidine side chain often constitute an essential part of some enzyme active sites.

In addition to the 20 'common' amino acids, some modified amino acids are also found in several proteins. These amino acids are normally altered via a process of post-translational modification (PTM) reactions (i.e. modified after protein synthesis is complete). Almost 200 such modified amino acids have been characterized to date. The more common such modifications are discussed separately in Section 2.5.

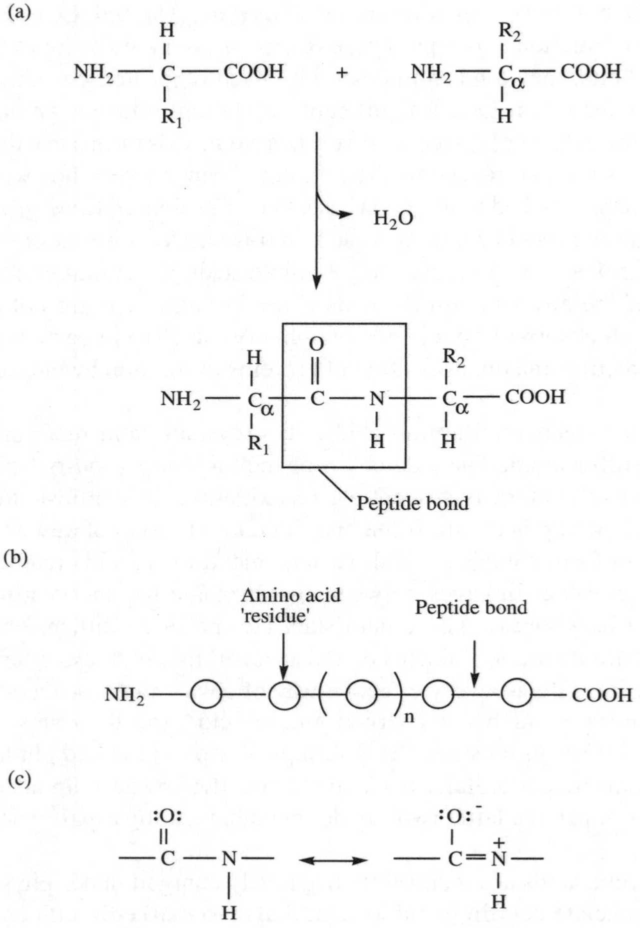

Figure 2.2 (a) Peptide bond formation. (b) Polypeptides consist of a linear chain of amino acids successively linked via peptide bonds. (c) The peptide bond displays partial double-bonded character

2.2.2 The peptide bond

Successive amino acids are joined together during protein synthesis via a 'peptide' (i.e. amide) bond (Figure 2.2). This is a condensation reaction, as a water molecule is eliminated during bond formation. Each amino acid in the resultant polypeptide is termed a 'residue', and the polypeptide chain will display a free amino (NH_2) group at one end and a free carboxyl (COOH) group at the other end. These are termed the amino and carboxyl termini respectively.

The peptide bond has a rigid, planar structure and is in the region of 1.33 Å in length. Its rigid nature is a reflection of the fact that the amide nitrogen lone pair of electrons is delocalized across the bond (i.e. the bond structure is a halfway house between the two forms illustrated in Figure 2.2c). In most instances, peptide groups assume a 'trans' configuration (Figure 2.2b). This minimizes steric interference between the R groups of successive amino acid residues.

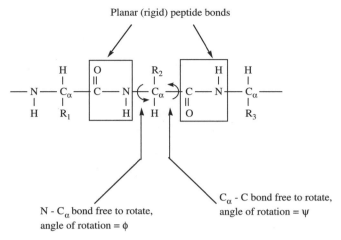

Figure 2.3 Fragment of polypeptide chain backbone illustrating rigid peptide bonds and the intervening N—C_α and C_α—C backbone linkages, which are free to rotate

Whereas the peptide bond is rigid, the other two bond types found in the polypeptide backbone (i.e. the N—C_α bond and the C_α—C bond, Figure 2.3) are free to rotate. The polypeptide backbone can thus be viewed as a series of planar 'plates' that can rotate relative to one another. The angle of rotation around the N—C_α bond is termed ϕ (phi) and that around the C_α—C bond is termed ψ (psi) (Figure 2.3). These angles are also known as rotation angles, dihedral angles or torsion angles. By convention, these angles are defined as being 180° when the polypeptide chain is in its fully extended, trans form. In principle, each bond can rotate to any value between −180° and +180°. However, the degrees of rotation actually observed are restricted due to the occurrence of steric hindrance between atoms of the polypeptide backbone and those of amino acid side chains.

For each amino acid residue in a polypeptide backbone, the actual ϕ and ψ angles that are physically possible can be calculated, and these angle pairs are often plotted against each other in a diagram termed a Ramachandran plot. Sterically allowable angles fall within relatively narrow bands in most instances. A greater than average degree of ϕ/ψ rotational freedom is observed around glycine residues, due to the latter's small R group – hence steric hindrance is minimized. On the other hand, bond angle freedom around proline residues is quite restricted due to this amino acid's unusual structure (Figure 2.1). The ϕ and ψ angles allowable around each C_α in a polypeptide backbone obviously exert a major influence upon the final three-dimensional shape assumed by the polypeptide.

2.2.3 Amino acid sequence determination

The amino acid sequence of a polypeptide may be determined directly via chemical sequencing or by physical fragmentation and analysis, usually by mass spectrometry. Direct chemical sequencing was the only method available until the 1970s. Insulin was the first protein to be sequenced by this approach (in 1953), requiring several years and several hundred grams of protein to complete. The method has been refined and automated over the years, such that, today, polypeptides containing 100 amino acids or more can be automatically sequenced within a few hours, using microgram to milligram levels of protein. The actual chemical sequencing procedure employed is termed the Edman degradation method.

Table 2.3 Representative organisms whose genomes have been or will soon be completely/ almost completely sequenced. Data taken largely from http://wit.integratedgenomics.com/GOLD/ eucaryoticgenomes.html and http://www.tigr.org/tdb/mdb/mdcomplete.html. Updated information is available on these sites

Organism	Classification	Genome size[a] (Mb)	Organism	Classification	Genome size[a] (Mb)
Aeropyrum pernix	Archaea	1.67	*Treponema pallidum*	Eubacteria	1.14
Archaeoglobus fulgidus	Archaea	2.18	*Vibrio chloerae*	Eubacteria	4.0
Pyrococcus horikoshii	Archaea	1.80	*Aspergillus nidulans*	Fungi	31.0
Pyrococcus furiosus	Archaea	2.10	*Candida albicans*	Fungi	15.0
Sulfolobus solfataricus	Archaea	2.99	*Neurospora crassa*	Fungi	47.0
Thermoplasma acidophilum	Archaea	1.56	*Schizosaccharomyces pombe*	Fungi	14.0
Aquifex aeolicus	Eubacteria	1.50	*Babesia bovis*	Protozoa	NL
Bacillus subtilis	Eubacteria	4.20	*Cryptosporidium parvum*	Protozoa	10.4
Bacillus anthracis	Eubacteria	4.50	*Leishmania major*	Protozoa	33.6
Bordetella pertussis	Eubacteria	3.88	*Arabidopsis thaliana* (thale cress)	Plant	70.0
Brucella suis	Eubacteria	3.30	*Hordeum vulgare* (barley)	Plant	5.0
Chlamydia pneumoniae	Eubacteria	1.23	*Gossypium hirsutum* (cotton)	Plant	NL
Clostridium tetani	Eubacteria	4.40	*Triticum aestivum* (wheat)	Plant	NL
Corynebacterium diphtheriae	Eubacteria	3.10	*Zea mays* (maize)	Plant	NL
E. coli	Eubacteria	5.23	*Danio rerio* (zebra fish)	Fish	NL
Lactobacillus acidophilus	Eubacteria	1.90	*Gallus gallus* (chicken)	Bird	NL
Listeria monocytogenes	Eubacteria	2.94	*Bos taurus* (cow)	Mammal	NL
Mycobacterium leprae	Eubacteria	2.80	*Canis familiaris* (dog)	Mammal	NL
Mycobacterium tuberculosis	Eubacteria	4.40	*Rattus norvegicus* (rat)	Mammal	NL
Neisseria meningitidis	Eubacteria	2.18	*Ovis aries* (sheep)	Mammal	NL
Pseudomonas aeruginosa	Eubacteria	6.3	*Sus scrofa* (pig)	Mammal	NL
Salmonella enterica	Eubacteria	NL	Ape	Primate	NL
Staphylococcus aureus	Eubacteria	2.80	*Homo sapiens*	Primate	NL
Streptococcus pneumoniae	Eubacteria	2.04			

[a]NL: not listed in source publication.

Table 2.4 The major primary sequence (protein and nucleic acid) databases and the web addresses from which they may be accessed

Database	Web address
Protein	
PIR	http://www-nbrf.georgetown.edu/
Swiss-Prot	http://www.ebi.ac.uk/swissprot/
MIPS	http://www.mips.biochem.mpg.de/
NRL-3D	http://www-nbrf.georgetown.edu/pirwww/ dbinfo/nrl3d.html
Tr EMBL	http://www.ebi.ac.uk/index.html
Owl	http://www.bis.med.jhmi.edu/Dan/ proteins/owl.html
Nucleic acid	
EMBL	http://www.ebi.ac.uk/embl/index.html/
GenBank	http://www.ncbi.nlm.nih.gov
DDBJ	http://www.ddbj.nig.ac.jp/

An alternative approach to amino acid sequence determination is to sequence its gene (Chapter 3). The amino acid sequence can be inferred from the nucleotide sequence obtained. This approach has gained favour in recent years. Refinements to DNA sequencing methodologies and equipment have made such sequence analysis both rapid and relatively inexpensive. The ongoing genome projects continue to generate enormous amounts of sequence data. By the early 2000s, substantial/complete sequence data for some 300 organisms were available (Table 2.3). As a result, the putative amino acid sequences of an enormous number of proteins (most of unknown function/structure) had been determined.

Upon its generation, sequence information is normally submitted to various databases. The major databases in which protein primary sequence data are available are listed in Table 2.4. Also included in this table are the major nucleic acid sequence databases, as amino acid sequence information can potentially be derived from these.

The Swiss-Prot database is probably the most widely used protein database. It is maintained collaboratively by the European Bioinformatics Institute (EBI) and the Swiss Institute for Bioinformatics. It is relatively easy to access and search via the World Wide Web (Table 2.4). A sample entry for human insulin is provided in Figure 2.4. Additional information detailing such databases is available via the web addresses provided in Table 2.4 and in the bioinformatics publications listed at the end of this chapter.

A polypeptide's amino acid sequence can thus be determined by direct chemical (Edman) or physical (mass spectrometry) means, or indirectly via gene sequencing. In practice, these methods are complementary to one another and can be used to cross-check sequence accuracy. If the target gene/messenger RNA (mRNA) has been previously isolated, then DNA sequencing is usually most convenient. However, this approach reveals little information regarding any PTMs present in the mature polypeptide, many of whom are of critical significance in the context of therapeutic proteins (discussed in Section 2.5).

General information about the entry	
Entry name	**INS_HUMAN**
Primary accession number	**P01308**
Secondary accession number(s)	None
Entered in SWISS-PROT in	Release 01, July 1986
Sequence was last modified in	Release 01, July 1986
Annotations were last modified in	Release 39, May 2000
Name and origin of the protein	
Protein name	INSULIN [Precursor]
Synonym(s)	None
Gene name(s)	INS
From	*Homo sapiens (Human)* [TaxID: *9606*]
Taxonomy	Eukaryota; Metazoa; Chordata; Craniata; Vertebrata; Euteleostomi; Mammalia; Eutheria; Primates; Catarrhini; Hominidae; Homo.

Features

SIGNAL	1	24		
CHAIN	25	54	INSULIN B CHAIN.	
PROPEP	57	87	C PEPTIDE.	
CHAIN	90	110	INSULIN A CHAIN.	
DISULFID	31	96	INTERCHAIN.	
DISULFID	43	109	INTERCHAIN.	
DISULFID	95	100		
VARIANT	34	34	H → D (IN PROVIDENCE; FAMILIAL HYPERPROINSULINEMIA). /FTID = VAR_003971.	
VARIANT	48	48	F → S (IN LOS-ANGELES; TYPE-II DIABETES MELLITUS). /FTID = VAR_003972.	
VARIANT	49	49	F → L (IN CHICAGO). /FTID = VAR_003973.	
VARIANT	89	89	R → H (IN FAMILIAL HYPERPROINSULINEMIA; IMPAIRS POSTTRANSLATIONAL CLEAVAGE). /FTID = VAR_003974.	Feature table viewer
VARIANT	89	89	R → L (IN KYOTO; FAMILIAL HYPERPROINSULINEMIA). /FTID = VAR_003975.	
VARIANT	92	92	V → L (IN WAKAYAMA). /FTID = VAR_003976.	
TURN	32	32		
HELIX	33	46		
STRAND	48	50		
HELIX	91	95		
TURN	96	97		
HELIX	102	108		
STRAND	109	109		

Sequence information

Length: **110 AA** [This is the length of the unprocessed precursor]	Molecular weight: **11981 Da** [This is the Mw of the unprocessed precursor]	CRC64: **C2C3B23B85E520E5** [This is a checksum on the sequence]

```
        10          20          30          40          50          60
         |           |           |           |           |           |
    MALWMRLLPL  LALLALWGPD  PAAAFVNQHL  CGSHLVEALY  LVCGERGFFY  TPKTRREAED
        70          80          90         100         110
         |           |           |           |           |                    P01308 in
    LQVGQVELGG  GPGAGSLQPL  ALEGSLQKRG  IVEQCCTSIC  SLYQLENYCN                FASTA format
```

Figure 2.4 Sample entry for human insulin as present in the Swiss-Prot database. Refer to text for further details. Reproduced from the Swiss-Prot database on the Uniprot website htt://www.ebi.uniprot.org/

2.2.4 Polypeptide synthesis

Full-scale polypeptide characterization usually requires modest/large (milligram to gram) amounts of the purified target polypeptide. Even larger quantities are then generally required if the polypeptide has a commercial application. In some cases a polypeptide can be obtained in sufficient quantities by direct extraction from its natural producer source. However, polypeptides may also be produced by direct chemical synthesis, as long as their amino acid sequence (and any PTMs) has been elucidated. Synthesis can be undertaken via a biological route (recombinant DNA technology), as is the case for virtually all modern therapeutic proteins.

2.3 Higher level structure

Thus far we have concentrated on the primary structure (amino acid sequence) of a polypeptide. Higher level protein structure can be described at various levels, i.e. secondary, tertiary and quaternary:

- Secondary structure can be described as the local spatial conformation of a polypeptide's backbone, excluding the constituent amino acid's side chains. The major elements of secondary structure are the α-helix and β-strands, as described below.

- Tertiary structure refers to the three-dimensional arrangement of all the atoms that contribute to the polypeptide.

- Quaternary structure refers to the overall spatial arrangement of polypeptide subunits within a protein composed of two or more polypeptides.

2.3.1 Secondary structure

By studying the backbone of most proteins, stretches of amino acids that adopt a regular, recurring shape usually become evident. The most commonly observed secondary structural elements are termed the α-helix and β-strands, which are usually separated by stretches largely devoid of regular, recurring conformation. The α-helix and β-sheets are commonly formed because they maximize formation of stabilizing intramolecular hydrogen bonds and minimize steric repulsion between adjacent side chain groups, while also being compatible with the rigid planar nature of the peptide bonds.

The α-helix contains 3.6 amino acid residues in a full turn (Figure 2.5). This approximates to a length of 0.56 nm along the long axis of the helix. The participating amino acid side chains protrude outward from the helical backbone. Amino acids most conducive with α-helix formation include alanine, leucine, methionine and glutamate. Proline, as well as the occurrence in close proximity of multiple residues with either bulky side groups or side groups of the same charge, tends to disrupt α-helical formation. The helical structure is stabilized by hydrogen bonding, with every backbone C=O group forming a hydrogen bond with the N—H group four residues ahead of it in the helix. Stretches of α-helix found in globular (i.e. tightly folded, approximately spherical) polypeptides can vary in length from a single helical turn to greater than 10 consecutive helical turns. The average length is about three turns.

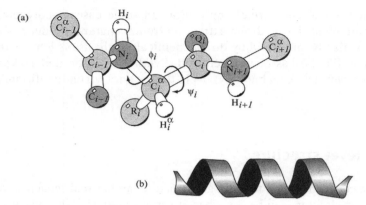

Figure 2.5   Ball-and-stick and ribbon representations of an α-helix. Reproduced from Sun, P. and Boyington. 1997. *Current Protocols in Protein Science* by kind permission of the publisher, John Wiley and Sons

Stretches of α-helix are most often positioned on the protein's surface, with one face of the helix facing the hydrophobic interior and the other facing the surrounding aqueous medium. The amino acid sequence of these helices is such that hydrophobic amino acid residues are positioned on one

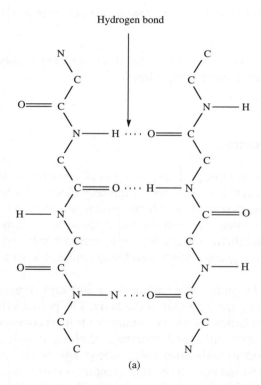

Figure 2.6   The β-sheet. (a) Two segments of β-strands (antiparallel) forming a β-sheet via hydrogen bonding. The β-strand is drawn schematically as a thick arrow. By convention the arrowhead points in the direction of the polypeptide's C terminus. (b) Schematic illustration of a two-strand β-sheet in parallel and antiparallel modes

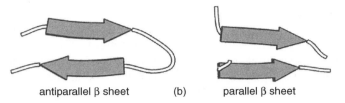

antiparallel β sheet   (b)   parallel β sheet

Figure 2.6 Continued

face of the helix, whereas hydrophobic amino acids line the other. The transmembrane sections of polypeptides that span biological membranes often display one (or more) α-helical stretches. In such instances, almost all the residues found in the helix display hydrophobic side chains.

β-strands represent the other major recurring structural element of proteins. β-strands usually are 5–10 amino acid residues in length, with the residues adopting an almost fully extended zigzag conformation. Single β-strands are rarely, if ever, found alone. Instead, two or more of these strands align themselves together to form a β-sheet. The β-sheet is a common structural element stabilized by maximum hydrogen bonding (Figure 2.6). The individual β-strands participating in β-sheet formation may all be present in the same polypeptide, or may be present in two polypeptides held in close juxtaposition. β-sheets are described as being parallel, antiparallel or mixed. A parallel sheet is formed when all the participating β-stretches are running in the same direction (e.g. from the amino terminus to the carboxy terminus; Figure 2.6). An antiparallel sheet is formed when successive strands have alternating directions (N-terminus to C-terminus followed by C-terminus to N-terminus, etc.). A β-sheet containing both parallel and antiparallel strands is termed a mixed sheet.

In terms of secondary structure, most proteins consist of several segments of α-helix and/ or β-strands separated from each other by various loop regions. These regions can vary in length and shape, and allow the overall polypeptide to fold into a compact tertiary structure. In addition to their obvious role in connecting stretches of regular secondary elements, loop regions themselves often participate/contribute directly to the polypeptide's biological function. The antigen-binding region of antibodies, for example, is largely constructed from six loop regions (Chapter 13). Such loops also often form the active site of enzymes (Chapter 12). One loop structure, termed a β-turn or β-bend, is a characteristic feature of many polypeptides (Figure 2.7).

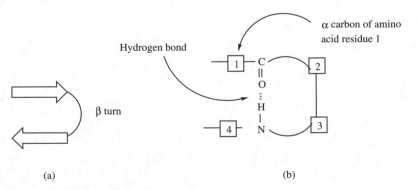

(a)   (b)

Figure 2.7 (a) The β-bend or β-turn is often found between two stretches of antiparallel β-strands. (b) It is stabilized in part by hydrogen bonding between the C=O bond and the NH groups of the peptide bonds at the neck of the turn

2.3.2 Tertiary structure

As mentioned previously, a polypeptide's tertiary structure refers to its exact three-dimensional structure, relating the relative positioning in space of all the polypeptide's constituent atoms to each other. The tertiary structure of small polypeptides (approximately 200 amino acid residues or less) usually forms a single discrete structural unit. However, when the three-dimensional structure of many larger polypeptides is examined, the presence of two or more structural subunits within the polypeptide becomes apparent. These are termed domains. Domains, therefore, are (usually) tightly folded subregions of a single polypeptide, connected to each other by more flexible or extended regions. As well as being structurally distinct, domains often serve as independent units of function. Cell surface receptors, for example, usually contain one or more extracellular domains (some or all of which participates in ligand binding), a transmembrane domain (hydrophobic in nature and serving to stabilize the protein in the membrane) and one or more intracellular domains that play an effector function (e.g. generation of second messengers). Many therapeutic proteins also display several domains. Tissue plasminogen activator (tPA), for example (Chapter 12), consists of five such domains.

2.3.3 Higher structure determination

There are three potential methods by which a protein's three-dimensional structure can be visualized: X-ray diffraction, NMR and electron microscopy. The latter method reveals structural information at low resolution, giving little or no atomic detail. It is used mainly to obtain the gross three-dimensional shape of very large (multi-polypeptide) proteins, or of protein aggregates such as the outer viral caspid. X-ray diffraction and NMR are the techniques most widely used to obtain high-resolution protein structural information, and details of both the principles and practice of these techniques may be sourced from selected references provided at the end of this chapter. The experimentally determined three-dimensional structures of some polypeptides are presented in Figure 2.8.

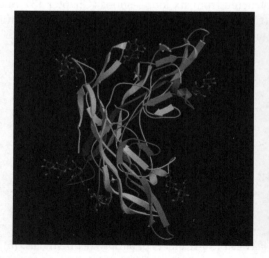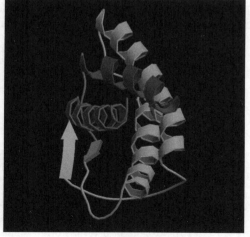

Figure 2.8 Three-dimensional structure of (a) human interleukin-4, as determined by NMR, and (b) human follicle-stimulating hormone, as determined by X-ray diffraction. Reproduced from protein data bank (www. rcsb.org/pdb, molecule ID numbers 1 ITM and 1 FL7 respectively)

2.4 Protein stability and folding

Upon biosynthesis, a polypeptide folds into its native conformation, which is structurally stable and functionally active. The conformation adopted ultimately depends upon the polypeptide's amino acid sequence, explaining why different polypeptide types have different characteristic conformations. We have previously noted that stretches of secondary structure are stabilized by short-range interactions between adjacent amino acid residues. Tertiary structure, on the other hand, is stabilized by interactions between amino acid residues that may be far apart from each other in terms of amino acid sequence, but which are brought into close proximity by protein folding. The major stabilizing forces of a polypeptide's overall conformation are:

- hydrophobic interactions

- electrostatic attractions

- covalent linkages.

Hydrophobic interactions are the single most important stabilizing influence of protein native structure. The 'hydrophobic effect' refers to the tendency of non-polar substances to minimize contact with a polar solvent such as water. Non-polar amino acid residues constitute a significant proportion of the primary sequence of virtually all polypeptides. These polypeptides will fold in such a way as to maximize the number of such non-polar residue side chains buried in the polypeptide's interior, i.e. away from the surrounding aqueous environment. This situation is most energetically favourable.

Stabilizing electrostatic interactions include van der Waals forces (which are relatively weak), hydrogen bonds and ionic interactions. Although nowhere near as strong as covalent linkages (Table 2.5), the large number of such interactions existing within a polypeptide renders them collectively quite strong.

Although polypeptides display extensive networks of intramolecular hydrogen bonds, such bonds do not contribute very significantly to overall conformational stability. This is because atoms hydrogen bonding with each other in a folded polypeptide can form energetically equivalent hydrogen bonds with water molecules if the polypeptide is in the unfolded state. Ionic attractions between (oppositely) charged amino acid side chains also contribute modestly to overall protein conformational stability. Such linkages are termed salt bridges, and, as one would expect, they are located primarily on the polypeptide surface.

Table 2.5 Approximate bond energies associated with various (non-covalent) electrostatic interactions, compared with a carbon–carbon single bond

Bond type	Bond strength (kJ mol^{-1})
Van der Waals forces	10
Hydrogen bond	20
Ionic interactions	86
Carbon–carbon bond	350

Disulfide bonds represent the major covalent bond type that can help stabilize a polypeptide's native three-dimensional structure. Intracellular proteins, although generally harbouring multiple cysteine residues, rarely form disulfide linkages, due to the reducing environment that prevails within the cell. Extracellular proteins, in contrast, are usually exposed to a more oxidizing environment, conducive to disulfide bond formation. In many cases the reduction (i.e. breaking) of disulfide linkages has little effect upon a polypeptide's native conformation. However, in other cases (particularly disulfide-rich proteins), disruption of this covalent linkage does render the protein less conformationally stable. In these cases the disulfide linkages likely serve to 'lock' functional/structurally important elements of domain/tertiary structure in place.

The description of protein structure as presented thus far may lead to the conclusion that proteins are static, rigid structures. This is not the case. A protein's constituent atoms are constantly in motion, and groups ranging from individual amino acid side chains to entire domains can be displaced via random motion by anything up to approximately 0.2 nm. A protein's conformation, therefore, displays a limited degree of flexibility, and such movement is termed 'breathing'.

Breathing can sometimes be functionally significant by, for example, allowing small molecules to diffuse in/out if the protein's interior. In addition to breathing, some proteins may undergo more marked (usually reversible) conformational changes. Such changes are usually functionally significant. Most often they are induced by biospecific ligand interactions (e.g. binding of a substrate to an enzyme or antigen binding to an antibody).

2.4.1 Structural prediction

Currently, there exists an enormous and growing deficit between the number of polypeptides whose amino acid sequence has been determined and the numbers of polypeptides whose three-dimensional structure has been resolved. Given the complexities of resolving three-dimensional structure experimentally, it is not surprising that scientists are continually attempting to develop methods by which they could predict higher order structure from amino acid sequence data. Although modestly successful secondary structure predictive approaches have been developed, no method by which tertiary structure may be predicted from primary data has thus far been developed.

Over 20 different methods of secondary structure prediction have been reported (Table 2.6). The approaches taken fall into two main categories:

Table 2.6 Some secondary structure predictive methods currently used. Refer to text for further details

Method	Basis of prediction
Chou and Fasman	Empirical statistical method
Garnier, Osguthorpe and Robson (GOR)	Empirical statistical method
EMBL profile neural network (PHD)	Empirical statistical method
Protein sequence analysis (PSA)	Empirical statistical method
Lim	Physicochemical criteria

1. Empirical statistical methods, which are based upon data generated from studying proteins of known three-dimensional structure and correlation of such proteins' primary amino acid sequences with structural features.

2. Methods based upon physicochemical criteria, such as fold compactness (i.e. the generation of a folded form displaying a tightly packed hydrophobic core and a polar surface).

Most such predictive methods are at best 50–70 per cent accurate. The relatively large inaccuracy stems from the fact that the folded (tertiary) structure imposes constraints upon the nature/extent of secondary structure within some regions of the polypeptide chain. Any generalized 'rules' relating secondary structure to amino acid sequence data, by nature, will not take such issues into consideration.

Accurate prediction of a polypeptide's three-dimensional structure from lower order structural information remains to be achieved. Tertiary structure prediction directly from amino acid sequence data remains in the distant future, although a technique known as threading will likely support some progress towards this goal. Three-dimensional structural analysis has shown that only a limited number of stable protein folds exist and, moreover, that many unrelated amino acid sequences can generate the same fold (a fold refers to a domain-like structure that is common to many proteins). By analysing databases containing polypeptide tertiary structure information, the various possible amino acid sequences that can give rise to any particular fold can potentially be determined.

Threading essentially entails comparing the sequence of the polypeptide whose three-dimensional structure you wish to predict with the database sequences known to generate specific fold patterns. Computer programs can then be used to estimate the probability of the target sequence adopting each known folding structure.

2.5   Protein post-translational modification

Many polypeptides undergo covalent modification after (or sometimes during) their ribosomal assembly. The most commonly observed such PTMs are listed in Table 2.7. Such modifications generally influence either the biological activity or the structural stability of the polypeptide. The majority of therapeutic proteins bear some form of PTM. Although glycosylation represents the most common such modification, additional PTMs important in a biopharmaceutical context include carboxylation, hydroxylation, sulfation and amidation; these PTMs are now considered further.

2.5.1   Glycosylation

Glycosylation (the attachment of carbohydrates) is one of the most common forms of PTM associated with eukaryotic proteins in general, particularly eukaryotic extracellular and cell surface proteins. In the case of some glycoproteins, removal of the sugar component has no detectable effect upon the biological properties (deglycosylated forms of the glycoprotein can be generated by including inhibitors of the glycosylation pathway, e.g. the antibiotic tunicamycin, in the cell growth media,

Table 2.7 Types of PTM that polypeptides may undergo. Refer to text for additional details

Modification	Example
Proteolytic processing	Various proteins become biologically active only upon their proteolytic cleavage (e.g. some blood factors)
Glycosylation	For some proteins glycosylation can increase solubility, influence biological half-life and/or biological activity
Phosphorylation	Influences/regulates biological activity of various polypeptide hormones
Acetylation	Function unclear
Acylation	May help some polypeptides interact with/anchor in biological membranes
Amidation	Influences biological activity/stability of some polypeptides
Sulfation	Influences biological activity of some neuropeptides and the proteolytic processing of some polypeptides
Hydroxylation	Important to the structural assembly of certain proteins
γ-Carboxyglutamate formation	Important in allowing some blood proteins to bind calcium
ADP-ribosylation	Regulates biological activity of various proteins
Disulfide bond formation	Helps stabilize conformation of some proteins

or by enzymatic degradation of the glycocomponent of preformed glycoproteins using glucosidase enzymes). However, in other cases the sugar component plays a direct role in the biological activity of the glycoprotein (Table 2.8). Native hCG (Chapter 11), for example, is a heavily glycosylated gonadotrophic hormone. Removal of its sugar components usually abolishes its ability to induce a biological response, although the hormone's binding affinity for its receptor remains unaltered, or is sometimes actually increased. Therapeutic proteins now approved for general medical use that are glycosylated are listed in Table 2.9.

Carbohydrate side chains are synthesized by a family of enzymes known as glycosyltransferases, located mainly in the endoplasmic reticulum. Two types of glycosylation can occur: N-linked and O-linked. In the case of N-linked glycosylation, the sugar chain (the oligosaccharide) is attached to the protein via the nitrogen atom of an asparagine (Asn) residue, whereas in O-linked systems the sugar chain is attached to the oxygen atom of hydroxyl groups, usually those of serine or threonine residues (Figure 2.9). Monosaccharides most commonly found in the sugar side chain(s) include mannose, galactose, glucose, fucose, N-acetylgalactosamine, N-acetylglucosamine, xylose and sialic acid. These can be joined together in various sequences and by a variety of glycosidic linkages. The carbohydrate chemistry of glycoproteins, therefore, is quite complex. The structure of two such example oligosaccharide chains is presented in Figure 2.10.

N-linked glycosylation is sequence specific, involving the transfer of a pre-synthesized oligosaccharide chain to an asn residue found in a characteristic sequence Asn–X–Ser, or Asn–X–Thr or Asn–X–Cys, where X represents any amino acid residue, with the exception of proline. An additional glycosylation determinant must also apply, as not all potential N-linked sites are glycosylated in some proteins. The pre-synthesized oligosaccharide side chain then undergoes

Table 2.8 The potential roles and effects of the glycocomponent of glycoproteins. Reproduced from Walsh, G. and Jefferies, R. (2006). *Nature Biotechnology* **24**, 1241–1252

Role/effect	Comment
Protein folding	Glycosylation can effect local protein secondary structure and help direct folding of the polypeptide chain
Protein targeting/trafficking	The glycocomponent can participate in the sorting/directing of a protein to its final destination
Ligand recognition/binding	The carbohydrate content of antibodies, for example, plays a role in antibody binding to monocyte F_c receptors and interaction with complement component $C1_q$
Biological activity	The carbohydrate side chain of gonadotrophins is essential to the activation of gonadotrophin signal transduction
Stability	Sugar side chains can potentially stabilize a glycoprotein in a number of ways, including enhancing its solubility, shielding hydrophobic patches on its surface, protection from proteolysis and by direct participation in intrachain stabilizing interactions
Regulates protein half-life	High levels of sialic acid (a family of acidic sugars that often caps sugar side chains) can increase a glycoprotein's plasma half-life. Exposure of galactose residues can decrease plasma half-life by promoting uptake via hepatic galactose residues. Yeast glycosylation is of a 'high mannose' type, which can also drive rapid removal from circulation via specific cell-surface mannose receptors
Immunogenicity	Some glycosylation motifs characteristic of plant-derived glycoproteins (often containing fucose and xylose residues) are highly immunogenic in mammals

additional glycosyltransferase-mediated trimming/modification. The determinants of O-linked glycosylation are even less well understood. Characteristic sequence recognition is not apparent in most cases, and three-dimensional structural features may be more important in such instances. Some glycosylated proteins will be characterized by one or more N-linked sugar side chains, others by one or more O-linked side chains, and still others by both N- and O-linked chains. Human EPO (Chapter 10), for example, displays three N-linked and one O-linked sugar side chain.

For any glycoprotein, the exact composition and structure of the carbohydrate side chain can vary slightly from one molecule of that glycoprotein to the next. This results in microheterogeneity which can be directly visualized by analytical techniques such as isoelectric focusing (Chapter 7). Also contributing to heterogeneity can be variable site glycosylation, in which some glycosylation sites remain unoccupied within a proportion of the glycoprotein molecules. The overall basis of heterogeneity is likely due to factors such as glycosyltransferase substrate specificity and the fact

Table 2.9 Approved therapeutic proteins (listed by trade name) that are glycosylated. These products are discussed in subsequent chapters. Reproduced from Walsh, G. and Jefferies, R. 2006. *Nature Biotechnology* **24**, 1241–1252

Product category	Specific products (by trade name)
Blood factors, anticoagulants and thrombolytics	Activase, Advate, Benefix, Bioclate, Helixate/Kogenate, Metalyse/TNKase, Novoseven, Recombinate, Refacto, Xigiris
Antibodies	Avastin, Bexxar, Erbitux, Herceptin, Humaspect, Humira, Mabcampath/Campath-H1, Mabthera/Rituxan, Mylotarg, Neutrospec, Oncoscint, Orthoclone OKT-3, Prostascint, Raptiva, Remicade, Simulect, Synagis, Xolair, Zenapax, Zevalin
Hormones	Gonal F, Luveris, Ovitrelle/Ovidrel, Puregon/Follistim, Thyrogen
EPO and colony-stimulating factors	Epogen/Procrit, Leukine, Neorecormon, Nespo/Aranesp
Interferons	Avonex, Rebif
Additional	Aldurazyme, Amevive, Cerezyme, Enbrel, Fabrazyme, Inductos, Infuse, Osigraft/OP-1 implant, Pulmozyme, Regranex, Replagal

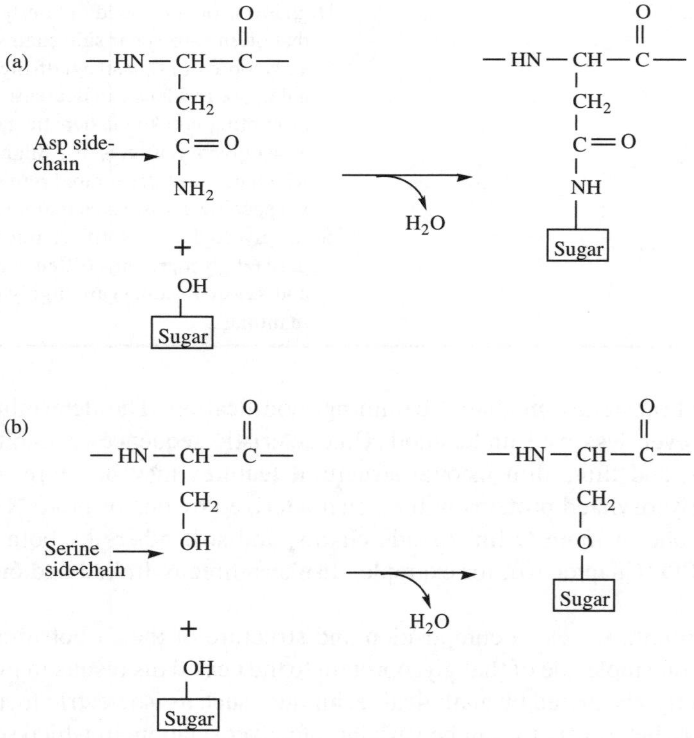

Figure 2.9   (a) N-linked versus (b) O-linked glycosylation. 'Sugar' represents an oligosaccharide chain, an example of which is provided in Figure 2.10

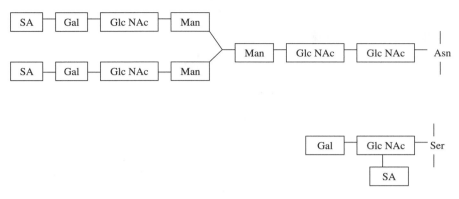

Figure 2.10 Structure of two sample oligosaccharide side chains (one N-linked the other O-linked) found in glycoproteins. Man: manose; Gal: galactose; SA: sailic acid; GlcNAc: *N*-acetyl glucosamine; GalNAc: *N*-acetyl galactosamine

that a proportion of the glycosylated proteins may be exported from the cell before they are fully processed by glycosyltransferases.

Virtually all therapeutic glycoproteins, even when produced naturally in the body, exhibit such heterogeneity; for example, two species of human interferon-γ (IFN-γ), one of molecular mass 20 kDa and the other of 25 kDa, differ from each other only in the degree and sites of (N-linked) glycosylation.

Furthermore, the glycosylation patterns obtained when human glycoproteins are expressed in non-human eukaryotic expression systems (e.g. animal cell culture) are usually somewhat different from the glycosylation pattern associated with the native human protein. The glycosylation pattern of human tPA produced in transgenic animals, for example, is different to the pattern obtained when the same gene is expressed in a recombinant mouse cell line. Both these patterns are, in turn, different to the native human pattern. The clinical significance, if any, of altered glycosylation patterns/microheterogeneity is not always predictable and is best determined by direct clinical trials. If the product is found to be safe and effective, then routine end-product quality control (QC) analysis for carbohydrate-based microheterogeneity is carried out more to determine batch-to-batch consistency (which is desirable) rather than to detect microheterogeneity *per se*.

2.5.2 Carboxylation and hydroxylation

γ-Carboxylation and β-hydroxylation are PTMs characteristic of a limited number of proteins, mainly a subset of proteins that function in the haemostatic process. γ-Carboxylation entails the enzymatic conversion of the side chains of specific glutamate residues in target proteins, forming γ-carboxyglutamate (conversion of 'Glu' residues to 'Gla' residues; Figure 2.11a). β-Hydroxylation usually entails the hydroxylation of target aspartate (Asp) residues yielding β-hydroxyaspartate (Asp → Hya; Figure 2.11b). Both PTMs help mediate the binding of calcium ions, which is important/essential to the effective functioning of blood factors VII, IX and X, as well as activated protein C and protein S of the anticoagulant system (Chapter 12).

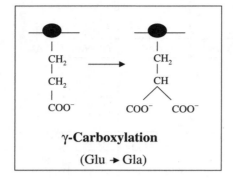

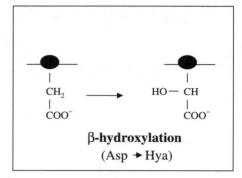

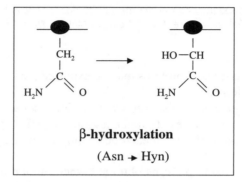

Figure 2.11 γ-Carboxylation of glutamate residues (Glu) yields γ-carboxyglutamate (Gla), whereas β-hydroxylation of aspartate (Asp) yields β-hydroxyaspartate (Hya) and β-hydroxylation of asparagine (Asn) yields β-hydroxyasparagine (Hyn)

2.5.3   Sulfation and amidation

Sulfation and amidation are two additional PTMs characteristic of a small number of biopharmaceuticals. Sulfation entails the enzyme-catalysed attachment of sulfate (SO_4^{2-}) groups to target polypeptides, usually via specific tyrosine side chains. Sulfation often plays a role in protein–protein interactions, and lack of sulfation tends to reduce a polypeptide's activity, as opposed to abolishing it completely. Notable therapeutic proteins that are sulfated in their natural state include the anticoagulant hirudin (Chapter 12) and blood factors VIII and IX. The recombinant forms of these proteins produced by genetic engineering generally have reduced levels/absence of sulfation, but yet they remain therapeutically effective.

Amidation refers to the replacement of a protein's C-terminal carboxyl group with an amide group (COOH → $CONH_2$). This PTM is usually characteristic of peptides (very short chains of amino acids), as opposed to the longer polypeptides, but one therapeutic polypeptide (salmon calcitonin, Chapter 11) is amidated, and amidation is required for full functional activity. Overall, the function(s) of amidation is not well understood, although in some cases at least it appears to contribute to peptide/polypeptide stability and/or activity.

Further reading

Books

Buxbaum, E. 2006. *Fundamentals of Protein Structure and Function*. Springer.

Carey, P. 1996. *Protein Engineering and Design*. Academic Press.

Cavanagh, J., Fairbrother, W.J., Palmer III, A.G., Skelton, N.J., and Rance, M. 2006. *Protein NMR Spectroscopy: Principles and Practice*, 2nd edition. Academic Press.

Creighton, T. 2006. *Proteins*. Oxford University Press.

Eidhammer, I., Jonassen, I., and Taylor, W.R. 2006. *Protein Bioinformatics*. Wiley.

Fersht, A. 1999. *Structure and Mechanism in Protein Science*. Freeman.

Higgins, D. 2000. *Bioinformatics*. Oxford University Press.

Rehm, H. 2006. *Protein Biochemistry and Proteomics*. Academic Press.

Walsh, C. 2005. *Posttranslational Modification of Proteins*. Roberts & Company Publishers.

Articles

Protein structure and folding

Al-Lazikani, B., Jung, J., Xiang, Z., and Honig, B. 2001. Protein structure prediction. *Current Opinion in Chemical Biology* **5**(1), 51–56.

Basharov, M. 2003. Protein folding. *Journal of Cellular and Molecular Medicine* **7**(3), 223–237.

Bonvin, A.M., Boelens, R., and Kaptein, R. 2005. NMR analysis of protein interactions. *Current Opinion in Chemical Biology* **9**(5), 501–508.

Brannigan, J. and Wilkinson, A. 2002. Protein engineering 20 years on. *Nature Reviews: Molecular Cell Biology* **3**(12), 964–970.

Chasse, G.A., Rodriguez, A.M., Mak, M.L., Deretey, E., Perczel, A., Sosa, C.P., Enriz, R.D., and Csizmadia I.G. 2001. Peptide and protein folding. *Journal of Molecular Structure: Theochem* **537**, 319–361.

Fiser, A. 2004. Protein structure modeling in the proteomics era. *Expert Review of Proteomics* **1**(1), 97–110.

Floudas, C.A., Fung, H.K., McAllister, S.R., Monnigmann, M., and Rajgaria, R. 2006. Advances in protein structure prediction and de novo protein design: a review. *Chemical Engineering Science* **61**(3), 966–988.

Gustafsson, C., Govindarajan S., and Minshull, J. 2003. Putting engineering back into protein engineering: bioinformatics approaches to catalyst design. *Current Opinion in Biotechnology* **14**(4), 366–370.

Koing, S. 2005. Functional protein analysis using mass spectrometry. *Current Organic Chemistry* **9**(9), 875–887.

Osguthorpe, D. 2000. *Ab initio* protein folding. *Current Opinion in Structural Biology* **10**(2), 146–152.

Petrey, D. and Honig, B. 2005. Protein structure prediction: inroads to biology. *Molecular Cell* **20**(6), 811–819.

Radford, S. 2000. Protein folding; progress made and promises ahead. *Trends in Biochemical Sciences* **25**(12), 611–618.

Wishart, D. 2005. NMR spectroscopy and protein structure determination: applications to drug discovery and development. *Current Pharmaceutical Biotechnology* **6**(2), 105–120.

Protein stability and post-translational modifications

Bayle, J.H. and Crabtree, G.R. 1997. Protein acetylation; more than chromatin modification to regulate transcription. *Chemistry and Biology* **4**(12), 885–888.

Bosshard, H.R., Marti, D.N., and Jelesarov, I. 2004. Protein stabilization by salt bridges: concepts, experimental approaches and clarification of some misunderstandings. *Journal of Molecular Recognition* **17**(1), 1–16.

Chi, E.Y., Krishnan, S., Randolph, T.W., and Carpenter, J.F. 2003. Physical stability of proteins in aqueous solution: mechanism and driving forces in nonnative protein aggregation. *Pharmaceutical Research* **20**(9), 1325–1336.

Frokjaer, S and Otzen, D. 2005. Protein drug stability: a formulation challenge. *Nature Reviews: Drug Discovery* **4**(4), 298–306.

Imperiali, B. and O'Connor, S. 1999. Effect of N-linked glycosylation on glycopeptide and glycoprotein structure. *Current Opinion in Chemical Biology* **3**(6), 643–649.

Lee, B. and Vasmatis, G. 1997. Stabilization of protein structures. *Current Opinion in Biotechnology* **8**, 423–428.

McLihnney, R. 1990. The fats of life; the importance and function of protein acylation. *Trends in Biochemical Sciences* **15**, 387–391.

Merry, T. 1999. Current techniques in protein glycosylation analysis – a guide to their application. *Acta Biochimica Polonica* **46**(2), 303–314.

Parodi, A. 2000. Protein glycosylation and its role in protein folding. *Annual Review of Biochemistry* **69**, 69–93.

Van den Burg, B. and Eijsink, V. 2002. Selection of mutations for increased protein stability. *Current Opinion in Biotechnology* **13**(4), 333–337.

Walsh, G. and Jefferis, R. 2006. Posttranslational modifications in the context of therapeutic proteins. *Nature Biotechnology* **24**, 1241–1252.

Yan, J.X., Packer, N.H., Gooley, A.A., and Williams, K.L. 1998. Protein phosphorylation; technologies for the identification of phosphoamino acids. *Journal of Chromatography A* **808**(1–2), 23–41.

3

Gene manipulation and recombinant DNA technology

3.1 Introduction

The biopharmaceutical sector is largely based upon the application of techniques of molecular biology and genetic engineering for the manipulation and production of therapeutic macromolecules. The majority of approved biopharmaceuticals (described from Chapter 8 onwards) are proteins produced in engineered cell lines by recombinant means. Examples include the production of insulin in recombinant *E. coli* and recombinant *S. cerevisiae*, as well as the production of EPO in an engineered (Chinese hamster ovary) animal cell line.

Terms such as 'molecular biology', 'genetic engineering' and 'recombinant DNA (rDNA) technology' are sometimes used interchangeably and often mean slightly different things to different people. Molecular biology, in its broadest sense, describes the study of biology at a molecular level, but focuses in particular upon the structure, function and interaction/relationship between DNA, RNA and proteins. Genetic engineering, on the other hand, describes the process of manipulating genes (outside of a cell's/organism's normal reproductive process). It generally involves the isolation, manipulation and subsequent reintroduction of stretches of DNA into cells and is usually undertaken in order to confer on the recipient cell the ability to produce a specific protein, such as a biopharmaceutical. 'rDNA technology' is a term used interchangeably with 'genetic engineering'. rDNA is a piece of DNA artificially created *in vitro* which contains DNA (natural or synthetic) obtained from two or more sources.

When developing a new protein biopharmaceutical, one of the earliest actions undertaken entails identifying and isolating the gene (or complementary DNA (cDNA); see later) coding for the target protein, the generation of an appropriate piece of rDNA containing the protein's coding sequence and the introduction of this rDNA into an appropriate host cell such that the target protein is made in large quantities by that engineered cell. The drug development process and the cell types generally chosen to produce recombinant proteins are described in Chapters 4 and 5 respectively. This chapter aims to provide an introductory overview of the approaches and techniques used to isolate the target gene, generate an rDNA sequence and introduce it into an appropriate producer cell. Before we look at these techniques, however, we will briefly review the basic biology and structure of nucleic acids.

Pharmaceutical biotechnology: concepts and applications   Gary Walsh
© 2007 John Wiley & Sons, Ltd   ISBN 978 0 470 01244 4 (HB)   978 0 470 01245 1 (PB)

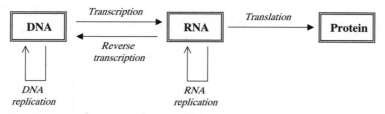

Figure 3.1   Schematic representation of the so-called central dogma of molecular biology. DNA replication is essential to the transmission of genetic information from one generation to the next in most life forms (i.e. in living forms whose genomes are DNA based). RNA replication is essential to the transmission of genetic information in the context of a small number of viruses whose genomes are RNA based. Transcription describes the copying of selected DNA sequences into RNA, and translation describes the conversion of the genetic information inherent in mRNA into a polypeptide of defined amino acid sequence. The process of reverse transcription is a central feature of certain viruses (retroviruses) containing an RNA-based genome which, as part of their life cycle, infect eukaryotic cells and convert their RNA-based genomes into a DNA-based one (see Box 14.1)

3.2   Nucleic acids: function and structure

Nucleic acids represent a prominent category of biomolecule present in living cells. The term incorporates both DNA and RNA. DNA represents the repository of genetic information (the genome) of most life forms. RNA replaces DNA as the repository of genetic information in some viruses. In most life forms, however, RNA plays a role in mediating the conversion of genetic information stored in specific DNA sequences (genes) into polypeptides. There are three subcategories of RNA, each playing a different role in the conversion of gene sequences into the amino acid sequence of polypeptides. Messenger RNA (mRNA) carries the genetic coding information from the gene to the ribosome, where the polypeptide is actually synthesized. Ribosomal RNA (rRNA), along with a number of proteins, forms the ribosome itself, and transfer RNA (tRNA) functions as an adaptor molecule, transferring a specific amino acid to a growing polypeptide chain on the ribosomal site of polypeptide synthesis. Therefore, nucleic acids, between them all, mediate the flow of genetic information via the processes of replication, transcription and translation as outlined in what has become known as the central dogma of molecular biology (Figure 3.1).

Structurally, nucleic acids are polymers in which the basic recurring monomer is a nucleotide (i.e. nucleic acids are polynucleotides). Nucleotides themselves consist of three components: a phosphate group, a pentose (five-carbon sugar) and a nitrogenous-containing cyclic structure known as a base (Figure 3.2). The nucleotide sugar associated with RNA is ribose, whereas that found in DNA is deoxyribose (Figure 3.3). In total, five different bases are found in nucleic acids. They are categorized as either purines (adenine and guanine, or A and G, found in both RNA and DNA) or pyrimidines (cytosine, thymine and uracil, or C, T and U). Cytosine is found in both RNA and DNA, whereas thymine is unique to DNA and uracil is unique to RNA (Figure 3.4).

The DNA or RNA polymer consists of a chain of nucleotides of specific base sequence, linked via phosphodiester bonds (Figure 3.5). RNA is a single-stranded polynucleotide, although RNA molecules tend to adopt higher order three-dimensional shapes. DNA, on the other hand, is a double-stranded molecule (Figure 3.6) that assumes a double helical structure. The two polynucleotide strands face each other in an antiparallel manner (Figure 3.6), with the hydrophilic sugar and phosphate residues facing outwards, towards the surrounding aqueous-based environment,

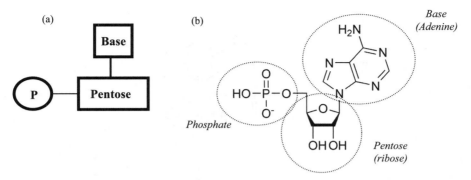

Figure 3.2 (a) The basic structure of a nucleotide. (b) The actual chemical structure of one representative nucleotide (adenylate, i.e. adenosine 5′-monophosphate)

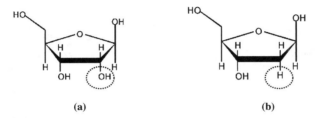

Figure 3.3 Chemical structure of (a) ribose and (b) 2′-deoxyribose, the nucleotide pentoses found in RNA and DNA respectively. The differences in chemical structure are highlighted by the dotted circles

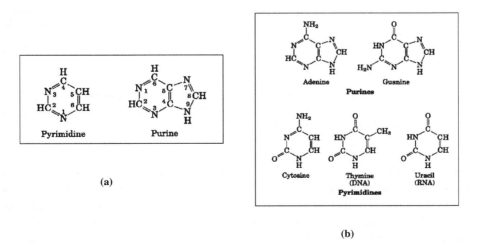

Figure 3.4 The five bases found in nucleic acids may be categorized as either pyrimidines or purines. Refer to text for details

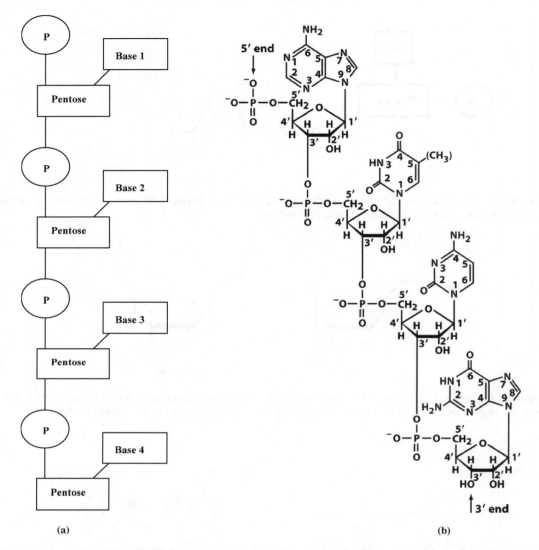

Figure 3.5   The basic polynucleotide structure as shown in (a) outline form and (b) in chemical detail. The 5′ end of the chain is defined by lacking a nucleotide attached to the first sugar's carbon number 5; the 3′ end lacks a nucleotide attached to the carbon number 3 of the last sugar in the backbone

and the more hydrophobic bases point inwards. The base sequence of each chain displays complementarity. Wherever thymine is found in one chain, adenine is found positioned opposite it in the other. Wherever guanine is found in one chain, cytosine is found positioned opposite it in the second chain. Complementarity provides an obvious mechanism to ensure the fidelity of DNA replication and to underline transcription. The double helical DNA structure is stabilized by (a) hydrogen bonding between complementary opposite bases (two hydrogen bonds between A and T, three hydrogen bonds between G and C; Figure 3.6) and (b) by hydrophobic stacking

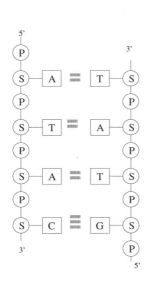

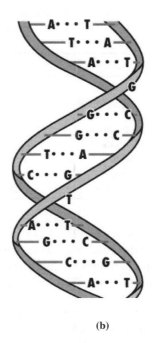

(b)

Figure 3.6 (a) DNA structure. The two complementary polynucleotide strands in DNA are antiparallel to each other in orientation (one runs 5′→3′, the other 3′→5′; see Figure 3.5). The two strands are held together by hydrogen bonds between opposite complementary bases, as well as hydrophobic interactions between stacked bases, as described in the text. (b) The double polynucleotide chain adopts a double helical structure

interactions between the planar, largely hydrophobic bases effectively stacked above each other along the length of each strand.

3.2.1 Genome and gene organization

The genome refers to the entire hereditary information present in an organism. As discussed earlier, this is usually encoded by double-stranded DNA (the genome of most plant viruses and some animal and bacterial viruses is RNA based). DNA-based genomes are largely or exclusively organized into chromosomes, each chromosome being a single DNA molecule housing multiple genes, as well as non-coding sequences.

Bacteria normally harbour a single, circular chromosome that tends to be tethered to the bacterial plasma membrane and tends to have few if any closely associated proteins. Many bacteria also contain extra-chromosomal DNA in the form of plasmids, as will be discussed later. Eukaryotes (plants, animals and yeasts) posses multiple linear chromosomes contained within a cell nucleus, and these chromosomes are normally closely associated with proteins termed histones (the protein–DNA complex is termed chromatin). Eukaryotes also invariably possess DNA sequences within mitochondria and in chloroplasts in plants. The (usually circular) DNA molecules are much

Table 3.1 The number of chromosomes found in selected species/cells, along with their predicted/estimated (approximate) number of genes

Cell/species	No. chromosomes	No. genes
E. coli	1	4400
S. cerevisiae	16	6200
Fern	1200	13600
Fruit fly	18	13000
Mouse	40	30000–35000
Rat	42	23000
Dog	78	19300
Human	46	30000–35000

shorter than chromosomal DNA, are often present in multiple copy number and tend to house genes coding for proteins required within these organelles. Human mitochondrial DNA, for example, is 6600 base pairs (6.6 kbp) in length and houses 37 genes. Such DNA molecules are believed to be vestiges of chromosomes from ancient bacteria that gained entry into early eukaryotic cells.

The genomes of different species are organized into different numbers of chromosomes, as is evident from Table 3.1. Chromosomes present in all cells contain both coding regions (i.e. genes, which are stretches of DNA that encode the specific amino acid sequence of a particular polypeptide or the exact nucleotide sequence of a tRNA or rRNA) and non-coding regions. Coding regions, as we will subsequently see, often represent only a small fraction of total genome sequences.

In close association with gene sequences are regulatory elements, i.e. stretches or regions of DNA that mark the beginning or end of a gene or a series of related genes or which regulate the level of gene expression (Figure 3.7). A characteristic regulatory sequence upstream (i.e. on the 5′ side) of a gene is termed the promoter region (P), which RNA polymerases (the enzymes responsible for transcribing the gene into RNA) identify and bind. Immediately adjacent to this is a characteristic sequence that represents the starting point for transcription (T_C). Immediately downstream of the gene is a transcriptional termination site (t_C). The intervening sequence, of course, represents the precise stretch of DNA that is copied into RNA and is often called the transcriptional unit. The gene sequence will often contain start and stop signals or sequences (T_L and t_L) that ultimately dictate the precise stretch of transcriptional unit actually translated into polypeptide (Figure 3.7). Other regulatory regions controlling gene expression can also be present, either upstream and/or downstream of the gene itself. In addition to genes and their associated

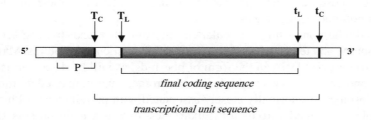

Figure 3.7 Generalized gene organization within the genome. Refer to text for details

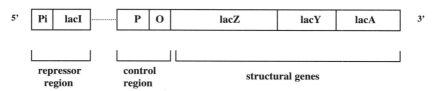

Figure 3.8 The *lac* operon houses three structural genes: *lacZ*, *lacY* and *lacA*. These code for three enzymes required for lactose metabolism (β-galactosidase, galactose permease and a transacetylase). Immediately upstream of these structural genes is a control region that houses a promoter (P) and operator (O) sequence. The operator represents a binding site for a 'repressor' protein that is in turn coded for by a repressor gene (*Pi*) found nearby but upstream of the *lac* operon. The repressor gene is in turn controlled by its own promoter. In the absence of the sugar lactose (or, more accurately, an isomer of lactose called 1,6-allolactose, which acts as an inducer) the repressor gene product is bound to the *lac* operator site, preventing transcription of the *lac* operon. In the presence of lactose (and hence the inducer), the inducer binds the repressor and the inducer–repressor complex disassociates from the operator, allowing transcription to go ahead. A polycystronic mRNA is produced, but the operon also houses translational start and stop sites that allow for independent ribosomal production of the three gene products

regulatory sequences, DNA molecules also invariably house additional non-coding sequences in the form of various kinds of repeat sequences. For example, genes account for only some 30 per cent of the total human genome sequence.

The detail and arrangement of gene structure is also normally different in prokaryotes and eukaryotes. In prokaryotes, genes of related function are often clustered together in operons, which are usually under the control of a single promoter/regulatory region. An example is the well-known 'lac operon' described in Figure 3.8. Transcribed operon mRNA thus usually contains coding sequence information for several polypeptides, and such mRNA is termed polycistronic. Although common in prokaryotes, the presence of polycistronic operons is infrequent in lower eukaryotes and essentially absent from higher eukaryotes, where virtually all protein-encoding genes are transcribed separately. Eukaryotic genes, however, usually contain coding sequences (exons) that are interrupted by non-coding intervening sequences (introns), and in many cases exons represent a minor proportion of the entire gene length (Figure 3.9, and also see Figure 4.3). For example, of the 30 per cent of the human genome believed to be taken up by genes, an estimated 28.5 per cent is accounted for by introns with only some 1.5 per cent being accounted for by exons.

mRNA transcripts in eukaryotes undergo substantial editing. The introns are enzymatically removed from (spliced out of) the primary transcript, and further characteristic modifications include the addition of a cap at the mRNA's 5′ end and the addition of a polyadenine nucleotide tail (poly A tail) at the molecule's 3′ terminus.

3.2.2 Nucleic acid purification

A prerequisite step to any rDNA work is the initial isolation of DNA or RNA from the source material (which can be microbial, plant, animal or viral). Numerous methodologies have been developed to achieve nucleic acid purification, and some of these methodologies have been adapted for use in a variety of commercially available purification kits. Although details vary, the general

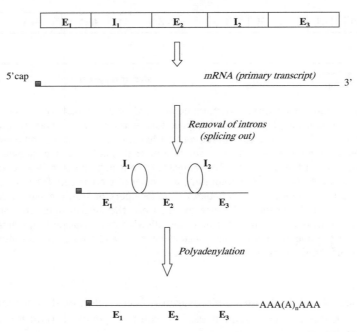

Figure 3.9 Overview of the transcription of eukaryote genes and subsequent mRNA editing. Prior to the completion of the synthesis of the primary transcript, a 5′ cap (7-methylguanosine) is enzymatically added at the 5′ end of the growing RNA chain. This helps to prevent mRNA degradation by nuclease enzymes. The non-coding introns (see main text) are enzymatically removed from the primary transcript (a process called splicing), yielding a mature mRNA sequence coding for the intended polypeptide. Finally, the mRNA 3′ end is also modified by the addition of a poly A tail, comprising 80–250 adenylate residues, which again likely helps protect the mRNA from degradation

approach adopted entails initial liberation of the nucleic acid by disruption of any cell wall present (or viral capsid) and of the cellular plasma membrane, followed by selective precipitation and often chromatography. In the context of plants and some microorganisms, initial disruption of the cell wall may require application of physical or other vigorous disruptive influences (see Chapter 6). This can potentially complicate DNA purification, particularly as it can cause physical shearing (fragmentation) of the extremely long DNA chromosome. The gentlest method of cell lysis usually involves incubation with cell-wall-degrading enzymes, and the addition of detergent will solubilize the plasma membrane. Following cellular disruption, initial purification steps normally entail solvent-based extraction/precipitation. For example, shaking in the presence of phenol (or a mixture of phenol and chloroform), followed by standing or centrifugation (to achieve phase separation) results in extraction of the (now denatured) proteins into the phenol phase and/or accumulation at the interphase, with nucleic acids remaining in the upper, aqueous phase. Further purification may be achieved by selective precipitation of the nucleic acids using ethanol or isopropanol as precipitant. If DNA is required, then the RNA present may now be removed by the addition of the enzyme ribonuclease, which selectively degrades RNA. On the other hand, if (eukaryotic) mRNA is required, then affinity-based purification may be undertaken using an oligo

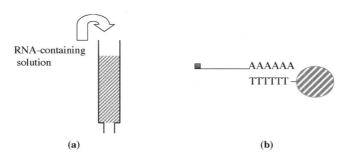

Figure 3.10 Affinity-based purification of mRNA. The unpurified mRNA-containing solution is percolated through a column packed with cellulose beads (a), to which a short chain of deoxythymidylate (an oligo dT chain) has been attached. Any mRNA present is retained in the column due to complementary base pairing between its 3′ poly A tail and the immobilized oligo dT, (b). Non-bound material can then be washed out of the column, with subsequent desorption of the mRNA by passing a low-salt buffer through the column. The mRNA collected may then be precipitated out of solution using ethanol, followed by its collection via centrifugation. An alternative, and now more commonly used variation, entails the direct addition of oligo (dT)-bound magnetic beads directly into the cell lysate and 'pulling out' the mRNA using a magnet. The method is rapid, thus minimizing contact time of the mRNA with degradative ribonucleases present naturally in the cytoplasm

(dT) column (Figure 3.10). Nucleic acids absorb UV light maximally at 260 nm (compared with 280 nm in the case of proteins); thus, absorbance at 260 nm can be used to quantify the amount of nucleic acid present and to follow the purification protocol. The ratio of absorbance at 260 nm versus 280 nm can also be used to determine how contaminated the nucleic acid preparation is with protein. The ratio $A_{260}/A_{280} \approx 1.8$ for pure DNA and 2.0 for pure RNA preparations; lower ratios usually indicate the presence of contaminant protein. DNA can also be detected and quantified by the addition of the chemical ethidium bromide. Ethidium bromide molecules intercalate (bind) in between DNA bases and fluoresce when illuminated with UV light.

3.2.3 Nucleic acid sequencing

The determination of the exact base sequence present in a stretch of nucleic acid (particularly in DNA) underpins much of modern molecular biology. Sequencing plays a central role in rDNA cloning experiments, as well as in determining genome data. Two approaches have been developed to sequence DNA: the Maxam–Gilbert chemical sequencing method and the Sanger–Coulson enzymatic sequencing method. Both involve the ultimate generation of a full set of fragments of the DNA strand to be sequenced, as illustrated in Figure 3.11. The methodologies employed ensure that the identity of the final (3′) base in each fragment is known. The fragments are then separated on the basis of their size by electrophoresis and, because the identity of the end base in each fragment is already known, the full sequence can simply be read from the ladder of fragments generated. Full details of sequencing methodologies are outside the scope of this book, but they are included in all core molecular biology and biochemistry student textbooks. RNA is sequenced by an enzyme-based method somewhat similar to the enzyme-based DNA method.

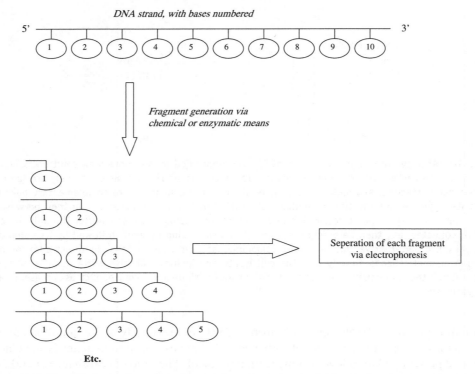

Figure 3.11 A simplified overview of the approaches adopted to both chemical and enzyme-based DNA sequencing. Refer to text for details

3.3   Recombinant production of therapeutic proteins

The evaluation of any protein as a potential biopharmaceutical and its subsequent routine medical use are dependent upon the availability of sufficient quantities of the target protein. In most instances this is best achieved via production by recombinant means (i.e. via genetic engineering). In addition to facilitating the production of any protein in substantial quantities, recombinant-based production can have a number of additional advantages over direct extraction from a naturally producing source, as described in Chapter 1. Production of any protein via rDNA technology entails the initial identification and isolation of a DNA sequence coding for the target protein. This sequence can be direct genomic DNA, but mRNA coding for the protein of interest can also act as a starting point. In the latter approach, the mRNA is enzymatically 'reverse transcribed' into cDNA. If the target therapeutic protein is eukaryotic (which is invariably the case) then the genomic DNA will contain both coding (exon) and non-coding (intron) sequences (Figure 3.9), whereas the cDNA will be a reflection of the exons only.

The desired gene/cDNA is normally amplified, sequenced and then introduced into an expression vector that facilitates its introduction and expression (transcription and translation) in an appropriate producer cell type. All recombinant therapeutic proteins approved to date are produced in *E. coli*, *S. cerevisiae* or in animal cell lines (mainly CHO or BHK cells). The general characteristics of these various producer cells and their advantages and disadvantages, along with factors taken into account when choosing one for biopharmaceutical production, are points considered in Chapter 5. In the remainder of this chapter we will review the basic molecular biology techniques that underpin the isolation, identification, cloning and expression of a target protein-encoding gene sequence. We will first overview the classical approach to cloning, which entails the generation of genomic libraries as described immediately below. We will then consider an alternative approach that has now come to the fore, and which is based upon the polymerase chain reaction (PCR) technique.

3.4 Classical gene cloning and identification

The basic approach to cloning a segment of DNA entails:

1. Initial enzyme-based fragmentation of intact genomic DNA (usually chromosomes isolated as described earlier in this chapter) so that it is broken down into manageable fragment sizes for further manipulation. Ideally all/most fragments will contain one gene.

2. Integration of the various fragments generated into cloning vectors, which are themselves small DNA molecules capable of self-replication. Typically, these are plasmids or viral DNAs and the composite or engineered DNA molecules generated are called rDNA.

3. Introduction of the vectors housing the DNA fragments into host cells.

4. Growing these cells on agar plates.

5. Screening/identification of the host cell colonies containing the rDNA molecules (i.e. screening the 'library' of clones generated) in order to identify the specific colony containing the target DNA fragment, i.e. the target gene (Figure 3.12).

We will now look at each of these stages separately. The initial fragmentation of genomic DNA is undertaken using enzymes known as restriction endonucleases (REs). Some 800 different REs have been identified thus far. These enzymes recognize, bind and cut DNA sequences which exhibit a defined base sequence (Table 3.2). These sequences normally exhibit a twofold symmetry around a specific point and are usually 4, 6 or 8 bp in length. Such areas are often termed palindromes. In general, the larger the recognition sequence the fewer such sequences present in a given DNA molecule and, hence, the smaller the number of DNA fragments that will be generated. Depending upon the specific RE utilized, DNA cleavage may yield blunt ends (e.g. BsaAI and EcoRV in Table 3.2) or staggered ends – the latter are often referred to as sticky ends.

An essential feature of the cloning vector used is that it must be capable of self-replication in the cell into which it is introduced, which is usually *E. coli*. Two of the most commonly used types of vector in conjunction with *E. coli* are plasmids and bacteriophage λ. Plasmids are circular extra-chromosomal DNA molecules, generally between 5000 and 350 0000 bp in length, that are found naturally in a wide range of bacteria. They generally house several

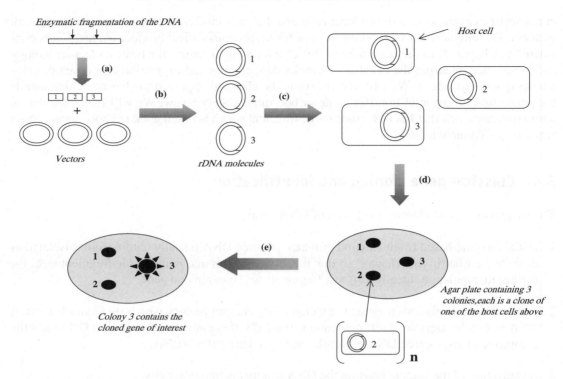

Figure 3.12   A basic overview of the DNA cloning process. Refer to text for specific details

genes, often including one or more genes whose product renders the plasmid-containing cell resistant to specific antibiotic(s). One plasmid often used in cloning experiments with *E. coli* is pUC18 (Figure 3.13). Bacteriophage ('phage') are viruses capable of infecting and replicating inside bacteria. Bacteriophage λ DNA is approximately 48 500 bp in length. Another vector type sometimes used are the bacterial artificial chromosomes (BACs), which are effectively very large plasmids used to clone very large stretches of DNA (usually DNA fragments above 100 000 bp).

Integration of the DNA fragments into the chosen vector is undertaken by 'opening up' the circular vector via treatment with the same RE as used to generate the DNA fragments for cloning, followed by co-incubation of the cleaved vector and the fragments under conditions that promote the annealing of complementary sticky ends. Some vectors may simply recircularize to reform their original structure, but pretreatment of the vector in various ways can prevent this from happening. Most of the recircularized plasmids will have incorporated a fragment of DNA to be cloned. The plasmids are then incubated with another enzyme, a DNA ligase, which catalyses the formation of phosphodiester bonds in the DNA backbone and thus will seal or 'ligate' the plasmid.

The next stage of the cloning process entails the introduction of the engineered vector into *E. coli* cells. This can be achieved by a number of different means. One approach (called transformation) involves co-incubation of the plasmids and cells in a solution of calcium chloride, initially at 0 °C, with subsequent increase in temperature to 42 °C. This temperature shock facilitates entry of plasmids into some cells.

Table 3.2 Some commercially available REs, their sources, DNA recognition sites and cleavage points

Restriction enzyme	Source	DNA recognition sequence and cleavage site[a]
BclI	*Bacillus caldolyticus*	5′-T↓GATCA-3′ 3′-ACTAG↑T-5′
BglII	Recombinant *E. coli* carrying *BglII* gene from *Bacillus globigii*	5′-A↓GATCT-3′ 3′-TCTAG↑A-5′
BsaAI	Recombinant *E. coli* carrying *BsaAI* gene from *Bacillus stearothermophilus* A	5′-PyAC↓GTPu-3′ 3′-PuTG↑CAPy-5′
BsaJI	*B. stearothermophilus* J	5′-C↓CNNGG-3′ 3′-GGNNC↑C-5′
BsiEI	*B. stearothermophilus*	5′-CGPuPy↓CG-3′ 3′-GC↑PyPuGC-5′
EcoRV	Recombinant *E. coli* carrying *EcoRV* gene from the plasmid J62 pIg 74	5′-GAT↓ATC-3′ 3′-CTA↑TAG-5′
MwoI	Recombinant *E. coli* carrying cloned *MwoI* gene from *Methanobacterium wolfeii*	5′-GCNNNNN↓NNGC-3′ 3′-CGNN↑NNNNNCG-5′
Tsp509I	*Thermus* sp.	5′-↓AATT-3′ 3′-TTAA↑-5′
XbaI	Recombinant *E. coli* carrying *XbaI* gene from *Xanthomonas badvii*	5′-T↓CTAGA-3′ 3′-AGATC↑T-5′
XhoI	Recombinant *E. coli* carrying *XhoI* gene from *X. holcicola*	5′-C↓TCGAG-3′ 3′-GAGCT↑C-5′

[a]G: guanine; C: cytosine; A: adenine; T: thymine; Pu: any purine; Py: any pyrimidine; N: either a purine or pyrimidine. Arrow indicates site of cleavage.

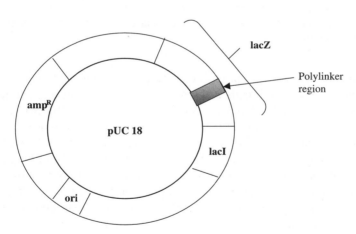

Figure 3.13 The plasmid pUC18 is often used for cloning purposes. It contains three genes: the ampicillin resistance gene (*amp*[R]), the *lacZ* gene, which codes for the enzyme β-galactosidase, and the *lacI* gene, which codes for a factor that controls the transcription of *lacZ*. Also present is an origin of replication (ori), essential for plasmid replication within the cell. Note the presence of a short stretch of DNA called the polylinker region located within the *lacZ* gene. The polylinker (also called a multiple cloning site) contains cleavage sites for 13 different REs. This allows genetic engineers great flexibility to insert a DNA fragment for cloning into this area. The polylinker has been designed and positioned within the *lacZ* gene so as not to prevent the expression of functional β-galactosidase. However, if a piece of DNA for cloning is introduced into the polylinker region, then the increased length does block β-galactosidase expression. The full sequence of the 2.69 kb plasmid is known and sequence analysis confirms the presence of multiple additional RE sites outside the polylinker region. There are at least six target sites for commonly used restriction enzymes within the *amp*[R] gene

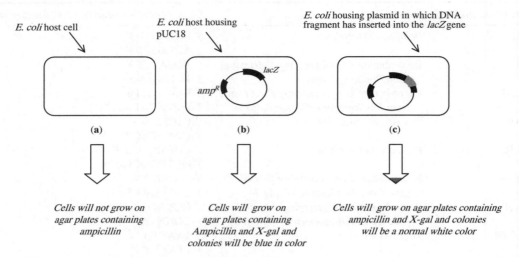

Figure 3.14   Identification of *E. coli* host cell clones containing rDNA using pUC18 vectors. After transformation the cells are spread on agar plates containing ampicillin and a chemical called X-gal. Any untransformed cells present will fail to grow on these plates as the host *E. coli* cells contain no ampicillin resistance gene. This step, therefore, identifies cells into which plasmid has successfully been transferred (cell types (b) and (c)). Cells containing plasmid into which no foreign DNA has been inserted (cell type (b)) will grow on ampicillin-containing plates as the plasmid contains the *amp*[R] gene. This cell type also produces functional β-galactosidase (Figure 3.13). This enzyme will cleave X-gal, liberating a product that is blue in colour. Colonies, therefore, appear blue. Cells containing a plasmid in which a DNA fragment has been inserted into the *lacZ* gene (cell type (c), i.e. the desired cells) do not produce β-galactosidase; therefore, colonies derived from these cells will be a normal white colour

The *E. coli* cells are next spread out on the surface of an agar plate and incubated under appropriate conditions in order to kill cells that have not taken up plasmid. Each individual cell will thus form a colony (clone of cells). Three main types of cell will be initially transferred onto these agar plates: (a) some cells will have failed to take up any plasmid; (b) some transformed cells may have a plasmid in which no foreign DNA had been inserted; (c) some cells will house a plasmid that does carry a fragment of the target DNA. These latter cells are the only ones of interest, and various strategies may be adopted to identify them. One such strategy is outlined in Figure 3.14. Once the various *E. coli* clones (colonies) containing vector into which DNA fragments have been successfully integrated (i.e. clones containing rDNA) have been identified, all that remains to be achieved is to pinpoint which colony harbours the rDNA fragment containing the gene of interest (see Figure 3.12).

Assuming you started off with whole genomic DNA, the procedure thus far has effectively generated a library of clones containing different genomic DNA fragments. The final task remaining, therefore, is to identify which specific clone/clones harbour the actual DNA fragment of interest (in our context this would be the fragment containing the gene coding for the desired therapeutic protein). This can be a major task, as libraries often consist of 10^9 or more clones. The most common means of achieving this is via sequence-based hybridization studies. The basic approach taken entails the use of a labelled (e.g. radioactive) probe that is a single-strand DNA fragment or an RNA fragment synthesized to have a base sequence complementary to a sequence within the gene of interest. Genome projects now mean that such sequence information is known for many proteins. Alternatively, likely base sequences can be deduced if a partial amino acid sequence of the protein is known. Hybridization studies are usually initiated by physically pressing a nitrocellulose paper onto the agar plates containing the recombinant colonies. A replica of the plate is thus created on the paper, as some cells from each colony adhere to it. Subsequent treatment of the paper with alkali lyses the cells, releasing and denaturing the DNA within. The DNA adsorbs tightly to the paper. The paper is then exposed to a solution containing the labelled DNA probe under conditions that allow it to anneal to the target DNA, if it is present. After washing (to remove unbound probe), any probe retained on the paper surface can de detected by an appropriate visualization technique (e.g. autoradiography if the probe is radiolabelled), and the positioning of the label on the paper surface pinpoints which colony on the agar surface houses the desired DNA fragment. Cells from the appropriate colony can then be grown up in larger amounts by submerged fermentation (see Chapter 5) in order to produce larger amounts of the desired (now cloned) gene. The cells can be collected, lysed and the vector therein recovered by standard microbiological techniques. The cloned gene can then be excised from the vector via treatment with an appropriate RE and purified by standard molecular techniques.

3.4.1 cDNA cloning

An alternative to cloning genomic DNA, as outlined in the sections above, entails beginning the process not with chromosomal fragments but with mRNA. This is often an approach taken when cloning eukaryotic genes in particular. As described in Figure 3.10, total eukaryotic cellular mRNA can be purified from the cell via an affinity-based mechanism. The mRNAs recovered in this way reflect only the polypeptide-encoding genes that are expressed in the cells at the time of their extraction. Incubation of the mRNA with the enzyme reverse transcriptase results in the conversion of the single-stranded mRNA into double-stranded DNA known as cDNA (see also Figure 3.1). These cDNA fragments can then be cloned to generate a cDNA library, and the desired cDNA clone can be identified by means similar to those already described in the context of cloning genomic DNA. cDNA libraries are smaller than genomic libraries as they are derived only from expressed genes. Non-coding regions in the genome (as well as quiescent genes and genes coding for rRNA and tRNA) are not represented in the library. Therefore, cDNA libraries are more manageable to work with, assuming that the gene of interest is being expressed.

3.4.2 Cloning via polymerase chain reaction

An enormous number of individual genes have been sequenced over the last two decades or more. Of latter years in particular, genome projects have also begun to make available the sequence of the entire complement of genes present in many species. The bulk of this sequence information has been made publicly available by its deposition in sequence databases. As a result, scientists who now wish to clone a particular gene will usually have prior access to partial/entire sequence information from the relevant organism or a closely related species. This sequence information allows them to obtain large amounts of the gene of interest by using the PCR technique (Figure 3.15). This

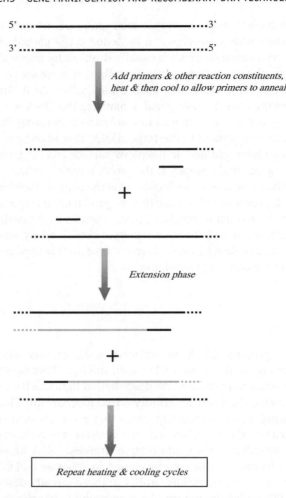

Figure 3.15   The PCR is initiated by separation of the double-stranded DNA into its two constituent strands. This is achieved by heating the sample (usually to 94 °C). Also present in the reaction mixture are: (a) two chemically synthesized oligonucleotide primers ('oligos') whose sequences are complementary to the sequences flanking the gene of interest; (b) the enzyme DNA polymerase, which can extend the primers to synthesize a new DNA strand of complementary sequence to the single stranded DNA template; (c) all the nucleoside precursors required for synthesis of the growing DNA strand (i.e. the deoxynucleoside triphosphates or dNTPs). Once strand separation has been achieved the reaction temperature is reduced, in order to allow primers to anneal to complementary sequences on each strand and allow the DNA polymerase to extend the primers. This extension phase is normally carried out at 74 °C. The DNA polymerase used is sourced from the thermophilic microorganism *Thermus aquaticus*; therefore, it is heat stable and not inactivated by PCR operational temperatures. This completes the first cycle of the PCR process; the result is a doubling of the amount of target DNA present in the reaction mixture. The cycle is then repeated and with each repeat comes a doubling of the amount of target DNA. After 25–30 cycle repeats, several hundred million copies of the target DNA have been generated

approach to cloning has now come to the fore, as it is faster and more convenient than the more classical methods described above. The process begins by extraction of total genomic DNA from the source of interest (e.g. human cells if you wish to clone a specific human gene). Oligonucleotide primers whose sequences flank the target gene/DNA segment are synthesized and used to amplify that portion of DNA selectively. Recognition sites for REs can be incorporated into the oligonucleotides to allow cloning of the amplified gene, as outlined earlier. Because the target gene sequence is the only segment of the extracted DNA to be amplified by the prior PCR step, the vast majority of clones in the library now generated should contain the desired gene. This can be confirmed by direct sequencing of the inserted DNA fragment from several of the colonies. Sequencing is important not only to prove definitively that the cloned DNA is the target gene, but also that its sequence perfectly matches the published sequence. The PCR process is prone to the introduction of sequence errors.

3.4.3 Expression vectors

The vectors described thus far have been designed to facilitate the cloning of genomic DNA/cDNA sequences, ultimately in order to identify and isolate a gene/cDNA coding for a particular polypeptide. The genetic construction of these vectors normally does not support the actual expression (i.e. transcription and translation) of the gene. Once the gene/cDNA coding for a potential target protein has been isolated, the goal usually becomes one of achieving high levels of expression of this target gene. This process entails ligation of the gene into a vector that will support high-level transcription and translation. In addition to the basic vector elements (e.g. an origin of replication and a selectable marker, such as an antibiotic resistance gene), expression vectors also contain all the genetic elements required to support transcription and translation, as described earlier in this chapter (e.g. promoters, translational start and stop signals, etc.; see Figures 3.7 and 3.8). A wide range of such expression vectors is now commercially available and, obviously, each is tailored to work best in a specific host cell type (e.g. bacterial, yeast, mammalian, etc.). The choice of exact host cell type in which to express a recombinant therapeutic protein will depend upon a number of factors, as described in Chapter 5.

3.4.4 Protein engineering

The advent of rDNA technology renders straightforward the manipulation of a protein's amino acid sequence. This process, termed site-directed mutagenesis or protein engineering, entails the controlled alteration of the nucleotide sequence coding for the polypeptide of interest such that specific, predetermined changes in amino acid sequence are introduced. Such changes can include insertions, deletions or substitutions. Site-directed mutagenesis is now most often undertaken by using a variant of the basic PCR method already described (Figure 3.15), known as 'overlap PCR', in which primers of altered nucleotide sequences are used for the PCR reactions.

Protein engineering facilitates a greater understanding of the link between a polypeptide's amino acid sequence and its structure. It also provides a powerful method of studying the relationship between structure and its function. As such, this technique will help greatly in achieving the much pursued, but still distant, objective of *de novo* protein design. Protein engineering is also

Table 3.3 Some approved biopharmaceuticals that have been altered by post-translational engineering

Engineered product	Effect of engineering
Insulin 'Detemir' (insulin with a fatty acid molecule attached; Chapter 11)	The insulin has a prolonged duration of action
PEGylated interferons (interferons to which polyethylene glycol molecules have been covalently attached; Chapter 8)	Increases product serum half-life, thereby reducing the frequency of dosage
Carbohydrate remodelled glucocerebrosidase (Cerezyme; Chapter 12)	Generation of product that is specifically targeted for uptake by macrophages

used to tailor structural or functional attributes of therapeutic and other commercially important proteins.

An increasing proportion of therapeutic proteins gaining approval for general medical use display an engineered amino acid sequence, altered in order to fulfil a predefined therapeutic goal better. Selected examples of biopharmaceuticals engineered in this way are presented in Table 1.3, and these, along with additional examples, will be discussed in various subsequent chapters. An alternative approach to protein engineering entails the covalent attachment or the alteration of specific molecules/groups to or on the polypeptide's backbone (Table 3.3). Specific examples of such post-translational engineering will again be provided in subsequent chapters.

Further reading

Books

Brown, T. 2006. *Gene Cloning and DNA Analysis, an Introduction*. Blackwell Science, UK.
Gellissen, G (ed.). 2005. *Production of Recombinant Proteins*. Wiley-VCH, Germany.
Primrose, S.B., Twyman, R.M., and Old, R.W. 2001. *Principles of Gene Manipulation*, 6th edn. Blackwell Science, UK.
Sambrook, J. and Russell, D. 2006. *Condensed Protocols from Molecular Cloning: A Laboratory Manual*. Cold Spring Harbor Laboratory Press, USA.

Articles

Abramowicz, M. 2003. The human genome project in retrospect. *Perspectives on Properties of the Human Genome Project* **50**, 231–261.
Arya, M., Shergill, I.S., Williamson, M., Gommersall, L., Arya, N., and Patel, H.R. 2005. Basic principles of real time quantitative PCR. *Expert Review of Molecular Diagnostics* **5**, 209–219.
Brannigan, J. and Wilkinson, A. 2002. Protein engineering 20 years on. *Nature Reviews Molecular Cell Biology* **3**, 964–970.
Christensen, A. 2001. Bacteriophage lambda-based expression vectors. *Molecular Biotechnology* **17**, 219–224.
Feuk, L., Carson, A.R., and Scherer, S.W. 2006. Structural variation in the human genome. *Nature Reviews Genetics* **7**, 85–97.

Marshall, S.A., Lazar G.A., Chirino A.J., and Desjarlais J.R. 2003. Rational design and engineering of therapeutic proteins. *Drug Discovery Today* **8**, 212–221.

Mergulhao, F.J.M., Monteiro, G.A., Cabral, J.M.S., and Taipa, M.A. 2004. Design of bacterial vector systems for the production of recombinant proteins in *Escherichia coli. Journal of Microbiology and Biotechnology.* **14**, 1–14.

Schumann, W. and Ferreira, L. 2004. Production of recombinant protein in *Escherichia coli. Genetics and Molecular Biology.* **27**, 442–453.

Shanklin, J. 2000. Exploring the possibilities of protein engineering. *Current Opinion in Plant Biology* **3**, 243–248.

Sodoyer, R. 2004. Expression systems for the production of recombinant pharmaceuticals. *Biodrugs* **18**, 51–62.

Steipe, B. (1999). Evolutionary approaches to protein engineering. *Combinatorial Chemistry in Biology* **243**, 55–86.

Wendland, J. 2003. PCR-based methods facilitate targeted gene manipulations and cloning procedures. *Current Genetics* **44**, 115–123.

Wurm, F. 2004. Production of recombinant protein therapeutics in cultivated mammalian cells. *Nature Biotechnology* **22**, 1393–1398.

4
The Drug development process

4.1 Introduction

In this chapter, the life history of a successful drug will be outlined (summarized in Figure 4.1).

A number of different strategies are adopted by the pharmaceutical industry in their efforts to identify new drug products. These approaches range from random screening of a wide range of biological materials to knowledge-based drug identification. Once a potential new drug has been identified, it is then subjected to a range of tests (both *in vitro* and in animals) in order to characterize it in terms of its likely safety and effectiveness in treating its target disease. The developer will also undertake manufacturing-related development work (development and initial optimization of upstream and downstream processing; Chapters 5 and 6), as well as investigating suitable potential routes of product administration.

After completing such preclinical trials, the developing company apply to the appropriate government-appointed agency, e.g. the Food and Drug Administration (FDA) in the USA, for approval to commence clinical trials (i.e. to test the drug in humans). Clinical trials are required to prove that the drug is safe and effective when administered to human patients, and these trials may take 5 years or more to complete. Once the drug has been characterized, and perhaps early clinical work is underway, the drug is normally patented by the developing company in order to ensure that it receives maximal commercial benefit from the discovery.

Upon completion of clinical trials, the developing company collates all the preclinical and clinical data they have generated, as well as additional pertinent information, e.g. details of the exact production process used to make the drug. They submit this information as a dossier (a multi-volume work) to the regulatory authorities. Regulatory scientific officers then access the information provided and decide (largely on criteria of drug safety and efficacy) whether the drug should be approved for general medical use.

If marketing approval is granted, the company can sell the product from then on. As the drug has been patented, they will have no competition for a number of years at least. However, in order to sell the product, a manufacturing facility is required, and the company will also have to gain manufacturing approval from the regulatory authorities. In order to gain a manufacturing licence, regulatory inspectors will review the proposed manufacturing facility. The regulatory authority will only grant the company a manufacturing licence if they are satisfied that every aspect of the manufacturing process is conducive to producing a safe and effective product consistently.

Pharmaceutical biotechnology: concepts and applications   Gary Walsh
© 2007 John Wiley & Sons, Ltd   ISBN 978 0 470 01244 4 (HB)   978 0 470 01245 1 (PB)

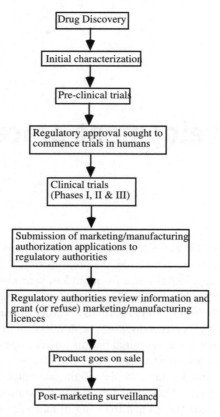

Figure 4.1   An overview of the life history of a successful drug. Patenting of the product is usually also undertaken, often during the initial stages of clinical trial work

Regulatory involvement does not end even at this point. Post-marketing surveillance is generally undertaken, with the company being obliged to report any subsequent drug-induced side effects/adverse reactions. The regulatory authority will also inspect the manufacturing facility on a regular basis in order to ensure that satisfactory manufacturing standards are maintained.

4.2   Discovery of biopharmaceuticals

The discovery of virtually all the biopharmaceuticals discussed in this text was a knowledge-based one. Continuing advances in the molecular sciences have deepened our understanding of the molecular mechanisms that underline health and disease. An understanding at the molecular level of how the body functions in health and of the deviations that characterize the development of a disease often renders obvious potential strategies likely to cure/control that disease. Simple examples illustrating this include the use of insulin to treat diabetes and the use of GH to treat certain forms of dwarfism (Chapter 11). The underlining causes of these types of disease are relatively straightforward, in that they are essentially promoted by the deficiency/absence of a single regulatory molecule. Other diseases, however, may be multifactorial and, hence, more complex. Examples include cancer and inflammation. Nevertheless, cytokines, such as interferons and interleukins, known to stimulate the immune response/regulate inflammation, have proven to be therapeutically useful in treating several such complex diseases (Chapters 8 and 9).

An understanding, at the molecular level, of the actions of various regulatory proteins, or the progression of a specific disease does not, however, automatically translate into pinpointing an effective treatment strategy. The physiological responses induced by the potential biopharmaceutical *in vitro* (or in animal models) may not accurately predict the physiological responses seen when the product is administered to a diseased human. For example, many of the most promising biopharmaceutical therapeutic agents (e.g. virtually all the cytokines, Chapter 8), display multiple activities on different cell populations. This makes it difficult, if not impossible, to predict what the overall effect administration of any biopharmaceutical will have on the whole body, hence the requirement for clinical trials.

In other cases, the widespread application of a biopharmaceutical may be hindered by the occurrence of relatively toxic side effects (as is the case with tumour necrosis factor α (TNF-α, Chapter 9). Finally, some biomolecules have been discovered and purified because of a characteristic biological activity that, subsequently, was found not to be the molecule's primary biological activity. TNF-α again serves as an example. It was first noted because of its cytotoxic effects on some cancer cell types *in vitro*. Subsequently, trials assessing its therapeutic application in cancer proved disappointing due not only to its toxic side effects, but also to its moderate, at best, cytotoxic effect on many cancer cell types *in vivo*. TNF's major biological activity *in vivo* is now known to be as a regulator of the inflammatory response.

In summary, the 'discovery' of biopharmaceuticals, in most cases, merely relates to the logical application of our rapidly increasing knowledge of the biochemical basis of how the body functions. These substances could be accurately described as being the body's own pharmaceuticals. Moreover, rapidly expanding areas of research, such as genomics and proteomics, will likely hasten the discovery of many more such products, as discussed below.

4.3   The impact of genomics and related technologies upon drug discovery

The term 'genomics' refers to the systematic study of the entire genome of an organism. Its core aim is to sequence the entire DNA complement of the cell and to map the genome arrangement physically (assign exact positions in the genome to the various genes/non-coding regions). Prior to the 1990s, the sequencing and study of a single gene represented a significant task. However, improvements in sequencing technologies and the development of more highly automated hardware systems now render DNA sequencing considerably faster, cheaper and more accurate. Modern sequencing systems can sequence thousands of bases per hour. Such innovations underpin the 'high-throughput' sequencing necessary to evaluate an entire genome sequence within a reasonable time-frame. By early 2006 some 364 genome projects had been completed (297 bacterial, 26 Archaeal and 41 Eucaryal, including the human genome) with in excess of 1000 genome sequencing projects ongoing.

From a drug discovery/development prospective, the significance of genome data is that they provide full sequence information of every protein the organism can produce. This should result in the identification of previously undiscovered proteins that will have potential therapeutic application, i.e. the process should help identify new potential biopharmaceuticals. The greatest pharmaceutical impact of sequence data, however, will almost certainly be the identification of numerous additional drug targets. It has been estimated that all drugs currently on the market target one (or more) of a maximum of 500 targets. The majority of such targets are proteins (mainly enzymes, hormones, ion channels and nuclear receptors). Hidden in the human genome sequence data is believed to be anywhere between 3000 and 10 000 new protein-based drug targets. Additionally, present in the sequence data of many human pathogens is sequence data of hundreds, perhaps thousands, of pathogen proteins that could serve as drug targets against those pathogens (e.g. gene products essential for pathogen viability or infectivity).

While genome sequence data undoubtedly harbours new drug leads/drug targets, the problem now has become one of specifically identifying such genes. Impeding this process is the fact that the biological function of many sequenced gene products remains unknown. The focus of genome research, therefore, is now shifting towards elucidating the biological function of these gene products, i.e. shifting towards 'functional genomics'.

Assessment of function is critical to understanding the relationship between genotype and phenotype and, of course, for the direct identification of drug leads/targets. The term 'function' traditionally has been interpreted in the narrow sense of what isolated biological role/activity the gene product displays (e.g. is it an enzyme and, if so, what specific reaction does it catalyse). In the context of genomics, gene function is assigned a broader meaning, incorporating not only the isolated biological function/activity of the gene product, but also relating to:

- where in the cell that product acts and, in particular, what other cellular elements does it influence/interact with;

- how do such influences/interactions contribute to the overall physiology of the organism.

The assignment of function to the products of sequenced genes can be pursued via various approaches, including:

- sequence homology studies;

- phylogenetic profiling;

- Rosetta stone method;

- gene neighbourhood method;

- knockout animal studies;

- DNA array technology (gene chips);

- proteomics approach;

- structural genomics approach.

With the exception of knockout animals, these approaches employ, in part at least, sequence structure/data interrogation/comparison. The availability of appropriate highly powerful computer programs renders these approaches 'high throughput'. However, even by applying these methodologies, it will not prove possible to identify immediately the function of all gene products sequenced.

Sequence homology studies depend upon computer-based (bioinformatic) sequence comparison between a gene of unknown function (or, more accurately, of unknown gene product function) and genes whose product has previously been assigned a function. High homology suggests likely related functional attributes. Sequence homology studies can assist in assigning a putative function to 40–60 per cent of all new gene sequences.

Phylogenetic profiling entails establishing a pattern of the presence or absence of the particular gene coding for a protein of unknown function across a range of different organisms whose genomes have been sequenced. If it displays an identical presence/absence pattern to an already characterized gene, then in many instances it can be inferred that both gene products have a related function.

The Rosetta stone approach is dependent upon the observation that sometimes two separate polypeptides (i.e. gene products X and Y) found in one organism occur in a different organism as a single fused protein XY. In such circumstances, the two protein parts (domains), X and Y, often display linked functions. Therefore, if gene X is recently discovered in a newly sequenced genome and is of unknown function but gene XY of known function has been previously discovered in a different genome, then the function of the unknown X can be deduced.

The gene neighbourhood method is yet another computation-based method. It depends upon the observation that two genes are likely to be functionally linked if they are consistently found side by side in the genome of several different organisms.

Knockout animal studies, in contrast to the above methods, are dependent upon phenotype observation. The approach entails the generation and study of mice in which a specific gene has been deleted. Phenotypic studies can sometimes yield clues as to the function of the gene knocked out.

4.4 Gene chips

Although sequence data provide a profile of all the genes present in a genome, they give no information as to which genes are switched on (transcribed) and, hence, which are functionally active at any given time/under any given circumstances. Gene transcription results in the production of RNA, either mRNA (usually subsequently translated into a polypeptide) or rRNA or tRNA (which have catalytic or structural functions; Chapter 3). The study of under which circumstances an RNA species is expressed/not expressed in the cell/organism can provide clues as to the biological function of the RNA (or, in the case of mRNA, the function of the final polypeptide product). Furthermore, in the context of drug lead/target discovery, the conditions under which a specific mRNA is produced can also point to putative biopharmaceuticals/drug targets. For example, if a particular mRNA is only produced by a cancer cell, that mRNA (or, more commonly, its polypeptide product) may represent a good target for a novel anti-cancer drug.

Levels of RNA (usually specific mRNAs) in a cell can be measured by well-established techniques such as northern blot analysis or by PCR analysis. However, the recent advent of DNA microarray technology has converted the identification and measurement of specific mRNAs (or

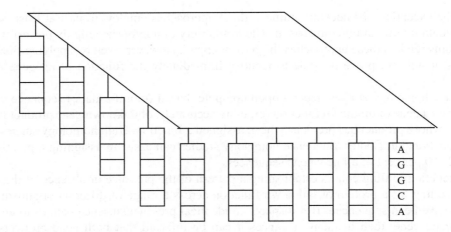

Figure 4.2 Generalized outline of a gene chip. In this example, short oligonucleotide sequences are attached to the anchoring surface (only the outer rows are shown). Each probe displays a different nucleotide sequence, and the sequences used are usually based upon genome sequence information. The sequence of one such probe is shown as AGGCA. By incubating the chip with, for example, total cellular mRNA under appropriate conditions, any mRNA with a complementary sequence (UCCGU in the case of the probe sequence shown) will hybridize with the probes. In reality, probes will have longer sequences than the one shown above

other RNAs if required) into a 'high-throughput' process. DNA arrays are also termed oligonucleotide arrays, gene chip arrays or, simply, chips.

The technique is based upon the ability to anchor nucleic acid sequences (usually DNA based) on plastic/glass surfaces at very high density. Standard gridding robots can put on up to 250 000 different short oligonucleotide probes or 10 000 full-length cDNA sequences per square centimetre of surface. Probe sequences are generally produced/designed from genome sequence data; hence, chip production is often referred to as 'downloading the genome on a chip'. RNA can be extracted from a cell and probed with the chip. Any complementary RNA sequences present will hybridize with the appropriate immobilized chip sequence (Figure 4.2). Hybridization is detectable as the RNA species are first labelled. Hybridization patterns obviously yield critical information regarding gene expression.

4.5 Proteomics

Although virtually all drug targets are protein based, the inference that protein expression levels can be accurately (if indirectly) detected/measured via DNA array technology is a false one, as:

- mRNA concentrations do not always directly correlate with the concentration of the mRNA-encoded polypeptide;

- a significant proportion of eukaryote mRNAs undergo differential splicing and, therefore, can yield more than one polypeptide product (Figure 4.3).

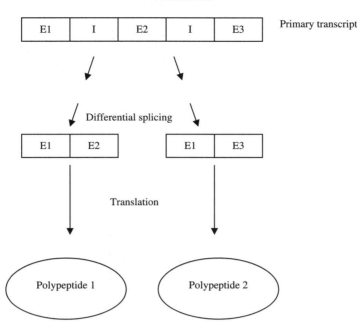

Figure 4.3 Differential splicing of mRNA can yield different polypeptide products. Transcription of a gene sequence yields a 'primary transcript' RNA. This contains coding regions (exons) and non-coding regions (introns). A major feature of the subsequent processing of the primary transcript is 'splicing', the process by which introns are removed, leaving the exons in a contiguous sequence. Although most eukaryotic primary transcripts produce only one mature mRNA (and hence code for a single polypeptide), some can be differentially spliced, yielding two or more mature mRNAs. The latter can, therefore, code for two or more polypeptides. E: exon; I: intron

Additionally, the cellular location at which the resultant polypeptide will function often cannot be predicted from RNA delection/sequences nor can detailed information regarding how the polypeptide product's functional activity will be regulated (e.g. via post-translational mechanisms such as phosphorylation, partial proteolysis, etc.). Therefore, protein-based drug leads/targets are often more successfully identified by direct examination of the expressed protein complement of the cell, i.e. its proteome. Like the transcriptome (total cellular RNA content), and in contrast to the genome, the proteome is not static, with changes in cellular conditions triggering changes in cellular protein profiles/concentrations. This field of study is termed proteomics.

Proteomics, therefore, is closely aligned to functional genomics and entails the systematic and comprehensive analysis of the proteins expressed in the cell and their function. Classical proteomic studies generally entailed initial extraction of the total protein content from the target cell/tissue, followed by separation of the proteins therein using two-dimensional electrophoresis (Chapter 7). Isolated protein 'spots' could then be eluted from the electrophoretic gel and subjected to further analysis; mainly to Edman degradation, in order to generate partial amino acid sequence data. The sequence data could then be used to interrogate protein sequence databanks in order to, for example, assign putative function by sequence homology searches (Figure 4.4). Two-dimensional electrophoresis, however, is generally capable of resolving no more than 2000 different proteins, and proteins expressed at low levels may not be detected at all if their gel concentration is below

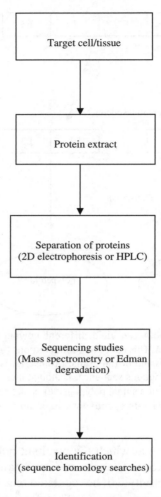

Figure 4.4   The proteomics approach. Refer to text for details

the (protein) staining threshold. The latter point can be particularly significant in the context of drug/target identification, as most such targets are likely to be kinases and other regulatory proteins that are generally expressed within cells at very low levels.

More recently, high-resolution chromatographic techniques (particularly reverse-phase and ion exchanged-based high-performance liquid chromatography (HPLC)) have been applied in the separation of proteome proteins and high-resolution mass spectrometry is being employed to aid high-throughput sequence determination.

4.6   Structural genomics

Related to the discipline of proteomics is that of structural genomics. The latter focuses upon the large-scale systematic study of gene product structure. While this embraces rRNA and tRNA, in practice the field focuses upon protein structure. The basic approach to structural genomics entails the cloning and recombinant expression of cellular proteins, followed by their purification and three-dimensional structural analysis. High-resolution determination of a protein's structure is amongst the most challenging of molecular investigations. By the year 2000, protein structure databanks housed in the region of 12 000 entries. However, such databanks are highly redundant, often containing multiple entries describing variants of the same molecule. For example, in excess of 50 different structures of 'insulin' have been deposited (e.g. both native and mutated/engineered forms from various species, as well as insulins in various polymeric forms and in the presence of various stabilizers and other chemicals). In reality, by the year 2000, the three-dimensional structure of approximately 2000 truly different proteins had been resolved.

Until quite recently, X-ray crystallography was the technique used almost exclusively to resolve the three-dimensional structure of proteins. As well as itself being technically challenging, a major limitation of X-ray crystallography is the requirement for the target protein to be in crystalline form. It has thus far proven difficult/impossible to induce the majority of proteins to crystallize. NMR is an analytical technique that can also be used to determine the three-dimensional structure of a molecule, and without the necessity for crystallization. For many years, even the most powerful NMR machines could resolve the three-dimensional structure of only relatively small proteins (less than 20–25 kDa). However, recent analytical advances now render it possible to analyse much larger proteins by this technique successfully.

The ultimate goal of structural genomics is to provide a complete three-dimensional description of any gene product. Also, as the structures of more and more proteins of known function are elucidated, it should become increasingly possible to link specific functional attributes to specific structural attributes. As such, it may prove ultimately feasible to predict protein function if its structure is known, and vice versa.

4.7 Pharmacogenetics

Pharmacogenetics relates to the emerging discipline of correlating specific gene DNA sequence information (specifically sequence variations) to drug response. As such, the pursuit will ultimately impinge directly upon the drug development process and should allow doctors to make better-informed decisions regarding what exact drug to prescribe to individual patients.

Different people respond differently to any given drug, even if they present with essentially identical disease symptoms. Optimum dose requirements, for example, can vary significantly. Furthermore, not all patients respond positively to a specific drug (e.g. IFN-β is of clinical benefit to only one in three multiple sclerosis patients; see Chapter 8). The range and severity of adverse effects induced by a drug can also vary significantly within a patient population base.

While the basis of such differential responses can sometimes be non-genetic (e.g. general state of health, etc.), genetic variation amongst individuals remains the predominant factor. Although all humans display almost identical genome sequences, some differences are evident. The most prominent widespread-type variations amongst individuals are known as single nucleotide polymorphisms (SNPs, sometimes pronounced 'snips'). SNPs occur in the general population at an average incidence of 1 in every 1000 nucleotide bases; hence, the entire human genome harbours 3 million or so. SNPs are not mutations; the latter arise more infrequently, are more diverse and are generally caused by spontaneous/mutagen-induced mistakes in DNA repair/replication. SNPs occurring in structural genes/gene regulatory sequences can alter amino acid sequence/expression levels of a protein and, hence, affect its functional attributes. SNPs largely account for natural physical variations evident in the human population (e.g. height, colour of eyes, etc.).

The presence of an SNP within the regulatory or structural regions of a gene coding for a protein that interacts with a drug could obviously influence the effect of the drug on the body. In this context, the protein product could, for example, be the drug target or perhaps an enzyme involved in metabolizing the drug.

The identification and characterization of SNPs within the human genomes is, therefore, of both academic and applied interest. Several research groups continue to map human SNPs, and over 1.5 million have thus far been identified.

By identifying and comparing SNP patterns from a group of patients responsive to a particular drug with patterns displayed by a group of unresponsive patients, it may be possible to identify specific SNP characteristics linked to drug efficacy. In the same way, SNP patterns/characteristics associated with adverse reactions (or even a predisposition to a disease) may be uncovered. This could usher a new era of drug therapy where drug treatment could be tailored to the individual patient. Furthermore, different drugs could be developed with the foreknowledge that each would be efficacious when administered to specific (SNP-determined) patient sub-types. A (distant) futuristic scenario could be visualized where all individuals could carry chips encoded with SNP details relating to their specific genome, allowing medical staff to choose the most appropriate drugs to prescribe in any given circumstance.

Linking specific genetic determinants to many diseases, however, is unlikely to be as straightforward as implied thus far. The progress of most diseases, and the relative effectiveness of allied drug treatment, is dependent upon many factors, including the interplay of multiple gene products. 'Environmental' factors such as patient age, sex and general health also play a prominent role.

The term 'pharmacogenomics' is one that has entered the 'genomic' vocabulary. Although sometimes used almost interchangeably with pharmacogenetics, it more specifically refers to studying the pattern of expression of gene products involved in a drug response.

4.8 Initial product characterization

The physicochemical and other properties of any newly identified drug must be extensively characterized prior to its entry into clinical trials. As the vast bulk of biopharmaceuticals are proteins, a summary overview of the approach taken to initial characterization of these biomolecules is presented. A prerequisite to such characterization is initial purification of the protein. Purification to homogeneity usually requires a combination of three or more high-resolution chromatographic steps (Chapter 6). The purification protocol is designed carefully, as it usually forms the basis of subsequent pilot- and process-scale purification systems. The purified product is then subjected to a battery of tests that aim to characterize it fully. Moreover, once these characteristics have been defined, they form the basis of many of the QC identity tests routinely performed on the product during its subsequent commercial manufacture. As these identity tests are discussed in detail in Chapter 7, only an abbreviated overview is presented here, in the form of Figure 4.5.

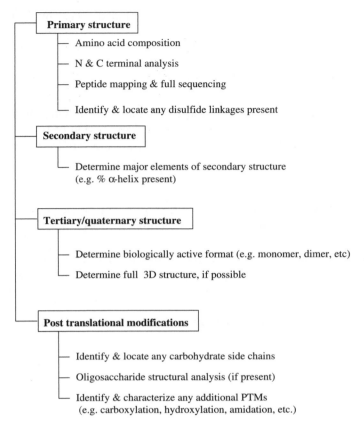

Figure 4.5 Task tree for the structural characterization of a therapeutic protein. A more detailed examination of many of these characterization studies is undertaken in chapter 7

In addition to the studies listed in Figure 4.5, stability characteristics of the protein with regard to e.g. temperature, pH and incubation with various potential excipients are studied. Such information is required in order to identify a suitable final product formulation, and to give an early indication of the likely useful shelf-life of the product (Chapter 6).

4.9 Patenting

The discovery and initial characterization of any substance of potential pharmaceutical application is followed by its patenting. The more detail given relevant to the drug's physicochemical characteristics, a method of synthesis and its biological effects, the better the chances of successfully securing a patent. Thus, patenting may not take place until preclinical trials and phase I clinical trials are completed. Patenting, once successfully completed, does not grant the patent holder an automatic right to utilize/sell the patented product; first, it must be proven safe and effective in subsequent clinical trials, and then be approved for general medical use by the relevant regulatory authorities.

4.9.1   What is a patent and what is patentable?

A patent may be described as a monopoly granted by a government to an inventor, such that only the inventor may exploit the invention/innovation for a fixed period of time (up to 20 years). In return, the inventor makes available a detailed technical description of the invention/innovation so that, when the monopoly period has expired, it may be exploited by others without the inventor's permission.

A patent, therefore, encourages innovation by promoting research and development. It can also be regarded as a physical asset, which can be sold or licensed to third parties for cash. Patents also represent a unique source of technical information regarding the patented product.

The philosophy underlining patent law is fairly similar throughout the world. Thus, although there is no worldwide patenting office, patent practice in different world regions is often quite similar. This is fortuitous, as there is a growing tendency towards world harmonization of patent law, fuelled by multinational trade agreements.

In order to be considered patentable, an invention/innovation must satisfy several criteria, the most important four of which are:

- novelty

- non-obviousness

- sufficiency of disclosure

- utility.

4.9.2   Patenting in biotechnology

Many products of nature (e.g. specific antibiotics, microorganisms, proteins, etc.) have been successfully patented. It might be argued that simply to find any substance naturally occurring on the Earth is categorized as a discovery and would be unpatentable because it lacks true novelty or any inventive step. However, if you enrich, purify or modify a product of nature such that you make available the substance for the first time in an industrially useful format, that product/process is generally patentable. In other words, patenting is possible if the 'hand of man' has played an obvious part in developing the product.

In the USA, purity alone often facilitates patenting of a product of nature (Table 4.1). The US Patent and Trademark Office (PTO) recognizes purity as a change in form of the natural material. For example, although vitamin B_{12} was a known product of nature for many years, it was only available in the form of a crude liver extract, which was of no use therapeutically. Development of a suitable

Table 4.1 Some products of nature that are generally patentable under US patent law. Additional patenting criteria (e.g. utility) must also be met. For many products, the patent will include details of the process used to purify the product. However, 'process' patents can be filed, as can 'use' patents. Refer to text for further details

---

A pure microbial culture
Isolated viruses
Specific purified proteins (e.g. EPO)
Purified nucleic acid sequences (including isolated genes, plasmids, etc.)
Other purified biomolecules (e.g. antibiotics, vitamins, etc.)

---

production (fermentation) and purification protocol allowed production of pure, crystalline vitamin B_{12} which could be used clinically. On this basis, a product patent was granted in the USA.

Using the same logic, the PTO has granted patents, for example, for pure cultures of specific microorganisms and for medically important proteins (e.g. Factor VIII purified from blood (Chapter 12) and EPO purified from urine (Chapter 10)).

Rapid technological advances in the biological sciences raises complex patenting issues, and patenting law as applied to modern biotechnology is still evolving.

In the late 1980s, the PTO confirmed they would consider issuing patents for non-human multicellular organisms, including animals. The first transgenic animal was patented in 1988 by Harvard University. The 'Harvard mouse' carried a gene that made it more susceptible to cancer and, hence, more sensitive in detecting possible carcinogens.

Another area of biotechnology patent law relates to the patenting of genes and DNA sequences. Thus far, patents have been issued for some human genes, largely on the basis of the use of their cloned products (e.g. EPO and tPA). A consensus has emerged that patent protection should only be considered for nucleotide sequences that can be used for specific purposes, e.g. for a sequence that can serve as a diagnostic marker or that codes for a protein product of medical value. This appears to be a reasonably approach, as it balances issues of public interest with encouraging innovation in the area.

The issue of patenting genetic material or transgenic plants/animals remains a contentious one. The debate is not confined to technical and legal argument: ethical and political issues, including public opinion, also impinge on the decision-making process. The increasing technical complexity and sophistication of the biological principles/processes upon which biotechnological innovations are based also render resolution of legal patenting issues more difficult.

A major step in clarifying European Union (EU)-wide law with regard to patenting in biotechnology stems from the introduction of the 1998 European Patent Directive. This directive (EU law) confirms that biological material (e.g. specific cells, proteins, genes, nucleotide sequences, antibiotics, etc.) that previously existed in nature are potentially patentable. However, in order actually to be patentable, they must (a) be isolated/purified from their natural environment and/or be produced via a technical process (e.g. rDNA technology in the case of recombinant proteins) and (b) they must conform to the general patentability principles regarding novelty, non-obviousness, utility and sufficiency of disclosure. The 'utility' condition, therefore, in effect prevents patenting of gene/genome sequences of unknown function. The directive also prohibits the possibility of patenting inventions if their exploitation would be contrary to public order or morality. Thus, it is not possible to patent:

- the human body;

- the cloning of humans;

- the use of human embryos for commercial purposes;

- modifying germ line identity in humans;

- modifying the genetic complement of an animal if the modifications cause suffering without resultant substantial medical benefits to the animal/to humans.

4.10   Delivery of biopharmaceuticals

An important issue that must be addressed during the preclinical phase of the drug development process relates to the route by which the drug will be delivered/administered. To date, the vast majority of biopharmaceuticals approved for general medical use are administered by direct injection (i.e. parenterally) usually by intravenous (i.v.), subcutaneous (s.c., i.e. directly under the skin) or intramuscular (i.m., i.e. into muscle tissue) routes. Administration via the s.c. or i.m. route is generally followed by slow release of the drug from its depot site into the bloodstream. Amongst the few exceptions to this parenteral route are the enzyme DNase, used to treat cystic fibrosis (Chapter 12), and platelet-derived growth factor (PDGF), used to treat certain skin ulcers (Chapter 10). However, neither of these products is required to reach the bloodstream in order to achieve its therapeutic effect. In fact, in each case the delivery system delivers the biopharmaceutical directly to its site of action (DNase is delivered directly to the lungs via aerosol inhalation, and PDGF is applied topically, i.e. directly on the ulcer surface, as a gel).

Parenteral administration is not perceived as a problem in the context of drugs which are administered infrequently, or as a once-off dose to a patient. However, in the case of products administered frequently/daily (e.g. insulin to diabetics), non-parenteral delivery routes would be preferred. Such routes would be more convenient, less invasive, less painful and generally would achieve better patient compliance. Alternative potential delivery routes include oral, nasal, transmucosal, transdermal or pulmonary routes. Although such routes have proven possible in the context of many drugs, routine administration of biopharmaceuticals by such means has proven to be technically challenging. Obstacles encountered include their high molecular mass, their susceptibility to enzymatic inactivation and their potential to aggregate.

4.10.1   Oral delivery systems

Oral delivery is usually the preferred system for drug delivery, owing to its convenience and the high level of associated patient compliance generally attained. Biopharmaceutical delivery via this route has proven problematic for a number of reasons:

- Inactivation due to stomach acid. Prior to consumption of a meal, stomach pH is usually below 2.0. Although the buffering action of food can increase the pH to neutrality, the associated stimulation of stomach acid secretion subsequently reduces the ambient pH back down to 3.0–3.5. Virtually all biopharmaceuticals are acid labile and are inactivated at low pH values.

- Inactivation due to digestive proteases. Therapeutic proteins would represent potential targets for digestive proteases such as pepsin, trypsin and chymotrypsin.

- Their (relatively) large size and hydrophilic nature renders difficult the passage of intact biopharmaceuticals across the intestinal mucosa.

- Orally absorbed drugs are subjected to first-pass metabolism. Upon entry into the bloodstream, the first organ encountered is the liver, which usually removes a significant proportion of absorbed drugs from circulation.

Given such difficulties, it is not unsurprising that bioavailabilities below 1 per cent are often recorded in the context of oral biopharmaceutical drug delivery. Strategies pursued to improve bioavailability include physically protecting the drug via encapsulation and formulation as microemulsions/microparticulates, as well as inclusion of protease inhibitors and permeability enhancers.

Encapsulation within an enteric coat (resistant to low pH values) protects the product during stomach transit. Microcapsules/spheres utilized have been made from various polymeric substances, including cellulose, polyvinyl alcohol, polymethylacrylates and polystyrene. Delivery systems based upon the use of liposomes and cyclodextrin-protective coats have also been developed. Included in some such systems also are protease inhibitors, such as aprotinin and ovomucoids. Permeation enhancers employed are usually detergent-based substances, which can enhance absorption through the gastrointestinal lining.

More recently, increasing research attention has focused upon the use of 'mucoadhesive delivery systems' in which the biopharmaceutical is formulated with/encapsulated in molecules that interact with the intestinal mucosa membranes. The strategy is obviously to retain the drug at the absorbing surface for a prolonged period. Non-specific (charge-based) interactions can be achieved by the use of polyacrylic acid, whereas more biospecific interactions are achieved by using selected lectins or bacterial adhesion proteins. Despite intensive efforts, however, the successful delivery of biopharmaceuticals via the oral route remains some way off.

4.10.2 Pulmonary delivery

Pulmonary delivery currently represents the most promising alternative to parenteral delivery systems for biopharmaceuticals. Delivery via the pulmonary route moved from concept to reality in 2006 with the approval of Exubera, an inhalable insulin product (Chapter 11). Although the lung is not particularly permeable to solutes of low molecular mass (e.g. sucrose or urea), macromolecules can be absorbed into the blood via the lungs surprisingly well. In fact, pulmonary macromolecular absorption generally appears to be inversely related to molecular mass, up to a mass of about 500 kDa. Many peptides/proteins delivered to the deep lung are detected in the blood within minutes, and bioavailabilities approaching/exceeding 50 per cent (relative to s.c. injection) have been reported for therapeutic proteins such as colony-stimulating factors and some interferons. Although not completely understood, such high pulmonary bioavailability may stem from:

- the lung's very large surface area;

- their low surface fluid volume;

- thin diffusional layer;

- relatively slow cell surface clearance;

- the presence of proteolytic inhibitors.

Additional advantages associated with the pulmonary route include:

- the avoidance of first-pass metabolism;

- the availability of reliable, metered nebulizer-based delivery systems capable of accurate dosage delivery, either in powder or liquid form;

- levels of absorption achieved without the need to include penetration enhancers which are generally too irritating for long-term use.

Although obviously occurring in practice, macromolecules absorbed via the pulmonary route must cross a number of biological barriers to get into the blood. These are:

- a protective monolayer of insoluble phospholipid, termed 'lung surfactant', and its underlying surface lining fluid, which lies immediately above the lung epithelial cells;

- the epithelial cells lining the lung;

- the interstitium (an extracellular space), and the basement membrane, composed of a layer of interstitial fibrous material;

- the vascular endothelium, i.e. the monolayer of cells that constitute the walls of the blood vessels.

Passage through the epithelium and endothelial cellular barriers likely represents the greatest challenge to absorption. Although the molecular details remain unclear, this absorption process appears to occur via one of two possible means: transcytosis or paracellular transport (Figure 4.6).

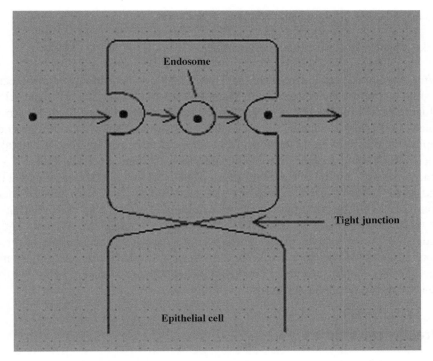

Figure 4.6 Likely mechanisms by which macromolecules cross cellular barriers in order to reach the blood-stream from (in this case) the lung. Transcytosis entails direct uptake of the macromolecule at one surface via endocytosis, travel of the endosome vesicle across the cell, with subsequent release on the opposite cell face via exocytosis. Paracellular transport entails the passage of the macromolecules through 'leaky' tight junctions found between some cells

4.10.3 Nasal, transmucosal and transdermal delivery systems

A nasal-based biopharmaceutical delivery route is considered potentially attractive as:

- it is easily accessible;

- nasal cavities are serviced by a high density of blood vessels;

- nasal microvilli generate a large potential absorption surface area;

- nasal delivery ensures the drug bypasses first-pass metabolism.

However, the route does display some disadvantages, including:

- clearance of a proportion of administered drug occurs due to its deposition upon the nasal mucous blanket, which is constantly cleared by ciliary action;

- the existence of extracellular nasal proteases/peptidases;

- low uptake rates for larger peptides/polypeptides.

Peptide/protein uptake rates across the nasal epithelia are dependent upon molecular mass. Relatively small peptides (such as oxytocin, desmopressin and luteinizing hormone-releasing hormone analogues) cross relatively easily, and several such products used medically are routinely delivered nasally. Larger molecules (of molecular mass greater than 10 kDa) generally do not cross the epithelial barrier without the concurrent administration of detergent-like uptake enhancers. Long-term use of enhancers is prohibited due to their damaging cellular effects.

Research efforts also continue to explore mucosal delivery of peptides/proteins via the buccal, vaginal and rectal routes. Again, bioavailabilities recorded are low, with modest increases observed upon inclusion of permeation enhancers. Additional barriers also exist relating, for example, to low surface areas, relatively rapid clearance from the mouth (buccal) cavity and the cyclic changes characteristic of vaginal tissue. Various strategies have been adopted in an attempt to achieve biopharmaceutical delivery across the skin (transdermal systems). Most have met with, at best, modest success thus far. Strategies employed include the use of a jet of helium or the application of a low voltage to accelerate proteins through the skin.

4.11   Preclinical studies

In order to gain approval for general medical use, the quality, safety and efficacy of any product must be demonstrated. Demonstration of conformance to these requirements, particularly safety and efficacy, is largely attained by undertaking clinical trials. However, preliminary data, especially safety data, must be obtained prior to the drug's administration to human volunteers. Regulatory authority approval to commence clinical trials is based largely upon preclinical pharmacological and toxicological assessment of the potential new drug in animals. Such preclinical studies can take up to 3 years to complete, and at a cost of anywhere between US$10 million and US$30 million. On average, approximately 10 per cent of potential new drugs survive preclinical trials.

In many instances, there is no strict set of rules governing the range of tests that must be undertaken during preclinical studies. However, guidelines are usually provided by regulatory authorities. The range of studies generally undertaken with regard to traditional chemical-based pharmaceuticals is summarized in Table 4.2. Most of these tests are equally applicable to biopharmaceutical products.

4.12   Pharmacokinetics and pharmacodynamics

Pharmacology may be described as the study of the properties of drugs and how they interact with/affect the body. Within this broad discipline exist (somewhat artificial) subdisciplines, including pharmacokinetics and pharmacodynamics.

Pharmacokinetics relates to the fate of a drug in the body, particularly its ADME, i.e. its absorption into the body, its distribution within the body, its metabolism by the body, and its excretion from the body.

Table 4.2 The range of major tests undertaken on a potential new drug during preclinical trials. The emphasis at this stage of the drug development process is upon assessing safety. Satisfactory pharmacological, and particularly toxicological, results must be obtained before any regulatory authority will permit commencement of human trials

---

Pharmacokinetic profile
Pharmacodynamic profile
Bioequivalence and bioavailability
Acute toxicity
Chronic toxicity
Reproductive toxicity and teratogenicity
Mutagenicity
Carcinogenicity
Immunotoxicity
Local tolerance

The results of such studies not only help to identify any toxic effects, but also point to the most appropriate method of drug administration, as well as the most likely effective dosage regime to employ. Generally, ADME studies are undertaken in two species, usually rats and dogs, and studies are repeated at various different dosage levels. All studies are undertaken in both males and females.

If initial clinical trials reveal differences in human versus animal model pharmacokinetic profiles, additional pharmacokinetic studies may be necessary using primates.

Pharmacodynamic studies deal more specifically with how the drug brings about its characteristic effects. Emphasis in such studies is often placed upon how a drug interacts with a cell/organ type, the effects and side effects it induces, and observed dose–response curves.

Bioavailability and bioequivalence are also usually assessed in animals. Such studies are undertaken as part of pharmacokinetic and/or pharmacodynamic studies. Bioavailability relates to the proportion of a drug that actually reaches its site of action after administration. As most biopharmaceuticals are delivered parenterally (e.g. by injection), their bioavailability is virtually 100 per cent. On the other hand, administration of biopharmaceuticals by mouth would, in most instances, yield a bioavailability at or near 0 per cent. Bioavailability studies would be rendered more complex if, for example, a therapeutic peptide was being administered intranasally.

Bioequivalence studies come into play if any change in product production/delivery systems was being contemplated. These studies would seek to identify whether such modifications still yield a product equivalent to the original one in terms of safety and efficacy. Modifications could include an altered formulation or method of administration, dosage regimes, etc.

4.12.1 Protein pharmacokinetics

A prerequisite to pharmacokinetic/pharmacodynamic studies is the availability of a sufficiently selective and sensitive assay. The assay must be capable of detecting and accurately quantifying the therapeutic protein in the presence of a complex soup of 'contaminant' molecules characteristic of tissue extracts/body fluids. As described in Chapter 7, specific proteins are usually detected and quantified either via immunoassay or bioassay. Additional analytical approaches occasionally used include liquid chromatography (e.g. HPLC) or the use of radioactively labelled protein.

The macromolecular structure of drugs and the fact that relatively minor structural alterations can potentially have a major influence upon bioactivity are often complicating factors. For example, an immunoassay may be blind to the oxidation of an amino acid residue, or very limited proteolytic processing, although such events can activate or decrease bioactivity.

As outlined previously, i.v. or s.c. administration is by far the most common delivery approach in the context of biopharmaceuticals. Whole-body distribution studies are undertaken mainly in order to assess tissue targeting and to identify the major elimination routes. The large molecular weight of therapeutic proteins, along with additional properties (e.g. charge), generally impairs their passage through biomembranes; hence, their initial distribution is usually limited to the volume of the extracellular space (mainly the plasma volume). Distribution volume usually subsequently increases, as the protein is taken up into tissue during its metabolism/elimination.

The metabolism/elimination of therapeutic proteins occurs via processes identical to those pertaining to native endogenous proteins. Ultimately this entails proteolytic degradation, with amino acid residues released either being incorporated into newly synthesized protein or being further degraded by standard metabolic pathways. Although the therapeutic protein may be subject to limited proteolysis in the blood, extensive and full metabolism occurs intracellularly, subsequent to product cellular uptake. Clearance of protein drugs from systemic circulation commences with passage across the capillary endothelia. The rate of passage depends upon the protein's physicochemical properties (e.g. mass and charge). Final product excretion is, in the main, either renal and/or hepatic mediated.

Many proteins of molecular mass <30 kDa are eliminated by the kidneys via glomerular filtration. In addition to size, filtration is also dependent upon the protein's charge characteristics. Owing to the presence of glycosaminoglycans, the glomerular filter is itself negatively charged, so negatively charged proteins are poorly filtered due to charge repulsion.

After initial filtration many proteins are actively reabsorbed (endocytosed) by the proximal tubules and subjected to lysosomal degradation, with subsequent amino acid reabsorption. Thus, very little intact protein actually enters the urine.

Uptake of protein by hepatocytes can occur via one of two mechanisms: (a) receptor-mediated endocytosis or (b) non-selective pinocytosis, again with subsequent protein proteolysis. Similarly, a proportion of some proteins are likely degraded within the target tissue, as binding to their functional cell surface receptors triggers endocytotic internalization of the receptor ligand complex (Figure 4.7).

Cellular uptake of some glycosylated therapeutic proteins occurs via specific sugar-binding cell surface receptors. Cell surface mannose receptors, for example, are capable of binding glycoproteins whose sugar side chains terminate in mannose, fucose, N-acetyl glucosamine or N-acetyl galactosamine. Evidence suggests that a liver-specific form of the mannose receptor mediates clearance of luteinizing hormone (LH, Chapter 11). The sugar side chains of many glycoproteins exhibit terminal sialic acid residues (sialic acid caps). The hepatic asialoglycoprotein receptor binds glycoproteins whose sialic acid caps have been removed, likely mediating their removal from general circulation.

Pharmacokinetic and indeed pharmacodynamic characteristics of therapeutic proteins can be rendered (even more) complicated by a number of factors, including:

- *The presence of serum-binding proteins.* Some biopharmaceuticals (including insulin-like growth factor (IGF), GH and certain cytokines) are notable in that the blood contains proteins that specifically bind them. Such binding proteins can function naturally as transporters or activators, and binding can affect characteristics such as serum elimination rates.

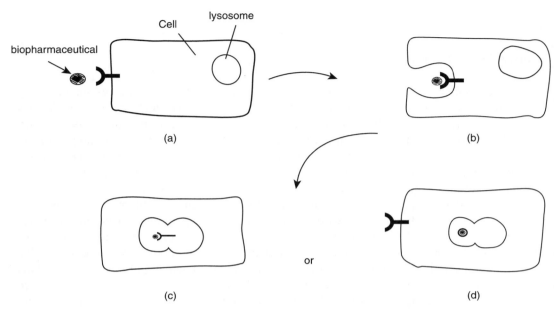

Figure 4.7 The process of receptor-mediated endocytosis. Binding of ligand, in this case a therapeutic protein, to its cell surface receptor (a) triggers invagination of the immediately surrounding area of plasma membrane, internalizing the receptor and its ligand in an intracellular vesicle (b). This is usually followed by fusion of the internalized vesicle with a lysosome and, therefore, degradation of both ligand and receptor by lysosomal hydrolases (c). In some cases, however, a variation can occur in which the ligand disassociated from the receptor (due to a lower pH in the vesicle), with subsequent budding off a small receptor-containing section of the vesicle, which returns the receptor to the cell surface. In this situation, only the ligand is available for degradation upon subsequent vesicular fusion with the lysosome (d)

- *Immunogenicity.* Many, if not most, therapeutic proteins are potentially immunogenic when administered to humans. The likelihood that non-human proteins (e.g. murine monoclonal antibodies; Chapter 13) are immunogenic in humans is an obvious one. However, human proteins can also be potentially immunogenic, as discussed in Box 4.1. Antibodies raised in this way can bind the therapeutic protein, neutralizing its activity and/or affecting its serum half-life.

- *Sugar profile of glycoproteins.* Expression of a therapeutic glycoprotein in different eukaryotic expression systems results in a product displaying differences in exact glycosylation detail (Chapter 2). The exact glycosylation pattern can influence protein activity and stability *in vivo*, and some sugar motifs characteristic of yeast-, insect- and plant-based expression systems are immunogenic in man.

4.12.2 Tailoring of pharmacokinetic profile

A number of different approaches may be used in order to alter a protein's pharmacokinetic profile. This can be desirable in order to achieve a predefined therapeutic goal, such as generating a faster- or slower-acting product, lengthening a product's serum half-life or altering a product's tissue

Box 4.1

Protein immunogenicity

Most traditional pharmaceuticals are relatively low molecular weight substances and generally escape the attention of the immune system. Proteins, on the other hand, are macromolecules and display molecular properties that can potentially trigger a vigorous immune response. During its formation our immune system develops tolerance to self-antigens. Such immunological tolerance is generally maintained throughout our lifetime by various regulatory mechanisms that either (a) prevent B- and T-lymphocytes from becoming responsive to self-antigens or (b) that inactivate such immune effector cells once they encounter self-antigens.

Based on the above principles, it might be assumed that a therapeutic protein obtained by direct extraction from human sources (e.g. some antibody preparations) or produced via recombinant expression of a human gene/cDNA sequence (e.g. recombinant human hormones or cytokines) would be non-immunogenic in humans whereas 'foreign' therapeutic proteins (e.g. non-engineered monoclonal antibodies) would stimulate a human immune response. This general principle holds in many cases, but not all. So why do therapeutic proteins of human amino acid sequences have the potential to trigger an immune response? Potential reasons can include:

- *Differences in post-translational modification (PTM) detail.* Human therapeutic proteins produced in several recombinant systems (e.g. yeast-, plant- and insect-based systems; Chapter 5) can display altered PTM detail, particularly in the context of glycosylation (Chapter 2). Some sugar residues/motifs characteristic of these systems can be highly immunogenic in humans.

- *Structural alteration of the protein during processing or storage.* Suboptimal product processing or formulation can result in partial degradation, denaturation, aggregation or precipitation of the therapeutic protein. Epitopes normally shielded from immune surveillance may be exposed as a result, triggering an immune response.

- *Some modes of administration.* In particular, s.c. injection may trigger protein aggregation or cause prolonged contact between the protein and immune system cells, thereby enhancing the potential for an immune response. An interesting example of this is provided by the recombinant human EPO-based product 'Eprex'. In the late 1990s the product's formulation was changed, with the removal of HSA as an excipient and its replacement with glycine and polysorbate 80. The product was being administered subcutaneously. The formulation change coincided with the product becoming immunogenic in a proportion of recipient humans. It is believed that the underlining immunogenicity was triggered by the association of multiple EPO molecules on polysorbate-generated micellar surfaces, with concurrent prolonged exposure to immune system cells. A switch from s.c. to i.v. administration relieved the problem.

- *Dosage levels and duration of treatment.* High dosage levels (well above normal physiological ranges), in particular if a product is administered on an ongoing and regular basis, may potentially contribute to breaking self-tolerance, particularly if combined with any of the circumstances outlined in the surrounding bulleted points.

- *Genetic and/or immunological factors.* Some individuals may display underlining or induced immunological abnormalities, rendering them more susceptible to breakdown of self-tolerance. For example, some blood factor and hormone preparations isolated by direct extraction from human serum or tissue stimulated an immunological response in a proportion of human patients receiving them. This may be triggered by some immune deficiency in the patients themselves, although the presence of product impurities or structural altered product forms may also be contributing factors.

Even if a biopharmaceutical triggers an immune response, it does not automatically follow that the response will be clinically significant or undesirable. In some instances, anti-product antibodies have no effect upon safety or efficacy. In other instances, antibody binding may alter the product's pharmacokinetic properties or directly neutralize the biopharmaceutical's biological activity. Even more seriously, antibodies raised against the product could potentially cross-react with the endogenous form of the protein, neutralizing it. Eprex provides an example of this latter phenomenon. Antibodies formed against the product cross-reacted with endogenous EPO, causing shutdown of (EPO-stimulated) red blood cell production, triggering antibody-mediated pure red cell aplasia.

A number of approaches may be adopted in an attempt to reduce or eliminate protein immunogenicity. Protein engineering (Chapter 3), for example, has been employed to humanize monoclonal antibodies (Chapter 13). An alternative approach entails the covalent attachment of polyethylene glycol (PEG) to the protein backbone. This can potentially shield immunogenic epitopes upon the protein from the immune system.

distribution profile. The approach taken usually relies upon protein engineering, be it alteration of amino acid sequence, alteration of a native post-translational modification (usually glycosylation) or the attachment of a chemical moiety to the protein's backbone (often the attachment of PEG, i.e. PEGylation). Specific examples of therapeutic proteins engineered in this way are discussed in detail within various subsequent chapters, and are summarized in Table 4.3.

4.12.3   Protein mode of action and pharmacodynamics

Different protein therapeutics bring about their therapeutic effect in different ways (Figure 4.8). Hormones and additional regulatory molecules invariably achieve their effect by binding to a specific cell surface receptor, with receptor binding triggering intracellular signal transduction event(s) that ultimately mediate the observed physiological effect(s). Many antibodies, on the other hand, bring about their effect by binding to their specific target molecule, which either inactivates/triggers destruction of the target molecule or (in the case of diagnostic applications) effectively tags the target molecules/cells. Therapeutic enzymes bring about their effect via a catalytic mechanism. The mode of action of many specific biopharmaceuticals will be outlined in

Table 4.3 Therapeutic proteins engineered in some way in order to alter pharmacokinetic or other pharmacological characteristics. Full details of specific products are provided in the chapter indicated

Protein	Engineering detail/rationale	Chapter
Chimaeric and humanized antibodies	Rendering murine antibodies more human in sequence, thereby decreasing their immunogenicity and increasing their serum half-life	13
Engineered tPA	Increasing serum half-life, allowing administration by a single i.v. injection as opposed to infusion over 90 min	12
Fast-acting insulins	Amino acid substitutions generate products that enter the bloodstream more quickly from the site of injection, facilitating product co-administration with a meal rather than administration 30–45 min prior to a meal	11
Long-acting insulins	Amino acid substitutions generate a product that enters the bloodstream very slowly from the site of injection, thereby maintaining basal insulin blood levels over an extended period	11
PEGylated interferons	Covalent attachment of PEG increases interferon half-life from 3 h to some 24 h, thereby generating a product whose dosage schedule requires once-weekly as opposed to daily administration	8
PEGylated Macugen	PEGylation increases the molecular weight of this aptamer from about 10 kDa to some 50 kDa, thereby increasing its half-life in the vitreous humor	14
Carbohydrate-remodelled EPO	EPO displaying additional sugar side chains, increasing its plasma half-life and thereby facilitating a once-weekly as opposed to three-times weekly injection schedule	10
Carbohydrate-remodelled glucocerebrosidase	Enzymatic removal of sialic acid caps exposes mannose residues, triggering selective product uptake by macrophages (the target cell type)	12

Chapters 8–14; however, it is important to note that the exact molecular detail underpinning the effects of many such drugs has not been fully characterized.

A significant element of preclinical studies, therefore, centres upon identification of a drug's mode of action at a molecular level, in addition to investigating the full range of resultant physiological effects. Pharmacodynamic studies will invariably include monitoring effects (and the timing of effects) of the therapeutic protein at different known drug concentrations and drug delivery schedules.

4.13   Toxicity studies

Toxicity studies are carried out on all putative new drugs, largely via testing in animals, in order to ascertain whether the product exhibits any short-term or long-term toxicity.

Acute toxicity is usually assessed by administration of a single high dose of the test drug to rodents. Both rats and mice (male and female) are usually employed. The test material is administered by two means, one of which should represent the proposed therapeutic method

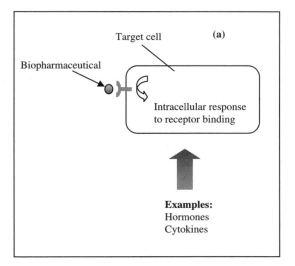

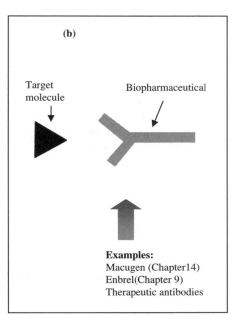

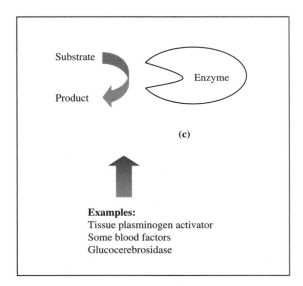

Figure 4.8 Overview of the mode of action of a biopharmaceutical. Some initiate their therapeutic effect by binding to a specific cell surface receptor, with subsequent signal transduction triggering the initiation of an intracellular response (e.g. activation of an intracellular enzyme or alteration of gene expression) (a). Other biopharmaceuticals function by binding to (and thus usually inactivating) a target molecule, perhaps whose overexpression is causing/exacerbating a medical condition (b). Therapeutic enzymes function by catalysing the conversion of specific target substrate molecule(s) into a product (c)

of administration. The animals are then monitored for 7–14 days, with all fatalities undergoing extensive post-mortem analysis.

Earlier studies demanded calculation of an LD_{50} value (i.e. the quantity of the drug required to cause death of 50 per cent of the test animals). Such studies required large quantities of animals, were expensive, and attracted much attention from animal welfare groups. Its physiological relevance to humans was often also questioned. Nowadays, in most world regions, calculation of the approximate lethal dose is sufficient.

Chronic toxicity studies also require large numbers of animals and, in some instances, can last for up to 2 years. Most chronic toxicity studies demand daily administration of the test drug (parenterally for most biopharmaceuticals). Studies lasting 1–4 weeks are initially carried out in order to, for example, assess drug levels required to induce an observable toxic effect. The main studies are then initiated and generally involve administration of the drug at three different dosage levels. The highest level should ideally induce a mild but observable toxic effect, whereas the lowest level should not induce any ill effects. The studies are normally carried out in two different species, usually rats and dogs, and using both males and females. All animals are subjected to routine clinical examination, and periodic analyses of, for example, blood and urine are undertaken. Extensive pathological examination of all animals is undertaken at the end of the study.

The duration of such toxicity tests varies. In the USA, the FDA usually recommends a period of up to 2 years, whereas in Europe the recommended duration is usually much shorter. Chronic toxicity studies of biopharmaceuticals can also be complicated by their likely stimulation of an immune response in the recipient animals. In the context of new chemical entities (NCEs, i.e. low molecular weight traditional chemicals), not only can the drug itself exhibit a toxic effect, but so potentially can drug breakdown products. As proteins are degraded to amino acids, any potentially toxicity associated with protein-based drugs is typically associated with the protein itself and not degradation products.

4.13.1   Reproductive toxicity and teratogenicity

All reproductive studies entail ongoing administration of the proposed drug at three different dosage levels (ranging from non-toxic to slightly toxic) to different groups of the chosen target species (usually rodents). Fertility studies aim to assess the nature of any effect of the substance on male or female reproductive function. The drug is administered to males for at least 60 days (one full spermatogenesis cycle). Females are dosed for at least 14 days before they are mated. Specific tests carried out include assessment of male spermatogenesis and female follicular development, as well as fertilization, implantation and early foetal development.

These reproductive toxicity studies complement teratogenicity studies, which aim to assess whether the drug promotes any developmental abnormalities in the foetus. (A teratogen is any substance/agent that can induce foetal developmental abnormalities. Examples include alcohol, radiation and some viruses.) Daily doses of the drug are administered to pregnant females of at least two species (usually rats and rabbits). The animals are sacrificed close to term and a full autopsy on the mother and foetus ensues. Post-natal toxicity evaluation often forms an extension of such studies. This entails administration of the drug to females both during and after pregnancy, with assessment of mother and progeny not only during pregnancy, but also during the lactation period. Therapeutic proteins rarely display any signs of reproductive toxicity or teratogenicity.

4.13.2   Mutagenicity, carcinogenicity and other tests

Mutagenicity tests aim to determine whether the proposed drug is capable of inducing DNA damage, either by inducing alterations in chromosomal structure or by promoting changes in nucleotide base sequence. Although mutagenicity tests are prudent and necessary in the case of chemical-based drugs, they are less so for most biopharmaceutical substances. In many cases, biopharmaceutical mutagenicity testing is likely to focus more so on any novel excipients added to the final product, rather than the biopharmaceutical itself. (Excipient refers to any substance other than the active ingredient that is present in the final drug formulation).

Mutagenicity tests are usually carried out *in vitro* and *in vivo*, often using both prokaryotic and eukaryotic organisms. A well-known example is the Ames test, which assesses the ability of a drug to induce mutation reversions in *E. coli* and *Salmonella typhimurium*.

Longer-term carcinogenicity tests are undertaken, particularly if (a) the product's likely therapeutic indication will necessitate its administration over prolonged periods (a few weeks or more) or (b) if there is any reason to suspect that the active ingredient or other constituents could be carcinogenic. These tests normally entail ongoing administration of the product to rodents at various dosage levels for periods of up to (or above) 2 years.

Some additional animal investigations are also undertaken during preclinical trials. These include immunotoxicity and local toxicity tests. Again, for many biopharmaceuticals, immunotoxicity tests (i.e. the product's ability to induce an allergic or hypersensitive response, or even a clinically relevant antibody response) are often impractical. The regulatory guidelines suggest that further studies should be carried out if a biotechnology drug is found capable of inducing an immune response. However, many of the most prominent biopharmaceuticals (e.g. cytokines) actually function to modulate immunological activities in the first place.

Prediction and preclinical assessment of the immunogenic potential of any biopharmaceutical in humans is by no means straightforward. The use of animal models is inappropriate, as the human protein will be automatically seen as foreign by their immune system, almost certainly stimulating an immune response. Some factors, such as the extent and nature of post-translational modifications, the mode and frequency of administration and whether the protein sequence is of human origin or not (Box 4.1), provide pointers but are by no means accurate predictors. One potential predictive approach entails the development and use of transgenic animals that are made immunotolerant for the human protein under development. Such animals can then provide some basis for the study of breaking immunological tolerance.

Many drugs, including many biopharmaceuticals, are administered to localized areas within the body by, for example, s.c. or i.m. injection. Local toxicity tests appraise whether there is any associated toxicity at/surrounding the site of injection. Predictably, these are generally carried out by s.c. or i.m. injection of product to test animals, followed by observation of the site of injection. The exact cause of any adverse response noted (i.e. active ingredient or excipient) is usually determined by their separate subsequent administration.

Preclinical pharmacological and toxicological assessment entails the use of thousands of animals. This is both costly and, in many cases, politically contentious. Attempts have been made to develop alternatives to using animals for toxicity tests, and these have mainly centred around animal cell culture systems. A whole range of animal and human cell types may be cultured, at least transiently, *in vitro*. Large-scale and fairly rapid screening can be undertaken by, for example, microculture of the target animal cells in microtitre plates, followed by addition of the drug and an indicator molecule.

The indicator molecule serves to assess the state of health of the cultured cells. The dye neutral red is often used (healthy cells assimilate the dye, dead cells do not). The major drawback to such systems is that they do not reflect the complexities of living animals and, hence, may not accurately reflect likely results of whole-body toxicity studies. Regulatory authorities are (rightly) slow to allow replacement of animal-based test protocols until the replacement system is proven to be reliable and is fully validated.

The exact range of preclinical tests that regulatory authorities suggest be undertaken for biopharmaceutical substances remains flexible. (Generally, only a subgroup of the standard tests for chemical-based drugs is appropriate. Biopharmaceuticals pose several particular difficulties, especially in relation to preclinical toxicological assessment. These difficulties stem from several factors (some of which have already been mentioned). These include:

- the species specificity exhibited by some biopharmaceuticals, e.g. GH and several cytokines, means that the biological activity they induce in man is not mirrored in test animals;

- for biopharmaceuticals, greater batch-to-batch variability exists compared with equivalent chemical-based products;

- induction of an immunological response is likely during long-term toxicological studies;

- lack of appropriate analytical methodologies in some cases.

In addition, tests for mutagenicity and carcinogenicity are not likely required for most biopharmaceutical substances. The regulatory guidelines and industrial practices relating to biopharmaceutical preclinical trials thus remain in an evolutionary mode, and each product is taken on a case-by-case basis. An overview of the main preclinical tests undertaken for a sample biopharmaceutical (Myozyme) is provided in Box 4.2.

4.13.3 Clinical trials

Clinical trials serve to assess the safety and efficacy of any potential new therapeutic 'intervention' in its intended target species. In our context, an intervention represents the use of a new biopharmaceutical. Examples of other interventions could be, for example, a new surgical procedure or a novel medical device. Veterinary clinical trials are based upon the same principles, but this discussion is restricted to investigations in humans. Clinical trials are also prospective rather than retrospective in nature, i.e. participants receiving the intervention are followed forward with time.

Clinical trials may be divided into three consecutive phases (Table 4.4). During phase I trials, the drug is normally administered to a small group of healthy volunteers. The aims of these studies are largely to establish:

- the pharmacological properties of the drug in humans (including pharmacokinetic and pharmacodynamic considerations);

Box 4.2

Preclinical assessment of Myozyme

Myozyme (tradename) is a recombinant form of the human lysosomal enzyme acid α-glucosidase (GAA) produced in a CHO cell line. It catalyses the degradation of lysosomal glycogen and it is approved for the treatment of Pompe disease (glycogen storage disease type II), an inherited disorder of glycogen metabolism caused by the absence or marked deficiency of lysosomal GAA. The product is administered by i.v. infusion at a dosage rate of 20 mg per kilogram body weight once every 2 weeks. The enzyme is taken up by various cells via endocytosis, triggered by binding of its carbohydrate component to cell surface sugar receptors. Internalization of endocytotic vesicles is followed by fusion with (and hence delivery to) lysosomes.

During product development, many of the initial non-clinical studies were undertaken in GAA knockout mice (i.e. mice devoid of a functional GAA gene), which serves as an animal model for Pompe's disease. The mice proved useful in assessing the pharmacodynamic effect of Myozyme on glycogen depletion and helped establish appropriate dosage regimens. The mice were also used to evaluate pharmacokinetics and biodistribution of GAA following its administration at clinically relevant doses.

Initial safety tests were carried out in beagle dogs and subsequently in cynomolgus monkeys. Single bolus i.v. doses of up to 100 mg kg^{-1} were used and were found to exert no negative effect upon general condition, blood pressure, heart or cardiovascular parameters, respiration rate or body temperature. No safety tests evaluating potential product effects upon the central nervous system were undertaken, as the protein is considered unlikely to cross the blood–brain barrier.

Repeat dose pharmacokinetic studies were undertaken in Sprague–Dawley rats and in monkeys. Biodistribution studies were carried out in both normal and knockout mice, with the majority of product distributed to the liver. No specific studies on product metabolism or excretion were undertaken, as the protein is almost certainly degraded via normal protein degradation mechanisms.

Toxicity was evaluated in mice, rats, dogs and monkeys following both acute and chronic product administration at various dosage levels (1–200 mg kg^{-1} range), with a proportion of animals displaying hypersensitivity/anaphylactic-like responses at high dosage levels.

Genotoxicity studies were not undertaken; this is normal practice for protein-based biopharmaceuticals, as proteins are unlikely to have mutagenic potential. No carcinogenicity studies were undertaken. Such studies are not normally required for therapeutic proteins unless there is some specific concern about carcinogenic potential. Reproductive toxicity studies evaluating product effect upon embryo–foetal development were undertaken in mice and revealed no concerns. Product antigenicity was evaluated, mainly as part of chronic toxicity studies. Antibody production against product was evaluated over a 26-week repeat administration period in monkeys. An enzyme-linked immunosorbent assay (ELISA)-based immunoassay (Chapter 7) was used to detect and quantify anti-product antibodies, with all monkey studies developing such antibodies.

Table 4.4 The clinical trial process. A drug must satisfactorily complete each phase before it enters the next phase. Note that the average duration listed here relates mainly to traditional chemical-based drugs. For biopharmaceuticals, the cumulative duration of all clinical trials is, on average, under 4 years

Trial phase	Evaluation undertaken (and usual number of patients)	Average duration (years)
I	Safety testing in healthy human volunteers (20–80)	1
II	Efficacy and safety testing in small number of patients (100–300)	2
III	Large-scale efficacy and safety testing in substantial numbers of patients (1000–3000)	3
IV	Post-marketing safety surveillance undertaken for some drugs that are administered over particularly long periods of time (number of patients varies)	Several

- the toxicological properties of the drug in humans (with establishment of the maximally tolerated dose);

- the appropriate route and frequency of administration of the drug to humans.

Thus, the emphasis of phase I trials largely remains upon assessing drug safety. If satisfactory results are obtained during phase I studies, the drug then enters phase II trials. These studies aim to assess both the safety and effectiveness of the drug when administered to volunteer patients (i.e. persons suffering from the condition the drug claims to cure/alleviate).

The design of phase II trials is influenced by the phase I results. Phase II studies typically last for anything up to 2 years, with anywhere between a few dozen and a hundred or more patients participating, depending upon the trial size.

If the drug proves safe and effective, phase III trials are initiated. (In the context of clinical trials, safe and effective are rarely used in the absolute sense. 'Safe' generally refers to a favourable risk:benefit ratio, i.e. the benefits should outweigh any associated risk. A drug is rarely 100 per cent effective in all patients. Thus, an acceptable level of efficacy must be defined, ideally prior to trial commencement. Depending upon the trial context, 'efficacy' could be defined as prevention of death/prolonging of life by a specific time-frame. It could also be defined as alleviation of disease symptoms or enhancement of the quality of life of sufferers (often difficult parameters to measure objectively). An acceptable incidence of efficacy should also be defined (particularly for phase II and III trials), e.g. the drug should be efficacious in, say, 25 per cent of all patients. If the observed incidence is below the minimal acceptable level, then clinical trials are normally terminated.

Phase III clinical trials are designed to assess the safety and efficacy characteristics of a drug in greater detail. Depending upon the trial size, usually hundreds if not thousands of patients are recruited, and the trial may last for up to 3 years. These trials serve to assess the potential role of the new drug in routine clinical practice; the phase III results will largely dictate whether or not the prospective drug subsequently gains approval for general medical use.

Even if a product gains marketing approval (on average, 10–20 per cent of prospective drugs that enter clinical trials are eventually commercialized), the regulatory authorities may demand further post-marketing surveillance studies. These are often termed 'phase IV clinical trials'. They aim to assess the long-term safety of a drug, particularly if the drug is administered to

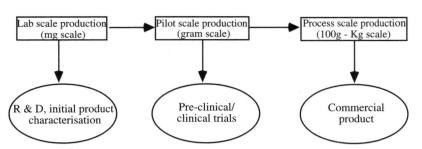

Figure 4.9 Scale-up of proposed biopharmaceutical production process to generate clinical trial material, and eventually commercial product. No substantive changes should be introduced to the production protocol during scale-up

patients for periods of time longer than the phase III clinical trials. The discovery of more long-term unexpected side effects can result in subsequent withdrawal of the product from the market.

Both preclinical and clinical trials are underpinned by a necessity to produce sufficient quantities of the prospective drug for its evaluation. Depending on the biopharmaceutical product, this could require from several hundred grams to over a kilogram of active ingredient. Typical production protocols for biopharmaceutical products are outlined in detail in Chapter 6. It is important that a suitable production process be designed prior to commencement of preclinical trials, that the process is amenable to scaling up and that, as far as is practicable, it is optimized (Figure 4.9). The material used for preclinical and clinical trials should be produced using the same process by which it is intended to undertake final-scale commercial manufacture. Extensive early development work is thus essential. Any significant deviation from the production protocol used to generate the trial material could invalidate all the clinical trial results with respect to the proposed commercialized product. (Changes in the production process could potentially change the final product characteristics, both for the active ingredient and the contaminant profile.)

4.13.4 Clinical trial design

Proper and comprehensive planning of a clinical trial is essential to the successful development of any drug. The first issue to be considered when developing a trial protocol is to define precisely what questions the trial results should be capable of answering. As discussed previously, the terms safety and efficacy are difficult to define in a therapeutic context. An acceptable meaning of these concepts, however, should be committed to paper prior to planning of the trial.

4.13.5 Trial size design and study population

A clinical trial must obviously have a control group, against which the test (intervention) group can be compared. The control group may receive: (a) no intervention at all; (b) a placebo (i.e. a substance such as saline, which will have no pharmacological or other effect); (c) the therapy most commonly used at that time to combat the target disease/condition.

The size of the trial will be limited by a number of factors, including:

- economic considerations (level of supporting financial resources);

- size of population with target condition;

- size of population with target condition after additional trial criteria have been imposed (e.g. specific age bracket, lack of complicating medical conditions, etc.);

- size of eligible population willing to participate in the trial.

Whereas a comprehensive phase III trial would normally require at least several hundred patients, smaller trials would suffice if, for example:

- the target disease is very serious/fatal;

- there are no existing acceptable alternative treatments;

- the target disease population is quite small;

- the new drug is clearly effective and exhibits little toxicity.

Choosing the study population is obviously critical to adequate trial design. The specific criteria of patient eligibility should be clearly predefined as part of the primary question the trial strives to answer.

A number of trial design types may be used (Table 4.5), each having its own unique advantages and disadvantages. The most scientifically pure is a randomized, double-blind trial (Box 4.3). However, in many instances, alternative trial designs are chosen based on ethical or other grounds. In most cases, two groups are considered: control and test. However, these designs can be adapted to facilitate more complex subgrouping. Clinical trial design is a subject whose scope is too broad to be undertaken in this text. The interested reader is referred to the 'Further reading' section at the end of this chapter.

Table 4.5   Some clinical trial design types. Additional information may be sourced from appropriate references provided in the 'Further reading' section

Randomized control studies (blinded or unblinded)
Historical control studies (unblinded)
Non-randomized concurrent studies (unblinded)
Cross-over trial design
Factorial design
Hybrid design
Large simple clinical trials

Box 4.3

Randomized control clinical trial studies

This trial design, which is the most scientifically desirable, involves randomly assigning participants into either control or test groups, with concurrent testing of both groups. (Randomness means that each participant has an equal chance of being assigned to one or the other group. This can be achieved by, for example, flipping a coin or drawing names from a hat). Randomness is important as it:

- removes the potential for bias (conscious or subconscious) and, thus, will produce comparable groups in most cases;

- guarantees the validity of subsequent statistical analysis of trial results.

The trial may also be unblinded or blinded. In an unblinded ('open') trial, both the investigators and participants know to which group any individual has been assigned. In a single blind trial, only the investigator is privy to this information, whereas neither the investigator nor the participants in double blind trials know to which group any individual is assigned. Obviously, the more blind the trial, the less scope for systematic error introduced by bias.

The most significant objection to the randomized control design is an ethical one. If a new drug is believed to be beneficial, then many feel it is ethically unsound to effectively deprive up to half the trial participants from receiving the drug.

One modification that overcomes the ethical difficulties is the use of historical control trials. In this instance, all the trial participants are administered the new drug and the results are compared with previously run trials in which a comparable group of participants was used. The control data are thus obtained from previously published or unpublished trial results. This trial design is non-randomized and non-concurrent. Although it bypasses ethical difficulties associated with withholding the new drug from any participant, it is vulnerable to bias. The trial designers have no influence over the criteria set for their control group. Furthermore, historical data can distort the result, as beneficial responses in the test subjects may be due not only to the therapeutic intervention, but generally improved patient management practices. This can be particularly serious if the control data are old (in some trials it was obtained 10–20 years previously).

4.14 The role and remit of regulatory authorities

The pharmaceutical industry is one of the most highly regulated industries known. Governments in virtually all world regions continue to pass tough laws to ensure that every aspect of pharmaceutical activity is tightly controlled. All regulations pertaining to the pharmaceutical industry are enforced by government-established regulatory agencies. The role and remit of some of the major world regulatory authorities is outlined below. In the context of this chapter, particular emphasis is placed upon their role with regard to the drug development process.

Table 4.6   The main product categories (and annual sales value) that the FDA regulate

US$500 billion worth of food
US$350 billion worth of medical devices
US$100 billion worth of drugs
US$50 billion worth of cosmetics and toiletries

4.14.1   The Food and Drug Administration

The FDA represents the American regulatory authority. Its mission statement defines its goal simply as being to 'protect public health'. In fulfilling this role, it regulates many products/consumer items (Table 4.6), the total annual value of which is estimated to be US$1 trillion. A more detailed list of the types of product regulated by the FDA is presented in Table 4.7. Its work entails inspecting/regulating almost 100 000 establishments in the USA (or those abroad who export regulated products for American consumption). The agency employs over 9000 people, of whom in the region of 4000 are concerned with enforcing drug law. The FDA's total annual budget is in the region of US$1 billion.

The FDA was founded in 1930. An act of Congress officially established it as a governmental agency in 1988, and it now forms a part of the US Department of Health and Human Services. The FDA Commissioner is appointed directly by the President (with the consent of the US Senate).

The FDA derives most of its statutory powers from the Federal Food, Drug and Cosmetic (FD & C) Act. This legislation was originally signed into law in 1930, but has been amended several times since. The agency interprets and enforces these laws. In order to achieve this, it draws up regulations based upon the legislation. Most of the regulations themselves are worded in general terms, and are supported by various FDA publications that explain/interpret these regulations in far greater detail. The publications include: 'Written Guidelines', 'Letters to Industry' and the 'Points to Consider' series of documents. As technological and other advances are made, the FDA further supplements their support publication list.

A partial organizational structure of the FDA is presented in Figure 4.10. The core activities of biopharmaceutical drug approval/regulation is undertaken mainly by the Center for Drug Evaluation and Research (CDER) and the Center for Biologics Evaluation and Research (CBER).

Table 4.7   A more detailed outline of the substances regulated by the FDA

Foods, nutritional supplements
Drugs: chemical-based, biologics, and biopharmaceuticals
Blood supply and blood products
Cosmetics/toiletries
Medical devices
All radioactivity-emitting substances
Microwave ovens
Advertising and promotional claims relating to the above product types

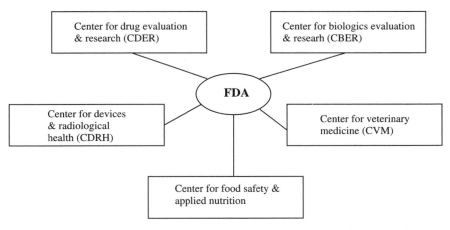

Figure 4.10 Partial organizational structure of the FDA, displaying the various centres primarily responsible for regulating drugs, devices and food

The major FDA responsibilities with regard to drugs include:

- assessing preclinical data to decide whether a potential drug is safe enough to allow commencement of clinical trials;

- protecting the interests and rights of patients participating in clinical trials;

- assessing preclinical and clinical trial data generated by a drug and deciding whether that drug should be made available for general medical use (i.e. if it should be granted a marketing licence);

- overseeing the manufacture of safe effective drugs (inspecting and approving drug manufacturing facilities on the basis of compliance to the principles of good manufacturing practice as applied to pharmaceuticals);

- ensuring the safety of the US blood supply.

In relation to the drug development process, the CDER has traditionally overseen and regulated the development and marketing approval of mainly chemical-based drugs. The CBER is more concerned with biologics. 'Biologic' has traditionally been defined in a narrow sense and has been taken to refer to vaccines and viruses, to blood and blood products, and to antiserum, toxins and antitoxins used for therapeutic purposes. Because of this, many established pharmaceutical products (e.g. microbial metabolites and hormones) have come under appraisal by the CDER, even though one might initially assume they would come under the biologics umbrella. The CDER has now also been assigned regulatory responsibility for the majority of products of pharmaceutical biotechnology (Table 4.8).

The criteria used by the CBER and CDER regulators in assessing product performance during the drug development process are similar, i.e. safety, quality and efficacy. However, the administrative details can vary in both name and content. Upon concluding preclinical trials, all

Table 4.8   Major biotechnology/biological-based drug types regulated by CDER and CBER

CDER regulated	CBER regulated
Monoclonal antibodies for *in vivo* use	Blood
Cytokines (e.g. interferons and interleukins)	Blood proteins (e.g. albumin and blood factors)
Therapeutic enzymes	Vaccines
Thrombolytic agents	Cell- and tissue-based products
Hormones	Gene therapy products
Growth factors	Antitoxins, venoms and antivenins
Additional miscellaneous proteins	Allenergic extracts

the data generated regarding any potential new drug are compiled in a dossier and submitted to the CDER or CBER in the form of an investigational new drug application (IND application). The FDA then assesses the application; if it does not object within a specific time-frame (usually 30 days), then clinical trials can begin. The FDA usually meet with the drug developers at various stages to be updated, and often to give informal guidance/advice. Once clinical trials have been completed, all the data generated during the entire development process are compiled in a multi-volume dossier.

The dossier submitted to the CDER is known as a new drug application (NDA), which, if approved, allows the drug to be marketed. If the drug is a CBER-regulated one, then a biologics licence application (BLA) is submitted.

4.14.2   The investigational new drug application

An investigational new drug is a new chemical-based, biologic or biopharmaceutical substance for which the FDA has given approval to undergo clinical trials. An IND application should contain information detailing preclinical findings, method of product manufacture and proposed protocol for initial clinical trials (Table 4.9).

In some instances, the FDA and drug sponsor (company/institution submitting the IND) will agree to hold a pre-IND meeting. This aims to acquaint the FDA officials with the background to/content of the IND application, and to get a feel for whether the IND application will be

Table 4.9   The major itemized points that must be included/addressed in an IND application to CDER or CBER

FDA Form 1571
Table of (IND) contents
Introduction
Proposed trial detail and protocol (general investigational plan)
Investigator's brochure
Chemistry (or biology, as appropriate) manufacturing, and control detail
Pharmacology and toxicology data
Any previous human experience regarding the drug substance
Any additional information

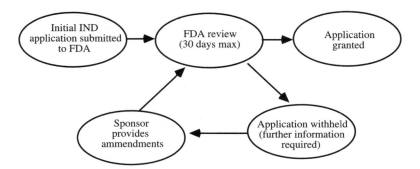

Figure 4.11 Outline of the IND application process

regarded as adequate (see below) by the FDA. An IND application can consist of up to 15 volumes of approximately 400 pages each. Once received by the FDA, it is studied to ensure that:

- it contains sufficient/complete information required;

- the information supplied supports the conclusion that clinical trial subjects would not be exposed to an unreasonable risk of illness/injury (the primary FDA role being to protect public health);

- the clinical investigator named is qualified to conduct the clinical trials;

- the sponsor's product brochures are not misleading/incomplete.

Based on their findings, the FDA may grant the application immediately or may require additional information which the sponsor then submits as IND amendments (Figure 4.11). Once clinical trials begin, the sponsor must provide the FDA with periodic updates, usually in the form of annual reports. Unscheduled reports must also be submitted under a variety of circumstances, including:

- if any amendments to the trial protocol are being considered;

- if any new scientific information regarding the product is obtained;

- if any unexpected safety observations are made.

During the clinical trial phase, the sponsor and FDA will meet on one or more occasions. A particularly important meeting is often the end of phase II meeting. This aims primarily to evaluate and agree upon phase III plans and protocols. This is particularly important, as phase III trials are the most costly and generate the greatest quantity of data, used later to support the drug approval application.

Table 4.10   An overview of the contents of a typical new NDA

Section 1	Index
Section 2	Overall summary
Section 3	Chemistry, manufacturing and control section
Section 4	Sampling, methods of validation, package and labelling
Section 5	Non-clinical pharmacology and toxicology data
Section 6	Human pharmacokinetics and bioavailability data
Section 7	Microbiology data
Section 8	Clinical data
Section 9	Safety update reports
Section 10	Statistical section
Section 11	Case report tabulations
Section 12	Case report forms
Sections 13 and 14	Patent information and certification
Section 15	Additional pertinent information

4.14.3   The new drug application

Upon completion of clinical trials, the sponsor will collate all the preclinical, clinical and other pertinent data (Table 4.10) and submit this to FDA in support of an application to allow the new drug to be placed upon the market. For CDER-related drugs, this submission document is termed an NDA.

The NDA must be an integrated document. It often consists of 200–300 volumes, which can represent over 120 000 pages. Several copies of the entire document, and sections thereof, are provided to the CDER. The FDA then classify the NDA based upon the chemical type of the drug and its therapeutic potential. Generally, drugs of high therapeutic potential (e.g. new drugs capable of curing/alleviating serious/terminal medical conditions) are appraised by the CDER in the shortest time.

After initial submission of the NDA, the FDA has 45 days in which to undertake a preliminary inspection of the document, to ensure that everything is in order. They then 'file' the NDA; or, if more information/better information management is needed, they refuse to file until such changes are implemented by the sponsor.

Once filed, an NDA undergoes several layers of review (Figure 4.12). A primary review panel generally consists of a chemist, microbiologist, pharmacologist, biostatistician, medical officer and biopharmaceutics scientist. Most hold PhDs in their relevant discipline. The team is organized by a project manager or consumer safety officer (CSO). The CSO initially forwards relevant portions of the NDA to the primary review panel member with the appropriate expertise.

Each reviewer then prepares a review report. This is forwarded to their supervisory officers, who undertake a second review. All of the reports are then sent to the division director who, in turn, recommends rejection or approval, or asks the sponsor to provide more information. On average, this entire process takes some 12 months.

Even when the NDA is approved and the product goes on sale, the sponsor must provide the FDA with further occasional reports. These can be in the form of scheduled annual reports, but also unscheduled reports are required in instances such as the occurrence of an unexpected adverse response to the drug.

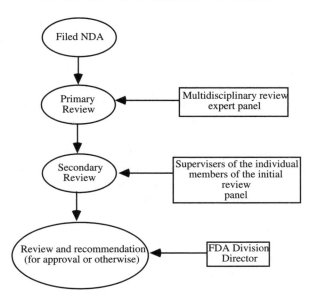

Figure 4.12 The CDER review process for a typical NDA. In addition to the review stages described, the FDA may also consult with a technical advisory committee. The members of the advisory committee are not routinely involved in IND or NDA assessment. The FDA is not obliged to follow any advice given by the advisory committee, but it generally does so

A similar general approach is taken by the CBER with regard to drugs being developed under their auspices. The CBER 'licensing process' for a new drug consists of three phases: the IND phase (already discussed), the pre-marketing approval phase (licensure phase) and the post-marketing surveillance phase. The pre-marketing approval phase (i.e. clinical trial phase) aims to generate data that prove the potency, purity and safety of the product. Upon completion of clinical trials, the sponsor collates the data generated and submits it to the FDA in the form of a BLA, which must provide a comprehensive description of both the product and product manufacture (including methods of QC analysis, product stability data, labelling data and, of course, safety and efficacy data).

A small number of biotechnology products are classified as medical devices and, hence, are regulated by the Center for Devices and Radiological Health (CDRH). The first approved biotech product to come under the auspices of the CDRH was OP-1 implant. Marketed by Stryker Biotech, OP-1 implant is a sterile powder composed of recombinant human oestrogenic protein-1 (OP-1) along with bovine collagen. It is used to treat fractured bones that fail to heal. The product is mixed with sterile saline immediately before application, and entails surgical insertion of the paste into the fracture.

4.14.4 European regulations

The overall philosophy behind granting a marketing authorization for a new drug is broadly similar in the USA and Europe. There are, however, major differences in the systems by which these philosophies are implemented in the two regions.

The EU is currently composed of 27 member states and a total population of 480 million (compared with 250 million in the USA and 125 million in Japan, the other two world pharmaceutical markets). The total European pharmaceutical market value stands at €70 billion, representing almost 33 per cent of world sales (the US and Japanese figures are 31 per cent and 21 per cent respectively). Annual expenditure on European pharmaceutical R&D stands at about €17 billion.

Prominent within the EU organizational structure is the European Commission. The major function of the European Commission is to propose new EU legislation and to ensure enforcement of existing legislation. European drug law, therefore, comes under the auspices of the Commission.

Two forms of legal instrument can be issued from the centralized European authorities: a 'regulation' and a 'directive'. A regulation is a strong legal instruction that, once passed, must be implemented immediately, and without modification, by national governments. A directive is a looser legal term, and provides an individual member state with 18 months to translate the flavour of that law into national law. Pharmaceutical law within the EU has been shaped by both directives and regulations, as discussed later.

4.14.5   National regulatory authorities

There are national regulatory authorities in all European countries. These authorities are appointed by the government of the country in question and are usually located in the national Ministry of Health. They serve to apply national and European law with regard to the drug development process. In many countries, different arms of the regulatory authorities are responsible for authorizing and assessing clinical trials, assessing the resultant drug dossier and deciding on that basis whether or not to grant a (national) marketing authorization/product licence. They are also often responsible for issuing manufacturing licences to companies.

In the past, a company wishing to gain a marketing licence within Europe usually applied for separate marketing authorizations on a country-by-country basis. This entailed significant duplication of effort as:

- the drug dossiers needed to be translated into various European languages;

- national laws often differed and, hence, different expectations/dossier requirements were associated with different countries;

- the time-scale taken for dossier assessment varied from country to country.

Attempts to harmonize European pharmaceutical laws were accelerated in the 1980s. From 1985 onwards, a substantial number of European pharmaceutical directives have been adopted. This entire legislation has been published in the form of a nine-volume series entitled 'The Rules Governing Medicinal Products in the European Union' (Table 4.11). These volumes form the basis of EU-wide regulation of virtually every aspect of pharmaceutical activity.

4.14.6   The European Medicines Agency and the new EU drug approval systems

In 1993, a significant advance in simplifying the procedures relating to drug marketing authorization applications in the EU was made. At that time, the legal basis of a new drug approval system

Table 4.11 The nine-volume series that comprises 'The Rules Governing Medicinal Products in the European Union'

Volume 1 – Pharmaceutical Legislation	Medicinal products for human use
Volume 2 – Notice to Applicants	Medicinal products for human use
Volume 3 – Guidelines	Medicinal products for human use
Volume 4 – Good Manufacturing Practices	Medicinal products for human and veterinary use
Volume 5 – Pharmaceutical Legislation	Veterinary medicinal products
Volume 6 – Notice to Applicants	Veterinary medicinal products
Volume 7 – Guidelines	Veterinary medicinal products
Volume 8 – Maximum Residue Limits	Veterinary medicinal products
Volume 9 – Pharmacovigilance	Medicinal products for human and veterinary use

was formed by an EU regulation and three directives. Central to this was the establishment of a European agency for the evaluation of medicinal products, the European Medicines Agency (EMEA), which is based in London and began work in January 1995. Two new marketing authorization (drug registration) procedures for human or veterinary drugs are now in place within the EU:

- a centralized procedure in which applications for a marketing licence are forwarded directly to the EMEA;

- a decentralized procedure based upon mutual acceptance or recognition of national authority decisions; disputes arising from this system are arbitrated by the EMEA.

The EMEA is comprised of:

- A management board, consisting of representatives of each EU member state and representatives of the European Commission and European Parliament. This group functions mainly to coordinate EMEA activities and manage its budget.

- Pre-authorization evaluation of medicines for human use unit. This unit provides scientific advice relating to quality, safety and efficacy issues, as well as relating to orphan drugs.

- Post-authorization evaluation of medicines for human use. This regulatory affairs unit is responsible for issues such as post-marketing surveillance of drugs.

- Veterinary medicines and inspections. This unit is responsible for veterinary marketing authorization procedures and inspections.

Central to the functioning of the EMEA are three committees:

- The Committee for Medicinal Products for Human use (CHMP). This committee is composed of 35 technical experts drawn from the various EU member countries. It is primarily responsible for formulating the EMEA's opinion on any medicinal product being considered for marketing approval under the centralized procedure.

- The Committee for Medicinal Products for Veterinary use (CVMP), whose structure and role is similar to the CPMP, except that it is concerned with animal medicines.

- The Committee for Orphan Medicinal Products (COMP), composed of a representative from each EU member state and EMEA and patient representatives. The COMP assesses applications relating to experimental medicines being granted 'orphan' status. Orphan medicines are those intended to treat rare diseases, and orphan designation results in a reduction of the fees charged by the EMEA when assessing marketing authorization applications.

Unlike the FDA, the EMEA itself does not directly undertake appraisals of drug dossiers submitted to support marketing authorization applications under the centralized procedure. Instead (as discussed in detail below), they forward the dossier to selected national EU regulatory bodies, who undertake the appraisal, and the EMEA makes a recommendation to approve (or not) the application based upon the national body's report. The overall role of the EMEA is thus to coordinate and manage the new system. The EMEA's annual budget is of the order of €120 million. The key objectives of the EMEA may be summarized as:

- protection of public and animal health by ensuring the quality, safety and efficacy of medicinal products for human and veterinary use;

- to strengthen the ideal of a single European market for human and veterinary pharmaceuticals;

- to support the European pharmaceutical industry as part of the EU industrial policy.

4.14.7   The centralized procedure

Under the centralized procedure, applications are accepted with regard to (a) products of biotechnology; (b) NCEs (drugs in which the active ingredient is new). Biotech products are grouped as 'list A' and NCEs as 'list B'. Marketing approval application for biotech products must be considered under the centralized procedure, whereas NCEs can be considered under centralized or decentralized mechanisms.

Upon receipt of a marketing authorization application under centralized procedures, the EMEA staff carry out an initial appraisal to ensure that it is complete and has been compiled in accordance with the appropriate EU guidelines (Box 4.4). This appraisal must be completed within 10 days, at which time (if the application is in order), it is given a filing date. The sponsor also pays an appropriate fee. The EMEA then has 210 days to consider the application. In the case of human drugs, the application immediately comes before the CPMP (which convenes for 2–3 days each month).

Two rappoteurs are then appointed (committee members who are responsible for getting the application assessed). The rappoteurs generally arrange to have the application assessed by their respective home national regulatory authorities. Once assessment is complete, the reports are presented via the rappoteur to the CPMP. After discussion, the CPMP issue an 'opinion' (i.e. a recommendation that the application be accepted, or not). This opinion is then forwarded to the European Commission, who have another 90 days to consider it. They usually accept the opinion

---

Box 4.4

The marketing authorization application in the EU

The dossier submitted to EU authorities, when seeking a manufacturing authorization, must be compiled according to specific EU guidelines. It generally consists of four parts as follows:

Part I (A)	Administrative data
Part I (B)	Summary of product characteristics
Part I (C)	Expert reports
Part II (A)	Composition of the product
Part II (B)	Method of product preparation
Part II (C)	Control of starting materials
Part II (D)	Control tests on intermediate products
Part II (E)	Control tests on finished product
Part II (F)	Stability tests
Part II (Q)	Bioavailability/bioequivalence and other information
Part III (A)	Single-dose toxicity
Part III (B)	Repeated-dose toxicity
Part III (C)	Reproductive studies
Part III (D)	Mutagenicity studies
Part III (E)	Carcinogenicity studies
Part III (F)	Pharmacodynamics
Part III (G)	Pharmacokinetics
Part III (H)	Local tolerance
Part III (Q)	Additional information
Part IV (A)	Clinical pharmacology
Part IV (B)	Clinical trial results
Part IV (Q)	Additional information

Part I (A) contains information that identifies the product, its pharmaceutical form and strength, route of administration and details of the manufacturer. The summary of product characteristics summarizes the qualitative and quantitative composition of the product, its pharmaceutical form, details of preclinical and clinical observations, as well as product particulars such as a list of added excipients, storage conditions, shelf life, etc. The expert reports contain written summaries of pharmaceutical, preclinical and clinical data.

Parts II, III and IV then make up the bulk of technical detail. They contain detailed breakdowns of all aspects of product manufacture and control (section II), preclinical data (section III) and clinical results (section IV).

---

and 'convert' it into a decision (they – and not the EMEA directly – have the authority to issue a marketing authorization). The single marketing authorization covers the entire EU. Once granted it is valid for 5 years, after which it must be renewed. The total time in which a decision must be taken is 300 days (210 + 90). The 300-day deadline only applies in situations where the CPMP do not require any additional information from the sponsor. In many cases, additional information is required; when this occurs, the 300-day 'clock' stops until the additional information is received.

4.14.8   Mutual recognition

The mutual recognition procedure is an alternative means by which a marketing authorization may be sought. It is open to all drug types except products of biotechnology. Briefly, if this procedure is adopted by a sponsor, then the sponsor applies for a marketing licence not to the EMEA, but to a specific national regulatory authority (chosen by the sponsor). The national authority then has 210 days to assess the application.

If adopted by the national authority in question, the sponsor can seek marketing licences in other countries on that basis. For this bilateral phase, other states in which marketing authorization are sought have 60 days in which to review the application. The theory, of course, is that no substantive difficulty should arise at this stage, as all countries are working to the same set of standards as laid down in 'The Rules Governing Medicinal Products in the European Union'.

A further 30 days is set aside in which any difficulties that arise may be resolved. The total application duration is 300 days. If one or more states refuse to grant the marketing authorization (i.e. mutual recognition breaks down), then the difficulties are referred back to the EMEA. The CPMP will then make a decision ('opinion'), which is sent to the European Commission. The Commission, taking into account the CPMP opinion, will make a final decision that is a binding.

4.14.9   Drug registration in Japan

The Japanese are the greatest consumers of pharmaceutical products per capita in the world. The Ministry of Health and Welfare in Japan has overall responsibility to implement Japanese pharmaceutical law. Within the department is the Pharmaceutical Affairs Bureau (PAB), which exercises this authority.

There are three basic steps in the Japanese regulatory process:

- approval ('shonin') must be obtained to manufacture or import a drug;

- a licence ('kyoka') must also be obtained;

- an official price for the drug must be set.

The PAB undertakes drug dossier evaluations, a process that normally takes 18 months. The approval requirements/process for pharmaceuticals (including biopharmaceuticals) are, in broad terms, quite similar to the USA. The PAB have issued specific requirements (Notification 243) for submission of recombinant protein drugs.

Applications are initially carefully checked by a single regulatory examiner to ensure conformance to guidelines. Subsequently, the application is reviewed in detail by a subcommittee of specialists. Clear evidence of safety, quality and efficacy are required prior to approval.

In addition, the Japanese normally insist that at least some clinical trials be carried out in Japan itself. This position is adopted due to, for example, differences in body size and metabolism of Japanese, compared with US and European citizens. Also, the quantity of active ingredient present in Japanese drugs is lower than in many other world regions. Hence, trials must be undertaken to prove product efficacy under intended Japanese usage conditions.

4.14.10 World harmonization of drug approvals

There is a growing trend in the global pharmaceutical industry towards internationalization. Increasingly, mergers and other strategic alliances are a feature of world pharmaceutical activity. Many companies are developing drugs that they aim to register in several world regions. Differences in regulatory practices and requirements in these different regions considerably complicate this process.

Development of harmonized requirements for world drug registrations would be of considerable benefit to pharmaceutical companies and to patients, for whom many drugs would be available much more quickly.

The International Conference on Harmonization (ICH) process (of technical requirements for registration of pharmaceuticals for human use) brings together experts from both the regulatory authorities and pharmaceutical industries based in the EU, USA and Japan. The aim is to achieve greater harmonization of technical/guidelines relating to product registration in these world regions, and considerable progress has been made in this regard over the last decade. The 'members' comprising the ICH are experts from the European Commission, the European Federation of Pharmaceutical Industries Association, the Ministry of Health and Welfare, Japan, the Japanese Pharmaceutical Manufacturers Association, the FDA and the Pharmaceutical Research and Manufacturers of America. The ICH process is supported by an ICH secretariat, based in Geneva, Switzerland.

4.15 Conclusion

The drug discovery and development process is a long and expensive one. A wide range of strategies may be adopted in the quest for identifying new therapeutic substances. Most biopharmaceuticals, however, have been discovered directly as a consequence of an increased understanding of the molecular mechanisms underlining how the body functions, both in health and disease.

Before any newly discovered drug is placed on the market, it must undergo extensive testing, in order to assure that it is both safe and effective in achieving its claimed therapeutic effect. The data generated by these tests (i.e. preclinical and clinical trials), are then appraised by independent, government-appointed regulatory agencies, who ultimately decide whether a drug should gain a marketing licence. While the drug development process may seem cumbersome and protracted, the cautious attitude adopted by regulatory authorities has served the public well in ensuring that only drugs of the highest quality finally come on to the market.

Further reading

Books

Adjei, H. 1997. *Inhalation Delivery of Therapeutic Peptides and Proteins*. J.A. Majors Company.
Ansel, H. 1999. *Pharmaceutical Dosage Forms and Drug Delivery Systems*. Lippincott Williams & Wilkins.
Askari, F. 2003. *Beyond the Genome – the Proteomics Revolution*. Prometheus Books.
Chakraborty, C. and Bhattacharya, A. 2005. *Pharmacogenomics*. Biotechnology Books, India.

Chow, S. and Liu, J. 2004. *Design and Analysis of Clinical Trials.* Wiley.

Crommelin, D. 2002. *Pharmaceutical Biotechnology.* Routledge.

De Jong, M. 1998. *FAQs on EU Pharmaceutical Regulatory Affairs.* Brookwood Medical Publications.

Desalli, R. 2002. *The Genomics Revolution.* Joseph Henry Press.

Everitt, B. and George, S. 2004. *Textbook of Clinical Trials.* Wiley.

Ferraiolo, B. 1992. *Protein Pharmacokinetics and Metabolism.* Plenum Publishers.

Goldberg, R. 2001. *Pharmaceutical Medicine, Biotechnology and European Patent Law.* Cambridge University Press.

Grindley, J. and Ogden, J. 2000. *Understanding Biopharmaceuticals. Manufacturing and regulatory Issues.* Interpharm Press.

Knight, H. 2001. *Patent Strategy.* Wiley.

Krogsgaard, L. 2002. *Textbook of Drug Design and Discovery.* Taylor and Francis.

McNally, E. (ed.). 2000. *Protein Formulation and Delivery.* Marcel Dekker.

Oxender, D. and Post, L. 1999. *Novel Therapeutics from Modern Biotechnology.* Springer Verlag.

Pennington, S. 2000. *Proteomics.* BIOS Scientific Publishers.

Pisano, D. and Mantus, D. 2003. FDA *Regulatory Affairs, a Guide for prescription Drugs, Medical Devices and Biologics.* CRC Press.

Poste, G. 1999. *The Impact of Genomics on Healthcare.* Royal Society of Medicine Press.

Rehm, H. 2005. *Protein Biochemistry and Proteomics.* Academic Press.

Scolnick, E. 2001. *Drug Discovery and Design.* Academic Press.

Venter, J. 2000. *From Genome to Therapy.* Wiley.

Articles and websites

Clinical trials, development and regulatory

Chirino, A.J., Ary, M.L., and Marshall, S.A. 2004. Minimizing the immunogenicity of protein therapeutics. *Drug Discovery Today* **9**, 82–90.

European Medicines Agency Website: http://www.emea.eu.int.

FDA Website: http://www.fda.gov.

Graffeo, A. 1994. The do's and don'ts of preclinical development. *Bio/Technology* **12**, 865.

Helms, P. 2002. Real world pragmatic clinical trials: what are they and what do they tell us? *Pediatric Allergy and Immunology* **13**(1), 4–9.

Hermeling, S., Crommelin, D.J.A., Schellekens, H., and Jiskoot, W. 2004. Structure–immunogenicity relationships of therapeutic proteins. Pharmaceutical research, **21**, 897–903.

Jefferys, D. and Jones, K. 1995. EMEA and the new pharmaceutical procedures for Europe. *European Journal of Clinical Pharmacology* **47**, 471–476.

Lubiniecki, A. 1997. Potential influence of international harmonization of pharmaceutical regulations on biopharmaceutical development. *Current Opinion in Biotechnology* **8**(3), 350–356.

Mahmood, I. and Green, M. 2005. Pharmacokinetic and pharmacodynamic considerations in the development of therapeutic proteins. *Clinical Pharmacokinetics* **44**(4), 331–347.

Pignatti, F., Aronsson, B., Vamvakas, S., Wade, G., Papadouli, I., Papaluca, M., Moulon, I., and Courtois, P.L. 2002. Clinical trials for registration in the European Union: the EMEA 5-year experience in oncology. *Critical Reviews in Oncology Haematology* **42**(2), 123–135.

Schwardt, O., Kolb, H., and Ernst, B. 2003. Drug discovery today. *Current Topics in Medicinal Chemistry* **3**(1), 1–9.

Shankar, G., Shores, E., Wagner, C., and Mire-Sluis, A. 2006. Scientific and regulatory considerations on the immunogenicity of biologics. *Trends in Biotechnology* **24**, 274–280.

Streatfield, S. 2005. Regulatory issues for plant-made pharmaceuticals and vaccines. *Expert Review of Vaccines* **4**(4), 591–601.

Tang, L., Persky, A.M., Hochhaus, G., and Meibohm, B. 2004. Pharmacokinetic aspects of biotechnology products. *Journal of Pharmaceutical Sciences* **93**(9), 2184–2204.

Walsh, G. 1999. Drug approval in Europe. *Nature Biotechnology*, **17**, 237–240.

Walsh, G. 2006. Drug approval in the European Union and United States. In *An Introduction to Molecular Biotechnology*, Wink, M. (ed.). Wiley-VCH; pp. 651–662.

Drug delivery

Davis, S. 2001. Nasal vaccines. *Advanced Drug Delivery Reviews* **51**, 21–42.

Hamman, J.H., Enslin, G.M., and Kotzé, A.F. 2005. Oral delivery of peptide drugs – barriers and developments. *Biodrugs* **19**(3), 165–177.

Kompella, U.B. and Lee, V.H.L. 2001. Delivery systems for penetration enhancement of peptide and protein drugs: design considerations. *Advanced Drug Delivery Reviews* **46**, 211–245.

Hussain, A., Arnold, J.J., Khan, M.A., and Ahsan, F. 2004. Absorption enhancers in pulmonary drug delivery. *Journal of Controlled Release* **94**(1), 15–24.

Orive, G., Hernandez, R.M., Gascon, A.R., Dominguez-Gil, A., and Pedraz, J.L. 2003. Drug delivery in biotechnology: present and future. *Current Opinion in Biotechnology* **14**(6), 659–664.

Patton, J. 1996. Mechanisms of macromolecule absorption by the lungs. *Advanced Drug Delivery Reviews* **19**, 3–36.

Patton, J.S., Bukar, J.G., and Eldon, M.A. 2004. Clinical pharmacokinetics and pharmacodynamics of inhaled insulin. *Clinical Pharmacokinetics* **43**(12), 781–801.

Quattrin, T. 2004. Inhaled insulin: recent advances in the therapy of type 1 and 2 diabetes. *Expert Opinion on Pharmacotherapy.* **5**(12), 2597–2604.

Shah, R.B., Ahsan, F., and Khan, M.A. 2002. Oral delivery of proteins: progress and prognostication. *Critical Reviews in Therapeutic Drug Carrier Systems* **19**(2), 135–169.

Shaikh, I.M., Jadhav, K.R., Ganga, S., Kadam, V.J., and Pisal, S.S. 2005. Advanced approaches in insulin delivery. *Current Pharmaceutical Biotechnology* **6**(5), 387–395.

Veronese, F.M. and Pasut, G. 2005. PEGylation, successful approach to drug delivery. *Drug Discovery Today* **10**, 1451–1458.

Genomics, proteomics and related technologies

Baba, Y. 2001. Development of novel biomedicine based on genome science. *European Journal of Pharmaceutical Sciences* **13**(1), 3–4.

Brenner, S. 2001. A tour of structural genomics. *Nature Reviews Genetics* **2**(10), 801–809.

Cunningham, M. 2000. Genomics and proteomics, the new millennium of drug discovery and development. *Journal of Pharmacological and Toxicological Methods* **44**(1), 291–300.

Debouck, C. and Goodfellow, P. 1999. DNA microarrays in drug discovery and development. *Nature Genetics* **21**, 48–50.

Gabig, M. and Wegrzyn, G. 2001. An introduction to DNA chips: principles, technology, applications and analysis. *Acta Biochimica Polonica* **48**(3), 615–622.

Jain, K. 2001. Proteomics: new technologies and their applications. *Drug Discovery Today* **6**(9), 457–459.

Jain, K. 2001. Proteomics: delivering new routes to drug discovery, part 2. *Drug Discovery Today.* **6**(16), 829–832.

Kassel, D. 2001. Combinatorial chemistry and mass spectrometry in the 21st century drug development laboratory. *Chemical Reviews* **101**(2), 255–267.

Lesley, S. 2001. High-throughput proteomics: protein expression and purification in the post-genomic world. *Protein Expression and Purification* **22**(2), 159–164.

McLeod, H. and Evans, W. 2001. Pharmacogenomics: unlocking the human genome for better drug therapy. *Annual Review of Pharmacology and Toxicology* **41**, 101–121.

Page, M.J., Amess, B., Rohlff, C., Stubberfield, C., and Parekh, R. 1999. Proteomics: a major new technology for the drug discovery process. *Drug Discovery Today* **4**(2), 55–62.

Ramsay, G. 1998. DNA chips: state of the art. *Nature Biotechnology* **16**(1), 40–44.

Roses, A. 2000. Pharmacogenetics and the practice of medicine. *Nature.* **405**, (6788), 857–865.

Schmitz, G., Aslanidis, C., and Lackner, K.J. 2001. Pharmacogenomics: implications for laboratory medicine. *Clinica Chemica Acta* **308**(1–2), 43–53.

Searls, D. 2000. Using bioinformatics in gene and drug discovery. *Drug Discovery Today* **5**(4), 135–143.

Terstappen, G. and Reggiani, A. 2001. *In silico* research in drug discovery. *Trends in Pharmacological Sciences* **22**(1), 23–26.

Wamg, J. and Hewick, R. 1999. Proteomics in drug discovery. *Drug Discovery Today* **4**(3), 129–133.

Wieczorek, S. and Tsongalis, G. 2001. Pharmacogenomics: will it change the field of medicine? *Clinica Chimica Acta* **308**(1–2), 1–8.

Wilgenbus, K. and Lichter, P. 1999. DNA chip technology *ante portas*. *Journal of Molecular Medicine* **77**(11), 761–768.

Wolf, C.R., Smith, G., and Smith, R.I. 2000. Science, medicine and the future – pharmacogenetics. *British Medical Journal* **320**(7240), 987–990.

Patenting

Allison, J. and Lemley, M. 2002. The growing complexity of the United States patenting system. *Boston University Law Review* **82**(1), 77–144.

Barton, J. 1991. Patenting life. *Scientific American* (March), 18–24.

Berks, A. 1994. Patent information in biotechnology. *Trends in Biotechnology* **12**, 352–364.

Farnley, S., Morey-Nase, P., and Sternfeld, D. 2004. Biotechnology – a challenge to the patent system. *Current Opinion in Biotechnology* **15**(3), 254–257.

Johnson, E. 1996. A benchside guide to patents and patenting. *Nature Biotechnology* **14**, 288–291.

Orr, R. and O'Neill, C. 2000. Patent review: therapeutic applications of antisense oligonucleotides, 1999–2000. *Current Opinion in Molecular Therapeutics* **2**(3), 325–331.

Schellekens, H. 2004. When biotech products go off patent. *Trends in Biotechnology* **22**(8), 406–410.

Yeh, J., and Fernandez, D. 2004. Patent protection strategies for biotechnological inventions. *Assay and Drug Development Technologies* **2**(6), 697–702.

5

Sources and upstream processing

5.1 Introduction

The manufacture of pharmaceutical substances is one of the most highly regulated and rigorously controlled of manufacturing processes. In order to gain a manufacturing licence, the producer must prove to the regulatory authorities that not only is the product itself safe and effective, but that all aspects of the proposed manufacturing process comply with the highest safety and quality standards.

This chapter aims to overview the manufacturing process of therapeutic proteins. It concerns itself with two major themes: (1) sources of biopharmaceuticals and (2) upstream processing. The additional elements of biopharmaceutical manufacturing, i.e. downstream processing and product analysis, are discussed in Chapters 6 and 7 respectively.

5.2 Sources of biopharmaceuticals

The bulk of biopharmaceuticals currently on the market are produced by genetic engineering using various recombinant expression systems. Although a wide range of potential protein production systems are available (Table 5.1), most of the recombinant proteins that have gained marketing approval to date are produced either in recombinant *E. coli* or in recombinant mammalian cell lines (Table 5.2). Such recombinant systems are invariably constructed by the introduction of a gene or cDNA coding for the protein of interest into a well-characterized strain of the chosen producer cell (Chapter 3). Examples include *E. coli* K12 and CHO strain K1 (CHO-K1). Each recombinant production system displays its own unique set of advantages and disadvantages, as described below.

5.2.1 *Escherichia coli* as a source of recombinant, therapeutic proteins

Many microorganisms represent attractive potential production systems for therapeutic proteins. They can usually be cultured in large quantities, inexpensively and in a short time, by standard

Pharmaceutical biotechnology: concepts and applications   Gary Walsh
© 2007 John Wiley & Sons, Ltd   ISBN 978 0 470 01244 4 (HB)   978 0 470 01245 1 (PB)

Table 5.1 Expression systems which are/could potentially be used for the production of recombinant biopharmaceutical products

E. coli (and additional prokaryotic systems, e.g. bacilli)
Yeast (particularly *S. cerevisiae*)
Fungi (particularly aspergilli)
Animal cell culture (particularly CHO and BHK cell lines)
Transgenic animals (focus thus far is upon sheep and goats)
Plant-based expression systems (various)
Insect cell culture systems

methods of fermentation. Production facilities can be constructed in any world region, and the scale of production can be varied as required.

The expression of recombinant proteins in cells in which they do not naturally occur is termed heterologous protein production (Chapter 3). The first biopharmaceutical produced by genetic engineering to gain marketing approval (in 1982) was recombinant human insulin (tradename 'Humulin'), produced in *E. coli*. An example of a more recently approved biopharmaceutical that is produced in *E. coli* is that of Kepivance, a recombinant keratinocyte growth factor used to treat oral mucositis (Chapter 10). Many additional examples are provided in subsequent chapters.

As a recombinant production system, *E. coli* displays a number of advantages. These include:

- *E. coli* has long served as the model system for studies relating to prokaryotic genetics. Its molecular biology is thus well characterized.

- High levels of expression of heterologous proteins can be achieved in recombinant *E. coli* (Table 5.3). Modern, high-expression promoters can routinely ensure that levels of expression of the recombinant protein reach up to 30 per cent total cellular protein.

- *E. coli* cells grow rapidly on relatively simple and inexpensive media, and the appropriate fermentation technology is well established.

These advantages, particularly its ease of genetic manipulation, rendered *E. coli* the primary biopharmaceutical production system for many years. However, *E. coli* also displays a number of drawbacks as a biopharmaceutical producer. These include:

Table 5.2 Some biopharmaceuticals currently on the market which are produced by genetic engineering in either *E. coli* or animal cells

Biopharmaceutical	Source	Biopharmaceutical	Source
tPA	*E. coli*, CHO	FSH	CHO
Insulin	*E. coli*	IFN-β	CHO
		EPO	CHO
IFN-α	*E. coli*	Glucocerebrosidase	CHO
IFN-γ	*E. coli*	Factor VIIa	BHK
IL-2	*E. coli*		
G-CSF	*E. coli*		
hGH	*E. coli*		

Table 5.3 Levels of expression of various biopharmaceuticals produced in recombinant *E. coli* cells

Biopharmaceutical	Level of expression (% of total cellular protein)
IFN-γ	25
Insulin	20
IFN-β	15
TNF	15
α_1-Antitrypsin	15
IL-2	10
hGH	5

- heterologous proteins accumulate intracellularly;

- inability to undertake post-translational modifications (particularly glycosylation) of proteins;

- the presence of lipopolysaccharide (LPS) on its surface.

The vast bulk of proteins synthesized naturally by *E. coli* (i.e. its homologous proteins) are intracellular. Few are exported to the periplasmic space or released as true extracellular proteins. Heterologous proteins expressed in *E. coli* thus invariably accumulate in the cell cytoplasm. Intracellular protein production complicates downstream processing (relative to extracellular production) as:

- additional primary processing steps are required, i.e. cellular homogenization with subsequent removal of cell debris by centrifugation or filtration;

- more extensive chromatographic purification is required in order to separate the protein of interest from the several thousand additional homologous proteins produced by the *E. coli* cells.

An additional complication of high-level intracellular heterologous protein expression is inclusion body formation. Inclusion bodies (refractile bodies) are insoluble aggregates of partially folded heterologous product. Because of their dense nature, they are easily observed by dark-field microscopy. Presumably, when expressed at high levels, heterologous proteins overload the normal cellular protein-folding mechanisms. Under such circumstances, it would be likely that hydrophobic patches normally hidden from the surrounding aqueous phase in fully folded proteins would remain exposed in the partially folded product. This, in turn, would promote aggregate formation via intermolecular hydrophobic interactions.

However, the formation of inclusion bodies displays one processing advantage: it facilitates the achievement of a significant degree of subsequent purification by a single centrifugation step. Because of their high density, inclusion bodies sediment even more rapidly than cell debris. Low-speed centrifugation thus facilitates the easy and selective collection of inclusion bodies directly after cellular homogenization. After collection, inclusion bodies are generally incubated with strong denaturants, such as detergents, solvents or urea. This promotes complete solubilization of the inclusion body (i.e. complete denaturation of the proteins therein). The denaturant is then removed by techniques such as dialysis or diafiltration. This facilitates refolding of the protein, a high percentage of which will generally fold into its native, biologically active, conformation.

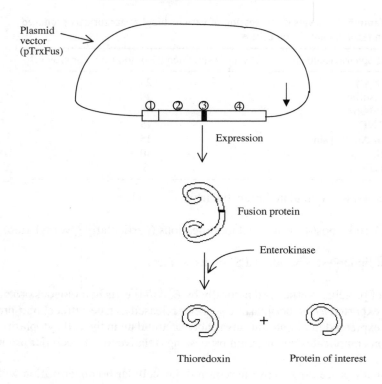

Key: ① = strong promoter

② = thioredoxin gene

③ = nucleotide sequence coding for the peptide sequence which
      serves as cleavage site for the protease (enterokinase)

④ = gene/cDNA coding for the protein of interest

Figure 5.1 High-level expression of a protein of interest in *E. coli* in soluble form by using the engineered 'thiofusion' expression system. Refer to text for specific details

Various attempts have been made to prevent inclusion body formation when expressing heterologous proteins in *E. coli*. Some studies have shown that a simple reduction in the temperature of bacterial growth (from 37 °C to 30 °C) can significantly decrease the incidence of inclusion body formation. Other studies have shown that expression of the protein of interest as a fusion partner with thioredoxin will eliminate inclusion body formation in most instances. Thioredoxin is a homologous *E. coli* protein, expressed at high levels. It is localized at the adhesion zones in *E. coli* and is a heat-stable protein. A plasmid vector has been engineered to facilitate expression of a fusion protein consisting of thioredoxin linked to the protein of interest via a short peptide sequence recognized by the protease enterokinase (Figure 5.1). The fusion protein is invariably expressed at high levels, while remaining in soluble form. Congregation at adhesion zones facilitates its selective release into the media by simple osmotic shock. This can greatly simplify its subsequent purification. After its release, the fusion protein is incubated with enterokinase, thus releasing the protein of interest (Figure 5.1).

Table 5.4 Proteins of actual or potential therapeutic use that are glycosylated when produced naturally in the body (or by hydridoma technology in the case of monoclonal antibodies). These proteins are discussed in detail in various subsequent chapters. See also Table 2.9

Most interleukins (IL-1 being an important exception)
IFN-β and -γ (most IFN-αs are unglycosylated)
CSFs
TNFs
Gonadotrophins (FSH, luteinizing hormone and hCG)
Blood factors (e.g. Factor VII, VIII and IX)
EPO
Thrombopoietin
tPA
α_1-Antitrypsin
Intact monoclonal antibodies

An alternative means of reducing/potentially eliminating inclusion body accumulation entails the high-level co-expression of molecular chaperones along with the protein of interest. Chaperones are themselves proteins that promote proper and full folding of other proteins into their biologically active, native three-dimensional shape. They usually achieve this by transiently binding to the target protein during the early stages of its folding and guiding further folding by preventing/correcting the occurrence of improper hydrophobic associations.

The inability of prokaryotes such as *E. coli* to carry out post-translational modifications (particularly glycosylation) can limit their usefulness as production systems for some therapeutically useful proteins. Many such proteins, when produced naturally in the body, are glycosylated (Table 5.4). However, the lack of the carbohydrate component of some glycoproteins has little, if any, negative influence upon their biological activity. The unglycosylated form of IL-2, for example, displays essentially identical biological activity to that of the native glycosylated molecule. In such cases, *E. coli* can serve as a satisfactory production system.

Another concern with regard to the use of *E. coli* is the presence on its surface of LPS molecules. The pyrogenic nature of LPS (Chapter 7) renders essential its removal from the product stream. Fortunately, several commonly employed downstream processing procedures achieve such a separation without any great difficulty.

5.2.2 Expression of recombinant proteins in animal cell culture systems

Technical advances facilitating genetic manipulation of animal cells now allow routine production of therapeutic proteins in such systems. The major advantage of these systems is their ability to carry out post-translational modification of the protein product. As a result, many biopharmaceuticals that are naturally glycosylated are now produced in animal cell lines. CHO and BHK cells have become particularly popular in this regard.

Although their ability to carry out post-translational modifications renders their use desirable/essential for producing many biopharmaceuticals, animal cell-based systems do suffer from a number of disadvantages. When compared with *E. coli*, animal cells display a very complex nutritional requirement, grow more slowly and are far more susceptible to physical damage. In industrial terms, this translates into increased production costs.

In addition to recombinant biopharmaceuticals, animal cell culture is used to produce various other biologically based pharmaceuticals. Chief amongst these are a variety of vaccines and hybridoma cell-produced monoclonal antibodies (Chapter 13). Earlier interferon preparations were also produced in culture by a particular lymphoblastoid cell line (the Namalwa cell line), which was found to synthesize high levels of several IFN-α's naturally (Chapter 8).

5.2.3   Additional production systems

5.2.3.1   Yeast

Attention has also focused upon a variety of additional production systems for recombinant biopharmaceuticals. Yeast cells (particularly *Saccharomyces cerevisiae*) display a number of characteristics that make them attractive in this regard. These characteristics include:

- their molecular biology has been studied in detail, facilitating their genetic manipulation;

- most are GRAS-listed organisms ('generally regarded as safe'), and they have a long history of industrial application (e.g. in brewing and baking);

- they grow relatively quickly in relatively inexpensive media, and their tough outer wall protects them from physical damage;

- suitable industrial-scale fermentation equipment/technology is already available;

- they possess the ability to carry out post-translational modifications of proteins.

The practical potential of yeast-based production systems has been confirmed by the successful expression of a whole range of proteins of therapeutic interest in such systems. However, a number of disadvantages relating to heterologous protein production in yeast have been recognized. These include:

- Although capable of glycosylating heterologous human proteins, the glycosylation pattern usually varies from the pattern observed on the native glycoprotein (when isolated from its natural source, or when expressed in recombinant animal cell culture systems).

- In most instances, expression levels of heterologous proteins remain less than 5 per cent of total cellular protein. This is significantly lower than expression levels typically achieved in recombinant *E. coli* systems.

Despite such potential disadvantages, several recombinant biopharmaceuticals now approved for general medical use are produced in yeast (*S. cerevisiae*)-based systems (Table 5.5). Interestingly, most such products are not glycosylated. The oligosaccharide component of glycoproteins produced in yeasts generally contains high levels of mannose. Such high mannose-

Table 5.5 Recombinant therapeutic proteins approved for general medical use that are produced in *S. cerevisiae*. All are subsequently discussed in the chapter indicated

Trade name	Description	Use	Chapter
Novolog	Engineered short-acting insulin	Diabetes mellitus	11
Leukine	Granulocyte macrophage CSF (GM-CSF)	Bone marrow transplantation	10
Recombivax, Comvax, Engerix B, Tritanrix-HB, Infanrmix, Twinrix, Primavax, Hexavax	All vaccine preparations containing rHBsAg as one component	Vaccination	13
Revasc, Refludan	Hirudin	Anticoagulant	12
Fasturtec	Urate oxidase	Hyperuricaemia	12
Regranex	PDGF	Diabetic ulcers	10

type glycosylation patterns generally trigger their rapid clearance from the blood stream. Such products, therefore, would be expected to display a short half-life when parenterally administered to humans, and some yeast sugar motifs can be immunogenic in humans.

5.2.3.2 Fungal production systems

Fungi have elicited interest as heterologous protein producers, as many have a long history of use in the production of various industrial enzymes such as α-amylase and glucoamylase. Suitable fermentation technology, therefore, already exists. In general, fungi are capable of high-level expression of various proteins, many of which they secrete into their extracellular media. The extracellular production of a biopharmaceutical would be distinctly advantageous in terms of subsequent downstream processing. Fungi also possess the ability to carry out post-translational modifications. Patterns of glycosylation achieved can, however, differ from typical patterns obtained when a glycoprotein is expressed in a mammalian cell line. Again, this can trigger a reduction is serum half-life or immunological complications in humans.

Most fungal host strains also naturally produce significant quantities of extracellular proteases, which can potentially degrade the recombinant product. This difficulty can be partially overcome by using mutant fungal strains secreting greatly reduced levels of proteases. Although researchers have produced a number of potential therapeutic proteins in recombinant fungal systems, no biopharmaceutical produced by such means has thus far sought/gained marketing approval.

5.2.3.3 Transgenic animals

The production of heterologous proteins in transgenic animals has gained much attention in the recent past. The generation of transgenic animals is most often undertaken by directly microinjecting exogenous DNA into an egg cell. In some instances, this DNA will be stably integrated into

Table 5.6   Proteins of actual/potential therapeutic use that have been produced in the milk of transgenic animals

Protein	Animal species	Expression level in milk (mg l^{-1})
tPA	Goat	6000
IL-2	Rabbit	0.5
Factor VIII	Pig	3
Factor IX	Sheep	1000
α_1-Antitrypsin	Goat	20 000
Fibrinogen	Sheep	5000
EPO	Rabbit	50
Antithrombin III	Goat	14 000
Human α-lactalbumin	Cow	2500
IGF-I	Rabbit	1000
Protein C	Pig	1000
GH	Rabbit	50

the genetic complement of the cell. After fertilization, the ova may be implanted into a surrogate mother. Each cell of the resultant transgenic animal will harbour a copy of the transferred DNA. As this includes the animal's germ cells, the novel genetic information introduced can be passed on from one generation to the next.

A transgenic animal harbouring a gene coding for a pharmaceutically useful protein could become a live bioreactor-producing the protein of interest on an ongoing basis. In order to render such a system practically useful, the recombinant protein must be easily removable from the animal, in a manner which would not be injurious to the animal (or the protein). A simple way of achieving this is to target protein production to the mammary gland. Harvesting of the protein thus simply requires the animal to be milked.

Mammary-specific expression can be achieved by fusing the gene of interest with the promoter-containing regulatory sequence of a gene coding for a milk-specific protein. Regulatory sequences of the whey acid protein (WAP), β-casein and α- and β-lactoglobulin genes have all been used to date to promote production of various pharmaceutical proteins in the milk of transgenic animals (Table 5.6).

One of the earliest successes in this regard entailed the production of human tPA in the milk of transgenic mice. The tPA gene was fused to the upstream regulatory sequence of the mouse WAP, the most abundant protein found in mouse milk. More practical from a production point of view, was the subsequent production of tPA in the milk of transgenic goats, again using the murine WAP gene regulatory sequence to drive expression (Figure 5.2). Goats and sheep have proven to be the most attractive host systems, as they exhibit a combination of attractive characteristics. These include:

- high milk production capacities (Table 5.7);

- ease of handling and breeding, coupled to well-established animal husbandry techniques.

A number of additional general characteristics may be cited that render attractive the production of pharmaceutical proteins in the milk of transgenic farm animals. These include:

- Ease of harvesting of crude product, which simply requires the animal to be milked.

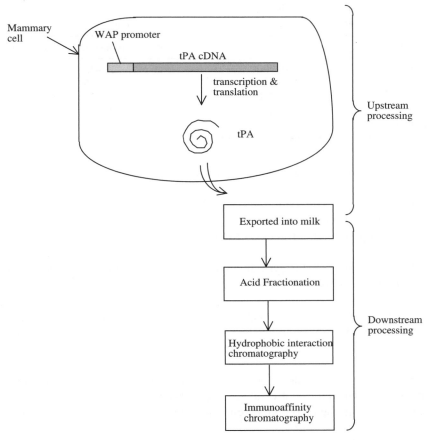

Figure 5.2 The production and purification of tPA from the milk of transgenic goats. (WAP promoter: murine whey acid promoter). The downstream processing procedure yielded in excess of an 8000-fold purification factor with an overall product yield of 25 per cent. The product was greater than 98 per cent pure, as judged by sodium dodecyl sulfate (SDS) electrophoresis

- Pre-availability of commercial milking systems, already designed with maximum process hygiene in mind.

- Low capital investments (i.e. relatively low-cost animals replace high-cost traditional fermentation equipment) and low running costs.

- High expression levels of proteins are potentially attained. In many instances, the level of expression exceeds 1 g protein/litre milk. In one case, initial expression levels of 60 g l^{-1} were observed, which stabilized at 35 g l^{-1} as lactation continued (the expression of the α_1-antitrypsin gene, under the influence of the ovine β-lactoglobulin promoter, in a transgenic sheep). Even at expression levels of 1 g l^{-1}, one transgenic goat would produce a similar quantity of product in 1 day as would be likely recoverable from a 50–100 l bioreactor system.

Table 5.7   Typical annual milk yields as well as time lapse between generation of the transgene embryo and first product harvest (first lactation) of indicated species

Species	Annual milk yield (l)	Time to first production batch (months)
Cow	6000–9000	33–36
Goat	700–800	18–20
Sheep	400–500	18–20
Pig	250–300	16–17
Rabbit	4–5	7

- Ongoing supply of product is guaranteed (by breeding).

- Milk is biochemically well characterized, and the physicochemical properties of the major native milk proteins of various species are well known. This helps rational development of appropriate downstream processing protocols (Table 5.8).

Despite the attractiveness of this system, a number of issues remain to be resolved before it is broadly accepted by the industry. These include:

- Variability of expression levels. Although in many cases the expression levels of heterologous proteins exceed 1 g l^{-1}, expression levels as low as 1.0 mg l^{-1} have been obtained in some instances.

- Characterization of the exact nature of the post-translational modifications the mammary system is capable of undertaking. For example, the carbohydrate composition of tPA produced in this system differs from the recombinant enzyme produced in murine cell culture systems.

- Significant time lag between the generation of a transgenic embryo and commencement of routine product manufacture. Once a viable embryo containing the inserted desired gene is generated, it must first be brought to term. This gestation period ranges from 1 month for rabbits to 9 months for cows. The transgenic animal must then reach sexual maturity before breeding (5 months for rabbits, 15 months for cows). Before they begin to lactate (i.e. produce the recombinant product), they must breed successfully and bring their offspring to term. The overall time

Table 5.8   Some physicochemical properties of the major (bovine) milk proteins

Protein	Caseins	β-Lactoglobulin	α-Lactalbumin	Serum albumin	IgG
Concentration (g l^{-1})	25 (total)	2–4	0.5–1.5	0.4	0.5–1.0
Mass (kDa)	20–25	18	14	66	150
Phosphorylated?	Yes	No	No	No	No
Isoelectric point	Vary	5.2	4.2–4.8	4.7–4.9	5.5–8.3
Glycosylated?	K-casein only	No	No	No	Yes

lag to routine manufacture can, therefore, be almost 3 years in the case of cows or 7 months in the case of rabbits. Furthermore, if the original transgenic embryo turns out to be male, a further delay is encountered, as this male must breed in order to pass on the desired gene to daughter animals, who will then eventually produce the desired product in their milk.

Another general disadvantage of this approach relates to the use of the microinjection technique to introduce the desired gene into the pronucleus of the fertilized egg. This approach is inefficient and time consuming. There is no control over issues such as if/where in the host genomes the injected gene will integrate. Overall, only a modest proportion of manipulated embryos will culminate in the generation of a healthy biopharmaceutical-producing animal.

A number of alternative approaches are being developed that may overcome some of these issues. Replication-defective retroviral vectors are available that will more consistently (a) deliver a chosen gene into cells and (b) ensure chromosomal integration of the gene. A second innovation is the application of nuclear transfer technology.

Nuclear transfer entails substituting the genetic information present in an unfertilized egg with donor genetic information. The best-known product of this technology is 'Dolly' the sheep, produced by substituting the nucleus of a sheep egg with a nucleus obtained from an adult sheep cell. (Genetically, therefore, Dolly was a clone of the original 'donor' sheep.) An extension of this technology applicable to biopharmaceutical manufacture entails using a donor cell nucleus previously genetically manipulated so as to harbour a gene coding for the biopharmaceutical of choice. The technical viability of this approach was proven in the late 1990s upon the birth of two transgenic sheep, 'Polly' and 'Molly'. The donor nucleus used to generate these sheep harboured an inserted (human) blood factor IX gene under the control of a milk protein promoter. Both produced significant quantities of human factor IX in their milk. The first (and only, at the time of writing) such product to gain approval anywhere in the world is 'Atryn', a recombinant human antithrombin that is produced in the milk of transgenic goats. Atryn is used to treat thromboembolism in surgery of people with congenital antithrombin deficiency.

In addition to milk, a range of recombinant proteins have been expressed in various other targeted tissues/fluids of transgenic animals. Antibodies and other proteins have been produced in the blood of transgenic pigs and rabbits. This mode of production, however, is unlikely to be pursued industrially for a number of reasons:

- Only relatively low volumes of blood can be harvested from the animal at any given time point.

- Serum is a complex fluid, containing a variety of native proteins. This renders purification of the recombinant product more complex.

- Many proteins are poorly stable in serum.

- The recombinant protein could have negative physiological side effects on the producer animal.

Therapeutic proteins have also been successfully expressed in the urine and seminal fluid of various transgenic animals. Again, issues of sample collection, volume of collected fluid and the appropriateness of these systems render unlikely their industrial-scale adoption. One system that does show industrial promise, however, is the targeted production of recombinant proteins in the egg white of transgenic birds. Targeted production is achieved by choice of an appropriate egg-white protein promoter sequence. A single egg white typically contains 4 g protein, of which approximately half is derived from the ovalbumin gene. Using an ovalbumin promoter-based expression system, a single hen could produce as much as 300 g recombinant product annually. Large quantities of product can, therefore, potentially accumulate in the egg, which can then be collected and processed with relative ease. The traditional use of eggs as incubation systems in the production of some viral vaccines would also render regulatory and some manufacturing issues more straightforward. Many of both the potential strengths and weaknesses of this system are encapsulated in recent findings relating to egg-produced antibodies.

Fully assembled and functional antibodies have recently been produced in eggs at levels of several milligrams per egg. When compared with the same antibodies produced by traditional CHO-based cell culture, some differences in product glycosylation detail were evident, although antigen binding affinity was not altered. The glycocomponent plays a number of roles, including influencing the ability of the antibody to interact with various immune effector cells, thereby triggering antibody-dependent cell cytoxicity (ADCC), which is important for effective therapeutic functioning. In this particular case, the altered glycosylation detail actually appeared to enhance ADCC. However, tests in mice showed the egg-derived antibody to have a much-reduced serum half-life (reduced from around 200 h to 100 h) and issues of antigenicity still remain to be assessed.

5.2.3.4   Transgenic plants

The production of pharmaceutical proteins using transgenic plants has also gained some attention over the last decade. The introduction of foreign genes into plant species can be undertaken by a number of means, of which *Agrobacterium*-based vector-mediated gene transfer is most commonly employed. *Agrobacterium tumefaciens* and *Agrobacterium rhizogenes* are soil-based plant pathogens. Upon injection, a portion of *Agrobacterium* Ti plasmid is translocated to the plant cell and is integrated into the plant cell genome. Using such approaches, a whole range of therapeutic proteins have been expressed in plant tissue (Table 5.9). Depending upon the specific promoters used, expression can be achieved uniformly throughout the whole plant or can be limited to, for example, expression in plant seeds.

Plants are regarded as potentially attractive recombinant protein producers for a number of reasons, including:

Table 5.9   Some proteins of potential/actual therapeutic interest that have been expressed (at laboratory level) in transgenic plants

Protein	Expressed in	Production levels achieved
EPO	Tobacco	0.003% of total soluble plant protein
HSA	Potato	0.02% of soluble leaf protein
Glucocerebrosidase	Tobacco	0.1% of leaf weight
IFN-α	Rice	Not listed
IFN-β	Tobacco	0.000 02% of fresh weight
GM-CSF	Tobacco	250 ng ml^{-1} extract
Hirudin	Canola	1.0% of seed weight
Hepatitis B surface antigen	Tobacco	0.007% of soluble leaf protein
Antibodies/antibody fragment	Tobacco	Various

- cost of plant cultivation is low;

- harvest equipment/methodologies are inexpensive and well established;

- ease of scale-up;

- proteins expressed in seeds are generally stable in the seed for prolonged periods of time (often years);

- plant-based systems are free of human pathogens (e.g. HIV).

However, a number of potential disadvantages are also associated with the use of plant-based expression systems, including:

- variable/low expression levels sometimes achieved;

- potential occurrence of post-translational gene silencing (a sequence-specific mRNA degradation mechanism);

- glycosylation patterns achieved differ significantly from native human protein glycosylation patterns, and plant glycoforms are invariably immunogenic in humans;

- the potential presence of biologically active plant metabolites (e.g. alkaloids) that would 'contaminate' the crude product;

- environmental/public concerns relating to potential environmental escape of genetically altered plants;

- seasonal/geographical nature of plant growth.

For these reasons, as well as the fact that additional tried-and-tested expression systems are already available, production of recombinant therapeutic proteins in transgenic plant systems has not as yet impacted significantly on the industry.

The most likely focus of future industry interest in this area concerns the production of oral vaccines in edible plants/fruit, such as tomatoes and bananas. Animal studies have clearly shown that ingestion of transgenic plant tissue expressing recombinant subunit vaccines (see Chapter 13 for a discussion of subunit vaccines) induces the production of antigen-specific antibody responses not only in mucosal secretions, but also in the serum. The approach is elegant, in that direct consumption of the plant material provides an inexpensive, efficient and technically straightforward mode of large-scale vaccine delivery, particularly in poorer world regions. However, several hurdles hindering the widespread application of this technology include:

- the immunogenicity of orally administered vaccines can vary widely;

- the stability of antigens in the digestive tract varies widely;

- the genetics of many potential systems remain poorly characterized, leading to inefficient transformation systems and low expression levels.

CaroRX and Merispace are (at the time of writing) two of the lead plant-produced biopharmaceuticals. Both are in phase II clinical trials. Neither is destined for parenteral administration. CaroRX is a recombinant antibody that targets *Streptococcus mutans*, a major causative agent of bacterial tooth decay. Binding prevents bacterial adherence to teeth, and the product is being developed for regular topical administration. Merispace is a recombinant mammalian gastric lipase enzyme produced in transgenic corn. It is intended to be used orally to counteract lipid malabsorption relating to exocrine pancreatic insufficiency caused by conditions such as cystic fibrosis and chronic pancreatitis.

In order to overcome environmental concerns in particular, some companies are investigating the use of engineered plant cell lines as opposed to intact transgenic plants in the context of biopharmaceutical production. One company (DowAgroSciences) gained approval in 2006 for a veterinary subunit vaccine against Newcastle disease in poultry produced by such means.

5.2.3.5   *Insect cell-based systems*

A wide range of proteins have been produced at laboratory scale in recombinant insect cell culture systems. The approach generally entails the infection of cultured insect cells with an engineered baculovirus (viral family that naturally infect insects) carrying the gene coding for the desired protein placed under the influence of a powerful viral promoter. Amongst the systems most commonly employed are:

- the silkworm virus *Bombyx mori* nuclear polyhedrovirus (BmNPV) in conjunction with cultured silkworm cells (i.e. *Bombyx mori* cells);

- the virus *Autographa californica* nuclear polyhedrovirus (AcNPV), in conjunction with cultured armyworm cells (*Spodoptera frugiperda* cells).

Baculovirus/insect cell-based systems are cited as having a number of advantages, including:

- High-level intracellular recombinant protein expression. The use of powerful viral promoters, such as promoters derived from the viral polyhedrin or P10 genes, can drive recombinant protein expression levels to 30–50 per cent of total intracellular protein.

- Insect cells can be cultured more rapidly and using less expensive media compared with mammalian cell lines.

- Human pathogens (e.g. HIV) do not generally infect insect cell lines.

However, a number of disadvantages are also associated with this production system, including:

- Targeted extracellular recombinant production generally results in low-level extracellular accumulation of the desired protein (often in the milligram per litre range). Extracellular production simplifies subsequent downstream processing, as discussed later in this chapter.

- Post-translational modifications, in particular glycosylation patterns, can be incomplete and/or can differ very significantly from patterns associated with native human glycoproteins.

Therapeutic proteins successfully produced on a laboratory scale in insect cell lines include hepatitis B surface antigen, IFN-γ and tPA. To date, no therapeutic product produced by such means has been approved for human use. Two veterinary vaccines, however, have: 'Bayovac CSF E2' and 'porcilis pesti' are both subunnit vaccines (Chapter 13) containing the E2 surface antigen protein of classical swine fever virus as active ingredient. The vaccines are administered to pigs in order to immunize against classical swine fever; an overview of their manufacture is provided in Figure 5.3.

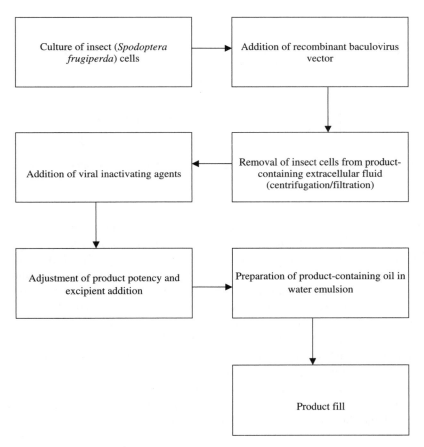

Figure 5.3 Generalized overview of the industrial-scale manufacture of recombinant E2 classical swine-fever-based vaccine, using insect cell culture production systems. Clean (uninfected) cells are initially cultured in 500–1000 l bioreactors for several days, followed by viral addition. Upon product recovery, viral inactivating agents such as beta propiolactone or 2-bromoethyl-imminebromide are added in order to destroy any free viral particles in the product stream. No chromatographic purification is generally undertaken, as the product is substantially pure; the cell culture medium is protein free and the recombinant product is the only protein exported in any quantity by the producer cells. Excipients added can include liquid paraffin and polysorbate 80 (required to generate an emulsion). Thiomersal may also be added as a preservative. The final product generally displays a shelf life of 18 months when stored refrigerated

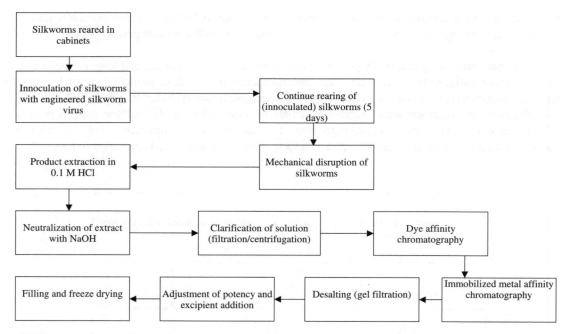

Figure 5.4   Overview of the industrial manufacture of the IFN-ω product 'Vibragen Omega'. Refer to text for details

An alternative insect cell-based system used to achieve recombinant protein production entails the use of live insects. Most commonly, live caterpillars or silkworms are injected with the engineered baculovirus vector, effectively turning the whole insect into a live bioreactor. One veterinary biopharmaceutical, Vibragen Omega, is manufactured using this approach, and an overview of its manufacture is outlined in Figure 5.4. Briefly, whole, live silkworms are introduced into pre-sterile cabinets and reared on laboratory media. After 2 days, each silkworm is inoculated with engineered virus using an automatic microdispenser. This engineered silkworm polyhedrosis virus harbours a copy of cDNA coding for feline IFN-ω. During the subsequent 5 days of rearing, a viral infection is established and, hence, recombinant protein synthesis occurs within the silkworms. After acid extraction, neutralization and clarification, the recombinant product is purified chromatographically. A two-step affinity procedure using blue sepharose dye affinity and copper chelate sepharose chromatography is employed. After a gel filtration step, excipients (sorbitol and gelatin) are added and the product is freeze-dried after filling into glass vials.

5.3   Upstream processing

Biopharm production can be divided into 'upstream' and 'downstream' processing (Figure 5.5). Upstream processing refers to the initial fermentation process that results in the initial generation of product, i.e. the product biosynthesis phase. Downstream processing refers to the actual purification of the protein product and generation of finished product format (i.e. filling into its final product containers,

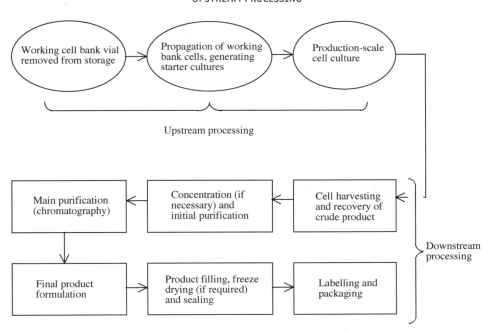

Figure 5.5 Overview of the production process for a biopharmaceutical product. Refer to text for specific details

freeze-drying if a dried product format is required), followed by sealing of the final product containers. Subsequent labelling and packaging steps represent the final steps of finished product manufacture.

Upstream processing is deemed to commence when a single vial of the working cell bank system (see later) is taken from storage and the cells therein cultured in order to initiate the biosynthesis of a batch of product. The production process is deemed complete only when the final product is filled in its final containers and those containers have been labelled and placed in their final product packaging.

5.3.1 Cell banking systems

Recombinant biopharmaceutical production cell lines are most often initially constructed by the introduction into these cells of a plasmid housing a nucleotide sequence coding for the protein of interest (Chapter 3). After culture, the resultant product-producing cell line is generally aliquoted into small amounts, which are placed in ampoules and subsequently immersed in liquid nitrogen. Therefore, the content of all the ampoules is identical, and the cells are effectively preserved for indefinite periods when stored under liquid nitrogen. This batch of cryopreserved ampoules forms a 'cell bank' system, whereby one ampoule is thawed and the cell therein cultured in order to seed, for example, a single production run. This concept is applied to both prokaryotic and eukaryotic biopharmaceutical-producing cells.

The cell bank's construction design is normally two tiered, consisting of a 'master cell bank' and a 'working cell bank' (Figure 5.6). The master cell bank is constructed first, directly from a culture of the newly constructed production cell line. It can consist of several hundred individually stored ampoules.

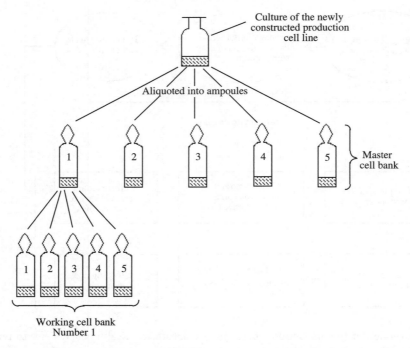

Figure 5.6 The master cell bank/working cell bank system. For simplicity, each bank shown above contains only five ampoules. In reality, each bank would likely consist of several hundred ampoules. Working cell bank number 2 will be generated from master cell bank vial number 2 only when working cell bank number 1 is exhausted and so on

These ampoules are not used directly to seed a production batch. Instead, they are used, as required, to generate a working cell bank. The generation of a single working cell bank normally entails thawing a single master cell bank ampoule, culturing of the cells therein and their subsequent aliquoting into multiple ampoules. These ampoules are then cryopreserved and form the working cell bank. When a single batch of new product is required, one ampoule from the working cell bank is thawed and used to seed that batch. When all the vials that compose the first working cell bank are exhausted, a second vial of the master cell bank is used to generate a second working cell bank, and so on.

The rationale behind this master cell bank/working cell bank system is to ensure an essentially indefinite supply of the originally developed production cells for manufacturing purposes. This is more easily understood by example. If only a single-tier cell bank system existed, containing 250 ampoules, and 10 ampoules were used per year to manufacture 10 batches of product, the cell bank would be exhausted after 25 years. However, if a two-tier system exists, where a single master cell bank ampoule is expanded as required, to generate a further 250 ampoule working cell bank, the entire master cell bank would not be exhausted for 6250 years

The upstream processing element of the manufacture of a batch of biopharmaceutical product begins with the removal of a single ampoule of the working cell bank. This vial is used to inoculate a small volume of sterile media, with subsequent incubation under appropriate conditions. This describes the growth of laboratory-scale starter cultures of the producer cell line. This starter culture is, in turn, used to inoculate a production-scale starter culture that is used to inoculate the production-scale bioreactor (Figure 5.7). The media composition and fermentation conditions required to

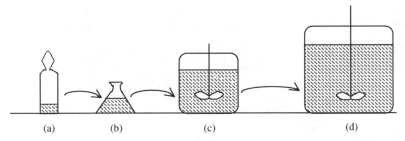

(a) (b) (c) (d)

Figure 5.7 Outline of the upstream processing stages involved in the production of a single batch of product. Initially, the contents of a single ampoule of the working cell bank (a) are used to inoculate a few hundred millilitres of media (b). After growth, this laboratory-scale starter culture is used to inoculate several litres/tens of litres of media present in a small bioreactor (c). This production-scale starter culture is used to inoculate the production-scale bioreactor (d), which often contains several thousands/tens of thousands litres of media. This process is equally applicable to prokaryotic or eukaryotic-based producer cell lines, although the bioreactor design, conditions of growth, etc., will differ in these two instances

promote optimal cell growth/product production will have been established during initial product development, and routine batch production is a highly repetitive, highly automated process. Bioreactors are generally manufactured from high-grade stainless steel and can vary in size from a few tens of litres to several tens of thousands of litres (Figure 5.8). At the end of the production-scale fermentation process, the crude product is harvested, which signals commencement of downstream processing.

Figure 5.8 Typical industrial-scale fermentation equipment as employed in the biopharmaceutical sector (a). Control of the fermentation process is highly automated, with all fermentation parameters being adjusted by computer (b). Photographs (a) and (b) courtesy of SmithKline Beecham Biological Services, s.a., Belgium. Photograph (c) illustrates the inoculation of a laboratory-scale fermenter with recombinant microorganisms used in the production of a commercial interferon preparation. Photograph (c) courtesy of Pall Life Sciences, Dublin, Ireland

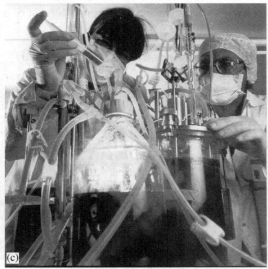

Figure 5.8   (*Continued*)

5.3.2 Microbial cell fermentation

Over half of all biopharmaceuticals thus far approved are produced in recombinant *E. coli* or *S. cerevisiae*. Industrial-scale bacterial and yeast fermentation systems share many common features, an overview of which is provided below. Most remaining biopharmaceuticals are produced using animal cell culture, mainly by recombinant BHK or CHO cells (or hybridoma cells in

Table 5.10 Various products (non-biopharmaceutical) of commercial significance manufactured industrially using microbial fermentation systems

Product type	Example	Example producer
Simple organic molecules	Ethanol	*S. cerevisiae*
		Pachysolen tannophilus
	Butanol	Some clostridia
		Clostridium acetobutylicum
		Clostridium saccharoacetobutylicum
	Acetone	*C. acetobutylicum*
		C. saccharoacetobutylicum
	Acetic acid	Various acetic acid bacteria
	Lactic acid	Lactobacilli
Amino acids	Lysine	*Corynebacterium glutamicum*
	Glutamic acid	*C. glutamicum*
Enzymes	Proteases	Various bacilli, e.g. *Bacillus licheniformis*
	Amylases	*Bacillus subtilis*
		Aspergillus oryzae
	Cellulases	*Trichoderma viride*
		Penicillium pinophilum
Antibiotics	Penicillin	*Penicillium chrysogenum*
	Bacitracin	*Bacillus licheniformis*

the case of some monoclonal antibodies; Chapter 13). While industrial-scale animal cell culture shares many common principles with microbial fermentation systems, it also differs in several respects, as described subsequently. Microbial fermentation/animal cell culture is a vast speciality area in its own right. As such, only a summary overview can be provided below and the interested reader is referred to the 'Further reading' section.

Microbial cell fermentation has a long history of use in the production of various biological products of commercial significance (Table 5.10). As a result, a wealth of technical data and experience have accumulated in the area. A generalized microbial fermenter design is presented in Figure 5.9. The impeller, driven by an external motor, serves to ensure even distribution of nutrients and cells in the tank. The baffles (stainless steel plates attached to the sidewalls) serve to enhance impeller mixing by preventing vortex formation. Various ports are also present through which probes are inserted to monitor pH, temperature and sometimes the concentration of a critical metabolite (e.g. the carbon source). Additional ports serve to facilitate addition of acid/base (pH adjustment) or, if required, addition of nutrients during the fermentation process.

Typically, the manufacture of a batch of biopharmaceutical product entails filling the production vessel with the appropriate quantity of purified water. Heat-stable nutrients required for producer cell growth are then added and the resultant medium is sterilized *in situ*. This can be achieved by heat, and many fermenters have inbuilt heating elements or, alternatively, outer jackets through which steam can be passed in order to heat the vessel contents. Heat-labile ingredients can be sterilized by filtration and added to the fermenter after the heat step. Media composition can vary

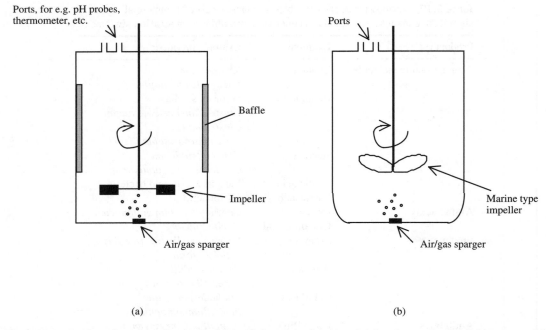

(a)   (b)

Figure 5.9   Design of a generalized microbial cell fermentation vessel (a) and an animal cell bioreactor (b). Animal cell bioreactors display several structural differences compared with microbial fermentation vessels. Note in particular: (i) the use of a marine-type impeller (some animal cell bioreactors–air lift fermenters–are devoid of impellers and use sparging of air–gas as the only means of media agitation); (ii) the absence of baffles; (iii) curved internal surfaces at the bioreactor base. These modifications aim to minimize damage to the fragile animal cells during culture. Note that various additional bioreactor configurations are also commercially available. Reprinted with permission from *Proteins; Biochemistry and Biotechnology* (2002), J. Wiley & Sons

from simple defined media (usually glucose and some mineral salts) to more complex media using yeast extract and peptone. Choice of media depends upon factors such as:

- Exact nutrient requirements of producer cell line to maximize cell growth and product production.

- Economics (total media cost).

- Extracellular or intracellular nature of product. If the biopharmaceutical is an extracellular product then the less complex the media composition the better, in order to render subsequent product purification as straightforward as possible.

Fermentation follows for several days subsequent to inoculation with the production-scale starter culture (Figure 5.7). During this process, biomass (i.e. cell mass) accumulates. In most cases, product accumulates intracellularly and cells are harvested when maximum biomass yields are achieved. This 'feed batch' approach is the one normally taken during biopharmaceutical manufacture, although reactors can also be operated on a continuous basis, where fresh nutrient media is continually added and a fraction of the media/biomass continually removed and processed. During fermentation, air (sterilized by filtration) is sparged into the tank to supply oxygen, and the fermenter is also operated at a temperature appropriate to optimal cell growth (usually between 25 and 37 °C, depending upon the producer cell type). In order to maintain this temperature, cooling rather than heating is required in some cases. Large-scale fermentations, in which cells grow rapidly and to a high cell density, can generate considerable heat due to (a) microbial metabolism and (b) mechanical activity, e.g. stirring. Cooling is achieved by passing the coolant (cold water or glycol) through a circulating system associated with the vessel jacket or sometimes via internal vessel coils.

5.3.3 Mammalian cell culture systems

Mammalian cell culture is more technically complex and more expensive than microbial cell fermentation. Therefore, it is usually only used in the manufacture of therapeutic proteins that show extensive and essential post-translational modifications. In practice, this usually refers to glycosylation, and the use of animal cell culture would be appropriate where the carbohydrate content and pattern are essential to the protein's biological activity, its stability or serum half-life. Therapeutic proteins falling into this category include EPO (Chapter 10), the gonadotrophins (Chapter 11), some cytokines (Chapters 8–10) and intact monoclonal antibodies (Chapter 13).

The culture of animal cells differs from that of microbial cells in several generalized respects, which include:

- they require more complex media;

- extended duration of fermentation due to slow growth of animal cells;

- they are more fragile than microbial cells due to the absence of an outer cell wall.

Basic animal cell culture media, such as Dulbecco's modified Eagle's medium, generally contain:

- most L-amino acids;

- many/most vitamins;

- salts (e.g. NaCl, KCl, $CaCl_2$);

- carbon source (often glucose);

- antibiotics (e.g. penicillin or streptomycin);

- supplemental serum;

- buffering agent (often CO_2 based).

Antibiotics are required to prevent microbial growth consequent to accidental microbial contamination. Supplemental serum (often bovine or foetal calf serum, or synthetic serum composed of a mixture of growth factors, hormones and metabolites typically found in serum) is required as a source of often ill-defined growth factors required by some animal cell lines.

The media constituents, several of which are heat labile, are generally dissolved in purified water and filter sterilized into the pre-sterile animal cell reactor. Reactor design (and operation) differs somewhat from microbial fermentations mainly with a view to minimizing damage to the more fragile cells during cell culture (Figure 5.9). Although the generalized reactor design presented in Figure 5.9 is commonly employed on an industrial scale, alternative reactor configurations are also available. These include hollow-fibre systems and the classical roller bottle systems. Roller bottles are still used in the industrial production of some vaccines, some EPO products and growth-hormone-based products. Roller bottles are cylindrical bottles that are partially filled with media, placed on their side and mechanically rolled during cell culture. This system is gentle on the cells, and the rolling action ensures homogeneity in the culture media and efficient oxygen transfer. The major disadvantage associated with applying roller bottle technology on an industrial scale is that many thousands of bottles are required to produce a single batch of product.

Different animal cell types display different properties pertinent to their successful culture. Those used to manufacture biopharmaceuticals are invariably continuous (transformed) cell lines. Such cells will grow relatively vigorously and easily in submerged culture systems, be they roller bottle or bioreactor based.

Unlike transformed cell lines, non-continuous cell lines generally:

- display anchorage dependence (i.e. will only grow and divide when attached to a solid substratum; continuous cell lines will grow in free suspension);

- grow as a monolayer;

- exhibit contact inhibition (physical contact between individual cells inhibits further division);

- display a finite lifespan, i.e. die, generally after 50–100 cell divisions, even when cultured under ideal conditions;

- display longer population doubling times and grow to lower cell densities than continuous cell lines;

- usually have more complex media requirements.

Many of these properties would obviously limit applicability of non-continuous cell lines in the industrial-scale production of recombinant proteins. However, such cell types are routinely cultured for research purposes, toxicity testing, etc.

The anchorage-dependent growth properties of such non-continuous cell lines impacts upon how they are cultured, both on laboratory and industrial scales. If grown in roller bottles/other low-volume containers, then cells grow attached to the internal walls of the vessel. Large-scale culture can be undertaken in submerged-type vessels, such as that described in Figure 5.9b in conjunction with the use of microcarrier beads. Microcarriers are solid or sometimes porous spherical particles approximately 200 μm in diameter manufactured from such materials as collagen, dextran or plastic. They display densities slightly greater than water, such that gentle mixing within the animal cell bioreactor is sufficient to maintain the beads in suspension and evenly distributed throughout the media. Anchorage-dependent cells can attach to and grow on the beads' outer surface/outer pores.

Overall, therefore, the routine manufacture of a biopharmaceutical product is initiated by large-scale culture of its producing cell line (upstream processing). Subsequent to this, the product is recovered, purified and formulated into final product format. These latter operations are collectively termed downstream processing and are described in Chapter 6.

Further reading

Books

Butler, M. 1996. *Animal Cell Culture and Technology. The basics.* IRL Press.
Flickinger, M. 1999. *The Encyclopedia of Bioprocess Technology.* Wiley.
Fresheny, I. (ed.). 2006. *Animal Cell Culture, a Practical Approach.* IRL Press.
Grindley, J. and Ogden, J. 2000. *Understanding Biopharmaceuticals. Manufacturing and Regulatory Issues.* Interpharm Press.
Merten, O.-W., Mattanovich, D., Lang, C., Larsson, G., Neubauer, P., Porro, D., Postma, P., Teixeira de Mattos, J., and Cole, J.A. *Recombinant Protein Production with Prokaryotic and Eukaryotic Cells: A Comparative View on Host Physiology?* Kluwer.
Oxender, D. and Post, L. 1999. *Novel Therapeutics from Modern Biotechnology.* Springer Verlag.
Ozturk, S. and Hu, W.S. (eds). 2006. *Cell Culture Technology for Pharmaceutical and Cell-based Therapies.* Taylor and Francis.
Vinci, V. and Parekh, S. 2003. *Handbook of Industrial Cell Culture.* Humana Press.

Articles

Baneyx, F. 1999. Recombinant protein expression in *E. coli. Current Opinion in Biotechnology* **10**, 411–421.
Butler, M. 2005. Animal cell cultures: recent achievements and perspectives in the production of biopharmaceuticals. *Applied Microbiology and Biotechnology* **68**, 283–291.
Carrio, M. and Villaverde, A. 2002. Construction and deconstruction of bacterial inclusion bodies. *Journal of Biotechnology* **96**(1), 3–12.
Chen, M., Liu, X., Wang, Z., Qi, Q., and Wang, P.G. 2005. Modification of plant *N*-glycan processing: the future of producing therapeutic protein by transgenic plants. *Medical Research Reviews* **25**(3), 343–360.
Datar, R.V. Cartwright, T., and Rosen, C.G. 1993. Process economics of animal cell and bacterial fermentations: a case study analysis of tissue plasminogen activator. *Bio/Technology* **11**, 340–357.
Dyck, M.K., Gagne, D., Ouellet, M., Senechal, J.-F., Belanger, E., Lacroix, D., Sirard, M.-A., and Pothier, F. 1999. Seminal vesicle production and secretion of growth hormone into seminal fluid. *Nature Biotechnology* **17**, 1087–1090.
Fischer, R., Emans, N., Schuster, F., Hellwig, S., and Drossard, J. 1999. Towards molecular farming in the future: using plant cell suspension cultures as bioreactors. *Biotechnology and Applied Biochemistry* **30**, 109–112.
Gomord, W., Chamberlain, P., Jefferis, R., and Faye, L. 2005. Biopharmaceutical production in plants: problems, solutions and opportunities. *Trends in Biotechnology* **23**(11), 559–565.
Grengross, T. 2004. Advances in the production of human therapeutic proteins in yeasts and filamentous fungi. *Nature Biotechnology* **22**(11), 1409–1414.
Helmrich, A. and Barnes, D. 1998. Animal cell culture equipment and techniques. *Methods in Cell Biology* **57**, 3–17.

Houdebine, L. 2000. Transgenic animal bioreactors. *Transgenic Research* **9**(4–5), 305–320.

Ivarie, R. 2006. Competitive bioreactor hens on the horizon. *Trends in Biotechnology* **24**(3), 99–101.

Li, H., Sethuraman, N., Stadheim, T.A., Zha, D., Prinz, B., Ballew, N., Bobrowicz, P., Choi, B.K., Cook, W.J., Cukan, M., Houston-Cummings, N.R., Davidson, R., Gong, B., Hamilton, S.R., Hoopes, J.P., Jiang, Y., Kim, N., Mansfield, R., Nett, J.H., Rios, S., Strawbridge, R., Wildt, S., and Gerngross, T.U. 2006. Optimization of humanized IgGs in glycoengineered *Pichia pastoris*. *Nature Biotechnology* **24**(2), 210–215.

Jarvis, D. 2003. Developing baculovirus–insect cell expression systems for humanized recombinant glycoprotein production. *Virology* **310**(1), 1–7.

Kjeldsen, T. 2002. Yeast secretory expression of insulin precursors. *Applied Microbiology and Biotechnology* **54**, 277–286.

Larrick, J. and Thomas, D. 2001. Producing proteins in transgenic plants and animals. *Current Opinion in Biotechnology* **12**(4), 411–418.

Mason, H.S., Warzecha, H., Mor, T., and Arntzen, C.J. 2002. Edible plant vaccines: applications for prophylactic and therapeutic molecular medicine. *Trends in Molecular Medicine* **8**(7), 324–329.

Punt, P.J., van Biezen, N., Conesa, A., Albers, A., Mangnus, J., and van den Hondel, C. 2002. Filamentous fungi as cell factories for heterologous protein production. *Trends in Biotechnology* **20**(5), 200–206.

Varley, J. and Birch, J. 1999. Reactor design for large scale suspension animal cell culture. *Cytotechnology* **29**(3), 177–205.

Wildt, S. and Gerngross, T. 2005. The humanization of *N*-glycosylation pathways in yeast. *Nature Reviews Microbiology* **3**(2), 119–128.

Zhu, L., van de Lavoir, M.C., Albanese, J., Beenhouwer, D.O., Cardarelli, P.M., Cuison, S., Deng, D.F., Deshpande, S., Diamond, J.H., Green, L., Halk, E.L., Heyer, B.S., Kay, R.M., Kerchner, A., Leighton, P.A., Mather, C.M., Morrison, S.L., Nikolov, Z.L., Passmore, D.B., Pradas-Monne, A., Preston, B.T., Rangan, V.S., Shi, M., Srinivasan, M., White, S.G., Winters-Digiacinto, P., Wong, S., Zhou, W., and Etches, R.J. 2005. Production of human monoclonal antibody in eggs of chimeric chickens. *Nature Biotechnology* **23**(9), 1159–1169.

6

Downstream processing

6.1 Introduction

Downstream processing serves to (a) recover the therapeutic protein from its producer cell source upon completion of the upstream processing phase, (b) purify the protein and (c) formulate the protein into final product format.

An overview of the steps normally undertaken during downstream processing is presented in Figure 6.1. Details of the exact steps undertaken during the downstream processing of any specific biopharmaceutical product are usually considered confidential by the manufacturer. Such details are thus rarely made generally available. However, a potential downstream processing procedure for recombinant tPA is presented in Figure 6.2, and other examples are provided at various stages through the remainder of this text.

Downstream processing is undertaken under clean-room conditions in order to protect the product stream from environmental contamination (Figure 6.3). In addition, the water used as solvent during downstream processing (and, indeed, often during upstream processing) is highly purified 'water for injections' (WFI). Standard potable (drinkable) water contains contaminants (e.g. microorganisms, dissolved organic and particulate matter, etc.) that could either react with the protein directly or that would have an adverse effect upon patient health if present in the final product. Generation of 'purified water' (often used to make up media for microbial fermentation) along with even purer WFI, is summarized in Figure 6.4.

All proteins retain their structural integrity and biological activity only over characteristic pH ranges. Proteins become denatured outside these ranges, losing their characteristic three-dimensional structure, and hence activity (Chapter 2). Most biopharmaceuticals are stable only at pH values approaching neutrality (approximately pH 5–8 for many). As such, downstream processing is carried out using not WFI *per se* as solvent but in buffer solutions made from WFI. A buffer is a solution that resists a change in its pH value even with the addition of small amounts of either acid or alkali, and hence effectively controls the pH environment of the protein.

Pharmaceutical biotechnology: concepts and applications   Gary Walsh
© 2007 John Wiley & Sons, Ltd   ISBN 978 0 470 01244 4 (HB)   978 0 470 01245 1 (PB)

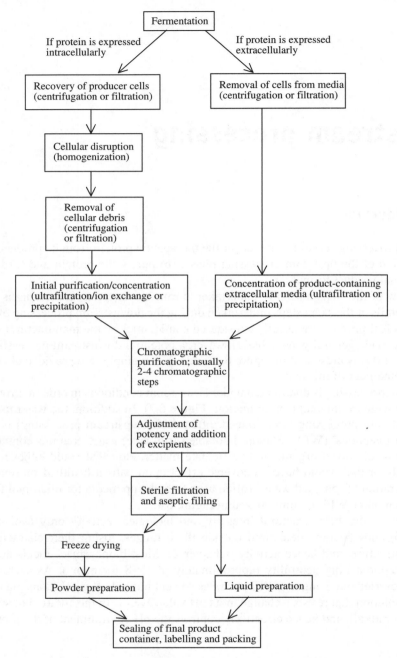

Figure 6.1 Overview of a generalized downstream processing procedure employed to produce a finished-product (protein) biopharmaceutical. QC also plays a prominent role in downstream processing. Qualty control personnel collect product samples during/after each stage of processing. These samples are analysed to ensure that various in-process specifications are met. In this way, the production process is tightly controlled at each stage

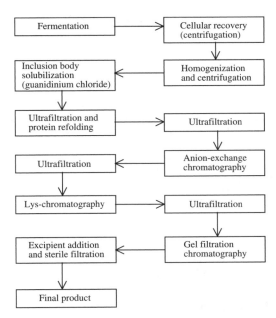

Figure 6.2 A likely purification procedure for tPA produced in recombinant *E. coli* cells. The heterologous product accumulates intracellularly in the form of inclusion bodies. In this particular procedure, an ultrafiltration step is introduced on several occasions to concentrate the product stream, particularly prior to application to chromatographic columns. Lysine affinity chromatography (Lys-chromatography) is employed, as tPA is known to bind immobilized lysine molecules. Adapted with permission from *Bio/Technology* (1993), **11**, 351

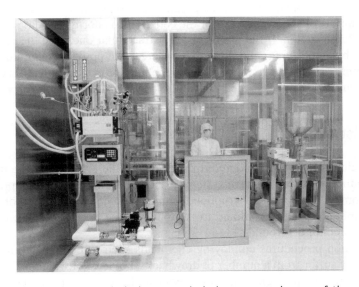

Figure 6.3 Photograph illustrating a typical pharmaceutical cleanroom and some of the equipment usually therein. Note the presence of a curtain of (transparent) heavy-gauge polyethylene strips (most noticeable directly in front of the operator). These strips box off a grade A laminar flow work station. Product filling into final product containers is undertaken within the grade A zone. The filling process is highly automated, requiring no direct contact between the operator and the product. This minimizes the chances of accidental product contamination by production personnel. Photograph courtesy of SmithKline Beecham Biological Services, s.a., Belgium

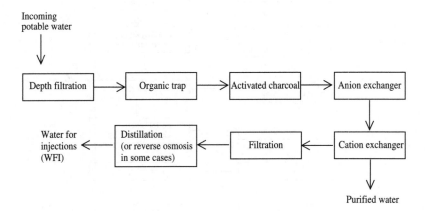

Figure 6.4 Overview of a generalized procedure by which purified water and WFI are generated in a pharmaceutical facility. Refer to text for specific details

6.2 Initial product recovery

The initial step of any downstream processing procedure involves recovery of the protein from its source. The complexity of this step depends largely upon whether the product is intracellular or extracellular (Figure 6.1). In general, animal cell-culture-derived biopharmaceuticals are secreted into the media (i.e. are produced as extracellular proteins), whereas the product accumulates intracellularly in many recombinant prokaryotic producer cell types. In both cases, upstream processing is followed by initial collection (harvest) of the cells. This is normally achieved by centrifugation, or sometimes microfiltration. If the product is an extracellular one, then the cell paste is generally inactivated (e.g. by autoclaving) and discarded, whereas the product-containing extracellular fluid is subject to further processing. In the case of intracellular product, cell recovery is followed by cellular disruption, in order to release the intracellular contents, including the protein of interest.

6.3 Cell disruption

Disruption of microbial cells is rendered difficult due to the presence of the microbial cell wall. Despite this, a number of very efficient systems exist that are capable of disrupting large quantities of microbial biomass (Table 6.1). Disruption techniques, such as sonication or treatment with the enzyme lysozyme, are usually confined to laboratory-scale operations, due either to equipment limitations or on economic grounds.

Protein extraction procedures employing chemicals such as detergents are effective in many instances, but they suffer from a number of drawbacks, not least of which is that they often induce protein denaturation and precipitation. This obviously limits their usefulness. Furthermore, even if the chemicals employed do not adversely affect the protein, their presence may adversely affect a subsequent purification step (e.g. the presence of detergent can prevent proteins from binding to a hydrophobic interaction column). In addition, the presence of such materials in the final preparation, even in trace quantities, may be unacceptable for medical reasons.

Disruption of microbial cells (and, indeed, some animal/plant tissue types) is most often achieved by mechanical methods, such as homogenization or by vigorous agitation with abrasives.

Table 6.1 Some chemical, physical and enzyme-based techniques that may be employed to achieve microbial cell disruption

---

Treatment with chemicals:
    detergents
    antibiotics
    solvents (e.g. toluene, acetone)
    chaotropic agents (e.g. urea, guanidine)
Exposure to alkaline conditions
Sonication
Homogenization
Agitation in the presence of abrasives (usually glass beads)
Treatment with lysozyme

During the homogenization process a cell suspension is forced through an orifice of very narrow internal diameter at extremely high pressures. This generates extremely high shear forces. As the microbial suspension passes through the outlet point, it experiences an almost instantaneous drop in pressure to normal atmospheric pressure. The high shear forces and subsequent rapid pressure drop act as very effective cellular disruption forces, and result in the rupture of most microbial cell types (Figure 6.5). In most cases a single pass through the homogenizer results in adequate cell breakage, but it is also possible to recirculate the material through the system for a second or third pass.

An efficient cooling system minimizes protein denaturation (denaturation would otherwise occur due to the considerable amount of heat generated during the homogenization process). Homogenizers capable of handling large quantities of cellular suspensions are now available, many of which can efficiently process several thousand litres per hour.

An additional method often employed to achieve microbial cell disruption, both at the laboratory level and on an industrial scale, involves cellular agitation in the presence of glass beads. In such bead mills, the microorganisms are placed in a chamber together with a quantity of glass beads of 0.2–0.3 mm in diameter. This mixture is then shaken/agitated vigorously, resulting

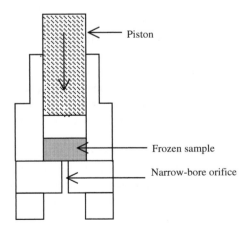

Figure 6.5 Diagrammatic representation of a cell homogenizer. This represents one of a number of instruments routinely used to rupture microbial cells, and in some cases animal/plant tissue

in numerous collisions between the microbial cells and the glass beads. It also results in the grinding of cells between the rotating beads. These forces promote efficient disruption of most microbial cell types. Operational parameters, such as ratio of cells to beads and the rate/duration of agitation, may be adjusted to achieve optimum disruption of the particular cells in question. Laboratory systems can homogenize several grams of microbial cells in minutes. Industrial-scale bead-milling systems can process in excess of 1000 l of cell suspension per hour. Cooling systems minimize protein inactivation by dissipating the considerable heat generated during this process.

Upon completion of the homogenization step, cellular debris and any remaining intact cells can be removed by centrifugation or by microfiltration. As mentioned previously, these techniques are also used to remove whole cells from the medium during the initial stages of extracellular protein purification.

6.4   Removal of nucleic acid

In some cases it is desirable/necessary to remove/destroy the nucleic acid content of a cell homogenate prior to subsequent purification of the released intracellular protein. Liberation of large amounts of nucleic acids often significantly increases the viscosity of the cellular homogenate. This generally renders the homogenate more difficult to process, particularly on an industrial scale. Significant increases in viscosity place additional demands upon the method of cell debris removal employed. Increased centrifugal forces for longer time periods may be required to collect (pellet) cell debris in such solutions efficiently. If a filtration system is employed to remove cellular debris, then the increased viscosity will also adversely affect flow rate and filter performance.

Effective nucleic acid removal is particularly important when purifying a protein destined for therapeutic use. Regulatory authorities generally insist that the nucleic acid content present in the final preparation be, at most, a few picograms per therapeutic dose (see Chapter 7).

Effective removal of nucleic acids during protein purification may be achieved by precipitation or by treatment with nucleases. A number of cationic (positively charged) molecules are effective precipitants of DNA and RNA; they complex with, and precipitate, the negatively charged nucleic acids. The most commonly employed precipitant is polyethylenimine, a long-chain cationic polymer. The precipitate is then removed, together with cellular debris, by centrifugation or filtration. The use of polyethylenimine during purification of proteins destined for therapeutic applications is often discouraged, however, as small quantities of unreacted monomer may be present in the polyethylenimine preparation. Such monomeric species may be carcinogenic. If polyethylenimine is utilized in such cases, then the subsequent processing steps must be shown to be capable of effectively and completely removing any of the polymer or its monomeric units that may remain in solution.

Nucleic acids may also be removed by treatment with nucleases, which catalyse the enzymatic degradation of these biomolecules. Indeed, nuclease treatment is quickly becoming the most popular method of nucleic acid removal during protein purification. This treatment is efficient, inexpensive and, unlike many of the chemical precipitants used, nuclease preparations themselves are innocuous and do not compromise the final protein product.

6.5 Initial product concentration

The next phase of downstream processing usually entails concentration of the crude protein product. This yields smaller product volumes, which are more convenient to work with and can be subsequently processed with greater speed. Concentration may be achieved by inducing product precipitation using salts, such as ammonium sulfate, or solvents, such as ethanol. The precipitate is then collected (usually by centrifugation, but potentially also by filtration) and the precipitate is redissolved in a small volume of processing buffer. Ion-exchange chromatography (in which proteins bind to charged beads immobilized in a column) can also potentially be used to concentrate protein solutions. This is discussed later in this chapter. Both of these methods also result in limited protein purification, as not all protein types present in the crude preparation will co-precipitate or bind to the ion exchanger along with the target protein. Ultrafiltration, however, is by far the most common method now used to achieve initial product concentration.

6.5.1 Ultrafiltration

As discussed previously, the technique of microfiltration is effectively utilized to remove whole cells or cell debris from solution. Membrane filters employed in the microfiltration process generally have pore diameters ranging from 0.1 to 10 μm. Such pores, while retaining whole cells and large particulate matter, fail to retain most macromolecular components, such as proteins. In the case of ultrafiltration membranes, pore diameters normally range from 1 to 20 nm. These pores are sufficiently small to retain proteins of low molecular mass. Ultrafiltration membranes with molecular mass cut-off points ranging from 1 to 300 kDa are commercially available. Membranes with molecular mass cut-off points of 3, 10, 30, 50, and 100 kDa are most commonly used.

Traditionally, ultrafilters have been manufactured from cellulose acetate or cellulose nitrate. Several other materials, such as polyvinyl chloride and polycarbonate, are now also used in membrane manufacture. Such plastic-type membranes exhibit enhanced chemical and physical stability when compared with cellulose-based ultrafiltration membranes. An important prerequisite in manufacturing ultrafilters is that the material utilized exhibits low protein adsorptive properties.

Ultrafiltration is generally carried out on a laboratory scale using a stirred-cell system (Figure 6.6a). The flat membrane is placed on a supporting mesh at the bottom of the cell chamber and the material to be concentrated is then transferred into the cell. Application of pressure, usually nitrogen gas, ensures adequate flow through the ultrafilter. Molecules of lower molecular mass than the filter cut-off pore size (e.g. water, salt and low molecular weight compounds) all pass through the ultrafilter, thus concentrating the molecular species present whose molecular mass is significantly greater than the molecular mass cut-off point. Concentration polarization (the build-up of a concentrated layer of molecules directly over the membrane surface that are unable to pass through the membrane) is minimized by a stirring mechanism operating close to the membrane surface. If unchecked, concentration polarization would result in a lowering of the flow rate. Additional ultrafilter formats used on a laboratory scale include cartridge systems, within which the ultrafiltration membrane is present in a highly folded format. In such cases, the pressure required to maintain a satisfactory flow rate through the membrane is usually generated by a peristaltic pump.

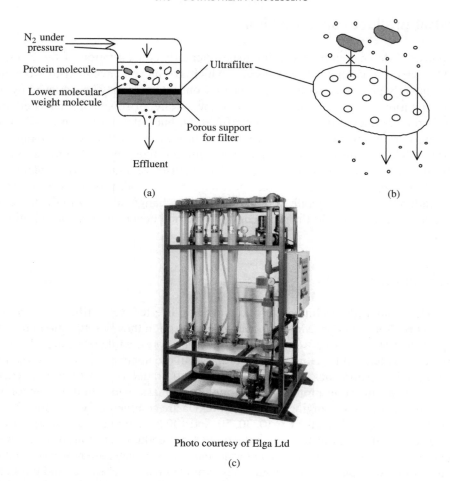

Figure 6.6 Ultrafiltration separates molecules based on size and shape. (a) Diagrammatic representation of a typical laboratory-scale ultrafiltration system. The sample (e.g. crude protein solution) is placed in the ultrafiltration chamber, where it sits directly above the ultrafilter membrane. The membrane, in turn, sits on a macroporous support to provide it with mechanical strength. Pressure is then applied (usually in the form of an inert gas), as shown. Molecules larger than the pore diameter (e.g. large proteins) are retained on the upstream side of the ultrafilter membrane. However, smaller molecules (particularly water molecules) are easily forced through the pores, thus effectively concentrating the protein solution (see also (b)). Membranes that display different pore sizes, i.e. have different molecular mass cut-off points, can be manufactured. (c) Photographic representation of an industrial-scale ultrafiltration system (photograph courtesy of Elga Ltd, UK)

Large-scale ultrafiltration systems invariably employ cartridge-type filters (Figure 6.6c). This allows a large filtration surface area to be accommodated in a compact area. Concentration polarization is avoided by allowing the incoming liquid to flow across the membrane surface at right angles, i.e. tangential flow. The ultrafiltration membrane may be pleated, with subsequent joining of the two ends to form a cylindrical cartridge. Alternatively, the membrane may be laid on a spacer mesh and this may then be wrapped spirally around a central collection tube, into which the filtrate can flow.

Another widely used membrane configuration is that of the hollow fibres. In this case, the hollow cylindrical cartridge casing is loaded with bundles of hollow fibres. Hollow fibres have an outward appearance somewhat similar to a drinking straw, although their internal diameters may be considerably smaller. In this configuration, the liquid to be filtered is pumped through the central core of the hollow fibres. Molecules of lower molecular mass than the membrane rated cut-off point pass through the walls of the hollow fibre. The permeate, which emerges from the hollow fibres along all of their length, is drained from the cartridge via a valve. The concentrate emerges from the other end of the hollow fibre and is collected by an outlet pipe; this is referred to as the retentate. The permeate is then normally discarded, whereas the retentate, containing the protein of interest, is processed further. The retentate may be recycled through the system if further concentration is required.

Ultrafiltration has become prominent as a method of protein concentration for a variety of reasons:

- the method is very gentle, having little adverse effect on bioactivity of the protein molecules;

- high recovery rates are usually recorded, with some manufacturers claiming recoveries of over 99 per cent;

- processing times are rapid when compared with alternative methods of concentration;

- little ancillary equipment is required.

One drawback relating to this filtration technique is its susceptibility to rapid membrane clogging. Viscous solutions also lead to rapid decreases in flow rates and prolonged processing times.

Ultrafiltration may also be utilized to achieve a number of other objectives. As discussed above, it may yield a limited degree of protein purification and may also be effective in depyrogenating solutions. This will be discussed further in Chapter 7. The technique is also widely used to remove low molecular mass molecules from protein solutions by *diafiltration*.

6.5.2 Diafiltration

Diafiltration is a process whereby an ultrafiltration system is utilized to reduce or eliminate low molecular mass molecules from a solution and is sometimes employed as part of biopharmaceutical downstream processing. In practice, this normally entails the removal of, for example, salts, ethanol and other solvents, buffer components, amino acids, peptides, added protein stabilizers or other molecules from a protein solution. Diafiltration is generally preceded by an ultrafiltration step to reduce process volumes initially. The actual diafiltration process is identical to that of ultrafiltration, except for the fact that the level of reservoir is maintained at a constant volume. This is achieved by the continual addition of solvent lacking the low molecular mass molecules that are to be removed. By recycling the concentrated material and adding sufficient fresh solvent to the system such that five times the original volume has emerged from the system as permeate, over 99 per cent of all molecules that freely cross the membrane will have been removed ('flushed out') from the solution. Removal of low molecular weight contaminants from protein solutions may also be achieved by other techniques, such as dialysis or gel-filtration chromatography. Diafiltration, however, is emerging as the method of choice, as it is quick, efficient and utilizes the same equipment as used in ultrafiltration.

6.6   Chromatographic purification

Once the protein is recovered from its producer source and concentrated it must be purified to homogeneity. In other words, all contaminant proteins and other potential contaminants of potential medical significance (discussed in Chapter 7) must be removed. Purification is generally achieved by column chromatography.

Column chromatography refers to the separation of different protein types from each other according to their differential partitioning between two phases: a solid stationary phase (the chromatographic beads, usually packed into a cylindrical column) and a mobile phase (usually a buffer). With the exception of gel filtration, all forms of chromatography used in protein purification protocols are adsorptive in nature. The protein mix is applied to the column (usually) under conditions that promote selective retention of the target protein. Ideally, this target protein should be the only one retained on the column, but this is rarely attained in practice. After sample application, the column is washed ('irrigated') with mobile phase in order to flush out all unbound material. The composition of the mobile phase is then altered in order to promote desorption of the bound protein. Fractions of eluate are collected in test tubes, which are then assayed for both total protein and for the protein of interest (Figure 6.7). The fractions containing the target protein are then pooled and subjected to the next step in the purification process.

Individual protein types possess a variety of characteristics that distinguish them from other protein molecules. Such characteristics include size and shape, overall charge, the presence of surface hydrophobic groups and the ability to bind various ligands. Quite a number of protein molecules may be similar to one another if compared on the basis of any one such characteristic. All protein types, however, present their own unique combination of characteristics, a protein chromatographic 'fingerprint'. Various chromatographic techniques have been developed that separate proteins from each other on the basis of differences in such characteristics (Table 6.2). Utilization of any one of these methods to exploit the molecular distinctiveness usually results in a dramatic increase in the purity of the protein of interest. A combination of methods may be employed to yield highly purified protein preparations.

In general, a combination of two to four different chromatographic techniques is employed in a typical downstream processing procedure. Gel-filtration and ion-exchange chromatography are amongst the most common. Affinity chromatography is employed wherever possible, as its high biospecificity facilitates the achievement of a very high degree of purification. Examples include the use of immunoaffinity chromatography to purify blood factor VIII and lysine affinity chromatography to purify tPA.

As with most aspects of downstream processing, the operation of chromatographic systems is highly automated and is usually computer controlled. Whereas medium-sized process-scale chromatographic columns (e.g. 5–15 l capacity) are manufactured from toughened glass or plastic, larger

1. Apply protein-containing sample
2. Irrigate with buffer (wash out unbound material)
3. Apply elution buffer and collect fractions

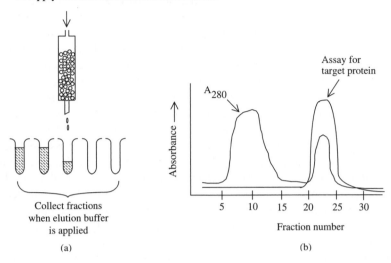

Collect fractions
when elution buffer
is applied

(a)

(b)

Figure 6.7 (a) Typical sequence of events undertaken during an (adsorption-based) protein purification chromatographic step. Note that the chromatographic beads are not drawn to scale, and in reality these display diameters <0.1 mm. Fractions collected during protein desorption are assayed for (i) total protein, usually by measuring absorbance at 280 nm, and (ii) target protein activity. (b) In the case illustrated, two major protein peaks are evident, only one of which contains the protein of interest. Thus, desorption and adsorption steps can result in selective purification

processing columns are available that are manufactured from stainless steel. Process-scale chromatographic separation is generally undertaken under low pressure, but production-scale high-pressure systems (i.e. process-scale HPLC) are sometimes used, as long as the protein product is not adversely affected by the high pressure experienced. An HPLC-based 'polishing step' is sometimes employed, e.g. during the production of highly purified insulin preparations (Chapter 11). Next, we will consider individually the most common forms of chromatography used to purify therapeutic proteins.

Table 6.2 Chromatographic techniques most commonly used in protein purification protocols. The basis of separation is listed in each case

Technique	Basis of separation
Ion-exchange chromatography	Differences in protein surface charge at a given pH
Gel-filtration chromatography	Differences in mass/shape of different proteins
Affinity chromatography	Based upon biospecific interaction between a protein and an appropriate ligand
Hydrophobic interaction chromatography	Differences in surface hydrophobicity of proteins
Chromatofocusing	Separates proteins on the basis of their isolectric points
Hydroxyapatite chromatography	Complex interactions between proteins and the calcium phosphate-based media; not fully understood

6.6.1   Size-exclusion chromatography (gel filtration)

Size-exclusion chromatography, also termed gel-permeation or gel-filtration chromatography, separates proteins on the basis of their size and shape. As most proteins fractionated by this technique are considered to have approximately similar molecular shape, separation is often described as being on the basis of molecular mass, although such a description is somewhat simplistic.

Fractionation of proteins by size-exclusion chromatography is achieved by percolating the protein-containing solution through a column packed with a porous gel matrix in bead form (Figure 6.8). As the sample travels down the column, large proteins cannot enter the gel beads and hence are quickly eluted. The progress of smaller proteins through the column is retarded, as such molecules are capable of entering the gel beads. The internal structure of the matrix beads could be visualized as a maze, through which proteins small enough to enter the gel must pass. Various possible routes through this maze are of varied distances. All proteins capable of entering the gel are thus not retained within the gel matrix for equal time periods. The smaller the protein, the more potential internal routes open to it and, thus, generally, the longer it is retained within the bead structure. Protein molecules, therefore, are usually eluted from a gel-filtration column in order of decreasing molecular size.

In most cases the gel matrices utilized are prepared by chemically cross-linking polymeric molecules such as dextran, agarose, acrylamide and vinyl polymers. The degree of cross-linking controls the average pore size of the gel prepared. Most gels synthesized from any one polymer type are thus available in a variety of pore sizes. The higher the degree of cross-linking introduced, the smaller the average pore size and the more rigid the resultant gel bead. Highly cross-linked gel matrices have pore sizes that exclude all proteins from entering the gel matrix. Such gels may be used to separate proteins from other molecules that are orders of magnitude smaller, and are often used to remove low molecular weight buffer components and salts from protein solutions (Figure 6.9).

Size-exclusion chromatography is rarely employed during the initial stages of protein purification. Small sample volumes must be applied to the column in order to achieve effective resolution. Application volumes are usually in the range of 2–5 per cent of the column volume. Furthermore, columns are easily fouled by a variety of sample impurities. Size-exclusion chromatography is thus often employed towards the end of a purification sequence, when the protein of interest is already relatively pure and is present in a small, concentrated volume. After sample application, the protein components are progressively eluted from the column by flushing with an appropriate buffer. In many cases, the eluate from the column passes through a detector. This facilitates immediate detection of protein-containing bands as they elute from the column. The eluate is normally collected as a series of fractions. On a preparative scale, each fraction may be a number of litres in volume. Although size-exclusion chromatography is an effective fractionation technique, it generally results in a significant dilution of the protein solution relative to the starting volume applied to the column. Column flow rates are also often considerably lower than flow rates employed with other chromatographic media. This results in long processing times, which, for industrial applications, has adverse process cost implications.

6.6.2   Ion-exchange chromatography

Several of the 20 amino acids that constitute the building blocks of proteins exhibit charged side chains. At pH 7.0, aspartic and glutamic acids have overall negatively charged acidic

(a)

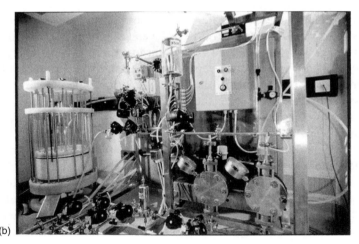

(b)

Figure 6.8 Chromatographic columns. The glass column illustrated in (a) is manufactured by Merck. A wide variety of columns (ranging in size from 1 ml to several litres, and constructed from glass/plastic or stainless steel) are available from this and a number of other manufacturers (e.g. Bio-Rad and Pharmacia Biotech). (b) Process-scale chromatographic system. This particular system is utilized by a UK-based biotech company in the manufacture of a (protein) drug for clinical trials. The actual column is positioned to the left of picture

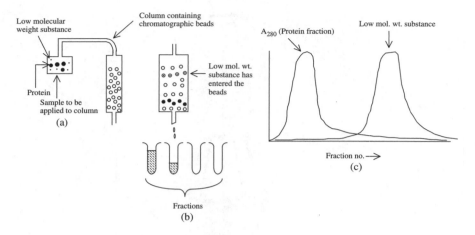

Figure 6.9 The application of gel-filtration chromatography to separate proteins from molecules of much lower molecular weight. The mobile phase (the 'running buffer') will be devoid of the molecular species to be removed from the protein. Highly cross-linked porous beads are used, which exclude all protein molecules. The lower molecular weight substances, however, can enter the beads; therefore, their progress down through the column will be retarded (a and b). The earlier fractions collected will contain the proteins, and the latter fractions will contain the low molecular weight contaminants (c). In practice, this 'group separations' application of gel-filtration chromatography is mainly used to separate proteins from salt (e.g. after an ammonium sulfate precipitation step) or for buffer exchange. *Note*: in practice, the chromatographic beads are tightly packed in the column. They are separated from each other in this diagram only for the purpose of clarity. Also, the drawing is not to scale; protein molecules are considerably smaller than individual beads

side groups, whereas lysine, arginine and histidine have positively charged basic side groups (Figure 6.10). Protein molecules, therefore, possess both positive and negative charges, largely due to the presence of varying amounts of these seven amino acids. (N-terminal amino groups and the C-terminal carboxy groups also contribute to overall protein charge characteristics.) The net charge exhibited by any protein depends on the relative quantities of these amino acids present in the protein, and on the pH of the protein solution. The pH value at which a protein molecule possesses zero overall charge is termed its isoelectric point (pI). At pH values above its pI, a protein will exhibit a net negative charge, whereas proteins will exhibit a net positive charge at pH values below the pI.

Ion-exchange chromatography is based upon the principle of reversible electrostatic attraction of a charged molecule to a solid matrix that contains covalently attached side groups of opposite charge (Figure 6.11). Proteins may subsequently be eluted by altering the pH or by increasing the salt concentration of the irrigating buffer. Ion-exchange matrices that contain covalently attached positive groups are termed anion exchangers. These will adsorb anionic proteins, e.g. proteins with a net negative charge. Matrices to which negatively charged groups are covalently attached are termed cation exchangers, adsorbing cationic proteins, e.g. positively charged proteins. Positively charged functional groups (anion exchangers) include species such as aminoethyl and diethylaminoethyl groups. Negatively charged groups attached to suitable matrices forming cation exchangers include sulfo- and carboxy-methyl groups (Table 6.3).

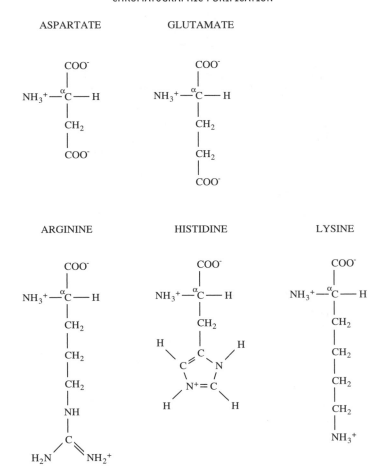

Figure 6.10 Structures of amino acids having overall net charges at pH 7.0. In proteins, the charges associated with the α-amino and α-carboxyl groups in all but the terminal amino acids are not present, as these groups are directly involved in the formation of peptide bonds

During the cation-exchange process, positively charged proteins bind to the negatively charged ion-exchange matrix by displacing the counter ion (often H^+), which is initially bound to the resin by electrostatic attraction. Elution may be achieved using a salt-containing irrigation buffer. The salt cation, often Na^+ of NaCl, in turn displaces the protein from the ion-exchange matrix. In the case of negatively charged proteins, an anion exchanger is obviously employed, with the protein adsorbing to the column by replacing a negatively charged counter ion.

The vast majority of purification procedures employ at least one ion-exchange step; it represents the single most popular chromatographic technique in the context of protein purification. Its popularity is based upon the high level of resolution achievable, its straightforward scale-up (for industrial application), together with its ease of use and ease of column regeneration. In addition,

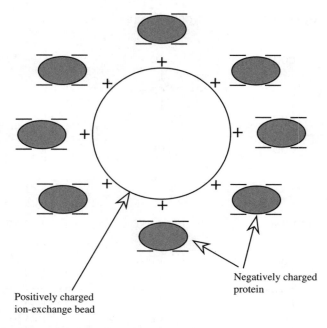

Negatively charged
protein

Positively charged
ion-exchange bead

Figure 6.11   Principle of ion-exchange chromatography, in this case anion exchange chromatography. The chromatographic beads exhibit an overall positive charge. Proteins displaying a nett negative charge at the pH selected for the chromatography will bind to the beads due to electrostatic interactions

it leads to a concentration of the protein of interest. It is also one of the least expensive chromatographic methods available. At physiological pH values most proteins exhibit a net negative charge. Anion-exchange chromatography, therefore, is most commonly used.

6.6.3   Hydrophobic interaction chromatography

Of the 20 amino acids commonly found in proteins, eight are classified as hydrophobic, due to the non-polar nature of their side chains (R groups, Figure 6.12). Most proteins are folded such

Table 6.3   Functional groups commonly attached to chromatographic beads in order to generate cation or anion exchangers

Group name	Group structure	Exchanger type
Diethylaminoethyl (DEAE)	$-O-(CH_2)_2-NH^+-(CH_2-CH_3)_2$	Anion
Quaternary ammonium (Q)	$-CH_2-NH^+-(CH_3)_3$	Anion
Quaternary aminoethyl (QAE)	$-O-(CH_2)_2-N^+(C_2H_5)_2-CH_2-CHOH-CH_3$	Anion
Carboxymethyl (CM)	$-O-CH_2-COO^-$	Cation
Methyl sulfonate (S)	$-CH_2-SO_3^-$	Cation
Sulfopropyl (SP)	$-CH_2-CH_2-CH_2SO_3^-$	Cation

that the majority of their hydrophobic amino acid residues are buried internally in the molecule and, hence, are shielded from the surrounding aqueous environment (Chapter 2). Internalized hydrophobic groups normally associate with adjacent hydrophobic groups. A minority of hydrophobic amino acids, however, are present on the protein surface and, hence, are exposed to the outer aqueous environment. Different protein molecules differ in the number and types of hydrophobic amino acid on their surface, and hence on their degree of surface hydrophobicity. Hydrophobic amino acids tend to be arranged in clusters or patches on the protein surface. Hydrophobic interaction chromatography fractionates proteins by exploiting their differing degrees of surface hydrophobicity. It depends on the occurrence of hydrophobic interactions between the hydrophobic patches on the protein surface and hydrophobic groups covalently attached to a suitable matrix.

The most popular hydrophobic interaction chromatographic beads (resins) are cross-linked agarose gels to which hydrophobic groups have been covalently linked. Specific examples include

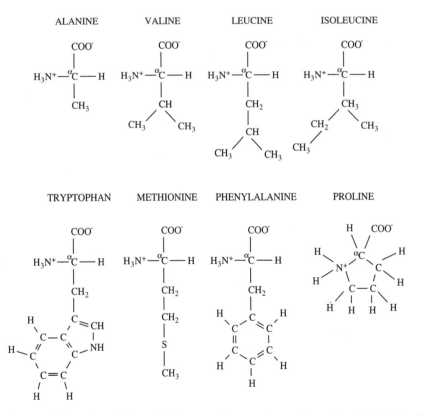

Figure 6.12 Structural formulae of the eight commonly occurring amino acids that display hydrophobic characteristics

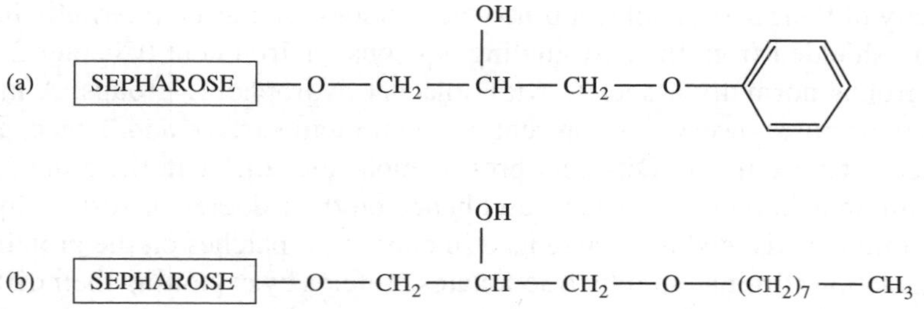

Figure 6.13 Chemical structure of (a) phenyl and (b) octyl sepharose, widely used in hydrophobic interaction chromatography

octyl- and phenyl-sepharose gels, which contain octyl and phenyl hydrophobic groups respectively (Figure 6.13).

Protein separation by hydrophobic interaction chromatography is dependent upon interactions between the protein itself, the gel matrix and the surrounding aqueous solvent. Increasing the ionic strength of a solution by the addition of a neutral salt (e.g. ammonium sulfate or sodium chloride) increases the hydrophobicity of protein molecules. This may be explained (somewhat simplistically) on the basis that the hydration of salt ions in solution results in an ordered shell of water molecules forming around each ion. This attracts water molecules away from protein molecules, which in turn helps to unmask hydrophobic domains on the surface of the protein.

Protein samples, therefore, are best applied to hydrophobic interaction columns under conditions of high ionic strength. As they percolate through the column, proteins may be retained via hydrophobic interactions. The more hydrophobic the protein, the tighter the binding. After a washing step, bound protein may be eluted by utilizing conditions that promote a decrease in hydrophobic interactions. This may be achieved by irrigation with a buffer of decreased ionic strength, inclusion of a suitable detergent, or lowering the polarity of the buffer by including agents such as ethanol or ethylene glycol.

Reverse-phase chromatography may also be used to separate proteins on the basis of differential hydrophobicity. This technique involves applying the protein sample to a highly hydrophobic column to which most proteins will bind. Elution is promoted by decreasing the polarity of the mobile phase. This is normally achieved by the introduction of an organic solvent. Elution conditions are harsh and generally result in denaturation of many proteins.

6.6.4 Affinity chromatography

Affinity chromatography is often described as the most powerful highly selective method of protein purification available. This technique relies on the ability of most proteins to bind specifically and reversibly to other compounds, often termed ligands (Figure 6.14). A wide variety of ligands may be covalently attached to an inert support matrix, and subsequently packed into a chromatographic column. In such a system, only the protein molecules that selectively bind to the immobilized ligand will be retained on the column. Washing the column with a suitable buffer will flush

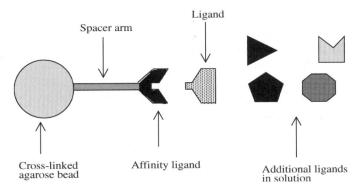

Spacer arm

Ligand

Cross-linked
agarose bead

Affinity ligand

Additional ligands
in solution

Figure 6.14 Schematic representation of the principle of biospecific affinity chromatography. The chosen affinity ligand is chemically attached to the support matrix (agarose bead) via a suitable spacer arm. Only those ligands in solution that exhibit biospecific affinity for the immobilized species will be retained

out all unbound molecules. An appropriate change in buffer composition, such as inclusion of a competing ligand, will result in desorption of the retained proteins.

Elution of bound protein from an affinity column is achieved by altering the composition of the elution buffer, such that the affinity of the protein for the immobilized ligand is greatly reduced. A variety of non-covalent interactions contribute to protein–ligand interaction. In many cases, changes in buffer pH, ionic strength, inclusion of a detergent or agents such as ethylene glycol (which reduce solution polarity) may suffice to elute the protein. In other cases, inclusion of a competing ligand promotes desorption. Competing ligands often employed include free substrates, substrate analogues or cofactors. Use of a competing ligand generally results in more selective protein desorption than does a generalized approach such as alteration of buffer pH or ionic strength. In some cases, a combination of such elution conditions may be required. Identification of optimal desorption conditions often requires considerable empirical study.

Affinity chromatography offers many advantages over conventional chromatographic techniques. The specificity and selectivity of biospecific affinity chromatography cannot be matched by other chromatographic procedures. Increases in purity of over 1000-fold, with almost 100 per cent yields, are often reported, at least on a laboratory scale. Incorporation of an affinity step could thus drastically reduce the number of subsequent steps required to achieve protein purification. This, in turn, could result in dramatic time and cost savings, which would be particularly significant in an industrial setting. Despite such promise, biospecific affinity chromatography does display some practical limitations:

- many biospecific ligands are extremely expensive and often exhibit poor stability;

- many of the ligand coupling techniques are chemically complex, hazardous, time consuming and costly;

- any leaching of coupled ligands from the matrix also gives cause for concern for two reasons, as (a) it effectively reduces the capacity of the system and (b) leaching of what are often noxious chemicals into the protein products is undesirable.

It is not normally prudent to employ biospecific affinity chromatography as an initial purification step, as various enzymatic activities present in the crude fractions may modify or degrade
the expensive affinity gels. However, it should be utilized as early as possible in the purification
procedure in order to accrue the full benefit afforded by its high specificity.

6.6.5   Immunoaffinity purifications

Immobilized antibodies may be used as affinity adsorbents for the antigens that stimulated their
production (Figure 6.15). Antibodies, like many other biomolecules, may be immobilized on a
suitable support matrix by a variety of chemical coupling procedures.

Immunoaffinity chromatography is amongst the most highly specific of all forms of biospecific chromatography. The high affinity with which an antibody normally binds its ligand can
often make subsequent ligand desorption from the column difficult. Desorption can require
conditions that result in partial denaturation of the bound protein. This is often achieved by
alteration of buffer pH or by employing chemical disrupting agents, such as urea or guanidine.
One of the most popular elution methods employed involves irrigation with a glycine–HCl
buffer at pH 2.2–2.8. In some cases, elution is more readily attainable at alkaline pH values. Specific examples have been documented in which protein elution was performed under
relatively mild conditions, such as a change of buffer system or an increase in ionic strength;
however, such examples are exceptional. The inclusion of an immunoaffinity step in the purification of recombinant blood factor VIII used to treat haemophilia (Chapter 12) is one example
of the industrial usage of this technique

6.6.6   Protein A chromatography

Most species of *Staphylococcus aureus* produce a protein known simply as protein A. This protein
consists of a single polypeptide chain of molecular weight 42 kDa. Protein A binds the F_c region
(the constant region) of immunoglobulin G (IgG; Chapter 13) obtained from human and many
other mammalian species with high specificity and affinity. Immobilization of protein A on chromatography beads provides a powerful affinity system that may be used to purify IgG. There is,
however, a considerable variation in the binding affinity of protein A for various IgG subclasses
obtained from different mammalian sources. In some cases another protein, protein G, may be
used instead of protein A. Most immunoglobulin molecules that bind to immobilized protein A do
so under alkaline conditions, and may subsequently be eluted at acidic pH values.

6.6.7   Lectin affinity chromatography

Lectin affinity chromatography may be used to purify a range of glycoproteins. Lectins are a
group of proteins synthesized by plants, vertebrates and a number of invertebrate species. Especially high levels of lectins are produced by a variety of plant seeds. Plant lectins are often termed

Table 6.4 Some lectins commonly used in immobilized format for the purification of glycoproteins. The sugar specificity is listed, as are the free sugars used to elute the bound glycoprotein

Lectin	Source	Sugar specificity	Eluting sugar
Con A	Jack bean seeds	α-D-Mannose, α-D-glucose	α-D-Methyl mannose
WG A	Wheat germ	N-Acetyl-β-D-glucosamine	N-Acetyl-β-D-glucosamine
PSA	Peas	α-D-Mannose	α-D-Methyl mannose
LEL	Tomato	N-Acetyl-β-D-glucosamine	N-Acetyl-β-D-glucosamine
STL	Potato tubers	N-Acetyl-β-D-glucosamine	N-Acetyl-β-D-glucosamine
PHA	Red kidney bean	N-Acetyl-D-galactosamine	N-Acetyl-D-galactosamine
ELB	Elderberry bark	Sialic acid or N-acetyl-D-galactosamine	Lactose
GNL	Snowdrop bulbs	α-1 → 3 Mannose	α-Methyl mannose
AAA	Freshwater eel	α-L-Fucose	L-Fucose

phytohaemagglutinins. All lectins have the ability to bind certain monosaccharides (such as α-D-mannose, α-D-glucose, D-N-acetyl galactosamine), and the sugar specificity for many is known (Table 6.4). Among the best-known and most widely used lectins are concanavalin A (Con A), soybean lectin (SBL) and wheat germ agglutinin (WGA).

Glycoproteins generally bind to lectin affinity columns at pH values close to neutrality. Desorption may be achieved in some cases by alteration of the pH of the eluting buffer. The most common method of desorption, however, involves inclusion of free sugar molecules for which the lectin exhibits a high affinity in this elution buffer, i.e. the inclusion of a competing ligand.

Although lectin affinity chromatography may be utilized to purify a variety of glycoproteins, it has not been widely employed for a number of reasons:

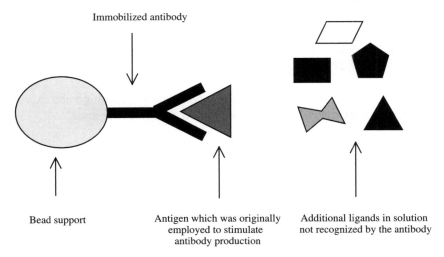

Immobilized antibody

Bead support

Antigen which was originally employed to stimulate antibody production

Additional ligands in solution not recognized by the antibody

Figure 6.15 Principle of immunoaffinity chromatography. Only antigen that is specifically recognized by the immobilized antibody will be retained on the column

- Most lectins are quite expensive.

- Crude protein sources containing one glycoprotein usually contain multiple glycoproteins. In most such instances, lectin-based affinity systems will result in the co-purification of several such glycoproteins.

- Limited application of this approach means that it has little track record, particularly on an industrial scale.

6.6.8   Dye affinity chromatography

The development of dye affinity chromatography may be attributed to the observation that some proteins exhibited anomalous elution characteristics when fractionated on gel-filtration columns in the presence of blue dextran. Blue dextran consists of a triazine dye (cibacron blue F3G-A) covalently linked to the high molecular weight sugar dextran. The discovery that some proteins bind the triazine dye soon led to its use as an affinity adsorbent by immobilization on an agarose matrix. A variety of other triazine dyes (Figure 6.16) also bind certain proteins and, hence, have also been used as affinity adsorbents. Dye affinity chromatography displays some positive general characteristics:

- The dyes are readily available in bulk and are relatively inexpensive.

- Chemical coupling of the dyes to the matrix is usually straightforward, often requiring no more than incubation under alkaline conditions at elevated temperature. The use of noxious coupling chemicals, such as cyanogen bromide, is avoided.

- The dye–matrix bead linkage is relatively resistant to chemical, physical and enzymatic degradation. In this way, ligand leakage from the column is minimized and is easily recognizable if it does occur – due to the dye colour.

- The protein binding capacity of immobilized dye adsorbents is also high and exceeds the binding capacity normally exhibited by natural biospecific adsorption ligands.

- Elution of bound protein is also relatively easily achieved.

A major potential disadvantage, however, is that it is not possible to predict accurately whether a specific protein will be retained on a dye affinity column, or what conditions will allow optimum binding/elution. Such information must be derived by empirical study. An understanding of the specific interactions that allow many apparently unrelated proteins to bind to dye affinity ligands, whereas other proteins are not retained, is usually lacking. The presence of negatively charged sulfonate groups lends triazine dyes an ion exchange character. These dyes also contain aromatic groups that can lend them some degree of hydrophobicity. Hydrophobic interactions, along with charged-based interactions, therefore, may play some role in protein adsorption. The dyes can also hydrogen bond with proteins.

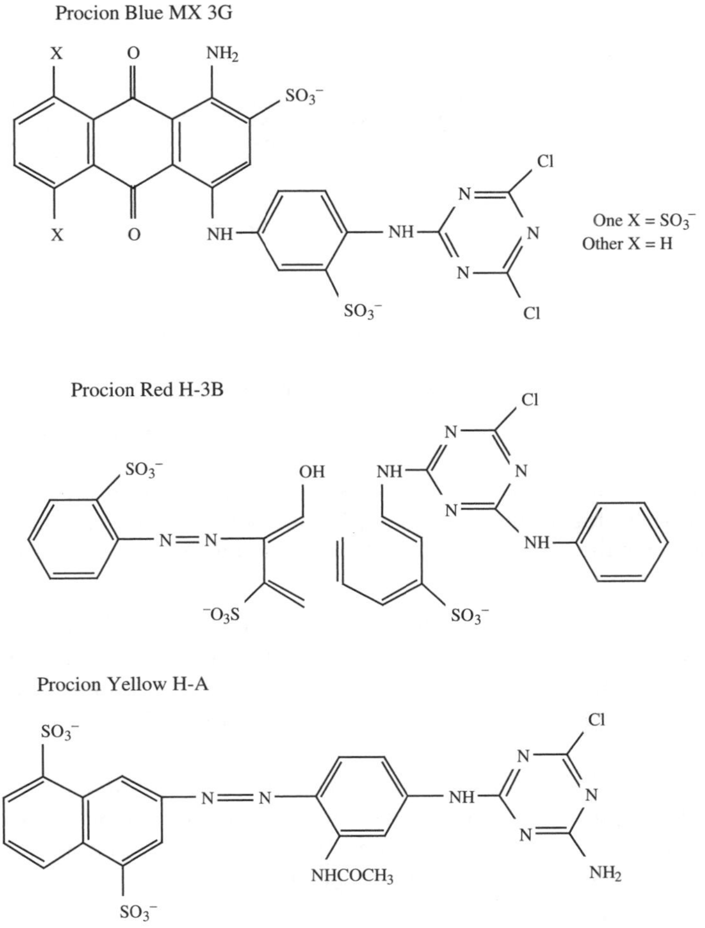

Procion Blue MX 3G

Procion Red H-3B

Procion Yellow H-A

Figure 6.16 Some mono- and di-chloro triazine dyes commonly used as affinity ligands in dye affinity chromatography

6.6.9 Metal chelate affinity chromatography

Metal chelate affinity chromatography is a pseudoaffinity protein purification technique first developed in the 1970s. The mode of adsorption relies upon the formation of weak coordinate bonds between basic groups on a protein surface with metal ions immobilized on chromatographic beads (Figure 6.17). The affinity medium is synthesized by covalent attachment of a metal chelator to the chromatographic bead via a spacer arm. Chelating agents, such as iminodiacetate, are capable of binding a number of metal ions (e.g. Fe, Co, Ni, Cu, Zn, Al), and binding effectively immobilizes the ion on the bead. The affinity gel is normally supplied without bound metal, so the gel can be 'charged' with the metal of choice (by flushing the column with a solution containing a salt of that metal, e.g. $CuSO_4$ in the case of copper). The metal ions most commonly used are Zn^{2+}, Ni^{2+} and Cu^{2+}. Basic groups on protein surfaces, most notably the side chain of histidine residues,

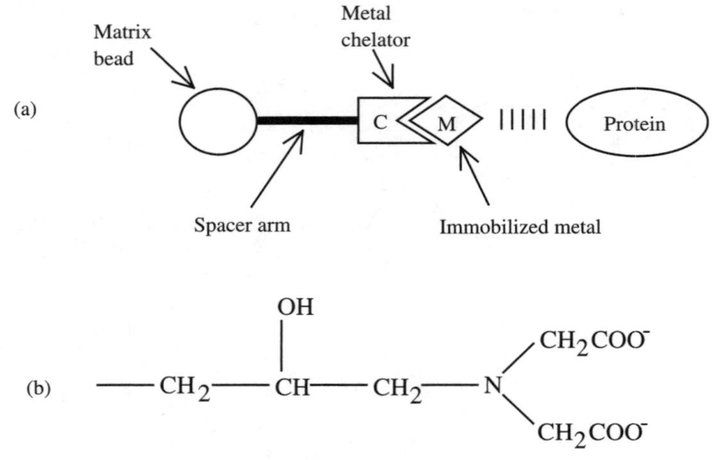

Figure 6.17 Schematic representation of the basic principles of metal chelate affinity chromatography. Certain proteins are retained on the column via the formation of coordinate bonds with the immobilized metal ion (a). The actual structure of the most commonly used metal chelator, iminodiacetic acid, is presented in (b)

are attracted to the metal ions, forming the weak coordinate bonds. Elution of bound proteins is undertaken by lowering the buffer pH (this causes protonation of the histidine residues, which are then unable to coordinate with the metal ion). Alternatively, a strong competitor complexing agent (e.g. the chelating agent ethylenediaminetetraacetic acid (EDTA)) can be added to the elution buffer.

Metal chelate affinity chromatography finds most prominent application in the affinity purification of recombinant proteins to which a histidine tag has been attached (described later). As protein binding occurs via the histidine residues, this technique is no more inherently useful for the purification of metalloproteins than for the purification of non-metalloproteins (a common misconception, given its name).

6.6.10   Chromatography on hydroxyapatite

Hydroxyapatite occurs naturally as a mineral in phosphate rock and also constitutes the mineral portion of bone. It may also be used to fractionate protein by chromatography.

Hydroxyapatite is prepared by mixing a solution of sodium phosphate (Na_2HPO_4) with calcium chloride ($CaCl_2$). A white precipitate known as brushite is formed. Brushite is then converted to hydroxyapatite by heating to 100°C in the presence of ammonia:

$$Ca_2HPO_4 \cdot 2H_2O \xrightarrow{100°C/NH_3} Ca_{10}(PO_4)_6(OH)_2$$

$$\text{brushite} \qquad\qquad\qquad\qquad \text{hydroxyapatite}$$

The underlying mechanism by which this substance binds and fractionates proteins is poorly understood. Protein adsorption is believed to involve interaction with both calcium and phosphate moieties of the hydroxyapatite matrix. Elution of bound species from such columns is normally achieved by irrigation with a potassium phosphate gradient.

6.6.11 Chromatofocusing

Fractionation by chromatofocusing separates proteins on the basis of their isoelectric points. This technique basically involves percolating a buffer of one pH through an ion-exchange column that is pre-equilibrated at a different pH. Owing to the natural buffering capacity of the exchanger, a continuous pH gradient may be set up along the length of the column. In order to achieve maximum resolution, a linear pH gradient must be constructed. This necessitates the use of an eluent buffer and exchanger that exhibit even buffering capacity over a wide range of pH values. The range of the pH gradient achieved will obviously depend on the pH at which the ion exchanger is pre-equilibrated and the pH of the eluent buffer. The sample is applied, usually in the running buffer, whose pH is lower than that of the pre-equilibrated column. After sample application, the column is constantly percolated with a specially formulated buffer that establishes an increasing pH gradient down the length of the column.

Upon sample application, negatively charged proteins immediately adsorb to the anion exchanger, while positively charged proteins flow down the column. Owing to the increasing pH gradient formed, such positively charged proteins will eventually reach a point within the column where the column pH equals their own pI values (their isoelectric points, i.e. the pH value at which the protein has an overall net charge of zero). Immediately upon further migration down the column such proteins become negatively charged, as the surrounding pH values increase above their pI values; hence, they bind to the column. Overall, therefore, upon initial application of the elution buffer, all protein species will migrate down the column until they reach a point where the column pH is marginally above their isoelectric points. At this stage they bind to the anion exchanger. Proteins of differing isoelectric points are thus fractionated on the basis of this parameter of molecular distinction.

The pH gradient formed is not a static one. As more elution buffer is applied, the pH value at any given point along the column is continually increasing. Thus, any protein that binds to the column will be almost immediately desorbed, as once again it experiences a surrounding pH value above its pI and becomes positively charged. Any such desorbed protein flows down the column until it reaches a further point where the pH value is marginally above its pI value, and it again rebinds. This process is repeated until the protein emerges from the column at its isoelectric point. To achieve the best results, the isoelectric point (pI value) of the required protein should ideally be in the middle of the pH gradient generated. Chromatofocusing can result in a high degree of protein resolution, with protein bands being eluted as tight peaks. This technique is particularly effective when used in conjunction with other chromatographic methods during protein purification. Most documented applications of this method still pertain to laboratory-scale procedures. Scaling up to industrial level is somewhat discouraged by economic factors, most notably the cost of the eluent required.

6.7 High-performance liquid chromatography of proteins

Most of the chromatographic techniques described thus far are usually performed under relatively low pressures, where flow rates through the column are generated by low-pressure pumps (i.e. low-pressure liquid chromatography). Fractionation of a single sample on such chromatographic columns typically requires a minimum of several hours to complete. Low flow rates are required because, as the protein sample flows through the column, the proteins are brought into contact with the surface of the chromatographic beads by direct (convective) flow. The protein molecules then rely entirely upon molecular diffusion to enter the porous gel beads. This is a slow process, especially when compared with the direct transfer of proteins past the outside surface of the gel beads by liquid flow. If a flow rate significantly higher than the diffusional rate is used, then 'protein band spreading' (and hence loss of resolution) will result. This occurs because any protein molecules that have not entered the bead will flow through the column at a faster rate than the (identical) molecules that have entered into the bead particles. Such high flow rates will also result in a lowering of adsorption capacity, as many molecules will not have the opportunity to diffuse into the beads as they pass through the column.

One approach that allows increased chromatographic flow rates without loss of resolution entails the use of microparticulate stationary-phase media of very narrow diameter. This effectively reduces the time required for molecules to diffuse in and out of the porous particles. Any reduction in particle diameter dramatically increases the pressure required to maintain a given flow rate. Such high flow rates may be achieved by utilizing high-pressure liquid chromatographic systems. By employing such methods, sample fractionation times may be reduced from hours to minutes.

The successful application of HPLC was made possible largely by (a) the development of pump systems that can provide constant flow rates at high pressure and (b) the identification of suitable pressure-resistant chromatographic media. Traditional soft gel media utilized in low-pressure applications are totally unsuited to high-pressure systems due to their compressibility.

In the context of protein purification/characterization, HPLC may be used for analytical or preparative purposes. Most analytical HPLC columns available have diameters ranging from 4 to 4.6 mm and lengths ranging from 10 to 30 cm. Preparative HPLC columns currently available have much wider diameters, typically up to 80 cm, and can be longer than 1 m (Figure 6.18). Various chemical groups may be incorporated into the matrix beads; thus, techniques such as ion-exchange, gel-filtration, affinity, hydrophobic interaction and reverse-phase chromatography are all applicable to HPLC.

Many small proteins, in particular those that function extracellularly (e.g. insulin, GH and various cytokines) are quite stable and may be fractionated on a variety of HPLC columns without significant denaturation or decrease in bioactivity. Preparative HPLC is used in industrial-scale purification of insulin and of IL2. In contrast, many larger proteins (e.g. blood factor VIII) are relatively labile, and loss of activity due to protein denaturation may be observed upon high-pressure fractionation.

At both preparative and analytical levels, HPLC exhibits several important advantages compared with low-pressure chromatographic techniques:

- HPLC offers superior resolution due to the reduction in bead particle size. The diffusional distance inside the matrix particles is minimized, resulting in sharper peaks than those obtained when low-pressure systems are employed.

- Owing to increased flow rates, HPLC systems also offer much improved fractionation speeds, typically in the order of minutes rather than hours.

- HPLC is amenable to a high degree of automation.

The major disadvantages associated with HPLC include cost and, to a lesser extent, capacity. Thus, for both technical and economic reasons, preparative HPLC is employed almost exclusively

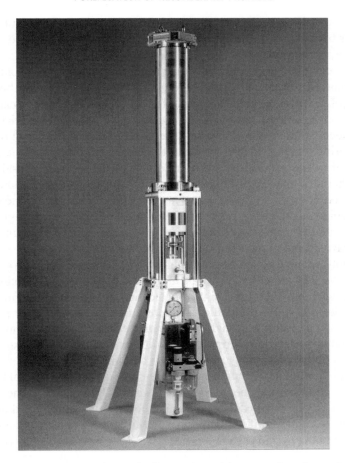

Figure 6.18 Preparative HPLC column (15 cm in diameter) used in processing of proteins required for therapeutic or diagnostic purposes. Column manufactured by Prochrom, Nancy, France (photograph courtesy of Affinity Chromatography Ltd)

in downstream processing of low-volume, extremely high-value proteins, mostly intended for therapeutic use, as opposed to proteins used for industrial preparations.

An alternative chromatographic system to HPLC is also available. Termed fast protein liquid chromatography (FPLC), this technique employs operating pressures significantly lower than those used in conventional HPLC systems. Lower pressures allow the use of matrix beads based on polymers such as agarose. FPLC columns are constructed of glass or inert plastic materials. Conventional HPLC columns are manufactured from high-grade stainless steel. In many cases, FPLC systems are economically more attractive than HPLC. Despite their operation at lower pressures, they still combine high resolution with enhanced speed of operation when compared with traditional low-pressure systems.

6.8 Purification of recombinant proteins

Proteins produced by recombinant DNA technology are usually purified by means identical to those available for purification of traditional non-recombinant proteins extracted directly from a natural producer source. In fact, purification of recombinant proteins can be somewhat more straightforward, as high expression levels of the target protein can be attained. This increases the ratio of target protein to contaminants.

Two features of recombinant production in particular can impact very significantly upon the approach subsequently taken to purify the recombinant product: inclusion body formation and the incorporation of purification tags. The processes of inclusion body formation, recovery and recombinant protein renaturation have been considered in Chapter 5. Once the recombinant protein has been refolded, additional purification (if required) follows traditional lines.

Genetic engineering techniques also facilitate the incorporation of specific peptide or protein tags to the protein of interest. A tag is chosen that confers on the resultant hybrid protein some pronounced physicochemical characteristic, facilitating its subsequent purification. Such a molecule is normally produced by fusing a DNA sequence that codes for the tag to one end of the genetic information encoding the protein of interest. Tags that allow rapid and straightforward purification of the hybrid protein by techniques such as ion-exchange, hydrophobic interaction or affinity chromatography have been designed and successfully employed.

Addition of a polyarganine (or polylysine) tag to the C-terminus of a protein confers on it a strong positive charge. The protein may then be more readily purified by cation-exchange chromatography. This approach has been used in the purification of various interferons and urogastrone on a laboratory scale at least. Addition of a tag containing a number of hydrophobic amino acids confers on the resultant molecule a strongly hydrophobic character, which allows its effective purification by hydrophobic interaction chromatography. A purification tag consisting of polyhistidine may be employed to purify proteins by metal chelate chromatography.

Tags that facilitate protein purification by affinity chromatography have also been developed. The gene coding for protein A may be fused to the gene or cDNA encoding the protein of interest. The resultant hybrid may be purified using a column containing immobilized IgG. Immunoaffinity purification may be employed if antibodies have been raised against the tag utilized.

Upon purification of the hybrid protein it is necessary to remove the tag, as the tag itself will be immunogenic. Removal of the tag is generally carried out by chemical or enzymatic means. This is achieved by designing the tag sequence such that it contains a cleavage point for a specific protease or chemical cleavage method at the protein–tag fusion junction. Sequence-specific proteases often employed to achieve tag removal include the endopeptidases trypsin, factor Xa and enterokinase. Exopeptidases, such as carboxypeptidase A, are also sometimes utilized. Generally speaking, endopeptidases, which cleave internal protein peptide bonds, are used to remove long tags, whereas exopeptidases are used most often to remove short tags. The exopeptidase carboxypeptidase A, for example, sequentially removes amino acids from the C-terminus of a protein until it encounters a lysine, arginine or proline residue. Chemical cleavage of specific peptide bonds relies on the use of chemicals such as cyanogen bromide or hydroxylamine.

Although several methods exist that can achieve tag removal, most such methods suffer from some inherent drawbacks. One essential prerequisite for any method is that the protein of interest must remain intact after the cleavage treatment. The required protein, therefore, should not contain any peptide bonds susceptible to cleavage by the specific method chosen. Chemical methods, for example, must generally be carried out under harsh conditions, often requiring high temperatures or extremes of pH. Such conditions can have a detrimental effect on normal protein functioning. Proteolytic removal of tags is also often less than 100 per cent efficient. Selective cleavage of the tag must be followed by subsequent separation of the tag from the protein of interest. This may require a further chromatographic step.

6.9 Final product formulation

High-resolution chromatography normally yields a protein that is 98–99 per cent pure. The next phase of downstream processing entails formulation into final product format. This generally involves:

- Addition of various excipients (substances other than the active ingredient(s) which, for example, stabilize the final product or enhance the characteristics of the final product in some other way).

- Filtration of the final product through a 0.22 μm absolute filter in order to generate sterile product, followed by its aseptic filling into final product containers.

- Freeze-drying (lyophilization) if the product is to be marketed in a powdered format.

The decision to market the product in liquid or powder form is often dictated by how stable the protein is in solution. This, in turn, must be determined experimentally, as there is no way to predict the outcome for any particular protein. Some proteins may remain stable for months (or even years) in solution, particularly if stabilizing excipients are added and the solution is refrigerated. Other proteins, particularly when purified, may retain biological activity for only a matter of hours or days when in aqueous solution.

6.9.1 Some influences that can alter the biological activity of proteins

A number of different influences can denature or otherwise modify proteins, rendering them less active/inactive. As all protein products are marketed on an activity basis, every precaution must be taken to minimize loss of biological activity during downstream processing and subsequent storage. Disruptive influences can be chemical (e.g. oxidizing agents, detergents, etc.), physical (e.g. extremes of pH, elevated temperature, vigorous agitation) or biological (e.g. proteolytic degradation). Minimization of inactivation can be achieved by minimizing the exposure of the product stream to such influences, and undertaking downstream processing in as short a time as possible. In addition, it is possible to protect the protein from many of these influences by the addition of suitable stabilizing agents. The addition of such agents to the final product is often essential in order to confer upon the product an acceptably long shelf life. During initial development, considerable empirical study is undertaken by formulators to determine what excipients are most effective in enhancing product stability.

A number of different molecular mechanisms can underpin the loss of biological activity of any protein. These include both covalent and non-covalent modification of the protein molecule, as summarized in Table 6.5. Protein denaturation, for example, entails a partial or complete alteration of the protein's three-dimensional shape. This is underlined by the disruption of the intramolecular forces that stabilize a protein's native conformation, namely hydrogen bonding, ionic attractions and hydrophobic interactions (Chapter 2). Covalent modifications of protein structure that can adversely affect its biological activity are summarized below.

Table 6.5 The various molecular alterations that usually result in loss of a protein's biological activity

Non-covalent alterations	
Partial/complete protein denaturation	
Covalent alterations	
Hydrolysis	Oxidation
Deamidation	Disulfide exchange
Imine formation	Isomerization
Raceimization	Photodecomposition

6.9.1.1   *Proteolytic degradation and alteration of sugar side-chains*

Proteolytic degradation of a protein is characterized by hydrolysis of one or more peptide (amide) bonds in the protein backbone, generally resulting in loss of biological activity. Hydrolysis is usually promoted by the presence of trace quantities of proteolytic enzymes, but can also be caused by some chemical influences.

Proteases belong to one of six mechanistic classes:

- serine proteases I (mammalian) or II (bacterial);

- cysteine proteases;

- aspartic proteases;

- metalloproteases I (mammalian) and II (bacterial).

The classes are differentiated on the basis of groups present at the protease active site known to be essential for activity, e.g. a serine residue forms an essential component of the active site of serine proteases. Both exo-proteases (which catalyse the sequential cleavage of peptide bonds beginning at one end of the protein), and endo-proteases (cleaving internal peptide bonds, generating peptide fragments) exist. Even limited endo-or exo-proteolytic degradation of biopharmaceuticals usually alters/destroys their biological activity.

Proteins differ greatly in their intrinsic susceptibility to proteolytic attack. Resistance to proteolysis seems to be dependent upon higher levels of protein structure (i.e. secondary and tertiary structure), as tight packing often shields susceptible peptide bonds from attack. Denaturation thus renders proteins very susceptible to proteolytic degradation.

A number of strategies may be adopted in order to minimize the likelihood of proteolytic degradation of the protein product, these include:

- minimizing processing times;

- processing at low temperature;

- use of specific protease inhibitors.

Table 6.6 Some of the most commonly employed protease inhibitions and the specific classes of proteases they inhibit

Inhibitor	Protease class inhibited
Phenylmethylsulfonyl fluoride	Serine proteases
	Some cysteine proteases
Benzamidine	Serine proteases
Pepstatin A	Aspartic proteases
EDTA	Metallo-proteases

Minimizing processing times obviously limits the duration during which proteases may come into direct contact with the protein product. Processing at low temperatures (often 4 °C) reduces the rate of proteolytic activity. Inclusion of specific proteolytic inhibitors in processing buffers, in particular homogenization buffers, can be very effective in preventing uncontrolled proteolysis. Although no one inhibitor will inhibit proteases of all mechanistic classes, a number of effective inhibitors for specific classes are known (Table 6.6). The use of a cocktail of such inhibitors is thus most effective. However, the application of many such inhibitors in biopharmaceutical processing is inappropriate due to their toxicity.

In most instances, instigation of precautionary measures protecting proteins against proteolytic degradation is of prime importance during the early stages of purification. During the later stages, most of the proteases present will have been removed from the product stream. A major aim of any purification system is the complete removal of such proteases, as the presence of even trace amounts of these catalysts can result in significant proteolytic degradation of the finished product over time.

As discussed in Chapter 2, many therapeutic proteins are glycosylated, and the sugar side chains can influence protein function, structure and stability. Chemical or enzymatic modification of a protein's glycocomponent, therefore, could affect its therapeutic properties. The presence of glycosidase enzymes in crude preparations, for example, could lead to partial degradation of sugar side chains. Generally, however, such eventualities may be effectively minimized by carrying out downstream processing at lower temperatures and as quickly as possible.

6.9.1.2 Protein deamidation

Deamidation and imide formation can also negatively influence a protein's biological activity. Deamidation refers to the hydrolysis of the side chain amide group of asparagine and/or glutamine, yielding aspartic acid and glutamic acid respectively (Figure 6.19). This reaction is promoted especially at elevated temperatures and extremes of pH. It represents the major route by which insulin preparations usually degrade. Imide formation occurs when the α-amino nitrogen of either asparagine, aspartic acid, glutamine or glutamic acid attacks the side chain carbonyl group of these amino acids. The resultant structures formed are termed aspartimides or glutarimides respectively. These cyclic imide structures are, in turn, prone to hydrolysis.

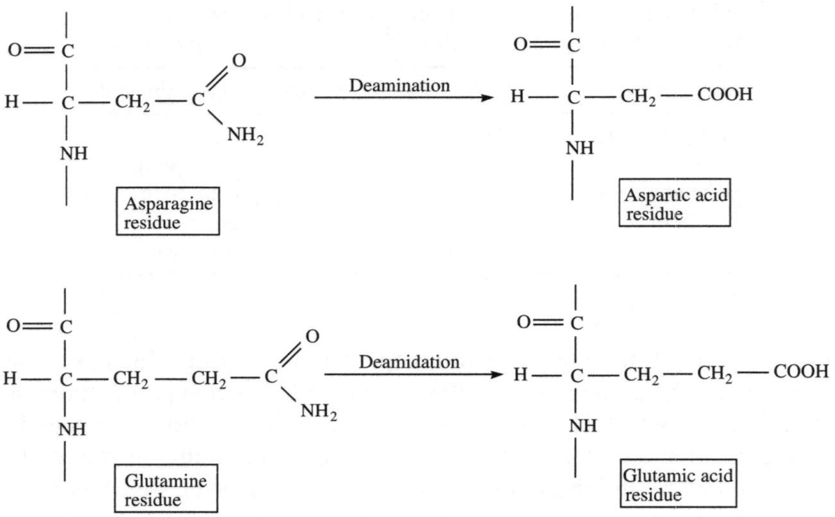

Figure 6.19 Deamidation of asparagine and glutamine, yielding aspartic acid and glutamic acid respectively. This process can often be minimized by reducing the final product pH to 4–5

6.9.1.3   Oxidation and disulfide exchange

The side chains of a number of amino acids are susceptible to oxidation by air. Although the side chains of tyrosine, tryptophan and histidine can be oxidized, the sulfur atoms present in methionine or cysteine are by far the most susceptible. Methionine can be oxidized by air or more potent oxidants, initially forming a sulfoxide and, subsequently, a sulfone (Figure 6.20). The sulfur atom of cysteine is readily oxidized, forming either a disulfide bond or (in the presence of potent oxidizing agents) sulfonic acid (Figure 6.20). Oxidation by air normally results only in disulfide bond formation. The oxidation of any constituent amino acid residue can (potentially) drastically reduce the biological activity of a polypeptide.

Oxidation of methionine is particularly favoured under conditions of low pH, and in the presence of various metal ions. Methionine residues on the surface of a protein are obviously particularly susceptible to oxidation. Those buried internally in the protein are less accessible to oxidant. hGH contains three methionine residues (at positions 14, 125 and 170). Studies have found that oxidation of methionine 14 and 125 (the more readily accessible ones) does not greatly effect hGH activity. However, oxidation of all three methionine residues results in almost total inactivation of the molecule.

Oxidation can be best minimized by replacing the air in the headspace of the final product container with an inert gas such as nitrogen, and/or the addition of antioxidants to the final product.

Disulfide exchange can also sometimes occur, and prompt a reduction in biological activity (Figure 6.21). Intermolecular disulfide exchange can result in aggregation of individual polypeptide molecules.

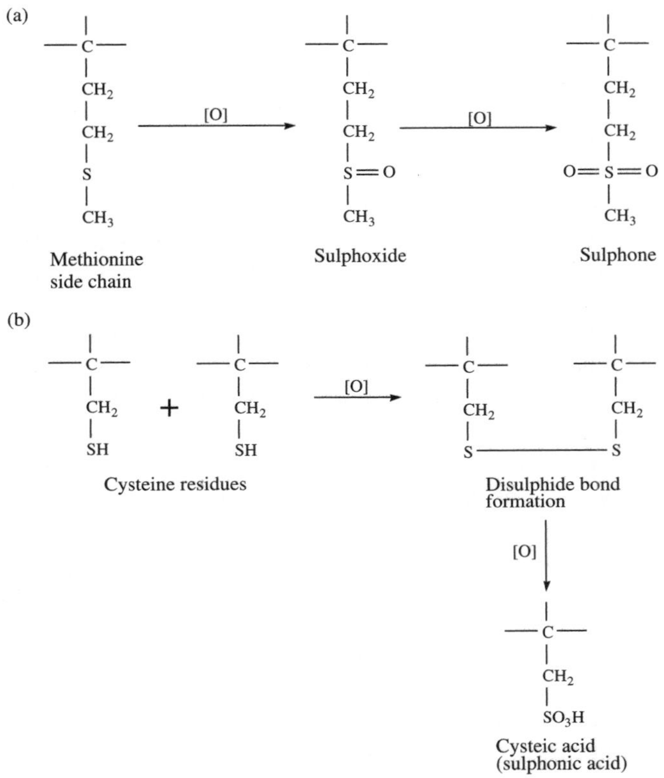

Figure 6.20 Oxidation of (a) methionine and (b) cysteine side chains, as can occur upon exposure to air or more potent oxidizing agents (e.g. peroxide, superoxide, hydroxyl radicals or hypochlorite). Refer to text for specific details

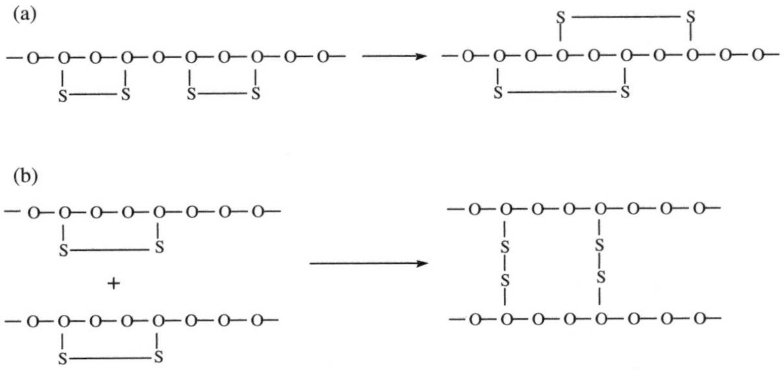

Figure 6.21 Diagram representing the molecular process of intrachain (a) and interchain (b) disulfide exchange. Refer to text for specific details. (—O— are amino acid residues in the polypeptide)

6.9.2   Stabilizing excipients used in final product formulations

A range of various substances may be added to a purified therapeutic protein in order to stabilize that product (Table 6.7). Such agents can stabilize proteins in a number of different ways, and some specific examples are outlined below.

Serum albumin addition has been shown to stabilize various different polypeptides (Table 6.8). HSA is often employed in the case of biopharmaceuticals destined for parenteral administration to humans. In many cases, it is used in combination with additional stabilizers, including amino acids (mainly glycine) and carbohydrates. Serum albumin itself is quite a stable molecule, capable of withstanding conditions of low pH or elevated temperature (it is stable for over 10 h at 60 °C). It also displays excellent solubility characteristics. It is postulated that albumin stabilizers exert their stabilizing influences by both direct and indirect means. Certainly, it helps decrease the level of surface adsorption of the active biopharmaceutical to the internal walls of final product containers. It also could act as an alternative target, for example, for traces of proteases or other agents that could be deleterious to the product. It may also function to stabilize the native conformation of many proteins directly. It has been shown to be an effective cryoprotectant for several biopharmaceuticals (e.g. IL-2, tPA and various interferon preparations), helping to minimize potentially detrimental effects of the freeze-drying process on the product.

However, the use of HSA in now discouraged due to the possibility of accidental transmission of blood-borne pathogens. The use of recombinant HSA would overcome such fears.

Various amino acids are also used as stabilizing agents for some biopharmaceutical products (Table 6.9). Glycine is most often employed, and it (as well as other amino acids) has been found to help stabilize various interferon preparations, as well as EPO, factor VIII, urokinase and arginase. Amino acids are generally added to final product at concentrations ranging from 0.5 per cent to 5 per cent. They appear to exert their stabilizing influence by various means, including reducing surface adsorption of product, inhibiting aggregate formation, and directly stabilizing

Table 6.7   Some major excipient groups that may be added to protein-based biopharmaceuticals in order to stabilize the biological activity of the finished product

Serum albumin
Various individual amino acids
Various carbohydrates
Alcohols and polyols
Surfactants

Table 6.8   Various biopharmaceutical preparations for which HSA has been described as a potential stabilizer

IFN-α and -β interferons	tPA
IFN-γ	Tumor necrosis factor
IL-2	Monoclonal antibody preparations
Urokinase	γ-Globulin preparations
EPO	Hepatitis B surface antigen

Table 6.9 Amino acids, carbohydrates and polyols that have found most application as stabilizers for some biopharmaceutical preparations

Amino acids	Carbohydrates	Polyols
Glycine	Glucose	Glycerol
Alanine	Sucrose	Mannitol
Lysine	Trehalose	Sorbitol
Threonine	Maltose	PEG

the conformation of some proteins, particularly against heat denaturation. The exact molecular mechanisms by which such effects are achieved remain to be elucidated.

Several polyols (i.e. molecules displaying multiple hydroxyl groups) have found application as polypeptide stabilizing agents. Polyols include substances such as glycerol, mannitol, sorbitol and PEG, as well as inositol (Table 6.9 and Figure 6.22). A subset of polyols is the carbohydrates, which are listed separately (and thus somewhat artificially) from polyols in Table 6.9. Various polyols have been found to stabilize proteins in solution directly, and carbohydrates in particular are also often added to biopharmaceutical products prior to freeze-drying in order to provide physical bulk to the freeze-dried cake.

Surfactants are well-known protein denaturants. However, when sufficiently dilute, some surfactants (e.g. polysorbate) exert a stabilizing influence on some protein types. Proteins display a tendency to aggregate at interfaces (air—liquid or liquid—liquid), a process that often promotes their denaturation. Addition of surfactant reduces surface tension of aqueous solutions and often increases the solubility of proteins dissolved therein. This helps reduce the rate of protein

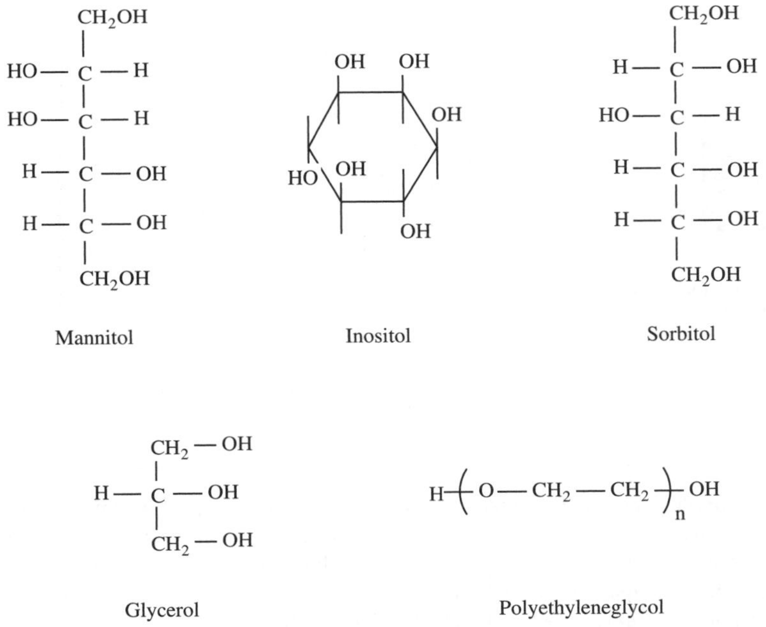

Figure 6.22 Structure of some polyols that are sometimes used to stabilize proteins

denaturation at interfaces. Polysorbate, for example, is included in some γ-globulin preparations, cytokines and in some monoclonal antibody-based products.

Although various polysorbates are used, the experience with an EPO-based product (trade-name Eprex) sounds a potential cautionary note in terms of formulation development, as outlined in Box 4.1.

6.9.3   Final product fill

An overview of a typical final product filling process is presented in Figure 6.23. The bulk final product first undergoes QC testing to ensure its compliance with bulk product specifications. Although implementation of good practices during manufacturing will ensure that the product carries a low microbial load, it will not be sterile at this stage. The product is then passed through a (sterilizing) 0.22 μm filter (Figure 6.24). The sterile product is housed (temporarily) in a sterile-product holding tank, from where it is aseptically filled into pre-sterile final product containers (usually glass vials). The filling process normally employs highly automated liquid filling systems. All items of equipment, pipework, etc. with which the sterilized product comes into direct contact must obviously themselves be sterile. Most such equipment items may be sterilized by autoclaving, and be aseptically assembled prior to the filling operation (which is undertaken under Grade A laminar flow conditions).

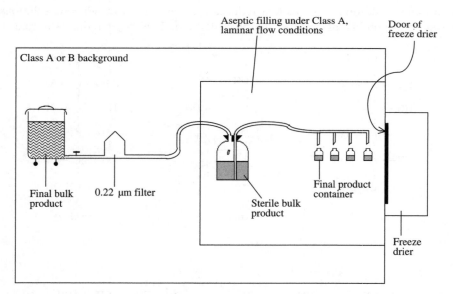

Figure 6.23   Final product filling. The final bulk product (after addition of excipients and final product QC testing), is filter sterilized by passing through a 0.22 μm filter. The sterile product is aseptically filled into (pre-sterile) final product containers under grade A laminar flow conditions. Much of the filling operation uses highly automated filling equipment. After filling, the product container is either sealed (by an automated aseptic sealing system), or freeze-dried first, followed by sealing

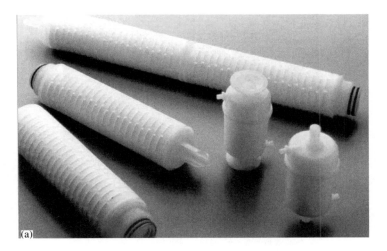

Figure 6.24 Photographic representation of a range of filter types and their stainless steel housing. Most filters used on an industrial scale are of a pleated cartridge design, which facilitates housing of maximum filter area within a compact space (a). These are generally housed in stainless steel housing units (b). Some process operations, however, still make use of flat (disc) filters, which are housed in a tripod-based stainless steel housing (c). All photographs courtesy of Pall Life Sciences, Ireland

The final product containers must also be pre-sterilized. This may be achieved by autoclaving or passage through special equipment that subjects the vials to a hot WFI rinse, followed by sterilizing dry heat and UV treatment.

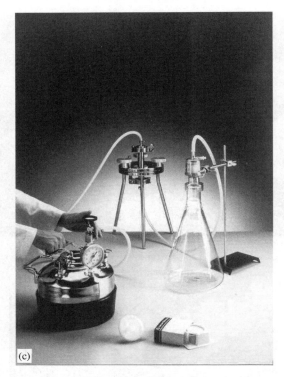

(c)

Figure 6.24 (*Continued*)

6.9.4 Freeze-drying

Freeze drying (lyophilization) refers to the removal of solvent directly from a solution while in the frozen state. Removal of water directly from (frozen) biopharmaceutical products via lyophilization yields a powdered product, usually displaying a water content of the order of 3 per cent. In general, removal of the solvent water from such products greatly reduces the likelihood of chemical/biological-mediated inactivation of the biopharmaceutical. Freeze-dried biopharmaceutical products usually exhibit longer shelf lives than products sold in solution. Freeze-drying is also recognized by the regulatory authorities as being a safe and acceptable method of preserving many parenteral products.

Freeze-drying is a relatively gentle way of removing water from proteins in solution. However, this process can promote the inactivation of some protein types, and specific excipients (cryoprotectants) are usually added to the product in order to minimize such inactivation. Commonly used cryoprotectants include carbohydrates (such as glucose and sucrose), proteins (such as HSA), and amino acids (such as lysine, arginine or glutamic acid). Alcohols/polyols have also found some application as cryoprotectants.

The freeze-drying process is initiated by the freezing of the biopharmaceutical product in its final product containers. As the temperature is decreased, ice crystals begin to form and grow. This results in an effective concentration of all the solutes present in the remaining liquid phase, including the protein and all added excipients. For example, the concentration of salts may increase to levels as high as 3 mol l^{-1}. Increased solute concentration alone can accelerate chemical reactions damaging to the protein product. In addition, such concentration effectively brings individual protein molecules into more intimate contact with each other, which can prompt protein–protein interactions and, hence, aggregation.

As the temperature drops still lower, some of the solutes present may also crystallize, thus being effectively removed from the solution. In some cases, individual buffer constituents can crystallize out of solution at different temperatures. This will dramatically alter the pH values of the remaining solution and, in this way, can lead to protein inactivation.

As the temperature is lowered further, the viscosity of the unfrozen solution increases dramatically until molecular mobility effectively ceases. This unfrozen solution will contain the protein, as well as some excipients, and (at most) 50 per cent water. As molecular mobility has effectively stopped, chemical reactivity also all but ceases. The consistency of this 'solution' is that of glass, and the temperature at which this is attained is called the glass transition temperature $T_{g'}$. For most protein solutions, $T_{g'}$ values reside between $-40\ °C$ and $-60\ °C$. The primary aim of the initial stages of the freeze-drying process is to decrease the product temperature below that of its $T_{g'}$ value and as quickly as possible in order to minimize the potential negative effects described above.

The next phase of the freeze-drying process entails the application of a vacuum to the system. When the vacuum is established, the temperature is increased, usually to temperatures slightly in excess of 0 °C. This promotes sublimation of the crystalline water, leaving behind a powdered cake of dried material. Once satisfactory drying has been achieved, the product container is sealed.

The drying chamber of industrial-scale freeze dryers usually opens into a cleanroom (Figure 6.23). This facilitates direct transfer of the product-containing vials into the chamber. Immediately prior to filling, rubber stoppers are usually partially inserted into the mouth of each vial in such a way as not to hinder the outward flow of water vapour during the freeze-drying process. The drying chamber normally contains several rows of shelves, each of which can accommodate several thousand vials (Figure 6.25). These shelves are wired to allow their electrical heating, cooling, and their upward or downward movement. After the freeze-drying cycle is complete (which can take 3 days or more), the shelves are then moved upwards. As each shelf moves up, the partially inserted rubber seals are inserted fully into the vial mouth as they come in contact with the base plate of the shelf immediately above them. After product recovery, the empty chamber is closed and is then heat-sterilized (using its own chamber-heating mechanism). The freeze-drier is then ready to accept its next load.

6.9.5 Labelling and packing

After the product has been filled (and sealed) in its final product container. QC personnel then remove representative samples of the product and carry out tests to ensure conformance to final product specification. The most important specifications will relate to product potency, sterility and final volume fill, as well as the absence of endotoxin or other potentially toxic substances. Detection and quantification of excipients added will also be undertaken. Product analysis is considered in Chapter 7.

Only after QC personnel are satisfied that the product meets these specifications will it be labelled and packed. These operations are highly automated. Labelling, in particular, deserves special attention. Mislabelling of product remains one of the most common reasons for product recall. This can occur relatively easily, particularly if the facility manufactures several different products, or even a single product at several different strengths. Information presented on a label should normally include:

Figure 6.25   Photographic representation of (a) laboratory-scale, (b) pilot-scale and (c) industrial-scale freeze-driers. Refer to text for details. Photograph courtesy of Virtis, USA

Figure 6.25   (*Continued*)

- name and strength/potency of the product;

- specific batch number of the product;

- date of manufacture and expiry date;

- storage conditions required.

Additional information often presented includes the name of the manufacturer, a list of excipients included and a brief summary of the correct mode of product usage.

When a batch of product is labelled and packed, and QC personnel are satisfied that labelling and packing are completed to specification, the QC manager will write and sign a 'Certificate of Analysis'. This details the predefined product specifications and confirms conformance of the actual batch of product in question to these specifications. At this point, the product, along with its Certificate of Analysis, may be shipped to the customer.

Further reading

Books

Ahmed, H. 2005. *Principles and Reactions of Protein Extraction, Purification and Characterization*. CRC Press.

Carpenter, J. 2002. *Rational Design of Stable Protein Formulations*. Kluwer.

Costantino, H. and Pikal, M. (eds). 2005. *Lyophilization of Biopharmaceuticals*. AAPS Press.

Desai, M. 2000. *Downstream Protein Processing Methods*. Humana Press.

Flickinger, M. 1999. *The Encyclopedia of Bioprocess Technology.* Wiley.

Frokjaer, S. 2000. *Pharmaceutical Formulation Development of Peptides and Proteins.* Taylor and Francis.

Grindley, J. and Ogden, J. 2000. *Understanding Biopharmaceuticals. Manufacturing and Regulatory Issues.* Interpharm Press.

Harris, E. 2000. *Protein Purification Applications.* Oxford University Press.

Merten, O.-W., Mattanovich, D., Lang, C., Larsson, G., Neubauer, P., Porro, D., Postma, P., Teixeira de Mattos, J., and Cole, J.A. *Recombinant Protein Production with Prokaryotic and Eukaryotic Cells: A Comparative View on Host Physiology?* Kluwer.

Janson, J. 1998. *Protein Purification.* Wiley.

Oxender, D. and Post, L. 1999. *Novel Therapeutics from Modern Biotechnology.* Springer Verlag.

Roe, S. 2001. *Protein Purification Techniques.* Oxford University Press.

Walsh, G. 2002. *Proteins: Biochemistry and Biotechnology.* Wiley.

Articles

Arakawa, T., Prestrelski, S.J., Kenney, W.C., and Carpenter, J.F. 2001. Factors effecting short-term and long-term stabilities of proteins. *Advanced Drug Delivery Reviews* **46**, 307–326.

Bernard, A.R., Lusti-Narasimhan, M., Radford, K.M., Hale, R.S., Sebille, E., and Graber, P. 1996. Downstream processing on insect cultures. *Cytotechnology* **20**(1–3), 239–257.

Cleland, J.L., Powell, M.F., and Shire, S.J. 1993. The development of stable protein formulations: a close look at protein aggregation, deamidation and oxidation. *Critical Reviews in Therapeutic Drug Carrier Systems* **10**(4), 307–377.

Frokjaer, S. and Otzen, D. 2005. Protein drug stability: a formulation challenge. *Nature Reviews Drug Discovery* **4**, 298–306.

Hedhammar, M., Gräslund, T., and Hober, S. 2005. Protein engineering strategies for selective protein purification. *Chemical Engineering and Technology* **28**, 1315–1325.

Keller, K., Friedmann, T., and Boxman, A. 2001. The bioseparation needs for tomorrow. *Trends in Biotechnology* **19**(11), 438–441.

Lee, J. 2000. Biopharmaceutical formulation. *Current Opinion in Biotechnology* **11**(1), 81–84.

Nikolov, Z. and Woodard, S. 2004. Downstream processing of recombinant proteins from transgenic feedstock. *Current Opinion in Biotechnology* **15**, 479–486.

Shire, S.J., Shahrokh, Z., and Liu J. 2004. Challenges in the development of high protein concentration formulations. *Journal of Pharmaceutical Sciences* **93**, 1390–1402.

Wang, W. 2000. Lyophilizatin and development of solid protein pharmaceuticals. *International Journal of Pharmaceutics* **203**(1–2), 1–60.

Wei, W. 1999. Instability, stabilization and formulation of liquid protein pharmaceuticals. *International Journal of Pharmaceutics* **185**(2), 129–188.

7

Product analysis

7.1 Introduction

All pharmaceutical finished products undergo rigorous QC testing in order to confirm their conformance to predetermined specifications. Potency testing is of obvious importance, ensuring that the drug will be efficacious when administered to the patient. A prominent aspect of safety testing entails analysis of product for the presence of various potential contaminants.

The range and complexity of analytical testing undertaken for recombinant biopharmaceuticals far outweighs that undertaken with regard to 'traditional' pharmaceuticals manufactured by organic synthesis. Not only are proteins (or additional biopharmaceuticals such as nucleic acids; Chapter 14) much larger and more structurally complex than traditional low molecular mass drugs, their production in biological systems renders the range of potential contaminants far broader (Table 7.1). Recent advances in analytical techniques render practical the routine analysis of complex biopharmaceutical products. An overview of the range of finished-product tests of recombinant protein biopharmaceuticals is outlined below. Explanation of the theoretical basis underpinning these analytical methodologies is not undertaken, as this would considerably broaden the scope of the text. Appropriate references are provided at the end of the chapter for the interested reader.

7.2 Protein-based contaminants

Most of the chromatographic steps undertaken during downstream processing are specifically included to separate the protein of interest from additional contaminant proteins. This task is not an insubstantial one, particularly if the recombinant protein is expressed intracellularly.

In addition to protein impurities emanating directly from the source material, other proteins may be introduced during upstream or downstream processing. For example, animal cell culture media are typically supplemented with bovine serum/foetal calf serum (2–25 per cent), or with a defined cocktail of various regulatory proteins required to maintain and stimulate growth of these cells. Downstream processing of intracellular microbial proteins often requires the addition of

Pharmaceutical biotechnology: concepts and applications   Gary Walsh
© 2007 John Wiley & Sons, Ltd   ISBN 978 0 470 01244 4 (HB)   978 0 470 01245 1 (PB)

Table 7.1 The range and medical significance of potential impurities present in biopharmaceutical products destined for parenteral administration

Impurity	Medical consequence
Microorganisms	Potential establishment of a severe microbial infection – septicaemia
Viral particles	Potential establishment of a severe viral infection
Pyrogenic substances	Fever response that, in serious cases, culminates in death
DNA	Significance is unclear – could bring about an immunological response
Contaminating proteins	Immunological reactions. Potential adverse effects if the contaminant exhibits an unwanted biological activity

endonuleases to the cell homogenate to degrade the large quantity of DNA liberated upon cellular disruption. (DNA promotes increased solution viscosity, rendering processing difficult. Viscosity, being a function of the DNA's molecular mass, is reduced upon nuclease treatment.)

Minor amounts of protein could also potentially enter the product stream from additional sources, e.g. protein shed from production personnel. Implementation of good manufacturing practice (GMP), however, should minimize contamination from such sources.

The clinical significance of protein-based impurities relates to (a) their potential biological activities and (b) their antigenicity. Whereas some contaminants may display no undesirable biological activity, others may exhibit activities deleterious to either the product itself (e.g. proteases that could modify/degrade the product) or the recipient patient (e.g. the presence of contaminating toxins).

Their inherent immunogenicity also renders likely and immunological reaction against protein-based impurities upon product administration to the recipient patient. This is particularly true in the case of products produced in microbial or other recombinant systems (i.e. most biopharmaceuticals). Although the product itself is likely to be non-immunogenic (usually being coded for by a human gene), contaminant proteins will be endogenous to the host cell, and hence foreign to the human body. Administration of the product can elicit an immune response against the contaminant. This is particularly likely if a requirement exists for ongoing, repeat product administration (e.g. administration of recombinant insulin). Immunological activation of this type could also potentially (and more seriously) have a sensitizing effect on the recipient against the actual protein product.

In addition to distinct gene products, modified forms of the protein of interest are also considered impurities, rendering desirable their removal from the product stream. Although some such modified forms may be innocuous, others may not. Modified product 'impurities' may compromise the product in a number of ways, e.g.:

- biologically inactive forms of the product will reduce overall product potency;

- some modified product forms remain biologically active, but exhibit modified pharmacokinetic characteristics (i.e. timing and duration of drug action);

- modified product forms may be immunogenic.

Altered forms of the protein of interest can be generated in a number of ways by covalent and non-covalent modifications (e.g. see Table 6.5).

7.3 Removal of altered forms of the protein of interest from the product stream

Modification of any protein will generally alter some aspect of its physicochemical characteristics. This facilitates removal of the modified form by standard chromatographic techniques during downstream processing. Most downstream procedures for protein-based biopharmaceuticals include both gel-filtration and ion-exchange steps (Chapter 6). Aggregated forms of the product will be effectively removed by gel filtration (because they now exhibit a molecular mass greater by several orders of magnitude than the native product). This technique will also remove extensively proteolysed forms of the product. Glycoprotein variants whose carbohydrate moieties have been extensively degraded will also likely be removed by gel-filtration (or ion-exchange) chromatography. Deamidation and oxidation will generate product variants with altered surface charge characteristics, often rendering their removal by ion exchange relatively straightforward. Incorrect disulfide bond formation, partial denaturation and limited proteolysis can also alter the shape and surface charge of proteins, facilitating their removal from the product by ion exchange or other techniques, such as hydrophobic interaction chromatography.

The range of chromatographic techniques now available, along with improvements in the resolution achievable using such techniques, renders possible the routine production of protein biopharmaceuticals which are in excess of 97–99 per cent pure. This level of purity represents the typical industry standard with regard to biopharmaceutical production.

A number of different techniques may be used to characterize protein-based biopharmaceutical products, and to detect any protein-based impurities that may be present in that product (Table 7.2). Analysis for non-protein-based contaminant is described in subsequent sections.

7.3.1 Product potency

Any biopharmaceutical must obviously conform to final product potency specifications. Such specifications are usually expressed in terms of 'units of activity' per vial of product (or per thera-

Table 7.2 Methods used to characterize (protein-based) finished product biopharmaceuticals. An overview of most of these methods is presented over the next several sections of this chapter

Non-denaturing gel electrophoresis
Denaturing (SDS) gel electrophoresis
Two-dimensional electrophoresis
Capillary electrophoresis
Peptide mapping
HPLC (mainly RP-HPLC)
Isoelectric focusing
Mass spectrometry
Amino acid analysis
N-terminal sequencing
Circular dichroism studies
Bioassays and immunological assays

peutic dose, or per milligram of product). A number of different approaches may be undertaken to determine product potency. Each exhibits certain advantages and disadvantages.

Bioassays represent the most relevant potency-determining assay, as they directly assess the biological activity of the biopharmaceutical. Bioassay involves applying a known quantity of the substance to be assayed to a biological system that responds in some way to this applied stimulus. The response is measured quantitatively, allowing an activity value to be assigned to the substance being assayed.

All bioassays are comparative in nature, requiring parallel assay of a 'standard' preparation against which the sample will be compared. Internationally accepted standard preparations of most biopharmaceuticals are available from organizations such as the World Health Organization (WHO) or the United States Pharmacopeia.

An example of a straightforward bioassay is the traditional assay method for antibiotics. This usually entailed measuring the zone of inhibition of microbial growth around an antibiotic-containing disc, placed on an agar plate seeded with the test microbe. Bioassays for modern biopharmaceuticals are generally more complex. The biological system used can be whole animals, specific organs or tissue types, or individual mammalian cells in culture.

Bioassays of related substances can be quite similar in design. Specific growth factors, for example, stimulate the accelerated growth of specific animal cell lines. Relevant bioassays can be undertaken by incubation of the growth-factor-containing sample with a culture of the relevant sensitive cells and radiolabelled nucleotide precursors. After an appropriate time period, the level of radioactivity incorporated into the DNA of the cells is measured. This is a measure of the bioactivity of the growth factor.

The most popular bioassay of EPO involves a mouse-based bioassay (EPO stimulates red blood cell production, making it useful in the treatment of certain forms of anaemia; Chapter 10). Basically, the EPO-containing sample is administered to mice along with radioactive iron (^{57}Fe). Subsequent measurement of the rate of incorporation of radioactivity into proliferating red blood cells is undertaken. (The greater the stimulation of red blood cell proliferation, the more iron taken up for haemoglobin synthesis.)

One of the most popular bioassay for interferons is termed the 'cytopathic effect inhibition assay'. This assay is based upon the ability of many interferons to render animal cells resistant to viral attack. It entails incubation of the interferon preparation with cells sensitive to destruction by a specific virus. That virus is then subsequently added, and the percentage of cells that survive thereafter is proportional to the levels of interferon present in the assay sample. Viable cells can assimilate certain dyes, such as neutral red. Addition of the dye followed by spectrophotometric quantitation of the amount of dye assimilated can thus be used to quantitate percentage cell survival. This type of assay can be scaled down to run in a single well of a microtitre plate. This facilitates automated assay of large numbers of samples with relative ease.

Although bioassays directly assess product potency (i.e. activity), they suffer from a number of drawbacks, including:

- *Lack of precision.* The complex nature of any biological system, be it an entire animal or individual cell, often results in the responses observed being influenced by factors such as metabolic status of individual cells, or (in the case of whole animals) subclinical infections, stress levels induced by human handling, etc.

- *Time*. Most bioassays take days, and in some cases week, to run. This can render routine bio-assays difficult, and impractical to undertake as a quick QC potency test during downstream processing.

- *Cost*. Most bioassay systems, in particular those involving whole animals, are extremely expensive to undertake.

Because of such difficulties alternative assays have been investigated, and sometimes are used in conjunction with, or instead of, bioassays. The most popular alternative assay system is the immunoassay.

Immunoassays employ monoclonal or polyclonal antibody preparations (Chapter 13) to detect and quantify the product (Box 7.1). The specificity of antibody–antigen interaction ensures good assay precision. The use of conjugated radiolabels (RIA) or enzymes (EIA) to allow detection of antigen–antibody binding renders such assays very sensitive. Furthermore, when compared with

---

Box 7.1

Immunoassays

In addition to their therapeutic use (Chapter 13), antibodies are frequently employed as diagnostic reagents because they exhibit extreme specificity in their recognition of a particular ligand, i.e. the antigen that stimulated their production. Antibody preparations are often used in the detection and quantification of a wide variety of specific analytes, including specific therapeutic proteins, and assays that employ antibodies in this way are termed immunoassays. The substance of interest is first employed as an antigen and injected into animals in order to elicit the production of antibodies against that particular molecule. Either monoclonal or polyclonal antibody preparations (Chapter 13) may be used in immunoassay systems.

Antibody molecules have no inherent characteristic that facilitates their direct detection in immunoassays. A second important step in developing a successful immunoassay, therefore, involves the incorporation of a suitable 'marker'. The marker serves to facilitate the rapid detection and quantification of antibody–antigen binding. Earlier immunoassay systems used radioactive labels as a marker (radioimmunoassay; RIA) although immunoassay systems using enzymes (enzyme immunoassays; EIA) subsequently have come to the fore. Yet additional immunoassay systems use alternative markers including fluorescent or chemiluminescent tags.

EIA systems take advantage of the extreme specificity and affinity with which antibodies bind antigens which stimulated their initial production, coupled to the catalytic efficiency of enzymes, which facilitates signal amplification as well as straightforward detection and quantification. In most such systems, the antibody is immobilized on the internal walls of the wells in a multi-well microtitre plate, which therefore serves as collection of reaction mini-test tubes.

Since their initial introduction over 30 years ago many variations on the basic enzyme immunoassay concept have been designed. One of the most popular EIA systems currently in use

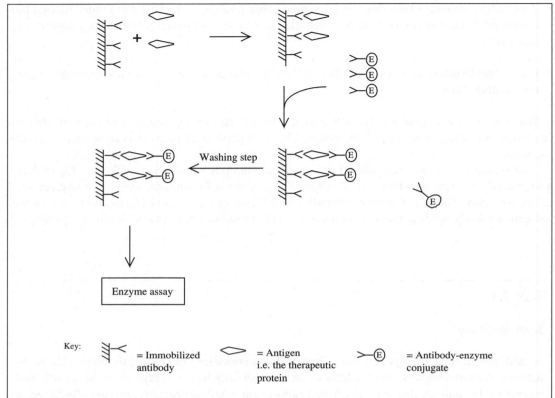

Figure 7.B1 Basic principle of the ELISA system

is that of the ELISA. The basic principle upon which the ELISA system is based is illustrated in Figure 7.B1. In this form it is also often referred to as the double antibody sandwich technique.

Antibodies raised against the antigen of interest (i.e. the therapeutic protein) are first adsorbed onto the internal walls of microtitre plate wells. The sample to be assayed is then incubated in the wells. Antigen present will bind to the immobilized antibodies. After an appropriate time, which allows antibody–antigen binding to reach equilibrium, the wells are washed.

A preparation containing a second antibody, which also recognizes the antigen, is then added. The second antibody will also bind to the retained antigen and the enzyme label is conjugated to this second antibody.

Subsequent to a further washing step, to remove any unbound antibody–enzyme conjugate, the activity of the enzyme retained is quantified by a straightforward enzyme assay. The activity recorded is proportional to the quantity of antigen present in the sample assayed. A series of standard antigen concentrations may be assayed to allow construction of a standard curve. The standard curve facilitates calculation of antigen quantities present in 'unknown' samples.

a bioassay, immunoassays are rapid (undertaken in minutes to hours), inexpensive, and straightforward to undertake.

The obvious disadvantage of immunoassays is that immunological reactivity cannot be guaranteed to correlate directly to biological activity. Relatively minor modifications of the protein product, although having a profound influence on its biological activity, may have little or no influence on its ability to bind antibody.

For such reasons, although immunoassays may provide a convenient means of tracking product during downstream processing, performing a bioassay on at the very least the final product is usually necessary to prove that potency falls within specification.

7.3.2 Determination of protein concentration

Quantification of total protein in the final product represents another standard analysis undertaken by QC. A number of different protein assays may be potentially employed (Table 7.3).

Detection and quantification of protein by measuring absorbency at 280 nm is perhaps the simplest such method. This approach is based on the fact that the side chains of the amino acids tyrosine and tryptophan absorb at this wavelength. The method is popular, as it is fast, easy to perform and is non-destructive to the sample. However, it is a relatively insensitive technique, and identical concentrations of different proteins will yield different absorbance values if their content of tyrosine and tryptophan vary to any significant extent. Hence, this method is rarely used to determine the protein concentration of the final product, but it is routinely used during downstream processing to detect protein elution off chromatographic columns, and hence track the purification process.

Table 7.3 Common assay methods used to quantitate proteins. The principle upon which each method is based is also listed

Method	Principle
Absorbance at 280 nm (A_{280}; UV method)	The side chain of selected amino acids (particularly tyrosine and tryptophan) absorbs UV at 280 nm
Absorbance at 205 nm (far-UV method)	Peptide bonds absorb UV at 190–220 nm
Biuret method	Binding of copper ions to peptide bond nitrogen under alkaline conditions generates a purple colour
Lowry method	Lowry method uses a combination of the Biuret copper-based reagent and the 'Folin–Ciocalteau' reagent, which contains phosphomolybdic-phosphotungstic acid. Reagents react with protein, yielding a blue colour that displays an absorbance maximum at 750 nm
Bradford method	Bradford reagent contains the dye Coomassie blue G-250 in an acidic solution. The dye binds to protein, yielding a blue colour that absorbs maximally at 595 nm
Bicinchonic acid method	Copper-containing reagent that, when reduced by protein, reacts with bicinchonic acid yielding a complex that displays an absorbance maximum at 562 nm
Peterson method	Essentially involves initial precipitation of protein out of solution by addition of trichloroacetic acid. The protein precipitate is redissolved in NaOH and the Lowry method of protein determination is then performed
Silver-binding method	Interaction of silver with protein – very sensitive method

Measuring protein absorbance at lower wavelengths (205 nm) increases the sensitivity of the assay considerably. Also, as it is the peptide bonds that are absorbing at this wavelength, the assay is subject to much less variation due to the amino acid composition of the protein.

The most common methods used to determine protein concentration are the dye-binding procedure using Coomassie brilliant blue, and the bicinchonic-acid-based procedure. Various dyes are known to bind quantitatively to proteins, resulting in an alteration of the characteristic absorption spectrum of the dye. Coomassie brilliant blue G-250, for example, becomes protonated when dissolved in phosphoric acid, and has an absorbance maximum at 450 nm. Binding of the dye to a protein (via ionic interactions) results in a shift in the dye's absorbance spectrum, with a new major peak (at 595 nm) being observed. Quantification of proteins in this case can thus be undertaken by measuring absorbance at 595 nm. The method is sensitive, easy and rapid to undertake. Also, it exhibits little quantitative variation between different proteins.

Protein determination procedures using bicinchonic acid were developed by Pierce Chemicals, who hold a patent on the product. The procedure entails the use of a copper-based reagent containing bicinchonic acid. Upon incubation with a protein sample, the copper is reduced. In the reduced state it reacts with bicinchonic acid, yielding a purple colour that absorbs maximally at 562 nm.

Silver also binds to proteins, an observation that forms the basis of an extremely sensitive method of protein detection. This technique is used extensively to detect proteins in electrophoretic gels, as discussed in the next section.

7.4 Detection of protein-based product impurities

SDS polyacrylamide gel electrophoresis (SDS-PAGE) represents the most commonly used analytical technique in the assessment of final product purity (Figure 7.1). This technique is well established and easy to perform. It provides high-resolution separation of polypeptides on the basis of their molecular mass. Bands containing as little as 100 ng of protein can be visualized by staining the gel with dyes such as Coomassie blue. Subsequent gel analysis by scanning laser densitometry allows quantitative determination of the protein content of each band (thus allowing quantification of protein impurities in the product).

The use of silver-based stains increases the detection sensitivity up to 100 fold, with individual bands containing as little as 1ng of protein usually staining well. However, because silver binds to protein non-stoichiometrically, quantitative studies using densitometry cannot be undertaken.

SDS-PAGE is normally run under reducing conditions. Addition of a reducing agent such as β-mercaptoethanol or dithiothreitol (DTT) disrupts interchain (and intrachain) disulfide linkages. Individual polypeptides held together via disulfide linkages in oligomeric proteins will thus separate from each other on the basis of their molecular mass.

The presence of bands additional to those equating to the protein product generally represent protein contaminants. Such contaminants may be unrelated to the product or may be variants of the product itself (e.g. differentially glycosylated variants, proteolytic fragment, etc.). Further characterization may include western blot analysis. This involves eluting the protein bands from the electrophoretic gel onto a nitrocellulose filter. The filter can then be probed using antibodies raised against the product. Binding of the antibody to the 'contaminant' bands suggests that they are variants of the product.

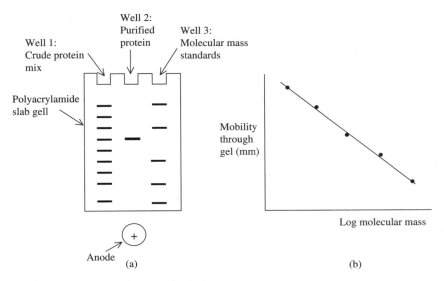

Figure 7.1 Separation of proteins by SDS-PAGE. Protein samples are incubated with SDS (as well as reducing agents, which disrupt disulfide linkages). The electric field is applied across the gel after the protein samples to be analysed are loaded into the gel wells. The rate of protein migration towards the anode is dependent upon protein size. After electrophoresis is complete, individual protein bands may be visualized by staining with a protein-binding dye (a). If one well is loaded with a mixture of proteins, each of known molecular mass, a standard curve relating distance migrated to molecular mass can be constructed (b). This allows estimation of the molecular mass of the purified protein

One concern relating to SDS-PAGE-based purity analysis is that contaminants of the same molecular mass as the product will go undetected as they will co-migrate with it. Two-dimensional electrophoretic analysis would overcome this eventuality in most instances.

Two-dimensional electrophoresis is normally run so that proteins are separated from each other on the basis of a different molecular property in each dimension. The most commonly utilized method entails separation of proteins by isoelectric focusing (see below) in the first dimension, with separation in the second dimension being undertaken in the presence of SDS, thus promoting band separation on the basis of protein size. Modified electrophoresis equipment that renders two-dimensional electrophoretic separation routine is freely available. Application of biopharmaceutical finished products to such systems allows rigorous analysis of purity.

Isoelectric focusing entails setting up a pH gradient along the length of an electrophoretic gel. Applied proteins will migrate under the influence of an electric field until they reach a point in the gel at which the pH equals the protein's isoelectric point pI (the pH at which the protein exhibits no overall net charge; only species with a net charge will move under the influence of an electric field). Isoelectric focusing thus separates proteins on the basis of charge characteristics.

This technique is also utilized in the biopharmaceutical industry to determine product homogeneity. Homogeneity is best indicated by the appearance in the gel of a single protein band, exhibiting the predicted pI value. Interpretation of the meaning of multiple bands, however, is less straightforward, particularly if the protein is glycosylated (the bands can also be stained for the presence of carbohydrates). Glycoproteins varying slightly in their carbohydrate content will vary in their sialic acid content and, hence, exhibit slightly different pI values. In such instances, isoelectric focusing analysis seeks to establish batch-to-batch consistency in terms of the banding pattern observed.

Isoelectric focusing also finds application in analysing the stability of biopharmaceuticals over the course of their shelf life. Repeat analysis of samples over time will detect deamidation or other degradative processes that alter protein charge characteristics.

7.4.1 Capillary electrophoresis

Capillary electrophoresis systems are also likely to play an increasingly prominent analytical role in the QC laboratory (Figure 7.2). As with other forms of electrophoresis, separation is based upon different rates of protein migration upon application of an electric field.

As its name suggests, in the case of capillary electrophoresis, this separation occurs within a capillary tube. Typically, the capillary will have a diameter of 20–50 μm and be up to a 1 m long (it is normally coiled to facilitate ease of use and storage). The dimensions of this system yield greatly increased surface area:volume ratio (compared with slab gels), hence greatly increasing the efficiency of heat dissipation from the system. This, in turn, allows operation at a higher current density, thus speeding up the rate of migration through the capillary. Sample analysis can be undertaken in 15–30 min, and on-line detection at the end of the column allows automatic detection and quantification of eluting bands.

The speed, sensitivity, high degree of automation and ability to quantitate protein bands directly render this system ideal for biopharmaceutical analysis.

Figure 7.2 Photograph of a capillary electrophoresis system (the HP-3D capillary electrophoresis system manufactured by Hewlett Packard). Refer to text for details. Photograph courtesy of Hewlett Packard GmbH, Germany

7.4.2 High-performance liquid chromatography

HPLC occupies a central analytical role in assessing the purity of low molecular mass pharmaceutical substances (Figure 7.3). It also plays an increasingly important role in analysis of macromolecules, such as proteins. Most of the chromatographic strategies used to separate proteins under 'low pressure' (e.g. gel filtration, ion exchange, etc.) can be adapted to operate under high pressure. Reverse-phase-, size-exclusion- and, to a lesser extent, ion-exchange-based HPLC chromatography systems are now used in the analysis of a range of biopharmaceutical preparations. On-line detectors (usually a UV monitor set at 220 or 280 nm) allows automated detection and quantification of eluting bands.

HPLC is characterized by a number of features that render it an attractive analytical tool. These include:

- excellent fractionation speeds (often just minutes per sample);

- superior peak resolution;

Figure 7.3 Photograph of a typical HPLC system (the Hewlett Packard HP1100 system). Photograph courtesy of Hewlett Packard GmbH, Germany

- high degree of automation (including data analysis);

- ready commercial availability of various sophisticated systems.

Reverse-phase HPLC (RP-HPLC) separates proteins on the basis of differences in their surface hydophobicity. The stationary phase in the HPLC column normally consists of silica or a polymeric support to which hydrophobic arms (usually alkyl chains, such as butyl, octyl or octadecyl groups) have been attached. Reverse-phase systems have proven themselves to be a particularly powerful analytical technique, capable of separating very similar molecules displaying only minor differences in hydrophobicity. In some instances a single amino acid substitution or the removal of a single amino acid from the end of a polypeptide chain can be detected by RP-HPLC. In most instances, modifications such as deamidation will also cause peak shifts. Such systems, therefore, may be used to detect impurities, be they related or unrelated to the protein product. RP-HPLC finds extensive application in, for example, the analysis of insulin preparations. Modified forms, or insulin polymers, are easily distinguishable from native insulin on reverse-phase columns.

Although RP-HPLC has proven its analytical usefulness, its routine application to analysis of specific protein preparations should be undertaken only after extensive validation studies. HPLC in general can have a denaturing influence on many proteins (especially larger, complex proteins). Reverse-phase systems can be particularly harsh, as interaction with the highly hydrophobic stationary phase can induce irreversible protein denaturation. Denaturation would result in the generation of artifactual peaks on the chromatogram.

Size-exclusion HPLC (SE-HPLC) separates proteins on the basis of size and shape. As most soluble proteins are globular (i.e. roughly spherical in shape), separation is essentially achieved on the basis of molecular mass in most instances. Commonly used SE-HPLC stationary phases include silica-based supports and cross-linked agarose of defined pore size. Size-exclusion systems are most often used to analyse product for the presence of dimers or higher molecular mass aggregates of itself, as well as proteolysed product variants.

Calibration with standards allows accurate determination of the molecular mass of the product itself, as well as any impurities. Batch-to-batch variation can also be assessed by comparison of chromatograms from different product runs.

Ion-exchange chromatography (both cation and anion) can also be undertaken in HPLC format. Though not as extensively employed as reverse-phase or size-exclusion systems, ion-exchange-based systems are of use in analysing for impurities unrelated to the product, as well as detecting and quantifying deamidated forms.

7.4.3   Mass spectrometry

Recent advances in the field of mass spectrometry now extend the applicability of this method to the analysis of macromolecules such as proteins. Using electrospray mass spectrometry, it is now possible to determine the molecular mass of many proteins to within an accuracy of ± 0.01 per cent. A protein variant missing a single amino acid residue can easily be distinguished from the native protein in many instances. Although this is a very powerful technique, analysis of the results obtained can sometimes be less than straightforward. Glycoproteins, for example, yield extremely complex spectra (due to their natural heterogeneity), making the significance of the findings hard to interpret.

7.5 Immunological approaches to detection of contaminants

Most recombinant biopharmaceuticals are produced in microbial or mammalian cell lines. Thus, although the product is derived from a human gene, all product-unrelated contaminants will be derived from the producer organism. These non-self proteins are likely to be highly immunogenic in humans, rendering their removal from the product stream especially important. Immunoassays may be conveniently used to detect and quantify non-product-related impurities in the final preparation (immunoassays generally may not be used to determine levels of product-related impurities, as antibodies raised against such impurities would almost certainly cross-react with the product itself).

The strategy usually employed to develop such immunoassays is termed the 'blank run approach'. This entails constructing a host cell identical in all respects to the natural producer cell, except that it lacks the gene coding for the desired product. This blank producer cell is then subjected to upstream processing procedures identical to those undertaken with the normal producer cell. Cellular extracts are subsequently subjected to the normal product purification process, but only to a stage immediately prior to the final purification steps. This produces an array of proteins that could copurify with the final product. These proteins (of which there may be up to 200 as determined by two-dimensional electrophoretic analysis) are used to immunize horses, goats or other suitable animals. Therefore, polyclonal antibody preparations capable of binding specifically to these proteins are produced. Purification of the antibodies allows their incorporation in radioimmunoassay or enzyme-based immunoassay systems, which may subsequently be used to probe the product. Such multi-antigen assay systems will detect the sum total of host-cell-derived impurities present in the product. Immunoassays identifying a single potential contaminant can also be developed.

Immunoassays have found widespread application in detecting and quantifying product impurities. These assays are extremely specific and very sensitive, often detecting target antigen down to parts per million levels. Many immunoassays are available commercially, and companies exist that will rapidly develop tailor-made immunoassay systems for biopharmaceutical analysis.

Application of the analytical techniques discussed thus far focuses upon detection of proteinaceous impurities. A variety of additional tests are undertaken that focus upon the active substance itself. These tests aim to confirm that the presumed active substance observed by electrophoresis, HPLC, etc. is indeed the active substance, and that its primary sequence (and, to a lesser extent, higher orders of structure) conform to licensed product specification. Tests performed to verify the product identity include amino acid analysis, peptide mapping, N-terminal sequencing and spectrophotometric analyses.

7.5.1 Amino acid analysis

Amino acid analysis remains a characterization technique undertaken in many laboratories, in particular if the product is a peptide or small polypeptide (molecular mass $\leq$10 kDa.). The strategy is simple. Determine the range and quantity of amino acids present in the product and compare the results obtained with the expected (theoretical) values. The results should be comparable.

The peptide/polypeptide product is usually hydrolysed by incubation with 6 mol l^{-1} HCl at elevated temperatures (110 °C), under vacuum, for extended periods (12–24 h). The constituent amino acids are separated from each other by ion-exchange chromatography and identified by comparison with standard amino acid preparations. Reaction with ninhydrin allows subsequent quantification of each amino acid present.

Although this technique is relatively straightforward and automated amino acid analysers are commercially available, it is subject to a number of disadvantages that limits its usefulness in biopharmaceutical analysis. These include:

- hydrolysis conditions can destroy/modify certain amino acid residues, in particular tryptophan, but also serine, threonine and tyrosine;

- the method is semi-quantitative rather than quantitative;

- sensitivity is at best moderate; low-level contaminants may go undetected (i.e. not significantly alter the amino acid profile obtained), particularly if the product is a high molecular mass protein.

These disadvantages, along with the availability of alternative characterization methodologies, limit application of this technique in biopharmaceutical analysis.

7.5.2 Peptide mapping

A major concern relating to biopharmaceuticals produced in high-expression recombinant systems is the potential occurrence of point mutations in the product's gene, leading to an altered primary structure (i.e. amino acid sequence). Errors in gene transcription or translation could also have similar consequences. The only procedure guaranteed to detect such alterations is full sequencing of a sample of each batch of the protein, which is a considerable technical challenge. Although partial protein sequencing is normally undertaken (see later), the approach most commonly used to detect alterations in amino acid sequence is peptide (fingerprint) mapping.

Peptide mapping entails exposure of the protein product to a reagent that promotes hydrolysis of peptide bonds at specific points along the protein backbone. This generates a series of peptide fragments. These fragments can be separated from each other by a variety of techniques, including one- or two-dimensional electrophoresis, and RP-HPLC in particular. A standardized sample of the protein product when subjected to this procedure will yield a characteristic peptide fingerprint, or map, with which the peptide maps obtained with each batch of product can subsequently be compared. If the peptides generated are relatively short, then a change in a single amino acid residue is likely to alter the peptide's physicochemical properties sufficiently to alter its position within the peptide map (Figure 7.4). In

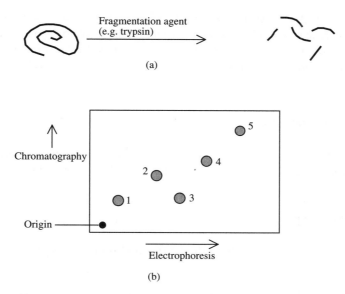

Figure 7.4 Generation of a peptide map. In this simple example, the protein to be analysed is treated with a fragmentation agent, e.g. trypsin (a). In this case, five fragments are generated. The digest is then applied to a sheet of chromatography paper (b) (at the point marked 'origin'). The peptides are then separated from each other in the first (vertical) dimension by paper chromatography. Subsequently, electrophoresis is undertaken (in the horizontal direction). The separated peptide fragments may be visualized by, for example, staining with ninhydrin. Two-dimensional separation of the peptides is far more likely to resolve each peptide completely from the others. In the case above, for example, chromatography (in the vertical dimension) alone would not have been sufficient to resolve peptides 1 and 3 fully. During biopharmaceutical production, each batch of the recombinant protein produced should yield identical peptide maps. Any mutation that alters the protein's primary structure (i.e. amino acid sequence) should result in at least one fragment adopting an altered position in the peptide map

this way, single (or multiple) amino acid substitutions, deletions, insertions or modifications can usually be detected. This technique plays an important role in monitoring batch-to-batch consistency of the product, and also obviously can confirm the identity of the actual product.

The choice of reagent used to fragment the protein is critical to the success of this approach. If a reagent generates only a few very large peptides, a single amino acid alteration in one such peptide will be more difficult to detect than if it occurred in a much smaller peptide fragment. On the other hand, generation of a large number of very short peptides can be counterproductive, as it may prove difficult to resolve all the peptides from each other by subsequent chromatography. Generation of peptide fragments containing an average of 7–14 amino acids is most desirable.

The most commonly utilized chemical cleavage agent is cyanogen bromide (it cleaves the peptide bond on the carboxyl side of methionine residues). V8 protease, produced by certain staphylococci, along with trypsin are two of the more commonly used proteolytic-based fragmentation agents.

Knowledge of the full amino acid sequence of the protein usually renders possible pre-determination of the most suitable fragmentation agent for any protein. The amino acid sequence of hGH, for example, harbours 20 potential trypsin cleavage sites. Under some circumstances it may be possible to use a combination of fragmentation agents to generate peptides of optimal length.

7.5.3   N-terminal sequencing

N-terminal sequencing of the first 20–30 amino acid residues of the protein product has become a popular quality control test for finished biopharmaceutical products. The technique is useful, as it:

- positively identifies the protein;

- confirms (or otherwise) the accuracy of the amino acid sequence of at least the N-terminus of the protein;

- readily identifies the presence of modified forms of the product in which one or more amino acids are missing from the N-terminus.

N-terminal sequencing is normally undertaken by Edman degradation (Figure 7.5). Although this technique was developed in the 1950s, advances in analytical methodologies now facilitate fast and automated determination of up to the first 100 amino acids from the N-terminus of most proteins, and usually requires a sample size of less than 1 μmol to do so (Figure 7.6).

Analogous techniques facilitating sequencing from a polypeptide's C-terminus remain to be satisfactorily developed. The enzyme carboxypeptidase C sequentially removes amino acids from the C-terminus, but often only removes the first few such amino acids. Furthermore, the rate at which it hydrolyses bonds can vary, depending on what amino acids have contributed to bond formation. Chemical approaches based on principles similar to the Edman procedure have been attempted. However, poor yields of derivatized product and the occurrence of side reactions have prevented widespread acceptance of this method.

7.5.4   Analysis of secondary and tertiary structure

Analyses such as peptide mapping, N-terminal sequencing or amino acid analysis yield information relating to a polypeptide's primary structure, i.e. its amino acid sequence. Such tests yield no information relating to higher-order structures (i.e. secondary and tertiary structure of polypeptides, along with quaternary structure of multi-subunit proteins). Although a protein's three-dimensional conformation may be studied in great detail by X-ray crystallography or NMR spectroscopy, routine application of such techniques to biopharmaceutical manufacture is impractical, both from a technical and an economic standpoint. Limited analysis of protein secondary and tertiary structure can, how-ever, be more easily undertaken using spectroscopic methods, particularly far-UV circular dichroism. More recently proton-NMR has also been applied to studying higher orders of protein structure.

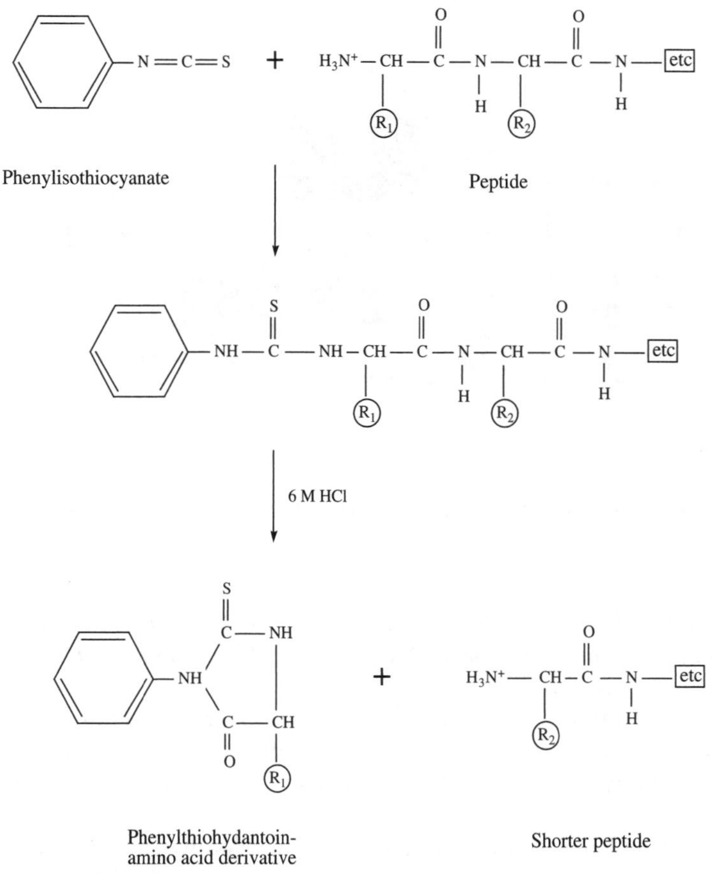

Figure 7.5 The Edman degradation method, by which the sequence of a peptide/polypeptide may be elucidated. The peptide is incubated with phenylisothiocyanate, which reacts specifically with the N-terminal amino acid of the peptide. Addition of 6 mol l^{-1} HCl results in liberation of a phenylthiohydantoin-amino acid derivative and a shorter peptide, as shown. The phenylthiohydantoin derivative can then be isolated and its constituent amino acid identified by comparison to phenylthiohydantoin derivatives of standard amino acid solutions. The shorter peptide is then subjected to a second round of treatment, such that its new amino terminus may be identified. This procedure is repeated until the entire amino acid sequence of the peptide has been established

7.6 Endotoxin and other pyrogenic contaminants

Pyrogens are substances that, when they enter the blood stream, influence hypothalamic regulation of body temperature, usually resulting in fever. Medical control of pyrogen-induced fever proves very difficult, and in severe cases results in patient death.

Pyrogens represent a diverse group of substances, including various chemicals, particulate matter and endotoxin (LPS), a molecule derived from the outer membrane of Gram-negative bacteria. Such Gram-negative organisms harbour 3–4 million LPS molecules on their surface, representing in the region of 75 per cent of their outer membrane surface area. Gram-negative bacteria

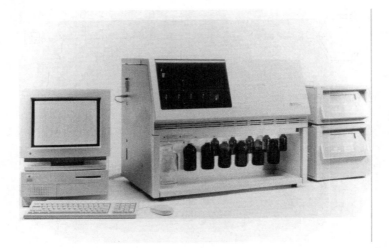

Figure 7.6 Photograph of a modern protein sequencing system. Photograph courtesy of Perkin Elmer Applied Biosystems Ltd, UK

clinically significant in human medicine include *E. coli*, *Haemophilus influenzae*, *Salmonella enterica*, *Klebsiella pneumoniae*, *Bordetella pertussis*, *Pseudomonas aeruginosa*, *Chlamydia psittaci* and *Legionella pneumophila*.

In many instances the influence of pyrogens on body temperature is indirect. For example, entry of endotoxin into the bloodstream stimulates the production of IL-1 (Chapter 9) by macrophages. It is the IL-1 that directly initiates the fever response (hence its alternative name, 'endogenous pyrogen').

Although entry of any pyrogenic substance into the bloodstream can have serious medical consequences, endotoxin receives most attention because of its ubiquitous nature. Therefore, it is the pyrogen most likely to contaminate parenteral (bio)pharmaceutical products. Effective implementation of GMP minimizes the likelihood of product contamination by pyrogens. For example, GMP dictates that chemical reagents used in the manufacture of process buffers be extremely pure. Such raw materials, therefore, are unlikely to contain chemical contaminants displaying pyrogenic activity. Furthermore, GMP encourages filtration of virtually all parenteral products through a 0.45 or 0.22 μm filter at points during processing and prior to filling in final product containers (even if the product can subsequently be sterilized by autoclaving). Filtration ensures removal of all particulate matter from the product. In addition, most final product containers are rendered particle free immediately prior to filling by an automatic pre-rinse using WFI. As an additional safeguard, the final product will usually be subject to a particulate matter test by QC before final product release. The simplest format for such a test could involve visual inspection of vial contents, although specific particle detecting and counting equipment is more routinely used.

Contamination of the final product with endotoxin is more difficult to control because:

- Many recombinant biopharmaceuticals are produced in Gram-negative bacterial systems; thus, the product source is also a source of endotoxin.

- Despite rigorous implementation of GMP, most biopharmaceutical preparations will be contaminated with low levels of Gram-negative bacteria at some stage of manufacture. These bacteria shed endotoxin into the product stream, which is not removed during subsequent bacterial filtration steps. This is one of many reasons why GMP dictates that the level of bioburden in the product stream should be minimized at all stages of manufacture.

- The heat stability exhibited by endotoxin (see Section 7.6.1) means that autoclaving of process equipment will not destroy endotoxin present on such equipment.

- Adverse medical reactions caused by endotoxin are witnessed in humans at dosage rates as low as 0.5 ng per kilogram body weight.

7.6.1 Endotoxin, the molecule

The structural detail of a generalized endotoxin (LPS) molecule is presented in Figure 7.7. As its name suggests, LPS consists of a complex polysaccharide component linked to a lipid (lipid A) moiety. The polysaccharide moiety is generally composed of 50 or more monosaccharide units linked by glycosidic bonds. Sugar moieties often found in LPS include glucose, glucosamine, mannose and galactose, as well as more extensive structures such as L-glycero-mannoheptose. The polysaccharide component of LPS may be divided into several structural domains. The inner (core) domains vary relatively little between LPS molecules isolated from different Gram-negative bacteria. The outer (O-specific) domain is usually bacterial-strain specific.

Most of the LPS biological activity (pyrogenicity) is associated with its lipid A moiety. This usually consists of six or more fatty acids attached directly to sugars such as glucosamine. Again, as is the case in relation to the carbohydrate component, lipid A moieties of LPS isolated from different bacteria can vary somewhat. The structure of *E coli*'s lipid A has been studied in the greatest detail; its exact structure has been elucidated and it can be chemically synthesized.

7.6.2 Pyrogen detection

Pyrogens may be detected in parenteral preparations (or other substances) by a number of methods. Two such methods are widely employed in the pharmaceutical industry.

Historically, the rabbit pyrogen test constituted the most widely used method. This entails parenteral administration of the product to a group of healthy rabbits, with subsequent monitoring of rabbit temperature using rectal probes. Increased rabbit temperature above a certain point suggests the presence of pyrogenic substances. The basic rabbit method, as outlined in the European Pharmacopoeia, entails initial administration of the product to three rabbits. The product is considered to have passed the test if the total (summed) increase of the temperature of all three animals is less than 1.15 °C. If the total increase recorded is greater than 2.65 °C then the product has failed. However, if the response observed falls between these two limits

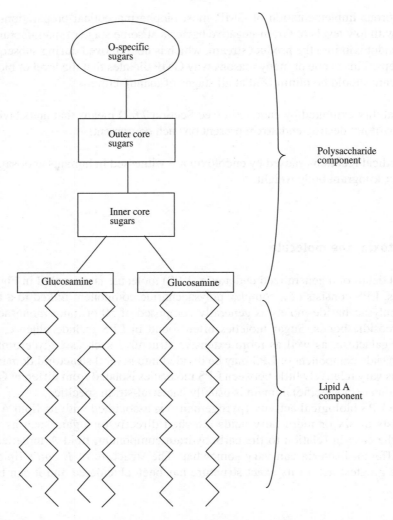

Figure 7.7   Structure of a generalized LPS molecule. LPS constitutes the major structural component of the outer membrane of Gram-negative bacteria. Although LPSs of different Gram-negative organisms differ in their chemical structure, each consists of a complex polysaccharide component, linked to a lipid component. Refer to text for specific details

the result is considered inconclusive, and the test must be repeated using a further batch of animals.

This test is popular because it detects a wide spectrum of pyrogenic substances. However, it is also subject to a number of disadvantages, including:

- it is expensive (there is a requirement for animals, animal facilities and animal technicians);

- excitation/poor handling of the rabbits can affect the results obtained, usually prompting a false positive result;

- subclinical infection/poor overall animal health can also lead to false positive results;

- use of different rabbit colonies/breeds can yield variable results.

Another issue of relevance is that certain biopharmaceuticals (e.g. cytokines such as 1L-1 and TNF; Chapter 9) themselves induce a natural pyrogenic response. This rules out use of the rabbit-based assay for detection of exogenous pyrogens in such products. Such difficulties have led to the increased use of an *in vitro* assay; the *Limulus* ameobocyte lysate (LAL) test. This is based upon endotoxin-stimulated coagulation of amoebocyte lysate obtained from horseshoe crabs. This test is now the most widely used assay for the detection of endotoxins in biopharmaceutical and other pharmaceutical preparations.

Development of the LAL assay was based upon the observation that the presence of Gram-negative bacteria in the vascular system of the American horseshoe crab, *Limulus polyphemus*, resulted in the clotting of its blood. Tests on fractionated blood showed that the factor responsible for coagulation resided within the crab's circulating blood cells, i.e. the amoebocytes. Further research revealed that the bacterial agent responsible of initiation of clot formation was endotoxin.

The endotoxin molecule activates a coagulation cascade quite similar in design to the mammalian blood coagulation cascade (Figure 7.8). Activation of the cascade also requires the presence of divalent cations such as calcium or magnesium. The final steps of this pathway entail the proteolytic cleavage of the polypeptide coagulogen, forming coagulin, and a smaller peptide fragment. Coagulin molecules then interact non-covalently, forming a 'clot' or 'gel'.

The LAL-based assay for endotoxin became commercially available in the 1970s. The LAL reagent is prepared by extraction of blood from the horseshoe crab, followed by isolation of its amoebocytes by centrifugation. After a washing step, the amoebocytes are lysed and the lysate dispensed into pyrogen-free vials. The assay is normally performed by making a series of 1:2 dilutions of the test sample using (pyrogen-free) WFI (and pyrogen-free test tubes; see later). A reference standard endotoxin preparation is treated similarly. LAL reagent is added to all tubes, incubated for 1 h, and these tubes are then inverted to test for gel (i.e. clot) formation, which would indicate presence of endotoxin.

More recently, a colorimetric-based LAL procedure has been devised. This entails addition to the LAL reagent of a short peptide, susceptible to hydrolysis by the LAL clotting enzyme. This synthetic peptide contains a chromogenic tag (usually *para*-nitroaniline, pNA) which is released free into solution by the clotting enzyme. This allows spectrophotometric analysis of the test sample, facilitating more accurate end-point determination.

The LAL system displays several advantages when compared with the rabbit test, most notably:

- sensitivity – endotoxin levels as low as a few picograms per millilitre of sample assayed will be detected;

- cost – the assay is far less expensive than the rabbit assay;

- speed – depending upon the format used, the LAL assay may be conducted within 15–60 min.

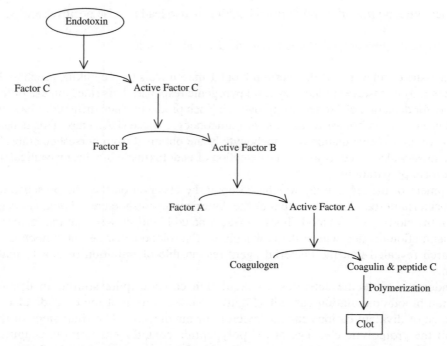

Figure 7.8 Activation of clot formation by endotoxin. The presence of endotoxin causes stepwise, sequential activation of various clotting factors present naturally within the amoebocytes of the American horseshoe crab. The net result is the generation of the polypeptide fragment coagulin, which polymerizes, thus forming a gel or clot

Its major disadvantage is its selectivity: it only detects endotoxin-based pyrogens. In practice, however, endotoxin represents the pyrogen that is by far the most likely to be present in pharmaceutical products. The LAL method is used extensively within the industry. It is used not only to detect endotoxin in finished parenteral preparations, but also in WFI and in biological fluids, such as serum or cerebrospinal fluid.

Before the LAL assay is routinely used to detect/quantify endotoxin in any product, its effective functioning in the presence of that product must be demonstrated by validation studies. Such studies are required to prove that the product (or, more likely, excipients present in the product) do not interfere with the rate/extent of clot formation (i.e. are neither inhibitors nor activators of the LAL-based enzymes). LAL enzyme inhibition could facilitate false-negative results upon sample assay. Validation studies entail, for example, observing the effect of spiking endotoxin-negative product with know quantities of endotoxin, or spiking endotoxin with varying quantities of product, before assay with the LAL reagents.

All ancillary reagents used in the LAL assay system (e.g. WFI, test tubes, pipette tips for liquid transfer, etc.) must obviously be endotoxin free. Such items can be rendered endotoxin free by heat. Its heat-stable nature, however, renders very vigorous heating necessary in order to destroy contaminant endotoxin. A single autoclave cycle is insufficient, with total destruction requiring three consecutive autoclave cycles. Dry heat may also be used (180 °C for 3 h or 240 °C for 1 h).

GMP requires that, where practicable, process equipment coming into direct contact with the biopharmaceutical product stream should be rendered endotoxin free (depyrogenated) before use. Autoclaving, steam or dry heat can effectively be used on many process vessels, pipework, etc., which are usually manufactured from stainless steel or other heat-resistant material. Such an approach is not routinely practicable in the case of some items of process equipment, such as chromatographic systems. Fortunately, endotoxin is sensitive to strongly alkaline conditions; thus, routine cleaning in place of chromatographic systems using 1 mol l^{-1} NaOH represents an effective depyrogenation step. Gentler approaches, such as exhaustive rinsing with WFI (until an LAL test shows the eluate to be endotoxin free), can also be surprisingly effective.

It is generally unnecessary to introduce specific measures aimed at endotoxin removal from the product during downstream processing. Endotoxin present in the earlier stages of production is often effectively removed from the product during chromatographic fractionation. The endotoxin molecule's highly negative charge often facilitates its effective removal from the product stream by ion-exchange chromatography. Gel-filtration chromatography also serves to remove endotoxin from the product. Although individual LPS molecules exhibit an average molecular mass of less than 20 kDa, these molecules aggregate in aqueous environments and generate supramolecular structures of molecular mass 100–1000 kDa.

The molecular mass of most biopharmaceuticals is considerably less than 100 kDa (Table 7.4). The proteins would thus elute from gel-filtration columns much later than contaminating endotoxin aggregates. Should the biopharmaceutical exhibit a molecular mass approaching or exceeding 100 kDa, then effective separation can still be achieved by inclusion of a chelating agent such as EDTA in the running buffer. This promotes depolymerization of the endotoxin aggregates into monomeric (20 kDa) form.

Additional techniques capable of separating biomolecules on the basis of molecular mass (e.g. ultrafiltration) may also be used to remove endotoxin from the product stream.

7.6.3 DNA

The clinical significance of DNA-based contaminants in biopharmaceutical products remains unclear. The concerns relating to the presence of DNA in modern biopharmaceuticals focus

Table 7.4 The molecular mass of some polypeptide biopharmaceuticals. Many are glycosylated, thereby exhibiting a range of molecular masses due to differential glycosylation

Protein	Molecular mass (kDa)	Protein	Molecular mass (kDa)	Protein	Molecular mass (kDa)
IFN-α	20–27	TNF-α	52[a]	EGF	6
IFN-β	20	GM-CSF	22	NGF	26
IFN-γ	20–25	G-CSF	21	Insulin	5.7
IL-2	15–20	EPO	36	hGH	22
IL-1	17.5	TPO	60	FSH	34
IL-12	30–35	IGF-1	7.6	LH	28.5

[a]Biologically active, trimeric form.
TPO: thrombopoietin; EGF: epidermal growth factor; NGF: nerve growth factor; LH: luteinizing hormone.

primarily upon the presence of active oncogenes in the genome of several producer cell types (e.g. monoclonal antibody production in hybridoma cell lines). Parenteral administration of DNA contaminants containing active oncogenes to patients is considered undesirable. The concern is that uptake and expression of such DNA in human cells could occur. There is some evidence to suggest that naked DNA can be assimilated by some cells at least, under certain conditions (Chapter 14). Guidelines to date state that an acceptable level of residual DNA in recombinant products is of the order of 10 pg per therapeutic dose.

DNA hybridization studies (e.g. the 'dot blot' assay) utilizing radiolabelled DNA probes allows detection of DNA contaminants in the product, to levels in the nanogram range. The process begins with isolation of the contaminating DNA from the product. This can be achieved, for example, by phenol and chloroform extraction and ethanol precipitation. The isolated DNA is then applied as a spot (i.e. a 'dot') onto nitrocellulose filter paper, with subsequent baking of the filter at 80 °C under vacuum. This promotes (a) DNA denaturation, yielding single strands, and (b) binding of the DNA to the filter.

A sample of total DNA derived from the cells in which the product is produced is then radiolabelled with ^{32}P using the process of nick translation. It is heated to 90 °C (promotes denaturation, forming single strands) and incubated with the baked filter for several hours at 40 °C. Lowering the temperature allows reannealing of single strands via complementary base-pairing to occur. Labelled DNA will reanneal with any complementary DNA strands immobilized on the filter. After the filter is washed (to remove non-specifically bound radiolabelled probe) it is subjected to autoradiography, which allows detection of any bound probe.

Quantification of the DNA isolated from the product involves concurrent inclusion in the dot blot assay of a set of spots, containing known quantities of DNA, and being derived from the producer cell. After autoradiography, the intensity of the test spot is compared with the standards.

In many instances there is little need to incorporate specific DNA removal steps during downstream processing. Endogenous nucleases liberated upon cellular homogenization come into direct contact with cellular DNA, resulting in its degradation. Commercial DNase's are sometimes added to crude homogenate to reduce DNA-associated product viscosity (Chapter 6). Most chromatographic steps are also effective in separating DNA from the product stream. Ion-exchange chromatography is particularly effective, as DNA exhibits a large overall negative charge (due to the phosphate constituent of its nucleotide backbone; Chapter 3).

7.6.4 Microbial and viral contaminants

Finished-product biopharmaceuticals, along with other pharmaceuticals intended for parenteral administration, must be sterile (the one exception being live bacterial vaccines). The presence of microorganisms in the final product is unacceptable for a number of reasons:

- Parenteral administration of contaminated product would likely lead to the establishment of a severe infection in the recipient patient.

- Microorganisms may be capable of metabolizing the product itself, thus reducing its potency. This is particularly true of protein-based biopharmaceuticals, as most microbes produce an array of extracellular proteases.

- Microbial-derived substances secreted into the product could adversely affect the recipient's health. Examples include endotoxin secreted from Gram-negative bacteria, or microbial proteins that would stimulate an immune response.

Terminal sterilization by autoclaving guarantees product sterility. Heat sterilization, however, is not a viable option in the case of biopharmaceuticals. Sterilization of biopharmaceuticals by filtration, followed by aseptic filling into a sterile final-product container, inherently carries a greater risk of product contamination. Finished-product sterility testing of such preparations thus represents one of the most critical product tests undertaken by QC. Specific guidelines relating to sterility testing of finished products are given in international pharmacopoeias.

Biopharmaceutical products are also subjected to screening for the presence of viral particles prior to final product release. Although viruses could be introduced, for example, via infected personnel during downstream processing, proper implementation of GMP minimizes such risk. Any viral particles found in the finished product are most likely derived from raw material sources. Examples could include HIV or hepatitis viruses present in blood used in the manufacture of blood products. Such raw materials must be screened before processing for the presence of likely viral contaminants.

A variety of murine (mouse) and other mammalian cell lines have become popular host systems for the production of recombinant human biopharmaceuticals. Moreover, most monoclonal antibodies used for therapeutic purposes are produced by murine-derived hybridoma cells. These cell lines are sensitive to infection by various viral particles. Producer cell lines are screened during product development studies to ensure freedom from a variety of pathogenic advantageous agents, including various species of bacteria, fungi, yeast, mycoplasma, protozoa, parasites, viruses and prions. Suitable microbiological precautions must subsequently be undertaken to prevent producer cell banks from becoming contaminated with such pathogens.

Removal of viruses from the product stream can be achieved in a number of ways. The physicochemical properties of viral particles differ greatly from most proteins, ensuring that effective fractionation is automatically achieved by most chromatographic techniques. Gel-filtration chromatography, for example, effectively separates viral particles from most proteins on the basis of differences in size.

In addition to chromatographic separation, downstream processing steps may be undertaken that are specifically aimed at removal or inactivation of viral particles potentially present in the product stream. Significantly, many are 'blanket' procedures, equally capable of removing known or potentially likely viral contaminants and any uncharacterized/undetected viruses. Filtration through a 0.22 μm filter effectively removes microbial agents from the product stream, but fails to remove most viral types. Repeat filtration through a 0.1 μm filter is more effective in this regard. Alternatively, incorporation of an ultrafiltration step (preferably at the terminal stages of downstream processing) also proves effective.

Incorporation of downstream processing steps known to inactivate a wide variety of viral types provides further assurance that the final product is unlikely to harbour active virus. Heating and irradiation are amongst the two most popular such approaches. Heating the product to between 40 and 60°C for several hours inactivates a broad range of viruses. Many biopharmaceuticals can be heated to such temperatures without being denatured themselves. Such an approach has been used extensively to inactivate blood-borne viruses in blood products. Exposure of product to controlled levels of UV radiation can also be quite effective, while having no adverse effect on the product itself.

7.6.5   Viral assays

A range of assay techniques may be used to detect and quantify viral contaminants in both raw materials and finished-biopharmaceutical products. No generic assay exists that is capable of detecting all viral types potentially present in a given sample. Viral assays currently available will detect only a specific virus, or at best a family of closely related viruses. The strategy adopted, therefore, usually entails screening product for viral particles known to be capable of infecting the biopharmaceutical source material. Such assays will not normally detect newly evolved viral strains, or uncharacterized/unknown viral contaminants. This fact underlines the importance of including at least one step in downstream processing that is likely to inactivate or remove viruses indiscriminately from the product. This acts as a safety net.

Current viral assays fall into one of three categories:

- immunoassays;

- assays based on viral DNA probes;

- bioassays.

Generation of antibodies that can recognize and bind to specific viruses is straightforward. A sample of live or attenuated virus, or a purified component of the viral caspid, can be injected into animals to stimulate polyclonal antibody production (or to facilitate monoclonal antibody production by hybridoma technology). Harvested antibodies are then employed to develop specific immunoassays that can be used to screen test samples routinely for the presence of that specific virus. Immunoassays capable of detecting a wide range of viruses are available commercially. The sensitivity, ease, speed and relative inexpensiveness of these assays render them particularly attractive.

An alternative assay format entails the use of virus-specific DNA probes. These can be used to screen the biopharmaceutical product for the presence of viral DNA. The assay strategy is similar to the dot blot assays used to detect host-cell-derived DNA contaminants, as discussed earlier.

Viral bioassays of various different formats have also been developed. One format entails incubation of the final product with cell lines sensitive to a range of viruses. The cells are subsequently monitored for cytopathic effects or other obvious signs of viral infection.

A range of mouse-, rabbit- or hamster-antibody production tests may also be undertaken. These bioassays entail administration of the product to a test animal. Any viral agents present will elicit production of antiviral antibodies in that animal. Serum samples (withdrawn from the animal approximately 4 weeks after product administration) are screened for the presence of antibodies recognizing a range of viral antigens. This can be achieved by enzyme immunoassay, in which immobilized antigen is used to screen for the virus-specific antibodies. These assay systems are extremely sensitive, as minute quantities of viral antigen will elicit strong antibody production. A single serum sample can also be screened for antibodies specific to a wide range of viral particles. Time and expense factors, however, militate against this particular assay format.

7.6.6   Miscellaneous contaminants

In addition to those already discussed, biopharmaceutical products may harbour other contaminants, some of which may be intentionally added to the product stream during the initial stages of downstream processing. Examples could include buffer components, precipitants (ethanol or other solvents, salts, etc.), proteolytic inhibitors, glycerol, anti-foam agents, etc. In addition to these, other contaminants may enter the product during downstream processing in a less controlled way. Examples could include metal ions leached from product-holding tanks/pipework, or breakdown products leaking from chromatographic media. The final product containers must also be chosen carefully. They must be chemically inert and be of suitable quality to eliminate the possibility of leaching of any substance from the container during product storage. For this reason, high-quality glass vials are often used.

In some instances it may be necessary to demonstrate that all traces of specific contaminants have been removed prior to final product filling. This would be true, for example, of many proteolytic inhibitors added during the initial stages of downstream processing to prevent proteolysis by endogenous proteases. Some such inhibitors may be inherently toxic, and many could (inappropriately) inhibit endogenous proteases of the recipient patient.

Demonstration of absence from the product of breakdown products from chromatographic columns may be necessary in certain instances. This is particularly true with regard to some affinity chromatography columns. Various chemical-coupling methods may be used to attach affinity ligands to the chromatographic support material. Some such procedures entail the use of toxic reagents, which, if not entirely removed after coupling, could leach into the product. In some cases ligands can also subsequently leach from the columns, particularly after sustained usage or overvigorous sanitation procedures. Improvements in the chemical stability of modern chromatographic media, however, have reduced such difficulties, and most manufacturers have carried out extensive validation studies regarding the stability of their product.

Sophisticated analytical methodologies facilitate detection of vanishingly low levels of many contaminants in biopharmaceutical preparations. The possibility exists, however, that uncharacterized contaminants may persist, remaining undetected in the final product. As an additional safety measure, finished products are often subjected to 'abnormal toxicity' or 'general safety' tests. Standardized protocols for such tests are outlined in various international pharmacopoeias. These normally entail parenteral administration of the product to at least five healthy mice. The animals are placed under observation for 48 h and should exhibit no ill effects (other than expected symptoms). The death or illness of one or more animals signals a requirement for further investigation, usually using a larger number of animals. Such toxicity testing represents a safety net, designed to expose any unexpected activities in the product that could compromise the health of the recipient.

7.6.7   Validation studies

Validation can be defined as 'the act of proving that any procedure, process, equipment, material, activity or system leads to the expected results'. Routine and adequate validation studies form a core principle of GMP as applied to (bio)pharmaceutical manufacture, as such studies help assure the overall safety of the finished product (Box 7.2).

All validation procedures must be carefully designed and fully documented in written format (Box 7.2). The results of all validation studies undertaken must also be documented, and retained in the plant files. As part of their routine inspection of manufacturing facilities, regulatory personnel will usually inspect a sample of these records, to ensure conformance to GMP.

Validation studies encompass all aspects of (bio)pharmaceutical manufacture. All new items of equipment must be validated before being routinely used. Initial validation studies should be

---

Box 7.2

Validation studies: a glossary of some important terms

Validation master plan	Document that serves as an overall guide for a facility's validation programme. It identifies all items/procedures, etc., that must be subjected to validation studies, describes the nature of testing in each instance and defines the responsibilities of those engaged in validation activities
Validation protocol	Document describing the specific item to be validated, the specific validation protocol to be carried out and acceptable results, as per acceptance criteria
Prospective validation	Validation undertaken prior to commencement of routine product manufacture
Concurrent validation	Validation undertaken while routine manufacture of product is also taking place
Retrospective validation	Validation carried out by review of historical records
Qualification	How an individual element of an overall validation programme performs. When validation of that specific element is complete, it is 'qualified'. When all elements are (satisfactorily) completed the system is validated
Design qualification	Auditing the design of a facility (or element of a facility, such as a cleanroom) to ensure that it is compliant with the specifications laid down and that it is, therefore, capable of meeting GMP requirements
Installation qualification	Auditing/testing to ensure that specific items of equipment have been correctly installed in accordance with the design specifications laid down
Operational qualification	Auditing/testing process that evaluates the system being tested to make sure that it is fully operational and will perform within operating specifications
Performance qualification	Demonstration that equipment/processes operate satisfactorily and consistently during the manufacture of actual product

comprehensive, with follow-up validation studies being undertaken at appropriate time intervals (e.g. daily, weekly or monthly). It is considered judicious to validate older items of equipment with increased frequency. Such studies can forewarn the manufacturer of impending equipment failure. Some validatory studies are straightforward, e.g. validation of weighing equipment simply entails weighing of standardized weights. Autoclaves may be validated by placing external temperature probes at various points in the autoclave chamber during a routine autoclave run. Validation studies should confirm that all areas within the chamber reach the required temperature for the required time.

Periodic validation of clean room air (HEPA) filters is also an essential part of GMP. After their installation, HEPA filters are subjected to a leak test. Particle counters are also used to validate cleanroom conditions. A particle counter is a vacuum-cleaner-like machine capable of sucking air from its surroundings at constant velocity and passing it through a counting chamber. The number of particles per cubic metre of air tested can easily be determined. Furthermore, passage of the air through a 0.2 μm filter housed in the counter will trap all airborne microorganisms. By placing the filter on the surface of a nutrient-agar-containing Petri dish, trapped microorganisms will grow as colonies, allowing determination of the microbial load per cubic metre of air.

In addition to equipment, many processes/procedures undertaken during pharmaceutical manufacture are also subject to periodic validation studies. Validation of biopharmaceutical aseptic filling procedures is amongst the most critical. The aim is to prove that the aseptic procedures devised are capable of delivering a sterile finished product, as intended.

Aseptic filling validation entails substituting a batch of final product with nutrient broth. The broth is subject to sterile filtration and aseptic processing. After sealing the final product containers, they are incubated at 30–37 °C, which encourages growth of any contaminant microorganisms. (Growth can be easily monitored by subsequently measuring the absorbance at 600 nm.) Absence of growth validates the aseptic procedures developed.

Contaminant-clearance validation studies are of special significance in biopharmaceutical manufacture. As discussed in Section 7.6.4, downstream processing must be capable of removing contaminants such as viruses, DNA and endotoxin from the product steam. Contaminant-clearance validation studies normally entail spiking the raw material (from which the product is to be purified) with a known level of the chosen contaminant and subjecting the contaminated material to the complete downstream processing protocol. This allows determination of the level of clearance of the contaminant achieved after each purification step, and the contaminant reduction factor for the overall process.

Viral clearance studies, for example, are typically undertaken by spiking the raw material with a mixture of at least three different viral species, preferably ones that represent likely product contaminants, and for which straightforward assay systems are available. Loading levels of up to 1×10^{10} viral particles are commonly used. The cumulative viral removal/inactivation observed should render the likelihood of a single viral particle remaining in a single therapeutic dose of product being greater that one in a million.

A similar strategy is adopted when undertaking DNA clearance studies. The starting material is spiked with radiolabelled DNA and then subjected to downstream processing. The level of residual DNA remaining in the product stream after each step can easily be determined by monitoring for radioactivity.

The quantity of DNA used to spike the product should ideally be somewhat in excess of the levels of DNA normally associated with the product prior to its purification. However, spiking of the product with a vast excess of DNA is counterproductive, in that it may render subsequent downstream processing unrepresentative of standard production runs.

For more comprehensive validation studies, the molecular mass profile of the DNA spike should roughly approximate to the molecular mass range of endogenous contaminant DNA in the crude product. Obviously, the true DNA clearance rate attained by downstream processing procedures (e.g. gel filtration) will depend to some extent on the molecular mass characteristics of the contaminant DNA.

Other manufacturing procedures requiring validation include cleaning, decontamination and sanitation (CDS) procedures developed for specific items of equipment/processing areas. Of particular importance is the ability of such procedures to remove bioburden. This may be assessed by monitoring levels of microbial contamination before and after application of CDS protocols to the equipment item in question.

Further reading

Books

Aguilar, M. 2003. *HPLC of Peptides and Proteins*. Humana Press.
Dass, C. 2000. *Principles and Practice of Biological Mass Spectrometry*. Wiley.
Kellner, R. 1999. *Micorcharacterization of Proteins*. Wiley.
Kinter, M. and Sherman, N. 2005. *Protein Sequencing and Identification Using Tandem Mass Spectrometry*. Wiley.
Ramstorp, M. 2000. *Contamination Control and Cleanroom Technology*. Wiley.
Rathore, A. and Sofer, G. (eds). 2005. *Process Validation in Manufacturing of Biopharmaceuticals: Guidelines, Current Practices, and Industrial Case Studies*. Taylor and Francis.
Rosenberg, I. 2004. *Protein Analysis and Purification*. Birkhauser.
Venn, R. 2000. *Principles and Practice of Bioanalysis*. Taylor and Francis.
Whyte, W. 2001. *Cleanroom Technology*. Wiley.
Wild, D. (ed.) 2005. *The Immunoassay Handbook*. Elsevier.

Articles

Dabbah, R. and Grady, L. 1998. Pharmacopoeial harmonization in biotechnology. *Current Opinion in Biotechnology* **9**, 307–311.
Darling, A. 2002. Validation of biopharmaceutical purification processes for virus clearance evaluation. *Molecular Biotechnology* **21**, 57–83.
Ding, J. and Ho, B. 2001. A new era in pyrogen testing. *Trends in Biotechnology* **18**(8), 277–280.
Domon, B. and Aebersold, R. 2006. Review – mass spectrometry and protein analysis. *Science* 312, 212–217.
Geisow, M.J. 1991. Characterizing recombinant proteins. *Bio/Technology* **9**, 921–924.
Geng, D., Shankar, G., Schantz, A., Rajadhyaksha, M., Davis, H., and Wagner, C. 2005. Validation of immunoassays used to assess immunogenicity to therapeutic monoclonal antibodies. *Journal of Pharmaceutical and Biomedical Analysis* **39**, 364–375.
Glennon, B. 1997. Control system validation in multipurpose biopharmaceutical facilities. *Journal of Biotechnology* **59**(1–2), 53–61.
Hu, S. and Dovichi, N. 2002. Capillary electrophoresis for the analysis of biopolymers. *Analytical Chemistry* **74**(12), 2833–2850.

Kakehi, K., Kinoshita, M., and Nakano, M. 2002. Analysis of glycoproteins and the oligosaccharides thereof by high performance capillary electrophoresis – significance in regulatory studies on biopharmaceutical products. *Biomedical Chromatography* **16**(2), 103–115.

Kett, V., McMahon, D., and Ward, K. 2004. Freeze-drying of protein pharmaceuticals – the application of thermal analysis. *Cryoletters* **25**, 389–404.

Larsen, M.R., Trelle, M.B., Thingholm, T.E., and Jensen, O.N. 2006. Analysis of posttranslational modifications of proteins by tandem mass spectrometry. *BioTechniques* **40**, 790–798.

Lin, S. and Hsu, S. 2005. Recent advances in capillary electrophoresis immunoassays. *Analytical Biochemistry* **341**, 1–15.

Mann, M., Hendrickson, R.C., and Pandey, A. 2001. Analysis of proteins and proteomes by mass spectrometry. *Annual Review of Biochemistry* **70**, 437–473.

Meager, A. 2006. Measurement of cytokines by bioassays: theory and applications. *Methods* **38**, 237–252.

Raetz, C. and Whitfield, C. 2002. Lipopolysaccharide endotoxins. *Annual Review of Biochemistry* **71**, 635–700.

Seamon, K. 1998. Specifications for biotechnology-derived protein drugs. *Current Opinion in Biotechnology* **9**, 319–325.

Chew, N. 1993. Validation of biopharmaceutical processes. *Pharmaceutical Technology Europe*, **5**(11), 34–39.

Tatford, O.C., Gomme, P.T., and Bertolini, J. 2004. Analytical techniques for the evaluation of liquid protein therapeutics. *Biotechnology and Applied Biochemistry* **40**, 67–81.

Tumanov, A. and Krestyaninov, P. 2002. Current status and prospects of bioassay. *Journal of Analytical Chemistry* **57**, 372–387.

Wuhrer, M., Koeleman, C.A., Hokke, C.H., and Deelder, A.M. 2005. Protein glycosylation analysis by liquid chromatography–mass spectrometry. *Journal of Chromatography B* **825**, 124–133.

8
The cytokines: The interferon family

8.1 Cytokines

Cytokines are a diverse group of regulatory proteins or glycoproteins whose classification remains somewhat diffuse (Table 8.1). These molecules are normally produced in minute quantities by the body. They act as chemical communicators between various cells, inducing their effect by binding to specific cell surface receptors, thereby triggering various intracellular signal transduction events. Over the next several chapters we consider various cytokines of therapeutic interest, focusing in particular upon those approved for clinical application.

Most cytokines act upon, or are produced by, leukocytes (white blood cells), which constitute the immune and inflammatory systems (Box 8.1). They thus play a central role in regulating both immune and inflammatory function and in related processes such as haematopoiesis (the production of blood cells from haematopoietic stem cells in the adult bone marrow), as well as in wound healing. Indeed, several immunosuppressive and anti-inflammatory drugs are now known to induce their biological effects by regulating production of several cytokines.

The term 'cytokine' was first introduced in the mid 1970s. It was applied to polypeptide growth factors controlling the differentiation and regulation of cells of the immune system. The interferons and interleukins represented the major polypeptide families classified as cytokines at that time. Additional classification terms were also introduced, including lymphokines (cytokines such as IL-2 and IFN-γ, produced by lymphocytes) and monokines (cytokines such as TNF-α, produced by monocytes). However, classification on the basis of producing cell types also proved inappropriate, as most cytokines are produced by a range of cell types (e.g. both lymphocytes and monocytes produce IFN-α).

Initial classification of some cytokines was also undertaken on the basis of the specific biological activity by which the cytokine was first discovered (e.g. TNF exhibited cytotoxic effects on some cancer cell lines; CSFs promoted the growth *in vitro* of various leukocytes in clumps or colonies). This, too, proved an unsatisfactory classification mechanism, as it was subsequently shown that most cytokines display a range of biological activities (e.g. the major biological function of TNF is believed to be as a regulator of both the immune and inflammatory response). More recently, primary sequence analysis of cytokines coupled to determination of secondary and tertiary structure reveal that most cytokines can be grouped into one of six families (Table 8.2).

Pharmaceutical biotechnology: concepts and applications   Gary Walsh
© 2007 John Wiley & Sons, Ltd   ISBN 978 0 470 01244 4 (HB)   978 0 470 01245 1 (PB)

Table 8.1   The major proteins/protein families that constitute the cytokine group of regulatory molecules[a]

The interleukins (IL-1 to IL-33)
The interferons (IFN-α, -β, -γ, -τ, -ω)
CSFs (G-CSF, M-CSF, GM-CSF)
TNFs (TNF-α, -β)
The neurotrophins (NGF, brain-derived neurotrophic factor (BDNF), NT-3, NT-4/5)
Ciliary neurotrophic factor (CNTF)
Glial cell-derived neurotrophic factor (GDNF)
EGF
EPO
Fibroblast growth factor (FGF)
Leukaemia inhibitory factor (LIF)
Macrophage inflammatory proteins (MIP-1α, -1β, -2)
PDGF
Transforming growth factors (TGF-α, -β)
TPO

[a]M-CSF: macrophage-colony-stimulating factor; NT: neurotrophin.

Box 8.1

Leukocytes, their range and function

Leukocytes (white blood cells) encompass all blood cells that contain a nucleus, and these cells basically constitute the cells of the immune system. They thus function to protect the body by inactivating and destroying foreign agents. Certain leukocytes are also capable of recognizing and destroying altered body cells, such as cancer cells. Most are not confined exclusively to blood, but can circulate/exchange between blood, lymph and body tissues. This renders them more functionally effective by facilitating migration and congregation at a site of infection.

Leukocytes have been subclassified into three families: mononuclear phagocytes, lymphocytes and granulocytes. These can be differentiated from each other on the basis of their interaction with a dye known as Romanowsky stain (Figure 8.B1).

Mononuclear phagocytes consist of monocytes and macrophages, and execute their defence function primarily by phagocytosis. Like all leukocytes, they are ultimately derived from bone marrow stem cells. Some such stem cells differentiate into monocytes, which enter the bloodstream from the bone marrow. From there, they migrate into most tissues in the body, where they settle and differentiate (mature) to become macrophages (sometimes called histocytes). Macrophages are found in all organs and connective tissue. They are given different names, depending upon in which organ they are located (hepatic macrophages are called Kupffer cells, central nervous system macrophages are called microglia, and lung macrophages are termed alveolar macrophages). All macrophages are effective scavenger cells, engulfing and destroying (by phagocytosis) any foreign substances they encounter. They also play an important role in other aspects of immunity by producing cytokines, and acting as antigen-presenting cells.

Lymphocytes are responsible for the specificity of the immune response. They are the only immune cells that recognize and respond to specific antigens, due to the presence on

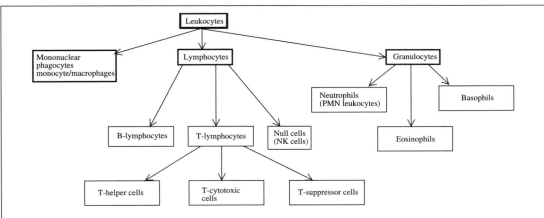

Figure 8.B1 The range of white blood cell types

their surface of high-affinity receptors. In addition to blood, lymphocytes are present in high numbers in the spleen and thymus. They may be subcategorized into antibody-producing B-lymphocytes, T-lymphocytes (which are involved in cell-mediated immunity) and null cells.

T-lymphocytes may be subcategorized on a functional basis into T-helper, T-cytoxic and T-suppressor cells. T-helper cells can produce various cytokines which can stimulate and regulate the immune response. T-cytotoxic cells can induce the lysis of cells exhibiting foreign antigen on their surface. As such, their major target cells are body cells infected by viruses or other intracellular pathogens (e.g. some protozoa). T-suppressor cells function to dampen or suppress an activated immune response, thus functioning as an important 'off' switch.

Most T-helper cells express a membrane protein termed CD4 on their surface. Most T-cytotoxic and T-suppressor cells produce a different cell surface protein, termed CD8. Monoclonal antibodies specifically recognizing CD4 or CD8 proteins can thus be used to differentiate between some T cell types.

Null cells are also known as 'large granular lymphocytes', but are best known as 'natural killer' (NK) cells. These represent a third lymphocyte subgroup. They are capable of directly lysing cancer cells and virally infected cells.

The third leukocyte cell type is termed granulocytes, due to the presence of large granules in their cytoplasm. Granulocytes, many of which can be activated by cytokines, play a direct role in immunity, and also in inflammation. Granulocytes can be subdivided into three cell types of which neutrophils (also known as polymorphonuclear leukocytes; PMN leukocytes) are the most abundant. Attracted to the site of infection, they mediate acute inflammation and phagocytose opsonized antigen efficiently due to the presence of an IgG F_c receptor on their surface. Eosinophils display a cell surface IgE receptor and, thus, seem to specialize in destroying foreign substances that specifically elicit an IgE response (e.g. parasitic worms). These cells also play a direct role in allergic reactions. Basophils also express IgE receptors. Binding of antigen–IgE complex prompts these cells to secrete their granule contents, which mediate hypersensitivity reactions.

Table 8.2   Cytokines, as grouped on a structural basis

Cytokine family	Members
'β-Trefoil' cytokines	FGFs
	IL-1
Chemokines	IL-8
	MIPs
'Cysteine knot' cytokines	NGF
	TGFs
	PDGF
EGF family	EGF
	TGF-α
Haematopoietins	IL-2–IL-7, IL-9, IL-13
	G-CSF
	GM-CSF
	LIF
	EPO
	CNTF
TNF family	TNF-α and -β

As a consequence of the various approaches adopted in naming and classifying cytokines, it is hardly surprising to note that many are known by more that one name. IL-1, for example, is also known as lymphocyte activating factor (LAF), endogenous pyrogen, leukocyte endogenous mediator, catabolin and mononuclear cell factor. This has led to even further confusion in this field.

During the 1980s, rapid developments in the areas of recombinant DNA technology and monoclonal antibody technology contributed to a greater depth of understanding of cytokine biology:

- Genetic engineering allowed production of large quantities of most cytokines. These could be used for structural and functional studies of the cytokine itself, and its receptor.

- Analysis of cytokine genes established the exact evolutionary relationships between these molecules.

- Detection of cytokine mRNA and cytokine receptor mRNA allowed identification of the full range of sources and target cells of individual cytokines.

- Hybridoma technology (Chapter 13) facilitated development of immunoassays capable of detecting and quantifying cytokines.

- Inhibition of cytokine activity *in vivo* by administration of monoclonal antibodies (and, more recently, by gene knockout studies) continues to elucidate the physiological and pathophysiological effect of various cytokines.

The cytokine family continues to grow, and often a decision to include a regulatory protein in this category is not a straightforward one. The following generalizations may be made with regard to most cytokines:

- They are very potent regulatory molecules, inducing their characteristic effects at nanomolar to picomolar concentrations.

- Most cytokines are produced by a variety of cell types, which may be leukocytes or non-leukocytes, e.g. IL-1 is produced by a wide range of cells, including leukocytes (such as monocytes, macrophages, NK cells, B- and T-lymphocytes) and non-leukocytes (such as smooth muscle cells, vascular endothelial cells (a single layer of cells lining blood vessels), fibroblasts (cells found in connective tissue that produce ground substance and collagen fibre precursors), astrocytes (non-neural cells found in the central nervous system) and chondrocytes (cells embedded in the matrix of cartilage)).

- Many cell types can produce more than one cytokine. Lymphocytes, for example, produce a wide range of interleukins, CSFs, TNF, IFN-αs and IFN-γ. Fibroblasts can produce IL-1, -6, -8, and -11, CSFs, IFN-β and TNF.

- Many cytokines play a regulatory role in processes other that immunity and inflammation. Neurotrophic factors, such as NGF and BDNF, regulate growth, development and maintenance of various neural populations in the central and peripheral nervous system. EPO stimulates the production of red blood cells from erythroid precursors in the bone marrow.

- Most cytokines are pleiotropic, i.e. can affect a variety of cell types. Moreover, the effect that a cytokine has on one cell type may be the same or different to its effect on a different cell type. IL-1, for example, can induce fever, hypotension and an acute phase response. G-CSF is a growth factor for neutrophils, but it is also involved in stimulating migration of endothelial cells and growth of haematopoietic cells. IFN-γ stimulates activation and growth of T- and B-lymphocytes, macrophages, NK cells, fibroblasts and endothelial cells. It also displays weak anti-proliferative activity with some cell types.

- Most cytokines are inducible, and are secreted by their producer cell, e.g. induction of IL-2 synthesis and release by T-lymphocytes is promoted by binding of IL-1 to its receptor on the surface of T cells. IFN-αs are induced by viral intrusion into the body. In general, potent cytokine inducers include infectious agents, tissue injury and toxic stimuli. The bodies main defence against such agents, of course, lies with the immune system and inflammation. Upon binding to target cells, cytokines can often induce the target cell to synthesize and release a variety of additional cytokines.

- In contrast, some cytokines (e.g. some CSFs and EPO) appear to be expressed constitutively. In yet other instances cytokines such as PDGF and TGF-β are stored in cytoplasmic granules and can be rapidly released in response to appropriate stimuli. Other cytokines (mainly ones with growth factor activity, e.g. TGF-β, FGF and IL-1) are found bound to the extracellular matrix in connective tissue, bone and skin. These are released, bringing about a biological response upon tissue injury.

- Many cytokines exhibit redundancy, i.e. two or more cytokines can induce a similar biological effect. Examples include TNF-α and -β, both of which bind to the same receptor and induce very similar, if not identical, biological responses. This is also true of the IFN-α family proteins and IFN-β, all of which bind the same receptor.

Although all cytokines are polypeptide regulatory factors, not all polypeptide regulatory factors are classified as cytokines. Classical polypeptide hormones, such as insulin, FSH and GH are not considered members of the cytokine family. The distinguishing features between these two groups is ill defined, and in many ways artificial. Originally, one obvious distinguishing feature was that hormones

Table 8.3 The cytokine receptor superfamilies. Refer to text for further details and to Table 8.1 for explanation of cytokine abbreviations

Receptor superfamily name	Alternative name	Main members
The haematopoietic receptor superfamily	The cytokine receptor superfamily	Receptors for: IL-2–IL-7, IL-9, IL-12, G-CSF, GM-CSF, EPO, LIF, CNTF, GH
The interferon receptor superfamily	Cytokine receptor type II family	Receptors for: IFN-α, -β, -γ, IL-10
The immunoglobulin superfamily	–	Receptors for: IL-1, IL-6, FGF, PDGF, M-CSF
PTK receptor superfamily	–	Receptors for: EGF, insulin, IGF-1
The nerve growth factor superfamily	–	Receptors for: NGF, TGF
The seven transmembrane spanning receptor superfamily	–	Receptors for various chemokines, including IL-8 and MIP
The complement control protein superfamily	–	IL-2 receptor (α-chain)

were produced by a multicellular, anatomically distinguishable gland (e.g. the pancreas, the pituitary, etc.) and functioned in a true endocrine fashion, affecting cells far distant from the site of their production. Many initially described cytokines are produced by white blood cells (which do not constitute a gland in the traditional sense of the word), and often function in an autocrine/paracrine manner.

However, even such distinguishing characteristics have become blurred. EPO, for example, is produced in the kidney and liver and acts in an endocrine manner, promoting production of red blood cells in the bone marrow. EPO could thus also be considered to be a true hormone.

8.1.1 Cytokine receptors

Recombinant DNA technology has also facilitated detailed study of cytokine receptors. Based upon amino acid sequence homology, receptors are usually classified as belonging to one of six known superfamilies (Table 8.3). Individual members of any one superfamily characteristically display 20–50 per cent homology. Conserved amino acids normally occur in discrete bands or clusters, which usually correspond to a discrete domain in the receptor. Most receptors exhibit multiple domains. In some cases a single receptor may contain domains characteristic of two or more superfamilies. For example, the IL-6 receptor contains domains characteristic of both the haematopoietic and immunoglobulin superfamilies, making it a member of both.

Some cytokine receptors are composed of a single transmembrane polypeptide (e.g. receptors for IL-8, -9 and -10). Many contain two polypeptide components (including the IL-3, -4, and -5 receptors), and a few contain three or more polypeptide components (e.g. the IL-2 receptor contains three polypeptide chains). In some instances a single cytokine may be capable of initiating signal transduction by binding two or more distinct receptors (e.g. IL-1 has two distinct receptors (types I and II), both of which are transmembrane glycoproteins).

In many cases where a receptor consists of multiple polypeptides, one of those polypeptides (which will be unique to that receptor) will interact directly with the ligand. The additional polypeptide(s), responsible for initiation of signal transduction, may be shared by a number of receptors (Figure 8.1).

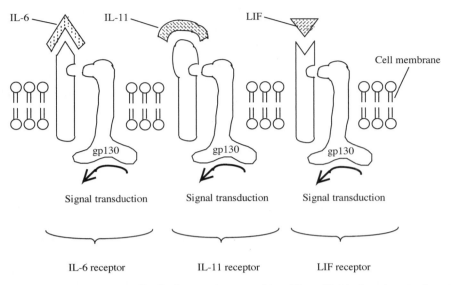

Figure 8.1 Cytokine receptors usually display a unique cytokine ('ligand')-binding domain, but they share additional receptor components that are normally responsible for signal transduction. This explains the molecular basis of pleiotropy. IL-6, IL-11 and LIF receptors, for example, are all composed of a distinct ligand-specific binding domain and a separate subunit (gp 130). gp 130 is responsible for initiating signal transduction and is identical in all three receptors. This is depicted schematically above

Some cytokine receptors can directly initiate signal transduction upon binding of ligand. In other cases additional elements are involved. For many receptors, the exact intracellular events triggered upon ligand binding remain to be elucidated. However, the molecular details of signal transduction pathways for others (e.g. the interferons) are now understood

8.1.2 Cytokines as biopharmaceuticals

Cytokines, in many ways, constitute the single most important group of biopharmaceutical substances. As coordinators of the immune and inflammatory response, manipulation of cytokine activity can have a major influence on the body's response to a variety of medical conditions. Administration of certain cytokines can enhance the immune response against a wide range of infectious agents and cancer cells. EPO has proven effective in stimulating red blood cell production in anaemic persons. Growth factors have obvious potential in promoting wound healing. And neurotrophic factors display some clinical promise in the abatement of certain neurodegenerative diseases.

A better understanding of the molecular principles underlining cytokine biology may also provide new knowledge-based strategies aimed at defeating certain viral pathogens. These pathogens appear to establish an infection successfully, at least in part, by producing specific proteins that thwart the normal cytokine-based immunological response. The cowpox virus, for example, produces an IL-1-binding protein, and the shope fibroma virus produces a TNF-binding protein. The Epstein–Barr virus, on the other hand, produces a protein homologous to IL-10.

A variety of medical conditions are now believed to be caused or exasperated by overproduction of certain cytokines in the body. A variety of pro-inflammatory cytokines, including IL-6, -8 and TNF, have been implicated in the pathogenesis of both septic shock and rheumatoid arthritis. Inhibiting the biological activity of such cytokines may provide effective therapies for such conditions. This may be achieved by administration of monoclonal antibodies raised against the target cytokine, or administration of soluble forms of its receptor that will compete with cell surface receptors for cytokine binding.

Some cytokines have already gained approval for medical use. Many more are currently undergoing clinical or preclinical trials. Over the next few chapters the biology and potential medical applications of these cytokines will be discussed in detail. The remainder of this chapter concerns itself with the prototypic cytokine family, namely the interferons.

8.2   The interferons

Interferons were the first family of cytokines to be discovered. In 1957, researchers observed that susceptible animal cells, if they were exposed to a colonizing virus, immediately became resistant to attack by other viruses. This resistance was induced by a substance secreted by virally infected cells which was named interferon. Subsequently, it has been shown that most species actually produce a whole range of interferons. Humans produce at least three distinct classes, IFN-α, IFN-β and IFN-γ (Table 8.4). These interferons are produced by a variety of different cell types and exhibit a wide range of biological effects, including:

- induction of cellular resistance to viral attack;

- regulation of most aspects of immune function;

- regulation of growth and differentiation of many cell types;

- sustenance of early phases of pregnancy in some animal species.

No one interferon will display all of these biological activities. Effects are initiated by the binding of the interferon to its specific cell surface receptor present in the plasma membrane of sensitive cells. IFN-α and -β display significant amino acid sequence homology (30 per cent), bind to the same receptor, induce similar biological activities and are acid stable. For these reasons, IFN-α and IFN-β are sometimes collectively referred to as type I interferons, or acid-stable interferons.

Table 8.4   Human interferons and the cells that produce them

Interferon family	Additional name	No. distinct interferons in family	Producing cells
IFN-α	Leukocyte interferon	>15	Lymphocytes
	B cell interferon		Monocytes
	Lymphoblast interferon		Macrophages
IFN-β	Fibroblast interferon	1	Fibroblasts
	IFN-β-1[a]		Some epithelial cells
IFN-γ	Immune interferon	1	T-lymphocytes
	T cell interferon		NK cells

[a]Originally a second cytokine was called IFN-β-2, but this was subsequently found to be actually IL-6.

IFN-γ is evolutionarily distinct from the other interferons; it binds to a separate receptor and induces a different range of biological activities. It is thus often referred to as type II interferon.

Owing to their biological activities most interferons are of actual or likely use in the treatment of many medical conditions, including:

- augmentation of the immune response against infectious agents (viral, bacterial, protozoal, etc.);

- treatment of some autoimmune conditions;

- treatment of certain cancer types.

Interferons may be detected and quantified using various bioassays or by immunoassay systems. Although such assays were available, subsequent purification, characterization and medical utilization of interferons initially proved difficult due to the tiny quantities in which these regulatory proteins are produced naturally by the body. By the early 1970s, advances in animal cell culture technology, along with the identification of cells producing increased concentrations of interferons, made some (mostly IFN-αs) available in reasonable quantities. It was not until the advent of genetic engineering, however, that all interferons could be produced in quantities sufficient to satisfy demand for both pure and applied purposes.

8.2.1 The biochemistry of interferon-α

For many years after its initial discovery it was assumed that IFN-α represented a single gene product. It is now known that virtually all species produced multiple, closely related IFN-αs. Purification studies from the 1970s on using high-resolution chromatographic techniques (mainly ion-exchange and gel-filtration chromatographies, immunoaffinity chromatography and isoelectric focusing) first elucidated this fact.

In humans, at least 24 related genes or pseudo-genes exist that code for the production of at least 16 distinct mature IFN-αs. These can be assigned to one of two families, i.e. type I and II. Humans are capable of synthesizing at least 15 type I IFN-αs and a single type II IFN-α.

Most mature type I IFN-αs contain 166 amino acids (one contains 165), whereas type II IFN-α is composed of 172 amino acids. All are initially synthesized containing an additional 23-amino-acid signal peptide. Based upon amino acid sequence data, the predicted molecular mass of all IFN-αs is in the 19–20 kDa range. SDS-PAGE analysis, however, reveals observed molecular masses up to 27 kDa. Isoelectric points determined by isoelectric focusing range between 5 and 6.5. The heterogeneity observed is most likely due to O-linked glycolylation, although several IFN-αs are not glycosylated. Some IFN-αs also exhibit natural heterogeneity due to limited proteolytic processing at the carboxyl terminus.

Individual IFN-αs generally exhibit in excess of 70 per cent amino acid homology with each other. They are rich in leucine and glutamic acid, and display conserved cysteines (usually at positions 1, 29, 99 and 139). These generally form two disulfide bonds in the mature molecule. Their tertiary structures are similar, containing several α helical segments, but appear devoid of β sheets.

Individual members of the IFN-α family each have an identifying name. In most cases the names were assigned by placing a letter after the 'α' (i.e. IFN-αA, IFN-αB, etc.). However, some exceptions exist which contain a number or a number and letter, e.g. IFN-α7, IFN-α8, IFN-α2B. Just to ensure total confusion, several are known by two different names, e.g. IFN-α7 is also known as IFN-αJ1.

8.2.2   Interferon-β

IFN-β, normally produced by fibroblasts, was the first interferon to be purified. Humans synthesize a single IFN-β molecule, containing 166 amino acid residues, that exhibits 30 per cent sequence homology to IFN-αs. The mature molecule exhibits a single disulfide bond and is a glycoprotein of molecular mass in excess of 20 kDa. The carbohydrate side chain is attached via an N-linked glycosidic bond to asparagine residue 80. The carbohydrate moiety facilitates partial purification by lectin affinity chromatography. Immunoaffinity chromatography using monoclonal antibodies raised against IFN-β, as well as dye affinity chromatography, has also been employed in its purification. IFN-β's tertiary structure is dominated by five α helical segments, three of which lay parallel to each other, with the remaining two being antiparallel to these.

8.2.3   Interferon-γ

IFN-γ is usually referred to as 'immune' interferon. It was initially purified from human peripheral blood lymphocytes. This interferon is produced predominantly by lymphocytes. Its synthesis by these cells is reduced when they come in contact with presented antigen. Additional cytokines, including IL-2 and -12, can also induce IFN-γ production under certain circumstances. A single IFN-γ gene exists, located on human chromosome number 12. It displays little evolutionary homology to type I interferon genes. The mature polypeptide contains 143 amino acids with a predicted molecular mass of 17 kDa. SDS-PAGE analysis reveals three bands of molecular mass 16–17, 20 and 25 kDa, arising because of differential glycosylation. The 20 kDa band is glycosylated at asparagine 97, whereas the 25 kDa species is glycosylated at asparagines 25 and 97. In addition, mature IFN-γ exhibits natural heterogeneity at its carboxyl terminus due to proteolytic-processing (five truncated forms have been identified). The molecule's tertiary structure consists of six α-helical segments linked by non-helical regions.

Gel-filtration analysis reveals bands of molecular mass 40–70 kDa. These represent dimers (and some multimers) of the IFN-γ polypeptide. Its biologically active form appears to be a homodimer in which the two subunits are associated in an antiparallel manner.

8.2.4   Interferon signal transduction

All interferons mediate their biological effect by binding to high-affinity cell surface receptors. Binding is followed by initiation of signal transduction, culminating in an altered level of expression of several interferon-responsive genes. Although both positive regulation and negative regulation exist, positive regulation (up-regulation) of gene expression has been studied in greatest detail thus far.

All interferon-stimulated genes are characterized by the presence of an associated interferon-stimulated response element (ISRE). Signal transduction culminates in the binding of specific regulatory factors to the ISRE, which stimulates RNA polymerase II-mediated transcription of the interferon-sensitive genes. The induced gene products then mediate the antiviral, immunomodulatory and other effects characteristically induced by interferons.

Table 8.5 Cell types which display an IFN-γ receptor on their surface

Haematopoietic cells	T-lymphocytes
	B-lymphocytes
	Macrophages
	Polymorphonuclear leukocytes
	Platelets
Somatic cells	Endothelial cells
	Epithelial cells
	Various tumour cells

8.2.5 The interferon receptors

The availability of large quantities of purified interferons facilitates detailed study of the interferon receptors. Binding studies using radiolabelled interferons can be undertaken, and photoaffinity cross-linking of labelled interferon to its receptor facilitates subsequent purification of the ligand–receptor complex. Recombinant DNA technology also facilitated direct cloning of interferon receptors. Binding studies using radiolabelled type I interferons reveals that they all compete for binding to the same receptor, whereas purified IFN-γ does not compete. Partial purification of the IFN-α receptor was undertaken by a number of means. One approach entailed covalent attachment of radiolabelled IFN-α to the receptor using bifunctional cross-linking agents, followed by purification of the radioactive complex. An alternative approach utilized an immobilized IFN-α ligand for affinity purification. The receptor has also been cloned, and the gene is housed on human chromosome number 21.

Studies have actually revealed two type I interferon receptor polypeptides. Sequence data from cloning studies place both in the class II cytokine receptor family. Both are transmembrane N-linked glycoproteins. Studies using isolated forms of each show that one polypeptide (called the α/β receptor) is capable of binding all type I interferons. The other one (the αβ receptor) is specific for IFN-α-B (a specific member of the IFN-α family). Both receptors are present on most cell types.

The IFN-γ receptor (the type II receptor) displays a more limited cellular distribution than that of the type I receptors (Table 8.5). This receptor is a transmembrane glycoprotein of molecular mass 50 kDa, which appears to function as a homodimer. The extracellular IFN-γ binding region consists of approximately 200 amino acid residues folded into two homologous domains. Initiation of signal transduction also requires the presence of a second transmembrane glycoprotein known as AF-1 (accessory factor 1), which associates with the extracellular region of the receptor.

The intracellular events triggered upon binding of type I or II interferons to their respective receptors are quite similar. The sequence of events, known as the JAK–STAT pathway, has been elucidated over the last few years. It has quickly become apparent that this pathway plays a prominent role in mediating signal transduction, not only for interferon, but also of many cytokines.

8.2.6 The JAK–STAT pathway

Cytokine receptors can be divided into two groups: those whose intracellular domains exhibit intrinsic protein tyrosine kinase (PTK) activity and those whose intracellular domains are devoid of such activity. Many of the latter group of receptors, however, activate intracellular soluble PTKs upon ligand binding.

Janus kinases (JAKs) represent a recently discovered family of PTKs that seem to play a central role in mediating signal transduction of many cytokines, and probably many non-cytokine regulatory molecules. These enzymes harbour two potential active sites and were thus named after Janus, a Roman god with two faces. It is likely that only one of those 'active' sites is functional. Four members of the JAK family have been best characterized to date: JAK1, JAK2, JAK3 and TYK2. They all exhibit molecular masses in the region of 130 kDa and approximately 40 per cent amino acid sequence homology. They appear to be associated with the cytoplasmic domain of many cytokine receptors, but remain catalytically inactive until binding of the cytokine to the receptor (Figure 8.2).

In most instances ligand binding appears to promote receptor dimerization, bringing their associated JAKs into close proximity (Figure 8.2). The JAKs then phosphorylate (and hence activate) each other (transphosphorylation). The activated kinases subsequently phosphorylate specific tyrosine residues on the receptor itself. This promotes direct association between one or more members of a family of cytoplasmic proteins (signal transducers and activators of transcription (STATs)) and the receptor. Once docked at the receptor surface, the STATs are in turn phosphorylated (and hence activated) by the JAKs (Figure 8.2). As described below, activated STATs then translocate to the nucleus, and directly regulate expression of interferon and other cytokine-sensitive genes.

As the term STAT suggests, these proteins (a) form an integral part of cytoplasmic signal transduction initiated by certain regulatory molecules and (b) activate transcription of specific genes in the nucleus. Thus far, at least six distinct mammalian STATs (STAT1–STAT6) have been identified which range in size from 84 to 113 kDa. Some may be differentially spliced, increasing the number of functional proteins in the family, e.g. STAT1 exists in two forms; STAT1α contains 750 amino acid residues and exhibits a molecular mass of 91 kDa (it is sometimes called STAT91). STAT1β is a splicing variant of the same gene product. It lacks the last 38 amino acid residues at the C-terminal of the protein and exhibits a molecular mass of 84 kDa (hence, it is sometimes called STAT84). Similar variants have been identified for STAT3 and STAT5. STATs have also been located in non-mammalian species such as the fruit fly. All STATs exhibit significant sequence homology and are composed of a number of functional domains (Figure 8.3). The SH2 domain functions to bind phosphotyrosine, thus docking the STAT at the activated receptor surface. As detailed below, this domain is also required for STAT interaction with JAKs (which then phosphorylate the STAT) and to promote subsequent dimerization of the STATs. An essential tyrosine is located towards the STAT C-terminus (around residue 700), which in turn is then phosphorylated by PTK.

STATs are differentially distributed in various cells/tissues. STATs 1, 2 and 3 seem to be present in most cell types, all be it at varying concentrations. Tissue distribution of STATs 4 and 5 is more limited.

Not surprisingly different ligands activate different members of the STAT family (Table 8.6). Some, such as STATs 1 and 3, are activated by many ligands, whereas others respond to far fewer ligands, e.g. STAT2 appears to be activated only by type I interferons.

STAT phosphorylation ensures its binding to the receptor, with subsequent disengagement from the receptor in dimeric form. STAT dimerization is believed to involve intermolecular associations between the SH2 domain of one STAT and phosphotyrosine of its partner. Dimerization appears to be an essential prerequisite for DNA binding. Dimers may consist of two identical STATs, but STAT1–STAT2 and STAT1–STAT3 heterodimers are also frequently formed in response to certain cytokines. The STAT dimers then translocate to the nucleus where they bind to specific

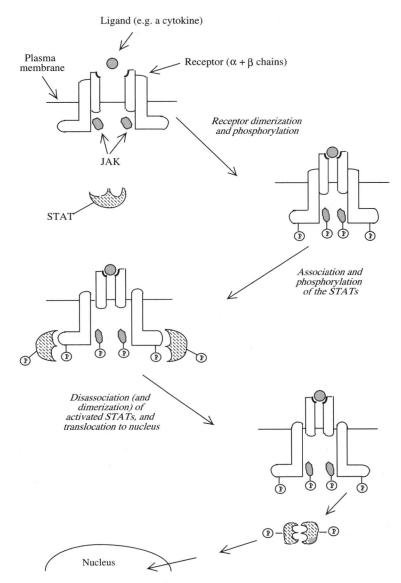

Figure 8.2 Simplified overview of the signal transduction process mediated by the JAK–STAT pathway. Refer to text for specific details

Figure 8.3 Schematic representation of the general domain structure of a STAT protein. A conserved ('C' or 'con') domain is located at the N-terminus, followed by the DNA-binding domain (D). Y represents a short sequence that contains the tyrosine residue phosphorylated by the Janus kinase. The carboxy terminus domain (T_r) represents a transcriptional activation domain

Table 8.6   Ligands which, upon binding to their cell surface receptors, are known to promote activation of one or more STATs. (The STATs activated are also shown.) This list, though representative, is not exhaustive

Ligand	STAT activated
IFN-α	1, 2, 3
IFN-γ	1
IL-2	1, 3, 5
IL-3	5
IL-6	1, 3
GM-CSF	5
EGF	1, 3
GH	1, 3, 5

DNA sequences. (STAT2-dependent signalling represents a partial exception. This STAT forms a complex with STAT1 and a non-STAT cytoplasmic protein (p48), and this complex translocates to the nucleus. Binding of this complex to the DNA is believed not to involve STAT2 directly.) STATs bind specific sequences of DNA that approach symmetry, or are palindromic (often TTCC X GGAA, where X can be different bases). These sequences are normally present in upstream regulatory regions of specific genes. Binding of the STAT complex enhances transcription of these genes, and the gene products mediate the observed cellular response to cytokine binding.

A number of proteins that inhibit the JAK–STAT function have also been identified. These include members of the so-called SOCS/Jab/Cis family and the PIAS family of regulatory proteins. Several appear to function by inhibiting the activation of various STATs, although the mechanisms by which this is achieved remain to be elucidated in detail. The JAK–STAT pathway likely does not function in isolation within the cell. JAKs are believed to activate elements of additional signalling pathways, and STATs are also likely activated by factors other than JAKs. As such, there may be considerable crosstalk between various JAK- and/or STAT-dependent signalling pathways.

8.2.7   The interferon JAK–STAT pathway

Binding of type I interferons to the IFN-α/β (type I) receptor results in the phosphorylation and, hence, activation of two members of the JAK family: Tyk2 and JAK1. These kinases then phosphorylate STAT1α (also called STAT91), STAT1β (STAT84) and STAT2 (STAT113). The three activated STATs disengage from the receptor and bind to the cytoplasmic protein p48. This entire complex translocates to the nucleus, where it interacts directly with upstream regulatory regions of interferon-sensitive genes. These nucleotide sequences are termed ISREs. This induces/augments expression of specific genes, as discussed later.

The essential elements of the signal transduction pathway elicited by IFN-γ are even more straightforward. IFN-γ binding to the type II receptor induces receptor dimerization with consequent activation of JAK1 and JAK2. The JAKs phosphorylate the receptor and subsequently the associated STAT1α. STAT1α is then released and forms a homodimer that translocates to the nucleus. It regulates expression of IFN-γ-sensitive genes by binding to a specific upstream regulatory sequence of the gene (the IFN-γ activated sequence, GAS). The PIAS-1 protein appears to play an inhibitory role in this pathway. By complexing with (phosphorylated) STAT-1 proteins, it inhibits DNA binding and transactivation.

8.2.8 The biological effects of interferons

Interferons induce a wide range of biological effects. Generally, type I interferons induce similar effects, which are distinct from the effects induced by IFN-γ. The most pronounced effect of type I interferons relates to their antiviral activity, as well as their anti-proliferative effect on various cell types, including certain tumour cell types. Anti-tumour effects are likely due not only to a direct anti-proliferative effect on the tumour cells themselves, but also due to the ability of type I interferons to increase NK and T-cytotoxic cell activity. These cells can recognize and destroy cancer cells.

Not all type I interferons induce exactly the same range of responses, and the antiviral to anti-proliferative activity ratio differs from one type I interferon to another. As all bind the same receptor, the molecular basis by which variation in biological activities is achieved is poorly understood as yet.

IFN-γ exhibits, at best, weak antiviral and anti-proliferative activity. When co-administered with type I interferons, however, it potentates these IFN-α/β activities. IFN-γ is directly involved in regulating most aspects of the immune and inflammatory responses. It promotes activation, growth and differentiation of a wide variety of cell types involved in these physiological processes (Table 8.7).

IFN-γ represents the main macrophage-activating factor, thus enhancing macrophage-mediated effects, including:

- destruction of invading microorganisms;

- destruction of intracellular pathogens;

- tumour cell cytotoxicity;

- increased major histocompatibility complex (MHC) antigen expression, leading to enhanced activation of lymphocytes via antigen presentation.

Table 8.7 Cell types participating in the immune, inflammatory or other responses whose activation, growth and differentiation are promoted by IFN-γ

Macrophages/monocytes
Polymorphonuclear neutrophils
T-lymphocytes
B-lymphocytes
NK cells
Fibroblasts
Endothelial cells

Binding of IFN-γ to its surface receptor on polymorphonuclear neutrophils induces increased expression of the gene coding for a neutrophil cell surface protein capable of binding the F_c portion (i.e. the constant region; see also Box 13.2) of IgG. This greatly increases the phagocytotic and cytotoxic activities of these cells.

IFN-γ also directly modulates the immune response by affecting growth, differentiation and function of both T- and B-lymphocytes. These effects are quite complex and are often influenced by additional cytokines. IFN-γ acts as a growth factor in an autocrine manner for some T cell sub-populations, and it is capable of suppressing growth of other T cell types. It appears to have an inhibitory effect on development of immature B-lymphocyte populations, but it may support mature B cell survival. It can both up-regulate and down-regulate antibody production under various circumstances.

All interferons promote increased surface expression of class I MHC antigens. Class II MHC antigen expression is stimulated mainly by IFN-γ (MHC proteins are found on the surface of various cell types. They play an essential role in triggering an effective immune response against not only foreign antigen, but also altered host cells). Although many interferons promote synergistic effects, some instances are known where two or more interferons can oppose each other's biological activities. IFN-αJ, for example, can inhibit the IFN-αA-mediated stimulation of NK cells.

The molecular basis by which interferons promote their characteristic effects, in particular antiviral activity, is understood at least in part. Interferon stimulation of the JAK–STAT pathway induces synthesis of at least 30 different gene products, many of which cooperate to inhibit viral replication. These antiviral gene products are generally enzymes, the most important of which are $2'$–$5'$ oligoadenylate synthetase (2,5-A_n synthetase) and the eIF-2α protein kinase.

These intracellular enzymes remain in an inactive state after their initial induction. They are activated only when the cell comes under viral attack, and their activation can inhibit viral replication in that cell. The 2,5-A_n synthetase acts in concert with two additional enzymes, i.e. an endoribonuclease and a phosphodiesterase, to promote and regulate the antiviral state (Figure 8.4).

Several active forms of the synthetase seem to be inducible in human cells; 40 kDa and 46 kDa variants have been identified that differ only in their carboxy terminus ends. They are produced as a result of differential splicing of mRNA transcribed from a single gene found on chromosome 11. A larger 85–100 kDa form of the enzyme has been detected, which may represent a heterodimer composed of the 40 and 46 kDa variants.

The synthetase is activated by double-stranded RNA (dsRNA). Although not normally present in human cells, dsRNA is often associated with commencement of replication of certain viruses. The activated enzyme catalyses the synthesis of oligonucleotides of varying length in which the sole base is adenine ($2'$–$5'A_n$). This oligonucleotide differs from oligonucleotides present naturally in the cell, in that the phosphodiester bonds present are $2'$–$5'$ bonds (Figure 8.5). The level of synthesis and average polymer length of the oligonucleotide products appear to depend upon the exact inducing interferon type, as well as on the growth state of the cell.

The sole biochemical function of $2'$–$5'A_n$ (and hence $2'$–$5'A_n$ synthetase) appears to be as an activator of a dormant endo-RNase, which is expressed constitutively in the cell. This RNase, known as RNase L or RNase F, cleaves all types of single-stranded RNA (ssRNA). This inhibits production of both viral and cellular proteins, thus paralyzing viral replication. Presumably, cellular destruction of the invading ssRNA will be accompanied by destruction of any additional viral components. Removal of dsRNA would facilitate deactivation of the endo-RNase, allowing translation of cellular mRNA to resume. A $2'$–$5'$ phosphodiesterase represents a third enzymatic

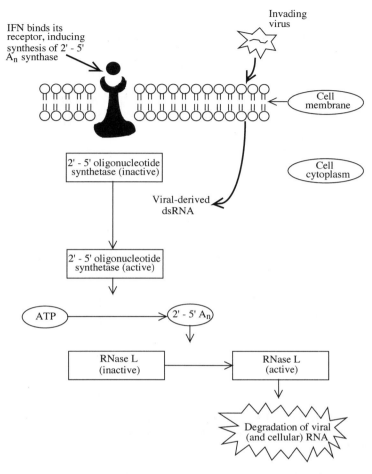

Figure 8.4 Outline of how the 2′–5′ synthetase system promotes its antiviral effect. The 2′–5′ phosphodiesterase 'off switch' is omitted for clarity. Refer to text for details

component of this system. It functions as an off switch, as it rapidly degrades the $2'-5'A_n$ oligonucleotides. Although this enzyme also appears to be expressed constitutively, interferon binding appears to increase its expression levels in most cells.

8.2.9 The eIF-2α protein kinase system

Intracellular replication of viral particles depends entirely upon successful intracellular transcription of viral genes with subsequent translation of the viral mRNA. Translation of viral or cellular mRNA is dependent upon ribosome formation. Normally, several constituent molecules interact with each other on the mRNA transcript, forming the smaller ribosomal subunit. Subsequent formation/attachment of the larger subunit facilitates protein synthesis.

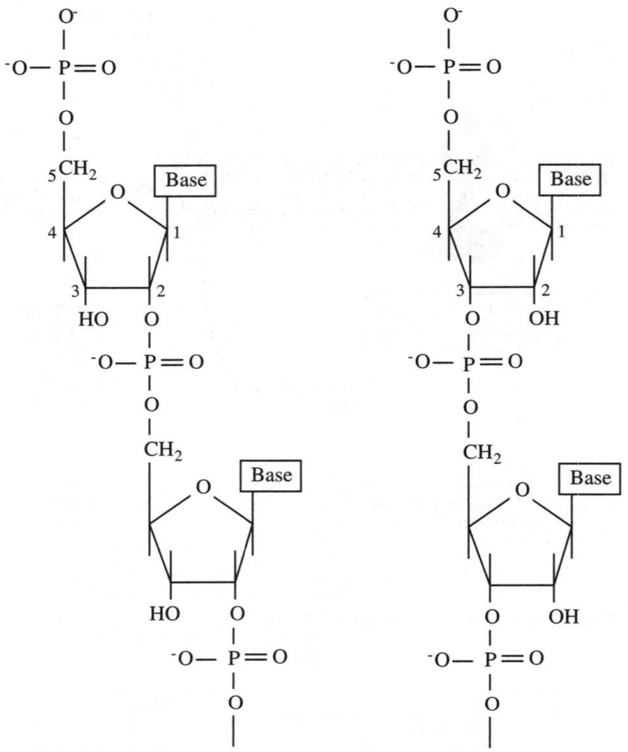

Figure 8.5 (a) Structural detail of the 2'–5' oligonucleotides (2'–5'A$_n$) generated by 2'–5'A$_n$ synthetase. Compare the 2'–5' phosphodiester linkages with the 3'–5' linkages characteristic of normal cellular oligonucleotides such as mRNA (b)

Exposure of cells to interferon normally results in the induction of a protein kinase termed eIF-2α protein kinase. The enzyme, which is synthesized in a catalytically inactive form, is activated by exposure to dsRNA. The activated kinase then phosplorylates its substrate, i.e. eIF-2α, which is the smallest subunit of initiation factor 2 (eIF$_2$). This, in turn, blocks construction of the smaller ribosomal subunit, thereby preventing translation of all viral (and cellular) mRNA (Figure 8.6).

Induction of eIF-2α protein kinase is dependent upon both interferon type and cell type.

IFN-α, -β and -γ are all known to induce the enzyme in various animal cells. However, in human epithelial cells the kinase is induced only by type I interferons, whereas none of the interferons seem capable of inducing synthesis of the enzyme in human fibroblasts. The purified kinase is highly selective for initiation factor eIF-2, which it phosphorylates at a specific serine residue.

Interferon, in particular type I interferon, is well adapted to its antiviral function. Upon entry into the body, viral particles are likely to encounter IFN-α/β-producing cells quickly, including macrophages and monocytes. This prompts interferon synthesis and release. These cells act like

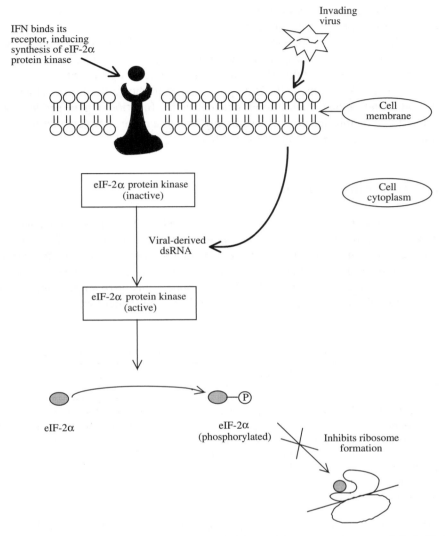

Figure 8.6   Outline of how the eIF-2α protein kinase system promotes an antiviral effect

sentries, warning other cells of the viral attack. Most body cells express the type I interferon receptor; thus, the released IFN-α or -β will induce an antiviral state in such cells.

The ability of interferons (especially type I interferons) to induce an antiviral state is unlikely
to be solely dependent upon the enzymatic mechanisms discussed above. Furthermore the $2'–5'A_n$
synthetase and eIF-2α kinase systems may play important roles in mediating additional interferon
actions. The ability of such systems to stall protein synthesis in cells may play a role in interferon-
induced alterations of cellular differentiation or cell cycle progression. They may also be involved
in mediating interferon-induced anti-proliferative effects on various transformed cells.

MHC antigens and β2-macroglobulin are amongst the best known proteins whose synthesis is also induced in a variety of cell types in response to various interferons.

Additional studies focus upon identification and characterization of gene products whose cellular levels are decreased in response to interferon binding. For example, studies using various human and animal cell lines found that IFN-α and -β can induce a significant decrease in the level of c-myc and c-fos mRNA in some cells. IFN-γ has also been shown to inhibit collagen synthesis in fibroblasts and chondrocytes. Such studies, elucidating the function of gene products whose cellular levels are altered by interferons, will eventually lead to a more complete picture of how these regulatory molecules induce their characteristic effects.

8.3 Interferon biotechnology

The antiviral and anti-proliferative activity of interferons, as well as their ability to modulate the immune and inflammatory response renders obvious their potential medical application. This has culminated in the approval for clinical use of several interferon preparations (Table 8.8). Ongoing clinical trials are likely to expand the medical uses of these regulatory molecules further over the next few years.

While at least some of these potential therapeutic applications were appreciated as far back as the late 1950s, initial therapeutic application was rendered impractical due to the extremely low

Table 8.8 Interferon-based biopharmaceuticals approved to date for general medical use

Product	Company	Indication
Intron A (rIFN-α-2b)	Schering Plough	Cancer, genital warts, hepatitis
PegIntron A (PEGylated rIFN-α-2b)	Schering Plough	Chronic hepatitis C
Viraferon (rIFN-α-2b)	Schering Plough	Chronic hepatitis B and C
ViraferonPeg (PEGylated rIFN-α-2b)	Schering Plough	Chronic hepatitis C
Roferon A (rhIFN-α-2a)	Hoffman-La-Roche	Hairy cell leukaemia
Actimmune (rhIFN-γ-1b)	Genentech	Chronic granulomatous disease (CGD)
Betaferon (rIFN-β-1b, differs from human protein in that Cys 17 is replaced by Ser)	Schering AG	MS
Betaseron (rIFN-β-1b, differs from human protein in that Cys 17 is replaced by Ser)	Berlex Laboratories and Chiron	Relapsing–remitting MS
Avonex (rhIFN-β-1a)	Biogen	Relapsing MS
Infergen (rIFN-α, synthetic type I interferon)	Amgen (USA) Yamanouchi Europe (EU)	Chronic hepatitis C
Rebif (rhIFN-β-1a)	Ares Serono	Relapsing–remitting MS
Rebetron (combination of ribavirin and rhIFN-α-2b)	Schering Plough	Chronic hepatitis C
Alfatronol (rhIFN-α-2b)	Schering Plough	Hepatitis B, C, and various cancers
Virtron (rhIFN-α-2b)	Schering Plough	Hepatitis B and C
Pegasys (Peginterferon α-2a)	Hoffman La Roche	Hepatitis C
Vibragen Omega (rFeline interferon omega)	Virbac	Vet. (reduce mortality/clinical signs of canine parvovirosis)

levels at which they are normally produced in the body. Large-scale purification from sources such as blood was non-viable. Furthermore, interferons exhibit species preference and, in some cases, strict species specificity. This rendered necessary the clinical use only of human-derived interferons in human medicine.

Up until the 1970s interferon was sourced (in small quantities) directly from human leukocytes obtained from transfused blood supplies. This 'interferon' preparation actually consisted of a mixture of various IFN-αs, present in varying amounts, and was only in the regions of 1 per cent pure. However, clinical studies undertaken with such modest quantities of impure interferon preparations produced encouraging results.

The production of interferon in significant quantities first became possible in the late 1970s, by means of mammalian cell culture. Various cancer cell lines were found to secrete interferons in greater than normal quantities, and were amenable to large-scale cell culture due to their transformed nature. Moreover, hybridoma technology facilitated development of sensitive interferon immunoassays. The Namalwa cell line (a specific strain of human lymphoblastoid cells) became the major industrial source of interferon. The cells were propagated in large animal cell fermenters (up to 8000 l), and subsequent addition of an inducing virus (usually the Sendai virus) resulted in production of significant quantities of leukocyte interferon. Subsequent analysis showed this to consist of at least eight distinct IFN-α subtypes.

Wellferon was the tradename given to one of the first such approved products. Produced by large-scale mammalian (lymphoblastoid) cell cultures, the crude preparation undergoes extensive chromatographic purification, including two immunoaffinity steps. The final product contains nine IFN-α subtypes.

Recombinant DNA technology also facilitated the production of interferons in quantities large enough to satisfy potential medical needs. The 1980s witnessed the cloning and expression of most interferon genes in a variety of expression systems. The expression of specific genes obviously yielded a product containing a single interferon (sub)type.

Most interferons have now been produced in a variety of expression systems, including *E. coli*, fungi, yeast and some mammalian cell lines, such as CHO cell lines and monkey kidney cell lines. Most interferons currently in medical use are recombinant human (rh) products produced in *E. coli*. *E. coli*'s inability to carry out post-translational modifications is irrelevant in most instances, as the majority of human IFN-αs, as well as IFN-β, are not normally glycosylated. Whereas IFN-γ is glycosylated, the *E. coli*-derived unglycosylated form displays a biological activity similiar to the native human protein.

The production of interferon in recombinant microbial systems obviously means that any final product contaminants will be microbial in nature. A high degree of purification is thus required to minimize the presence of such non-human substances. Most interferon final product preparations are in the region of 99 per cent pure. Such purity levels are achieved by extensive chromatographic purification. While standard techniques such as gel filtration and ion exchange are extensively used, reported interferon purification protocols have also entailed use of various affinity techniques using, for example, anti-interferon monoclonal antibodies, reactive dyes or lectins (for glycosylated interferons). Hydroxyapatite, metal-affinity and hydrophobic interaction chromatography have also been employed in purification protocols. Many production columns are run in HPLC (or FPLC) format, yielding improved and faster resolution. Immunoassays are used to detect and quantify the interferons during downstream processing, although the product (in particular the finished product) is also usually subjected to a relevant bioassay. The production and medical uses of selected interferons are summarized in the sections below.

8.3.1 Production and medical uses of interferon-α

Clinical studies undertaken in the late 1970s with multicomponent, impure IFN-α preparations clearly illustrated the therapeutic potential of such interferons as an anti-cancer agent. These studies found that IFN-α could induce regression of tumours in significant numbers of patients suffering from breast cancer, certain lymphomas (malignant tumour of the lymph nodes) and multiple myeloma (malignant disease of the bone marrow). The interferon preparations could also delay recurrence of tumour growth after surgery in patients suffering from osteogenic sarcoma (cancer of connective tissue involved in bone formation).

The first recombinant interferon to become available for clinical studies was IFN-α2a, in 1980. Shortly afterwards the genes coding for additional IFN-αs were cloned and expressed, allowing additional clinical studies. The antiviral, anti-tumour and immunomodulatory properties of these interferons assured their approval for a variety of medical uses. rhIFN-αs manufactured/marketed by a number of companies (Table 8.8) are generally produced in *E. coli*.

Clinical trials have shown the recombinant interferons to be effective in the treatment of various cancer types, with rhIFN-α2a and -α2b both approved for treatment of hairy cell leukaemia. This is a rare B-lymphocyte neoplasm for which few effective treatments were previously available. Administration of the recombinant interferons promotes significant regression of the cancer in up to 90 per cent of patients.

Schering Plough's rhIFN-α2b (Intron A) was first approved in the USA in 1986 for treatment of hairy cell leukaemia, but is now approved for use in more than 50 countries for up to 16 indications (Table 8.9). The producer microorganism is *E. coli*, which harbours a cytoplasmic expression vector (KMAC-43) containing the interferon gene. The gene product is expressed intracellulary. Intron A manufacturing facilities are located in New Jersey and in Brinny, Co. Cork, Ireland.

Upstream processing (fermentation) and downstream processing (purification and formulation) are physically separated, by being undertaken in separate buildings. Fermentation is generally undertaken in specially designed 42 000 l stainless steel vessels. After recovery of the product from the cells, a number of chromatographic purification steps are undertaken, essentially within

Table 8.9 Some of the indications (i.e. medical conditions) for which Intron A is approved. Note that the vast majority are either forms of cancer or viral infections

Hairy cell leukaemia	Laryngeal papillomatosis[a]
Renal cell carcinoma	Mycosis fungoides[b]
Basal cell carcinoma	Condyloma acuminata[c]
Malignant melanoma	Chronic hepatitis B
AIDS-related Kaposi's sarcoma	Hepatitis C
Multiple myeloma	Chronic hepatitis D
Chronic myelogenous leukaemia	Chronic hepatitis, non-A, non-B/C hepatitis
Non-Hodgkin's lymphoma	

[a]Benign growths (papillomas) in the larynx.
[b]A fungal disease.
[c]Genital warts.

a large cold-room adapted to function under cleanroom conditions. Crystallization of the IFN-α2b is then undertaken as a final purification step. The crystalline product is redissolved in phosphate buffer, containing glycine and human albumin as excipients. After aseptic filling, the product is normally freeze-dried. Intron A is usually sold at five commercial strengths (3, 5, 10, 25, and 50 million IU/vial).

More recently, a number of modified recombinant interferon products have also gained marketing approval. These include PEGylated interferons (PEG IntronA and Viraferon Peg (Table 8.8 and Box 8.2) and the synthetic interferon product Infergen. PEGylated interferons are generated by reacting purified IFN-αs with a chemically activated form of PEG. Activated methoxypolyethylene glycol is often used, which forms covalent linkages with free amino groups on the interferon molecule. Molecular mass analysis of PEGylated interferons (e.g. by mass spectroscopy, gel

---

Box 8.2

Product case study: ViraferonPeg

ViraferonPeg (tradename) is a PEGylated form of interferon alfa-2b (IFNα-2b) approved for medical use in the EU since 2000. It differs from native human IFNα-2b only by the presence of covalently attached PEG. ViraferonPeg is indicated for the treatment of chronic hepatitis C in adults, and is usually administered in combination with the antiviral drug ribavirin. It is produced via recombinant DNA technology in an engineered *E. coli* cell line carrying the human IFNα-2b gene. After cell fermentation the interferon is purified from the bacterial culture via crystallization and multiple chromatographic steps. As part of the downstream processing, the interferon is incubated with chemically activated PEG (methoxypolyethylene glycol, mPEG), which spontaneously forms a covalent linkage via selected protein amino acid groups. The majority of interferon molecules are monoPEGylated with minor quantities of unPEGylated and diPEGylated product also being produced.

The product is presented in lyophilized format and contains sodium phosphate, sucrose and polysorbate as excipients. It is usually administered as once-weekly s.c. injections, typically for periods of 6 months.

Pharmacokinetic studies indicate a plasma half-life of 13–25 h (compared with 4 h for the unPEGylated molecule) with maximum serum concentrations attained within 15–44 h. The main clinical study to establish initial safety and efficacy was a multicentre double blind, randomized trial involving 1200 patients split into four groups (three treated with increasing concentrations of ViraferonPeg and the fourth being treated with unPEGylated IFNα-2b (tradename Intron A). The primary efficacy measure was a composite of viral response, assessed in terms of reduction in serum viral particle load and normalized liver enzyme function, and the PEGylated product proved most effective. A subsequent trial showed that a combination of ribavirin and ViraferonPeg to be more effective than a combination of ribavirin and unPEGylated product.

Common side effects noted include injection site reactions, weakness, dizziness, weight loss and flu-like symptoms, with depression being the most common reason for treatment discontinuation. ViraferonPeg is manufactured and marketed by Schering Plough.

filtration or SDS-PAGE) indicates that the approved PEGylated products consist predominantly of monoPEGylated interferon molecules, with small amounts of both free and diPEGylated species also being present.

The intrinsic biological activity of PEGylated and non-PEGylated interferons is essentially the same. The PEGylated product, however, displays a significantly prolonged plasma half-life (13–25 h, compared with 4 h for unpegylated species). The prolonged half-life appears to be due mainly to slower elimination of the molecule, although PEGylation also appears to decrease systemic absorption from the site of injection following subcutaneous administration, as discussed in Chapter 4.

Infergen (interferon alfacon-1 or consensus interferon) is an engineered interferon recently approved for the treatment of hepatitis C (Table 8.8). The development of infergen entailed initial sequence comparisons between a range of IFN-αs. The product's amino acid sequence reflects the most frequently occurring amino acid residue in each corresponding position of these native interferons. A DNA sequence coding for the product was synthesized and inserted into *E. coli*. The recombinant product accumulates intracellularly as inclusion bodies.

Large-scale manufacture entails an initial fermentation step. After harvest, the *E. coli* cells are homogenized and the inclusion bodies recovered via centrifugation. After solubilization and re-folding, the interferon is purified to homogeneity by a combination of chromatographic steps. The final product is formulated in the presence of a phosphate butter and sodium chloride. It is presented as a 30 μg ml^{-1} solution in glass vials and displays a shelf life of 24 months when stored at 2–8 °C. When compared on a mass basis, the synthetic interferon displays higher antiviral, antiproliferative and cytokine-inducing activity than do native type I interferons.

Ongoing clinical trials continue to assess the efficacy of recombinant interferon preparations in treating a variety of cancers. Some trials suggest that treatments are most effective when administered in the early stages of cancer development. rhIFN-αs have also proven effective in the treatment of various viral conditions, most notably viral hepatitis. Hepatitis refers to an inflammation of the liver. It may be induced by toxic substances, immunological abnormalities, or by viruses (infectious hepatitis). The main viral causative agents are:

- hepatitis A virus (hepatitis A);

- hepatitis B virus (hepatitis B, i.e. classical serum hepatitis);

- hepatitis C virus (hepatitis C, formerly known as classical non-A, non-B hepatitis);

- hepatitis D virus (hepatitis D, i.e. delta hepatitis);

- hepatitis E virus (hepatitis E, i.e. endemic non-A, non-B hepatitis);

- hepatitis GB agent.

This disease may be acute (rapid onset, often accompanied by severe symptoms, but of brief duration) or chronic (very long duration).

Hepatitis A is common, particularly in areas of poor sanitation, and is transmitted by food or drink contaminated by a sufferer/carrier. Clinical symptoms include jaundice and are usually mild. A full recovery is normally recorded. Hepatitis B is transmitted via infected blood. Symptoms of acute hepatitis B include fever, chill, weakness and jaundice. Most suffers recover from such infection, although acute liver failure and death sometimes occur. Some 5–10 per cent of suffers go on to develop chronic hepatitis B. Acute hepatitis C is usually mild and asymptomatic. However, up to 90 per cent of infected persons go on to develop a chronic form of the condition. Hepatitis D is unusual in that it requires the presence of hepatitis B in order to replicate. It thus occurs in some persons concomitantly infected with hepatitis B virus. Its clinical symptoms are usually severe, and can occur in acute or chronic form.

Chronic forms of hepatitis (in particular B, C and D) can result in liver cirrhosis and/or hepatocellular carcinoma. This occurs in up to 20 per cent of chronic hepatitis B sufferers and in up to 30 per cent of chronic hepatitis C sufferers. The scale of human suffering caused by hepatitis on a worldwide basis is enormous. Approximately 5 per cent of the global population suffer from chronic hepatitis B. An estimated 50 million new infections occur each year. Over 1.5 million of the 300 million carriers worldwide die annually from liver cirrhosis and hepatocellular carcinoma.

IFN-α2b is now approved in the USA for the treatment of hepatitis B and C. Clinical studies undertaken with additional IFN-α preparations indicate their effectiveness in managing such conditions, and several such products are also likely to gain regulatory approval.

IFN-α2a, when administered three times weekly for several weeks/months, was found effective in treating several forms of hepatitis. Remission is observed in 30–45 per cent of patients suffering from chronic hepatitis B, and a complete recovery is noted in up to 20 per cent of cases. The drug induces sustained remission in up to 30 per cent of patients suffering from chronic hepatitis C, but can ease clinical symptoms of this disease in up to 75 per cent of such patients. Ongoing studies also indicate its efficacy in treating chronic hepatitis D, though relapse is frequently observed upon cessation of therapy. The drug is normally administered by i.m. or s.c. (directly beneath the skin) injection. Peak plasma concentrations of the interferon are observed more quickly upon i.m. injection (4 h versus 7.5 h). The elimination half-life of the drug ranges from 2.5 to 3.5 h.

IFN-α preparations have also proven efficacious in the treatment of additional viral-induced medical conditions. rhIFN-α2b and IFN-αn3 are already approved for the treatment of sexually transmitted genital warts, caused by a human papilloma virus. Although this condition is often unresponsive to various additional therapies, direct injection of the interferon into the wart causes its destruction in up to 70 per cent of patients. Another member of the papilloma family is associated with the development of benign growths in the larynx (laryngeal papillomatosis). This condition can be successfully treated with IFN-α preparations, as can certain papilloma-related epithelial cell cancers, such as cervical intraepithelial neoplasm (epithelial cells are those that cover all external surfaces of the body and line hollow structures, with the exception of blood and lymph vessels). IFN-α's ability to combat a range of additional virally induced diseases, including acquired immune deficiency syndrome (AIDS), is currently being appraised in clinical trials.

8.3.2 Medical uses of interferon-β

rhIFN-β has found medical application in the treatment of relapsing–remitting multiple sclerosis (MS), a chronic disease of the nervous system. This disease normally presents in young adults (more commonly women) aged 20–40 years. It is characterized by damage to the myelin sheath, which surrounds neurons of the central nervous system, and in this way compromises neural function. Clinical presentations include shaky movement, muscle weakness, paralysis, defects in speech, vision and other higher mental functions. The most predominant form of the condition is characterized by recurrent relapses followed by remission. MS appears to be an autoimmune disease, in which elements of immunity (mainly lymphocytes and macrophages) cooperate in the destruction of the myelin. What triggers onset of the condition is unknown, although genetic and environmental factors (including viral infection) have been implicated.

IFN-β preparations approved for medical use to date include Betaferon, Betaseron, Avonex and Rebif (Table 8.8 and Box 8.3). The former two products are produced in recombinant *E. coli* cells, whereas the latter two are produced in CHO cell lines. Manufacture using *E. coli* generates a non-glycosylated product, although lack of native glycosylation does not negatively affect its therapeutic efficacy. Typically, IFN-β-based drugs reduce the frequency of

---

Box 8.3

Product case study: Rebif

Rebif (tradename) is an rIFN-β-1a first approved for medical use in the EU in 1998 and subsequently in the USA in 2002. It is indicated for the treatment of patients with relapsing–remitting MS, to decrease the frequency of clinical exacerbations and delay the accumulation of physical disability.

Rebif is produced via recombinant DNA technology in a CHO cell line. It displays an identical amino acid sequence to that of native human IFN-β-1a and, like the native product, is glycosylated. After cell culture the interferon is purified using a series of chromatographic steps (affinity, ion-exchange, gel-filtration and reverse-phase liquid chromatography). It is formulated as a sterile solution in pre-filled syringes and contains mannitol, HSA, sodium acetate, acetic acid and sodium hydroxide as excipients. It is administered subcutaneously three times weekly.

Product pharmacokinetics were evaluated in healthy volunteers and a single s.c. injection (60 μg) resulted in a peak serum concentration $C_{max} = 5.1$ IU ml^{-1}, with a median time of peak serum concentration $T_{max} = 16$ h. The serum elimination half-life $t_{1/2}$ was of the order of 70 h.

The product was evaluated in two multicentre clinical studies that established safety and efficacy. The first (the 'PRISMS' study) was a randomized, double blind placebo-controlled study involving 187 patients. The primary efficacy end-point was the number of clinical relapses recorded. The mean number of relapses over 2 years for the placebo group patients was 2.56, whereas that of the Rebif-treated group was 1.73, a relative reduction of 32 per cent. Rebif treatment also provided positive relative outcomes for several secondary end-points, including the proportion of patients with sustained disability progression.

The most common side effects noted included injection site reactions and flu-like symptoms with serious potential side effects including depression, as well as liver and blood abnormalities. Rebif is manufactured and marketed by Serono Inc.

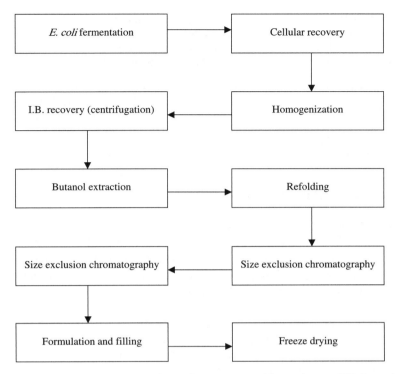

Figure 8.7 Overview of the manufacture of Betaferon, a recombinant human IFN-β produced in *E. coli*. The product differs from native human IFN-β in that it is unglycosylated and cysteine residue 17 had been replaced by a serine residue. *E. coli* fermentation is achieved using minimal salts/glucose media and product accumulates intracellularly in inclusion body (IB) form. During downstream processing, the Ibs are solubilized in butanol, with subsequent removal of this denaturant to facilitate product refolding. After two consecutive gel-filtration steps, excipients are added, the product is filled into glass vials and freeze-dried. It exhibits a shelf life of 18 months when stored at 2–8 °C

relapses by about 30 per cent in many patients. In some instances, a sustained reduction in the accumulation of MS brain lesions (as measured by magnetic resonance imaging) is also observed. However, there is little evidence that IFN-β significantly alters overall progression of the disease. A summary overview of the production of one such product (Betaferon) is presented in Figure 8.7.

The molecular mechanism by which IFN-β induces its therapeutic effect is complex and not fully understood. It is believed that the pathology of MS is linked to the activation and proliferation of T-lymphocytes specific for epitopes found on specific myelin antigens. Upon migration to the brain, these lymphocytes trigger an inflammatory response mediated by the production of pro-inflammatory cytokines, most notably IFN-γ, IL-1, IL-2 and TNF-α. The inflammatory response, in addition to other elements of immunity (e.g. antibodies and complement activation), results in the destruction of myelin surrounding neuronal axons. IFN-β likely counteracts these effects, in part at least, by inhibiting production of IFN-γ and TNF-α and hence mediating down-regulation of the pro-inflammatory response.

Table 8.10 Some pathogens (bacterial, fungal and protozoal) whose phagocytic-mediated destruction is impaired in persons suffering from CGD. Administration of IFN-γ, in most cases, enhances the phagocyte's ability to destroy these pathogens. These agents can cause hepatic and pulmonary infections, as well as genitourinary tract, joint and other infections

Staphylococcus aureus	*Plasmodium flaciparum*
Listeria monocytogenes	*Leishmania donovani*
Chlamydia psittaci	*Toxoplasma gondii*
Aspergillus fumigatus	

8.3.3   Medical applications of interferon-γ

The most notable medical application of IFN-γ relates to the treatment of CGD, a rare genetic condition with a population incidence of between 1 in 250 00 and 1 in 1 000 000. Phagocytic cells of patients suffering from CGD are poorly capable/incapable of ingesting or destroying infectious agents such as bacteria or protozoa. As a result, patients suffer from repeated infections (Table 8.10), many of which can be life threatening.

Phagocytes from healthy individuals are normally capable of producing highly reactive oxidative substances, such as hydrogen peroxide and hypochlorous acid, which are lethal to pathogens. Production of these oxidative species occurs largely via a multicomponent NADPH oxidase system (Figure 8.8). CGD is caused by a genetic defect in any component of this oxidase system that compromises its effective functioning.

In addition to recurrent infection, CGD sufferers also exhibit abnormal inflammatory responses which include granuloma formation at various sites of the body (granuloma refers to a tissue outgrowth that is composed largely of blood vessels and connective tissue). This can lead to obstruction of various ducts, e.g. in the urinary and digestive tracts.

Traditionally, treatment of CGD entailed prophylactic administration of antimicrobial agents in an attempt to prevent occurrence of severe infection. However, affected individuals still experience life-threatening infections, requiring hospitalization and intensive medical care, as often as once a year. Attempts to control these infections rely on strong antimicrobial agents and leukocyte transfusions.

Long-term administration of IFN-γ to CGD patients has proven effective in treating/moderating the symptoms of this disease. The recombinant human IFN-γ used therapeutically is produced in *E. coli*, and is termed IFN-γ1b. It displays identical biological activity to native human IFN-γ, although it lacks the carbohydrate component. The product, usually sold in liquid form, is manufactured by Genentech, who market it under the tradename Actimmune. The product is administered on an ongoing basis, usually by s.c. injection three times weekly. In clinical trials, its administration, when compared with a control group receiving a placebo, resulted in a reduction in the:

- incidence of life-threatening infections by 50 per cent or more;

- incidence of total infections by 50 per cent or more;

- number of days of hospitalization by threefold (and even when hospitalization was required, the average stay was cut in half).

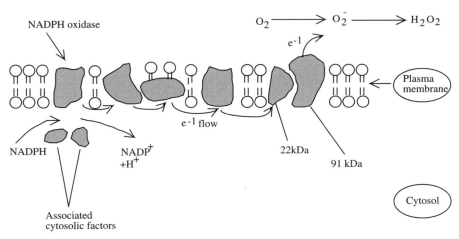

Figure 8.8 Production of reactive oxygen species by phagocytes. In addition to degrading foreign substances via phagocytosis, phagocytes secrete reactive oxygen species into their immediate environment. This can kill microorganisms (and indeed damage healthy tissue) in the vicinity, thus helping control the spread of infection. The reactive oxygen species are produced by an NADPH oxidase system, the main feature of which is a plasma membrane-based electron transport chain. NADPH represents the electron donor. The first membrane carrier is NADPH oxidase, which also requires interaction with at least two cytosolic proteins for activation. The electrons are passed via a number of carriers, including a flavoprotein, to cytochrome b_{558}. This is a haem protein consisting of two subunits (22 and 91 kDa). The cytochrome, in turn, passes the electrons to oxygen, generating a superoxide anion (O^-). The superoxide can be converted to hydrogen peroxide (H_2O_2) spontaneously, or enzymatically, by superoxide dismutase. A genetic defect affecting any element of this pathway will result in a compromised ability/inability to generate reactive oxygen species, normally resulting in CGD. Over 50 per cent of CGD sufferers display a genetic defect in the 91 kDa subunit of cytochrome b_{558}

The molecular basis by which IFN-γ induces these effects is understood, at least in part. In healthy individuals this cytokine is a potent activator of phagocytes. It potentiates their ability to generate toxic oxidative products (via the NADPH oxidase system), which they then use to kill infectious agents. In CGD sufferers, IFN-γ boosts flux through the NADPH oxidative system. As long as the genetic defect has not totally inactivated a component of the system, this promotes increased synthesis of these oxidative substances. IFN-γ also promotes increased expression of IgG F_c receptors on the surface of phagocytes. This would increase a phagocyte's ability to destroy opsonized infectious agents via phagocytes (Figure 8.9).

Additional molecular mechanisms must also mediate IFN-γ effects, as it promotes a marked clinical improvement in some CGD patients without enhancing phagocyte activity. IFN-γ's demonstrated ability to stimulate aspects of cellular and humoral immunity (e.g. via T- and B-lymphocytes), as well as NK cell activity, is most likely responsible for these observed improvements.

IFN-γ may also prove valuable in treating a variety of other conditions, and clinical trials for various indications are currently underway. This cytokine shows promise in treating leishmaniasis, a disease common in tropical and subtropical regions. The causative agent is a parasitic protozoan of the genus *Leishmania*. The disease is characterized by the presence of these protozoa inside certain immune cells, particularly macrophages. IFN-γ appears to stimulate the infected macrophage to produce nitric oxide, which is toxic for the parasite.

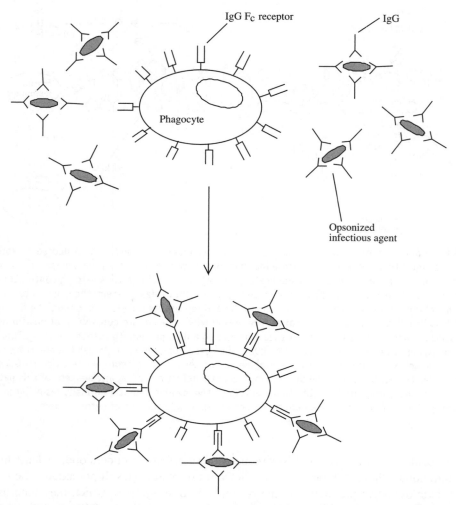

Figure 8.9   Increased expression of IgG F$_c$ receptors on phagocytes results in enhanced phagocytosis. These receptors will retain opsonized (i.e. antibody-coated) infectious agents at their surface by binding the F$_c$ portion of the antibody. This facilitates subsequent phagocytosis

Additional studies illustrate that IFN-γ stimulates phagocytic activity in humans suffering from various cancers, AIDS and lepromatous leprosy (leprosy is caused by the bacterium *Mycobacterium leprae*. Lepromatous leprosy is a severe contagious form of the disease leading to disfigurement). IFN-γ may thus prove useful in treating such conditions.

8.3.4   Interferon toxicity

Like most drugs, administration of interferons can elicit a number of unwanted side effects. Unfortunately, in some instances the severity of such effects can limit the maximum recommended therapeutic

Table 8.11 Side-effects sometimes associated with therapeutic administration of IFN-αs. In most cases, only minor side effects are noted. However, more serious effects, necessitating cessation of treatment, may occur in up to 17 per cent of patients

Minor side effects	Serious potential side effects

Range of flu-like symptoms, e.g.
    fever
    headache
    chills
	Anorexia
	Strong fatigue
	Insomnia
	Cardiovascular complications
	Autoimmune reactions
	Hepatic decompression

dose to a level below that which might have maximum therapeutic effect. Administration of interferons (in addition to many other cytokines) characteristically induces flu-like symptoms in many recipients. Such symptoms are experienced by most patients within 8 h of IFN-α administration. However, they are usually mild and are alleviated by concurrent administration of paracetamol. Tolerance of such effects also normally develops within the first few weeks of commencing treatment.

In some instances more severe side effects are noted (Table 8.11), whereas in a few cases very serious side effects, such as induction of autoimmune reactions and central nervous system or cardiovascular disturbances, render necessary immediate withdrawal of treatment.

Administration of IFN-β also characteristically causes flu-like symptoms. More serious side effects, however, are sometimes noted, including:

- hypersensitivity reactions;

- menstrual disorders;

- anxiety and emotional liability;

- depression, which in rare instances may prompt suicidal thoughts.

The only common side effect associated with IFN-γ is the characteristic flu-like symptoms. However, in rare instances and at high doses, adverse clinical reactions have been noted. These have included heart failure, central nervous system complications (confusion disorientation, Parkinsonian-like symptoms), metabolic complications (e.g. hyperglycaemia), and various other symptoms.

Prediction of the range or severity of side effects noted after administration of any interferon preparation is impossible. Careful monitoring of the patients, particularly in the earliest stages of treatment, soon reveals the onset of any side effects that might warrant suspension of treatment.

8.3.5 Additional interferons

In the last few years additional members of the interferon family has been discovered. Amino acid sequence analysis of a protein called trophoblastin (which is found in many ruminants) revealed it was closely related to IFN-α. This result was surprising because, in sheep and several other ruminants, the primary function (and until recently the only known function) of trophoblastin is to sustain the corpus luteum during the early stages of pregnancy. The 172 amino acid protein is produced by the trophoblast (an outer layer of cells that surrounds the cells which constitute the early embryo) for several days immediately preceding implantation. In many ruminants, therefore, trophoblastin plays an essentially similar role to hCG in humans (Chapter 11).

If amino acid analysis hinted that trophoblastin was in fact an interferon, functional studies have proven it. These studies show that trophoblastin:

- displays the same antiviral activity as type I interferons;

- displays anti-proliferative activity against certain tumour cells, *in vitro* at least;

- binds the type I interferon receptor.

Trophoblastin, therefore, has been named interferon-tau (IFN-τ), and is classified as a type I interferon. There are at least three or four functional IFN-τ genes in sheep and cattle. The molecule displays a molecular mass of 19 kDa and an isoelectric point of 5.5–5.7, in common with other type I interferons. Interestingly, the molecule can also promote inhibition of reverse transcriptase activity in cells infected with the HIV virus.

IFN-τ is currently generating considerable clinical interest. It induces effects similar to type I interferon, but it appears to exhibit significantly lower toxicity. Thus, it may prove possible to use this interferon safely at dosage levels far greater than the maximum dosage levels applied to currently used type I interferons. This, however, can only be elucidated by future clinical trials.

IFN-ω represents an additional member of the interferon (type I) family. This 170 amino acid glycoprotein exhibits 50–60 per cent amino acid homology to IFN-αs, and appears even more closely related to IFN-τ.

IFN-ω genes have been found in humans, pigs and a range of other mammals, but not in dogs or rodents. The interferon induces its antiviral, immunoregulatory and other effects by binding the type I interferon receptor, although the exact physiological relevance of this particular interferon remains to be elucidated. Recently, a recombinant form of feline IFN-ω has been approved within the EU for veterinary use. Its approved indication is for the reduction of mortality and clinical symptoms of parvoviral infections in young dogs. The recombinant product is manufactured using a novel expression system that entails direct inoculation of silkworms with an engineered silkworm nuclear polyhedrosis virus housing the feline IFN-ω gene, as overviewed in Figure 5.4.

8.4   Conclusion

Interferons represent an important family of biopharmaceutical products. They have a proven track record in the treatment of selected medical conditions, and their range of clinical applications continues to grow. It is also likely that many may be used to greater efficacy in the future by their application in combination with additional cytokines.

Although it is premature to speculate upon the likely medical applications of IFN-τ, the reduced toxicity exhibited by this molecule will encourage its immediate medical appraisal. The classification of tau as an interferon also raises the intriguing possibility that other interferons may yet prove useful in the treatment of some forms of reproductive dysfunctions in veterinary and human medicine.

Further reading

Books

Abbas, A. 2003. *Cellular and Molecular Immunology*. W.B. Saunders.

Aggarwal, B. 1998. *Human Cytokines*. Blackwell Science.

Estrov, Z. 1993. *Interferons, Basic Principles and Clinical Applications*. R.G. Landes.

Fitzgerald, K. 2001. *The Cytokine Facts Book*. Academic Press.

Karupiah, G. 1997. *Gamma Interferon in Antiviral Disease*. Landes Bioscience.

Korholz, D. 2003. *Cytokines and Colony Stimulating Factors*. Humana Press, NJ, USA.

Mire-Sluis, A. 1998. *Cytokines*. Academic Press

Mantovani, A. 2000. *Pharmacology of Cytokines*. Oxford University Press

Pieters, T. 2005. *Interferon*. Routledge, London, UK.

Reder, A. 1996. *Interferon Therapy of Multiple Sclerosis*. Marcel Dekker.

Stuart-Harris, R. 2005. *Clinical Applications of the Interferons*. Hodder Arnold, NC, USA.

Articles

Cytokines: general

Aringer, M., Frucht, D., and O'Shea, J.J. 1999. Interleukin/interferon signalling: a 1999 perspective. *The Immunologist* **7**(5), 139–146.

Baggiolini, M., Dewald, B., and Moser, B. 1997. Human chemokines: an update. *Annual Review of Immunology* **15**, 675–705.

Elenkov, I.J., Iezzoni, D.G., Daly, A., Harris, A.G., and Chrousos, G.P. 2005. Cytokine dysregulation, inflammation and well-being. *Neuroimmunomodulation* **12**(5), 255–269.

Hideshima, T., Podar, K., Chauhan, D., and Anderson K. 2005. Cytokines and signal transduction. *Best Practice and Research Clinical Haematology* **18**(4), 509–524.

Ihle, J. 1996. STATs: signal transducers and activators of transcription. *Cell* **84**, 331–334.

Lau, F. and Horvath, C. 2002. Mechanisms of type I interferon cell signalling and STAT-mediated transcriptional responses. *Mount Sinai Journal of Medicine* **69**(3), 156–168.

Liu, L., Damen, J.E., Ware, M., Hughes, M., and Krystal, G. 1997. SHIP, a new player in cytokine-induced signalling. *Leukaemia* **11**(2), 181–184.

McInnes, I. and Liew, F. (2005) Cytokine networks – towards new therapies for rheumatoid arthritis. *Nature Clinical Practice Rheumatology* **1**(1), 31–39.

Mire-Sluis, A. 1999. Cytokines: from technology to therapeutics. *Trends in Biotechnology* **17**, 319–325.

O'Shea, J.J., Gadina, M., and Schreiber, R.D. 2002. Cytokine signalling in 2002: new surprises in the JAK/STAT pathway. *Cell* **109**, S121–S131.

Piscitelli, S.C., Reiss, W.G., Figg, W.D., and Petros, W.P. 1997. Pharmacokinetic studies with recombinant cytokines: scientific issues and practical considerations. *Clinical Pharmacokinetics* **32**(5), 368–381.

Proost, P., Wuyts, A., and van Damme, J. 1996. The role of chemokines in inflammation. *International Journal of Clinical and Laboratory Research* **26**(4), 211–223.

Schooltink, H. and Rose, J. 2002. Cytokines as therapeutic drugs. *Journal of Interferon and Cytokine Research* **22**(5), 505–516.

Taniguchi, T. 1995. Cytokine signalling through non-receptor protein tyrosine kinases. *Science* **268**, 251–255.

Takatsu, K. 1997. Cytokines involved in B cell differentiation and their sites of action. *Proceedings of the Society for Experimental Biology and Medicine* **215**(2), 121–133.

Yoshimura, A. 2005. Negative regulation of cytokine signalling. *Clinical Reviews in Allergy and Immunology* **28**(3), 205–220.

Interferons

Alberti, A. 1999. Interferon alfacon-1. A novel interferon for the treatment of chronic hepatitis C. *Biodrugs* **12**(5), 343–357.

Bekisz, J., Schmeisser, H., Hernandez, J., Goldman, N.D., and Zoon, K.C. 2004. Human interferons alpha, beta and omega. *Growth Factors* **22**(4), 243–251.

Belardelli, F., Ferrantini M., Proietti E., and Kirkwood J.M. 2002. Interferon-alpha in tumor immunity and immunotherapy. *Cytokine and Growth Factor Reviews* **13**(2), 119–134.

Boehm, U., Klamp, T., Groot, M., and Howard, J.C. 1997. Cellular responses to interferon-gamma. *Annual Review of Immunology* **15**, 749–795.

Brassard, D.L., Grace, M.J., and Bordens, R.W. 2002. Interferon-alpha as an immunotherapeutic protein. *Journal of Leukocyte Biology* **71**(4), 565–581.

Cencic, A. and La Bonnardiere, C. 2002. Trophoblastic interferon-gamma: current knowledge and possible role in early pig pregnancy. *Veterinary Research* **33**(2), 139–157.

Chelmonska-Soyta, A. 2002. Interferon-tau and its immunological role in ruminant reproduction. *Archivum Immunologiae et Therapiae Experimentalis* **50**(1), 47–52.

Colonna, M., Krug, A., and Cella, M. 2002. Interferon-producing cells: on the front line in immune responses against pathogens. *Current Opinion in Immunology* **14**(3), 373–379.

Foster, G. 2004. PEGylated interferons: chemical and clinical differences. *Alimentary Pharmacology and Therapeutics* **20**(8), 825–830.

Haria, M. and Benfield, P. 1995. Interferon-α-2a. *Drugs* **50**(5), 873–896.

Jonasch, E. and Haluska, F. 2001. Interferon in oncological practice: review of interferon biology, clinical applications and toxicities. *Oncologist* **6**(1), 34–55.

Katre, N. 1993. The conjugation of proteins with polyethylene glycol and other polymers – altering properties of proteins to enhance their therapeutic potential. *Advanced Drug Delivery Reviews* **10**(1), 91–114.

Kirkwood, J. 2002. Cancer immunotherapy: the interferon-alpha experience. *Seminars in Oncology* **29**(3), 18–26.

Leaman, D.W., Leung, S., Li, X., and Stark, G.R. 1996. Regulation of stat-dependent pathways by growth factors and cytokines. *FASEB Journal* **10**(14), 1578–1588.

Leon, M. and Zuckerman, S. 2005. Gamma interferon: a central mediator in atherosclerosis. *Inflammation Research* **54**(10), 395–411.

Li, J. and Roberts, M. 1994. Interferon-τ and interferon-α interact with the same receptors in bovine endometrium. *Journal of Biological Chemistry* **269**(18), 13 544–13 550.

Lyseng-Williamson, K. and Plosker, G. 2002. Management of relapsing–remitting multiple sclerosis – defining the role of subcutaneous recombinant interferon-β-1a (Rebif). *Disease Management and Health Outcomes* **10**(5), 307–325.

Meager, A. 2002. Biological assays for interferons. *Journal of Immunological Methods* **261**(1–2), 21–36.

Pestka, S. and Langer, J. 1987. Interferons and their actions. *Annual Review of Biochemistry* **56**, 727–777.

Platanias, L. 2005. Mechanisms of type-I and type-II interferon-mediated signalling. *Nature Reviews Immunology* **5**(5), 375–386.

Rio, J. and Montalban, X. 2005. Interferon-β_{1b} in the treatment of multiple sclerosis. *Expert Opinion on Pharmacotherapy* **6**(16), 2877–2886.

Schroder, K., Hertzog, P.J., Ravasi, T., and Hume, D.A. 2004. Interferon gamma: an overview of signals, mechanisms and function. *Journal of Leukocyte Biology* **75**(2), 163–189.

Simko, R. and Nagy, K. 1996. Interferon-alpha in childhood hematological malignancies. *Postgraduate Medical Journal* **72**(854), 709–713.

Sleijfer, S., Bannink, M., van Cool, A.R., Kruit, W.H.J., and Stoter, G. 2005. Side effects of interferon-α therapy. *Pharmacy World and Science* **27**(6), 423–431.

Soos, J. and Johnson, H. 1999. Interferon-τ. Prospects for clinical use in autoimmune disorders. *BioDrugs* **11**(2), 125–135.

Todd, P. and Goa, K. 1992. Interferon gamma-1b. *Drugs* **43**(1), 111–122.

Tossing, G. 2001. New developments in interferon therapy. *European Journal of Medical Research* **6**(2), 47–65.

Veronese, F.M. and Pasut, G. 2005. PEGylation, successful approach to drug delivery. *Drug Discovery Today* **10**, 1451–1458.

Wang, Y.-S., Youngster, S., Grace, M., Bausch, J., Bordens, R., and Wyss, D.F. 2002. Structural and biological characterization of pegylated recombinant interferon alpha-2b and its therapeutic implications. *Advanced Drug Delivery Reviews* **54**(4), 547–570

Woo, M. and Burnakis, T. 1997. Interferon-alpha in the treatment of chronic viral hepatitis-B and viral hepatitis C. *Annals of Pharmacotherapy* **31**(3), 330–337.

Younes, H. and Amsden, B. 2002. Interferon-gamma therapy: evaluation of routes of administration and delivery systems. *Journal of Pharmaceutical Sciences* **91**(1), 2–17.

Young, H. and Hardy, K. 1995. Role of interferon-γ in immune cell regulation. *Journal of Leukocyte Biology* **58**, 373–379.

Zavyalov, V. and Zavyalova, G. 1997. Interferons alpha/beta and their receptors – place in the hierarchy of cytokines. *APMIS* **105**(3), 161–186

9
Cytokines: Interleukins and tumour necrosis factor

9.1 Introduction

The interleukins represent another large family of cytokines, with at least 33 different constituent members (IL-1 to IL-33) having been characterized thus far. Most of these polypeptide regulatory factors are glycosylated (a notable exception being IL-1) and display a molecular mass ranging from 15 to 30 kDa. A few interleukins display a higher molecular mass, e.g. the heavily glycosylated, 40 kDa, IL-9.

Most of the interleukins are produced by a number of different cell types. At least 17 different cell types are capable of producing IL-1, and IL-8 is produced by at least 10 distinct cell types. On the other hand, IL-2, -9 and -13 are produced only by T-lymphocytes.

Most cells capable of synthesizing one interleukin are capable of synthesizing several, and many prominent producers of interleukins are non-immune system cells (Table 9.1). Regulation of interleukin synthesis is exceedingly complex and only partially understood. In most instances, induction or repression of any one interleukin is prompted by numerous regulators (mostly additional cytokines). IL-1, for example, promotes increased synthesis and release of IL-2 from activated T-lymphocytes. It is highly unlikely that cells capable of synthesizing multiple interleukins concurrently synthesize them all at high levels.

Nearly all of the interleukins are soluble molecules (one form of IL-1 is cell associated). They promote their biological response by binding to specific receptors on the surface of target cells. Most interleukins exhibit paracrine activity (i.e. the target cells are in the immediate vicinity of the producer cells), although some display autocrine activity (e.g. IL-2 can stimulate the growth and differentiation of the cells that produce it). Other interleukins display more systematic endocrine effects (e.g. some activities of IL-1).

The signal transduction mechanisms by which most interleukins prompt their biological response are understood, in outline at least. In many cases, interleukin cell surface receptor binding is associated with intracellular tyrosine phosphorylation events. In other cases, serine and threonine residues of specific intracellular substrates are also phosphorylated. For some interleukins,

Pharmaceutical biotechnology: concepts and applications   Gary Walsh
© 2007 John Wiley & Sons, Ltd   ISBN 978 0 470 01244 4 (HB)   978 0 470 01245 1 (PB)

Table 9.1 Many cell types are capable of producing a whole range of interleukins. T-lymphocytes are capable of producing all the interleukins, with the possible exception of IL-7 and IL-15. Many cell types producing multiple interleukins can also produce additional cytokines. For example, both macrophages and fibroblasts are capable of producing several interleukins, CSFs and PDGF

Cell type	Interleukins produced
Macrophages	IL-1, IL-6, IL-10, IL-12
Eosinophils	IL-3, IL-5
Vascular endothelial cells	IL-1, IL-6, IL-8
Fibroblasts	IL-1, IL-6, IL-8, IL-11
Keratinocytes	IL-1, IL-6, IL-8, IL-10

receptor binding triggers alternative signal transduction events, including promoting an increase in intracellular calcium concentration, or inducing hydrolysis of phosphatidylethanolamine with release of diacyl glycerol.

The sum total of biological responses induced by the interleukins is large, varied and exceedingly complex. These cytokines regulate a variety of physiological and pathological conditions, including:

- normal and malignant cell growth;

- all aspects of the immune response;

- regulation of inflammation.

Several interleukins, particularly those capable of modulating transformed cell growth, as well as those exhibiting immunostimulatory properties, enjoy significant clinical interest. As with other cytokines, the advent of recombinant DNA technology facilitates production of these molecules in quantities sufficient to meet actual/potential medical needs. Thus far, only two interleukin-based products have gained approval for general medical use ('Proleukin', an IL-2 and 'Neumega', an IL-11; see later) and these interleukins, along with IL-1, form the focus of much of this chapter. Literature detailing additional interleukins is cited in the 'Further reading' section, to which the interested reader is referred.

9.2 Interleukin-2

IL-2, also known as T-cell growth factor, represents the most studied member of the interleukin family. It was the first T-cell growth factor to be identified and it plays a central role in the immune response. It is produced exclusively by T-lymphocytes (especially T-helper cells), in response to activation by antigen and mitogens.

Human IL-2 is a single-chain polypeptide containing 133 amino acids. It is a glycoprotein, the carbohydrate component being attached via an O-linked glycosidic bond to threonine

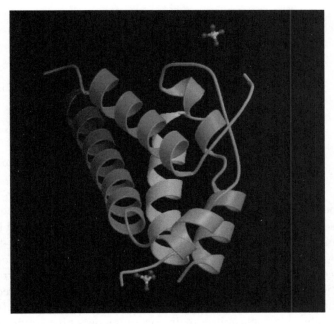

Figure 9.1 Three-dimensional structure of IL-2. Structural details courtesy of the Protein Data Bank, http://www.rcsb.org/pdb/

residue no. 3. The mature molecule displays a molecular mass ranging from 15 to 20 kDa, depending upon the extent of glycosylation. The carbohydrate moiety is not required for biological activity.

X-ray diffraction analysis shows the protein to be a globular structure, consisting of four α-helical stretches interrupted by bends and loops. It appears devoid of any β-conformation and contains a single stabilizing disulfide linkage involving cysteine numbers 58 and 105 (Figure 9.1).

IL-2 induces its characteristic biological activities by binding a specific receptor on the surface of sensitive cells. The high-affinity receptor complex consists of three membrane-spanning polypeptide chains (α, β and γ; Table 9.2).

Table 9.2 Summary of the polypeptide constituents of the high-affinity human IL-2 receptor

Receptor polypeptide constituent	Additional names	Molecular mass (kDa)
α	P55	
	Tac	55
	CD25	
β	P75	
	CD122	75
γ	P64	64

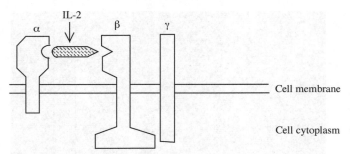

Figure 9.2   Schematic diagram of the high affinity IL-2 receptor

The α chain binds IL-2 with low affinity, with binding being characterized by high subsequent association–disassociation rates. The γ subunit does not interact directly with IL-2. It is sometimes known as γc (common), as it also appears to be a constituent of the IL-4, -7, -9, -13 and -15 receptors.

Heterodimeric complexes consisting of αγ or βγ can bind IL-2 with intermediate affinity. The heterotrimeric αβγ complex represents the cytokine's true high-affinity receptor (Figure 9.2). The exact intracellular signal transduction events triggered by IL-2 are not fully elucidated. The α receptor chain exhibits a cytoplasmic domain containing only nine amino acids, and is unlikely to play a role in intracellular signalling. Mutational studies reveal that the 286 amino acid β chain intracellular domain contains at least two regions (a serine-rich region and an acidic region) essential for signal transduction. The β chain is phosphorylated on a specific tyrosine residue subsequent to IL-2 binding, probably via association of a protein tyrosine kinase essential for generation of intracellular signals. The direct role played by the γ chain, although unclear, is likely critical. A mutation in the gene coding for this receptor constituent results in severe combined immunodeficiency (X-SCID).

Interestingly, prolonged elevated levels of IL-2 promote the shedding of the IL-2 receptor α subunit from the cell surface. Initially, it was suspected that these circulating soluble receptor fragments, capable of binding IL-2, might play a role in inducing immunosuppression (by competing for IL-2 with the cell surface receptor). However, the affinity of IL-2 for the intact receptor (αβγ) is far greater than for the α subunit, rendering this theory unlikely.

The IL-2 receptor is associated with a number of cell types, mainly cells playing a central role in the immune response (Table 9.3). Binding of IL-2 to its receptor induces growth and

Table 9.3   The range of cells expressing the IL-2 cell surface receptor. IL-2-stimulated growth and differentiation of these cells forms the molecular basis by which many aspects of the immune response are activated. It thus acts in an autocrine and paracrine manner to mobilize the immune response

T-lymphocytes	B-lymphocytes
NK cells	LAK cells
Monocytes	Macrophages
Oligodendrocytes	

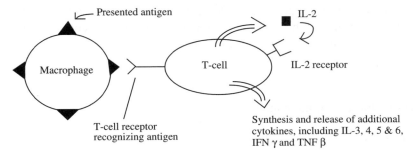

Figure 9.3 Activation of T-cells by interaction with macrophage-displayed antigen. Activation results in IL-2 production, which acts in an autocrine manner to stimulate further T-cell growth and division. IL-2 thus represents the major regulatory molecule responsible for stimulation of cell-mediated immunity. Note that it was initially believed that binding of presented antigen alone was insufficient to trigger T-cell activation. It was thought that co-stimulation with IL-1 was required. However, the assay used to detect the 'co-stimulation' was found not to be specific for IL-1 alone. The role of IL-1 as a co-stimulator of T-cell activation is now believed to be minimal at most

differentiation of these cells. This cytokine, therefore, behaves as a central molecular switch, activating most aspects of the immune response.

Quiescent T-lymphocytes are stimulated largely by direct binding to an antigen fragment presented on the surface of a macrophage in the context of MHC complex (Figure 9.3). This results in the induction of expression of at least 70 genes whose products are collectively important in immune stimulation. These products include:

- Several cytoplasmic proteins capable of inducing T-cell growth (i.e. several cellular proto-oncogenes, including C-*fos* and C-*myc*).

- Various cytokines, most notably IL-2.

- Cytokine receptors, most notably the IL-2 receptor α subunit. (The T-lymphocytes appear to constitutively express the β and γ IL-2 receptor polypeptides. Induction of the α gene leads to formation of a high-affinity αβγ receptor complex, thereby rendering the activated T cell highly sensitive to IL-2.)

IL-2 acts as a critical autocrine growth factor for T-cells, and the magnitude of the T-cell response is largely dependent upon the level of IL-2 produced. IL-2 also serves as a growth factor for activated B-lymphocytes. In addition to promoting proliferation of these cells, IL-2 (as well as some other interleukins) stimulates enhanced antibody production and secretion. In this way, it effectively potentates the humoral immune response.

A third biological activity of IL-2 pertinent to immunostimulation is its ability to promote the growth of NK cells. It also promotes further differentiation of NK cells, forming lymphokine-activated killer (LAK) cells, which exhibit an enhanced ability to kill tumour cells or virally infected cells directly. NK cells express the β and γ IL-2 receptor subunits only; thus, their stimulation by IL-2 requires elevated concentrations of this cytokine. NK cells are also activated by a variety of additional cytokines, including all interferons and TNF.

The immunopotentative effects of IL-2 rendered it an obvious target for clinical application. At the simplest level, it was hoped that administration of exogenous IL-2 could enhance the immune response to a number of clinical conditions, including:

- cancers,

- T-cell and other forms of immunodeficiency,

- infectious diseases.

9.2.1 Interleukin-2 production

As was the case for most other cytokines, medical appraisal/use of IL-2 was initially impractical due to the minute quantities in which it is normally produced. Some transformed cell lines, most notably the Jurkat leukaemia cell line, produces IL-2 in increased quantities, and much of the IL-2 used for initial characterization studies was obtained from this source. Large-scale IL-2 production was made possible by recombinant DNA technologies. Although the IL-2 gene/cDNA has now been expressed in a wide variety of host systems, it was initially expressed in *E. coli*, and most products being clinically evaluated are obtained from that source. As mentioned previously, the absence of glycosylation on the recombinant product does not alter its biological activity.

Proleukin is the tradename given to the recombinant IL-2 preparation manufactured by Chiron and approved for the treatment of certain cancers. It is produced in engineered *E. coli* and differs from native human IL-2 in that it is non-glycosylated, lacks an N-terminal alanine residue, and cysteine 125 has been replaced by a serine residue. After extraction and chromatographic purification, the product is formulated in a phosphate buffer containing mannitol and low levels of the detergent SDS. The product displays typical IL-2 biological activities, including enhancement of lymphocyte mitogenesis and cytotoxicity, induction of LAK/NK cell activity and the induction of IFN-γ production.

9.2.2 Interleukin-2 and cancer treatment

The immunostimulatory activity of IL-2 has proven beneficial in the treatment of some cancer types. An effective anti-cancer agent would prove not only medically valuable, but also commercially very successful. In the developed world, an average of one-in-six deaths is caused by cancer. In the USA alone, the annual death toll from cancer stands in the region of half a million people.

There exists direct evidence that the immune system mounts an immune response against most cancer types. Virtually all transformed cells express (a) novel surface antigens not expressed by normal cells or (b) express, at greatly elevated levels, certain antigens present normally on the cell at extremely low levels. These 'normal' expression levels may be so low that they have gone unnoticed by immune surveillance (and thus have not induced immunological tolerance).

The appearance of any such cancer-associated antigen should thus be capable of inducing an immune response, which, if successful, should eradicate the transformed cells. The exact elements of immunity responsible for destruction of transformed cells remain to be fully characterized. Both a humoral and cell-mediated response can be induced, although the T-cell response appears to be the most significant.

Cytotoxic T cells may play a role in inducing direct destruction of cancer cells, in particular those transformed by viral infection (and who express viral antigen on their surface). *In vitro* studies have shown that cytotoxic T-lymphocytes obtained from the blood of persons suffering from various cancer types are capable of destroying those cancer cells.

NK cells are capable of efficiently lysing various cancer cell types and, as already discussed, IL-2 can stimulate differentiation of NK cells forming LAK cells, which exhibit enhanced tumouricidal activity. Macrophages, too, probably play a role. Activated macrophages have been shown to lyse tumour cells *in vitro*, while leaving untransformed cells unaffected. Furthermore, these cells produce TNF and various other cytokines which can trigger additional immunological responses. The production of antibodies against tumour antigens (and the subsequent binding of the antibodies to those antigens) marks the transformed cells for destruction by NK/LAK cells and macrophages – all of which exhibit receptors capable of binding the F_c portion of antibodies.

Although immune surveillance is certainly responsible for the detection and eradication of some transformed cells, the prevalence of cancer indicates that this surveillance is nowhere near 100 per cent effective. Some transformed cells obviously display characteristics that allow them to evade this immune surveillance. The exact molecular details of how such 'tumour escape' is achieved remains to be confirmed, although several mechanisms have been implicated, including:

- Most transformed cells do not express class II MHC molecules and express lower than normal levels of class I MHC molecules. This renders their detection by immune effector cells more difficult. Treatment with cytokines, such as IFN-γ, can induce increased class I MHC expression, which normally promotes increased tumour cell susceptibility to immune destruction.

- Transformed cells expressing tumour-specific surface antigens that closely resemble normal surface antigens may not induce an immune response. Furthermore, some tumour antigens, although not usually expressed in adults, were expressed previously during the neonatal period (i.e. just after birth) and are thus believed by the immune cells to be 'self'.

- Some tumours secrete significant quantities of cytokines and additional regulatory molecules that can suppress local immunological activity. TGF-β (produced by many tumour types), for example, is capable of inhibiting lymphocyte and macrophage activity.

- Antibody binding to many tumour antigens triggers the immediate loss of the antibody–antigen complex from the transformed cell surface, either by endocytosis or extracellular shedding.

- The glycocalyx (carbohydrate-rich outer cell coat) can possibly shield tumour antigens from the immune system.

Whatever the exact nature of tumour escape, it has been demonstrated, both *in vitro* and *in vivo*, that immunostimulation can lead to enhanced tumour detection and destruction. Several approaches to cancer immunotherapy have thus been formulated, many involving application of IL-2 as the primary immunostimulant.

Experiments conducted in the early 1980s showed that lymphocytes incubated *in vitro* with IL-2 could subsequently kill a range of cultured cancer cell lines, including melanoma and colon cancer cells. These latter cancers do not respond well to conventional therapies. Subsequent investigations showed that cancer cell destruction was mediated by IL-2-stimulated NK cells (i.e. LAK cells). Similar responses were seen in animal models upon administration of LAK cells activated *in vitro* using IL-2.

Clinical studies have shown this approach to be effective in humans. LAK cells originally purified from a patient's own blood, activated *in vitro* using IL-2 and reintroduced into the patient along with more IL-2, promoted complete tumour regression in 10 per cent of patients suffering from melanoma or renal cancer. Partial regression was observed in a further 10–25 per cent of such patients. Administration of high doses of IL-2 alone could induce similar responses, but significant side effects were noted (discussed later).

IL-2-stimulated cytotoxic T cells appear even more efficacious than LAK cells in promoting tumour regression. The approach adopted here entails removal of a tumour biopsy, followed by isolation of T-lymphocytes present within the tumour. These tumour-infiltrating lymphocytes (TILs) are cytotoxic T-lymphocytes that apparently display a cell surface receptor which specifically binds the tumour antigen in question. They are thus tumour-specific cells. Further activation of these TILs by *in vitro* culturing in the presence of IL-2, followed by reintroduction into the patient along with IL-2, promoted partial/full tumour regression in well over 50 per cent of treated patients.

Further studies have shown additional cancer types, most notably ovarian and bladder cancer, non-Hodgkin's lymphoma and acute myeloid leukaemia, to be at least partially responsive to IL-2 treatment. However, a persistent feature of clinical investigations assessing IL-2 effects on various cancer types is variability of response. Several trials have yielded conflicting results, and no reliable predictor of clinical response is available.

9.2.3 Interleukin-2 and infectious diseases

Although antibiotics have rendered possible the medical control of various infectious agents (mainly bacterial), numerous pathogens remain for which no effective treatment exists. Most of these pathogens are non-bacterial (e.g. viral, fungal and parasitic, including protozoal). In addition, the overuse/abuse of antibiotics has hastened the development of antibiotic-resistant 'super bacteria', which have become a serious medical problem.

The most difficult microbial pathogens to treat are often those that replicate within host cells (e.g. viruses and some parasites). For example, during the complex life cycle of the malaria protozoan in humans, the parasite can infect (and destroy) liver cells and erythrocytes. Over 2 million people die each year from malaria, with at least 200–300 million people being infected at any given time. Some such agents have even evolved to survive and replicate within macrophages subsequent to uptake via phagocytosis. This is often achieved on the basis that the phagocytosed microbe is somehow capable of preventing fusion of the phagocytosed vesicle with lysozomes. Examples of pathogens capable of survival within macrophages include:

- Mycobacteria (e.g. *M. tuberculosis*, the causative agent of tuberculosis, and *M. leprae*, which causes leprosy).

- *Listeria monocytogenes*, a bacterium that, when transmitted to humans, causes listeriosis. Listeriosis is characterized by flu-like symptoms, but can cause swelling of the brain and induce abortions.

- *Legionella pneumophila*, the bacterium that causes legionnaire's disease.

The immunological response raised against intracellular pathogens is largely a T-cell response. IL-2's ability to stimulate T-cells may render it useful in the treatment of a wide range of such conditions. Clinical trials assessing its efficacy in treating a range of infectious diseases, including AIDS, continue. A related potential medical application of IL-2 relates to its possible use as adjuvant material in vaccination programmes (Chapter 13).

9.2.4   Safety issues

Like all other cytokines, administration of IL-2 can induce side effects that can be dose limiting. Serious side effects, including cardiovascular, hepatic or pulmonary complications, usually necessitate immediate termination of treatment. Such side effects may be induced not only directly by IL-2, but also by a range of additional cytokines whose synthesis is augmented by IL-2 administration. These cytokines, which can include IL-3, -4, -5 and -6, as well as TNF and IFN-γ, also likely play a direct role in the overall therapeutic benefits accrued from IL-2 administration.

9.2.5   Inhibition of interleukin-2 activity

A variety of medical conditions exist that are caused or exacerbated by the immune system itself. These are usually treated by administering immunosuppressive agents. Examples include:

- Autoimmune diseases in which immunological self-tolerance breaks down and the immune system launches an attack on self-antigens.

- Tissue/organ transplantation in which the donor is not genetically identical to the recipient (i.e. in cases other than identical twins). The recipient will mount an immune response against the transplanted tissue, culminating in tissue rejection unless immunosuppressive agents are administered.

Selective immunosuppression in individuals suffering from the above conditions is likely best achieved by preventing the synthesis or functioning of IL-2. Cyclosporin A, one of the foremost immunosuppressive agents currently in use, functions by preventing IL-2 synthesis. A number of alternative approaches are now being considered or tested directly in clinical trials. These include:

- Administration of soluble forms of the IL-2 receptor, which would complete with the native (cell surface) receptor for binding of IL-2.

- Administration of monoclonal antibodies capable of binding the IL-2 receptor. (It must be confirmed that the antibody used is not itself capable of initiating signal transduction upon binding the receptor.) See the products 'Zenapax' and 'Simulect'; Chapter 13.

- Administration of IL-2 variants that retain the ability to bind the receptor, but which fail to initiate signal transduction.

- Administration of IL-2 coupled to bacterial or other toxins. Binding of the cytokine to its receptor brings the associated toxin into intimate contact with the antigen-activated T cells (and other cells, including activated B cells), leading to the destruction of these cells. This would induce selective immunotolerance to whatever specific antigen activated the B/T cells. One product (tradename Ontak; Box 9.1) that is based upon this principle is now in general medical use.

Box 9.1

Product case study: Ontak

Ontak (tradename, also known as denileukin diftitox) is an engineered fusion protein produced by recombinant means in *E. coli*. It gained approval for general medical use in the USA in 1999 and is indicated for the treatment of patients with cutaneous T-cell lymphoma (CTCL). The fusion product is composed of diphtheria toxin fragments A and B directly linked to IL-2. It displays a molecular weight of 58 kDa and is purified using reverse-phase chromatography and diafiltration. The product is formulated as a sterile frozen solution containing citric acid, EDTA and polysorbate as excipients. It is presented in single-use vials. The product is normally administered i.v. daily for five consecutive days every 3 weeks over the treatment period.

The IL-2 portion of the fusion protein facilitates product interaction with cells displaying cell surface IL-2 receptors, found in high levels on some leukaemia and lymphoma cells, including CTCL cells. Binding appears to trigger internalization of the receptor–fusion protein complex (Figure 9.B1). Sufficient quantities of the latter escape immediate cellular destruction to allow diphtheria toxin-mediated inhibition of cellular protein synthesis. Cell death usually results within hours.

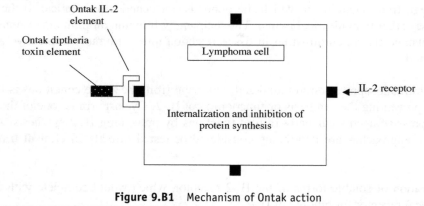

Figure 9.B1   Mechanism of Ontak action

Pharmacokinetic studies indicate the product displays two-compartment behaviour, with a distribution phase (half-life 2–5 min) and a terminal phase (half-life 70–80 min). Development of antibodies to the product significantly impacts upon clearance rates.

The major clinical study underpinning product approval was a randomized, double blind study in which 71 CTCL patients were administered the product at one of two dosage levels (9 or 18 μg kg^{-1} day^{-1}); overall, 30 per cent of patients experienced an objective tumour response. Serious side effects potentially associated with product administration include acute hypersensitivity-type reactions, vascular leak syndrome and visual impairment. Additional adverse reactions include flu-like symptoms, headache, hyper- or hypo-tension, as well as digestive upset. Ontak is manufactured by Seragen Inc. and is distributed by Ligand Pharmaceuticals.

9.3 Interleukin-1

IL-1 is also known as lymphocyte-activating factor (LAF), endogenous pyrogen and catabolin. It displays a wide variety of biological activities and has been appraised clinically in several trials.

Two distinct forms of IL-1 exist: IL-1α and IL-1β. Although different gene products, and exhibiting only 20 per cent amino acid sequence homology, both of these molecules bind the same receptor and induce similar biological activities. The genes coding for IL-1α and -1β both reside on human chromosome number 2, and display similar molecular organization, both containing seven exons.

IL-1α and -1β are expressed as large (30 kDa) precursor molecules from which the mature polypeptide is released by proteolytic cleavage. Neither IL-1α and -1β possess any known secretory signal peptide, and the molecular mechanism by which they exit the cell remains to be characterized. Neither interleukin appears to be glycosylated.

IL-1α is initially synthesized as a 271 amino acid precursor, with the mature form containing 159 amino acids (17.5 kDa). This molecule appears to remain associated with the extracellular face of the cell membrane. IL-1β, initially synthesized as a 269 amino acid precursor, is released fully from the cell. The mature form released contains 153 amino acids and displays a molecular mass in the region of 17.3 kDa.

X-ray diffraction analysis reveals the three-dimensional structure of both IL-1 molecules to be quite similar. Both are globular proteins, composed of six strands of antiparallel β pleated sheet forming a 'barrel' that is closed at one end by a further series of β sheets.

A wide range of cells are capable of producing IL-1 (Table 9.4). Different cell types produce the different IL-1s in varying ratios. In fibroblasts and endothelial cells, both are produced in roughly similar ratios, whereas IL-1β is produced in larger quantities than IL-1α in monocytes. Activated macrophages appear to represent the major cellular source for IL-1.

The IL-1s induce their characteristic biological activities by binding to specific cell surface receptors present on sensitive cells. Two distinct receptors, type I and II, have been identified. Both IL-1α and IL-1β can bind both receptors. The type I receptor is an 80 kDa transmembrane glycoprotein. It is a member of the IgG superfamily. This receptor is expressed predominantly on fibroblasts, keratinocytes, hepatocytes and endothelial cells. The type II receptor is a 60 kDa transmembrane glycoprotein, expressed mainly on B-lymphocytes, bone marrow cells and polymorphonuclear leukocytes. It displays a very short (29 amino acid) intracellular domain,

Table 9.4 The range of cells capable of producing IL-1

T-lymphocytes	Vascular endothelial cells
B-lymphocytes	Fibroblasts
Monocytes/macrophages	Astrocytes
NK cells	Dendritic cells
Large granular lymphocytes	Microglia
Keratinocytes	Glioma cells
Chondrocytes	

and some studies suggest that IL-1s can induce a biological response only upon binding to the type I receptor.

The exact IL-1-mediated mechanism(s) of signal transduction remain to be clarified. A number of different signal transduction pathways have been implicated, including involvement of G-proteins. IL-1 has also been implicated in activation of protein kinase C by inducing the hydrolysis of phosphatidylethanolamine.

9.3.1 The biological activities of interleukin-1

IL-1 mediates a wide variety of biological activities:

- It is a pro-inflammatory cytokine, promoting the synthesis of various substances such as eicosanoids, as well as proteases and other enzymes involved in generating inflammatory mediators. This appears to be its major biological function.

- It plays a role in activating B-lymphocytes, along with additional cytokines, and may also play a role in activating T-lymphocytes.

- Along with IL-6, it induces synthesis of acute-phase proteins in hepatocytes.

- It acts as a co-stimulator of haematopoietic cell growth/differentiation.

The relative prominence of these various biological activities depends largely upon the quantities of IL-1 produced in any given situation. At low concentrations, its effects are largely paracrine, e.g. induction of local inflammation. At elevated concentrations, it acts more in an endocrine manner, inducing systematic effects, such as the hepatic synthesis of acute-phase proteins, but also induction of fever (hence the name, endogenous pyrogen) and cachexia (general body wasting, such as that associated with some cancers). Many of these biological activities are also promoted by TNF, which is another example of cytokine redundancy.

In addition to IL-1α and -1β, a third IL-1-like protein has been identified, termed IL-1 receptor antagonist (IL-1Rα). As its name suggests, this molecule appears to be capable of binding to the IL-1 receptors without triggering an intracellular response.

9.3.2 Interleukin-1 biotechnology

IL-1 continues to be a focus of clinical investigation. This stems from its observed:

- immunostimulatory effects;

- ability to protect/restore the haematopoietic process during, or subsequent to, chemotherapy or radiation therapy;

- anti-proliferative effects against various human tumour cell lines grown *in vitro*, or in animal models.

Most of these effects are most likely mediated not only directly by IL-1, but also by various additional cytokines (including IL-2) induced by IL-1 administration.

The observed effects prompted initiation of clinical trials assessing IL-1's efficacy in treating:

- bone marrow suppression induced by chemo/radiotherapy;

- various cancers.

The initial findings of some such trials (involving both IL-1α and IL-1β) proved disappointing. No significant anti-tumour response was observed in many cases, although side effects were commonly observed. Virtually all patients suffered from fevers, chills and other flu-like symptoms. More serious side effects, including capillary leakage syndrome and hypotension, were also observed and were dose limiting.

IL-1 thus displays toxic effects comparable to administration of TNF (see later), or high levels of IL-2. However, several clinical studies are still underway, and this cytokine may yet prove therapeutically useful, either on its own or, more likely, when administered at lower doses with additional therapeutic agents.

Because of its role in mediating acute/chronic inflammation, (downward) modulation of IL-1 levels may prove effective in ameliorating the clinical severity of these conditions. Again, several approaches may prove useful in this regard, including administration of:

- anti-IL-1 antibodies;

- soluble forms of the IL-1 receptor;

- the native IL-1 receptor antagonist.

Kineret is the tradename given to a recently approved product based on the latter strategy. Indicated in the treatment of rheumatoid arthritis, the product consists of a recombinant form of the human IL-1 receptor antagonist. The 17.3 kDa, 153 amino acid product is produced in engineered *E. coli* and differs from the native human molecule in that it is non-glycosylated and contains an additional N-terminal methionine residue (a consequence of its prokaryotic expression system).

The purified product is presented as a solution and contains sodium citrate, EDTA, sodium chloride and polysorbate 80 as excipients. A daily (s.c.) injection of 100 mg is recommended for patients with rheumatoid arthritis. This inflammatory condition is (not surprisingly) characterized by the presence of high levels of IL-1 in the synovial fluid of affected joints. In addition to its pro-inflammatory properties, IL-1 also mediates additional negative influences on the joint/bone, including promoting cartilage degradation and stimulation of bone resorption.

An additional approach to IL-1 down-regulation could entail development of inhibitors of the proteolytic enzymes that release the active interleukin from its inactive precursor. Moreover, such inhibitors could probably be taken orally and, thus, would be suitable to treat chronic inflammation (the alternatives outlined above would be administered parenterally).

The enzyme that releases active IL-1β from its 31 kDa precursor has been identified and studied in detail. Termed IL-1β converting enzyme (ICE), it is a serine protease whose only known physiological substrate is the inactive IL-1β precursor. ICE cleaves this precursor between Asp 116 and Ala 117, releasing the active IL-1β.

ICE is an oligomeric enzyme (its active form may be a tetramer). It contains two distinct polypeptide subunits, p20 (20 kDa) and p10 (10 kDa). These two subunit types associate very closely, and the protease's active site spans residues from both. p10 and p20 are proteolytically derived from a single 45 kDa precursor protein.

9.4   Interleukin-11

IL-11 is an additional cytokine that has gained approval for general medical use. This cytokine, produced largely by IL-1-activated bone marrow stromal cells and fibroblasts, functions as a haematopoietic growth factor. It stimulates (a) thrombopoiesis (the production of platelets formed by the shedding of fragments of cytoplasm from large bone marrow cells called megakaryocytes; Chapter 10) and (b) growth/differentiation of bone marrow cells, derived from stem cells that have become committed to differentiate into macrophages.

IL-11 is a 23 kDa, 178 amino acid polypeptide. Its receptor appears to be a single-chain 150 kDa transmembrane protein. Binding of IL-11 results in tyrosine phosphorylation of several intracellular proteins, which, in turn, somehow promote the observed biological activities of IL-11.

The rationale for assessing IL-11 as a potential therapeutic agent relates mainly to its ability to induce platelet synthesis. Platelet counts can be significantly lowered due to certain disease conditions and in patients undergoing cancer chemotherapy. Trials in patients receiving chemotherapy illustrated that s.c. administration of IL-11, at levels above 25 μg per kilogram body weight per day, did stimulate platelet production, at least partially offsetting the chemotherapeutic effects. When administered at levels of 50 μg kg^{-1} day^{-1}, a significant increase in the bone marrow megakaryocyte population was noted. At levels in excess of 75 μg kg^{-1} day^{-1}, side effects, including fatigue and oedema (tissue swelling due to accumulation of fluid), were also noted.

Neumega is the tradename given to the IL-11-based product approved for the prevention of thrombocytopenia. The product is produced in engineered E. coli cells and is presented as a purified product in freeze-dried format. Excipients include phosphate buffer salts and glycine. It is reconstituted (with water for injections) to a concentration of 5 mg ml^{-1} before s.c. administration.

9.5 Tumour necrosis factors

The TNF family of cytokines essentially consists of two related regulatory factors: TNF-α (cachectin) and TNF-β (lymphotoxin). Although both molecules bind the same receptor and induce very similar biological activities, they display limited sequence homology. The human TNF-α and -β genes are located adjacent to each other on chromosome 6, being separated by only 1100 base pairs. Both contain three introns, and their expression appears to be coordinately regulated. TNF-α, sometimes referred to simply as TNF, has been studied in significantly greater detail than lymphotoxin.

9.5.1 Tumour necrosis factor biochemistry

TNF-α is also known as cachectin, macrophage cytotoxic factor, macrophage cytotoxin and necrosin. As some of these names suggest, activated macrophages appear to represent the most significant cellular source of TNF-α, but it is also synthesized by many other cell types (Table 9.5). Producer cells do not store TNF-α, but synthesize it *de novo* following activation.

Human TNF-α is initially synthesized as a 233 amino acid polypeptide that is anchored in the plasma membrane by a single membrane-spanning sequence. This TNF pro-peptide, which itself displays biological activity, is usually proteolytically processed by a specific extracellular metalloprotease. Proteolytic cleavage occurs between residues 76 (Ala) and 77 (Val), yielding the mature (soluble) 157 amino acid TNF-α polypeptide. Mature human TNF-α appears to be devoid of a carbohydrate component, and contains a single disulfide bond.

Monomeric TNF is biologically inactive; the active form is a homotrimer in which the three monomers associate non-covalently about a threefold axis of symmetry, forming a compact bell-shaped structure. X-ray crystallographic studies reveal that each monomer is elongated and characterized by a large content of antiparallel β pleated sheet, which closely resembles subunit proteins of many viral caspids (Figure 9.4).

A number of stimuli are known to act as inducers of TNF production (Table 9.6). Bacterial LPS represents the most important inducer, and TNF mediates the pathophysiological effects of this molecule. TNF biosynthesis is regulated by both transcriptional and post-transcriptional mechanisms. Macrophages appear to express TNF-α mRNA constitutively, which is translated only

Table 9.5 The major cellular sources of human TNF-β. As is evident, TNF-α synthesis is not restricted to cells of the immune system, but is undertaken by a wide variety of different cells in different anatomical locations, including the brain

Macrophages	B- and T-lymphocytes
NK cells	Polymorphonuclear leukocytes
Eosinophils	Astrocytes
Hepatic Kupffer cells	Langerhan's cells
Glomerular mesangial cells	Brain microglial cells
Fibroblasts	Various transformed cell lines

Figure 9.4   Three-dimensional structure of TNF-α. Structural details courtesy of the Protein Data Bank, http://www.rcsb.org/pdb/

upon their activation. After activation, the rate of gene expression may increase only threefold, although cellular TNF-α mRNA levels may increase 100-fold and secretion of soluble TNF-α may increase 10 000-fold.

9.5.2   Biological activities of tumour necrosis factor-α

The availability of large quantities of recombinant TNF-α facilitates rigorous investigation of the effects of this cytokine on cells *in vitro*, as well as its systemic *in vivo* effects. Like most cytokines, TNF-α exhibits pleiotropic effects on various cell types. The major biological responses induced by this cytokine include:

- activation of certain elements of both non-specific and specific immunity, in particular in response to Gram-negative bacteria;

Table 9.6   Major physiological inducers of TNF-α production

LPS	Bacterial enterotoxin
Mycobacteria	Various viruses
Fungi	Parasites
Antibody–antigen complexes	Various cytokines (e.g. IL-1)
Inflammatory mediators	TNF-α (autocrine activity)

- induction/regulation of inflammation;

- selective cytotoxic activity against a range of tumour cells;

- mediation of various pathological conditions, including septic shock, cachexia and anorexia.

The exact range of biological effects induced by TNF-α is dependent upon a number of factors, most notably the level at which TNF-α is produced. At low concentrations, TNF-α acts locally in a paracrine and autocrine manner, predominantly influencing white blood cells and endothelial cells. Under such circumstances, TNF-α's major activity relates to regulation of immunity and inflammation. In some situations, however, very large quantities of TNF-α may be produced (e.g. during severe Gram-negative bacterial infections). In such instances, TNF-α enters the blood stream and acts in an endocrine manner. Systemic effects of TNF-α (systemic means relating to the whole body, not just a specific area or organ), which include severe shock, are largely detrimental. Prolonged elevated systemic levels of TNF-α induce additional effects on, for example, whole-body metabolism. Many of TNF-α's biological effects are augmented by interferon-γ.

9.5.3 Immunity and inflammation

At low concentrations, TNF-α activates a range of leukocytes that mediate selected elements of both specific and non-specific immunity. These TNF-α actions include:

- activation of various phagocytic cells, including macrophages, neutrophils and polymorphonuclear leukocytes;

- enhanced toxicity of eosinophils and macrophages towards pathogens;

- exerting antiviral activity somewhat similar to class I interferons, and increasing surface expression of class I MHC molecules on sensitive cells;

- enhancing proliferation of IL-2-dependent T-lymphocytes.

In addition, TNF-α influences immunity indirectly by promoting synthesis and release of a variety of additional cytokines, including interferons, IL-1, IL-6, IL-8 and some CSFs.

TNF-α also plays a prominent role in mediating the inflammatory response; indeed, this may be its major normal physiological role. It promotes inflammation by a number of means, including:

- Promoting activation of neutrophils, eosinophils and other inflammatory leukocytes.

- Induction of expression of various adhesion molecules on the surface of vascular endothelial cells. These act as docking sites for neutrophils, monocytes and lymphocytes, facilitating their accumulation at local sites of inflammation.

- Displays chemotactic effects, especially for monocytes and polymorphonuclear leukocytes.

- Enhances vascular leakness by promoting a reorganization of the cytoskeleton of endothelial cells.

- Induction of synthesis of various lipid-based inflammatory mediators, including some prostaglandins and platelet-activating factor, by macrophages and other cells. Many of these promote sustained vasodilation and increased vascular leakage.

- Induction of synthesis of pro-inflammatory cytokines, such as IL-1 and IL-8.

TNF-α, therefore, promotes various aspects of immunity and inflammation. Blockage of its activity, e.g. by administration of anti-TNF-α antibodies, has been shown to compromise the body's ability to contain and destroy pathogens.

TNF-α exhibits cytotoxic effects on a wide range of transformed cells. Indeed, initial interest in this molecule (and its naming) stems from its anti-tumour activity. These investigations date back to the turn of the century, when an American doctor, William Coley, noted that tumours regressed in some cancer patients after the patient had suffered a severe bacterial infection. Although such observations prompted pioneering scientists to treat some cancer patients by injection with live bacteria, the approach was soon abandoned because many patients died due to the resulting uncontrolled bacterial infections, before they could be 'cured' of the cancer. The active tumour agent turned out to be TNF-α (high circulating levels of which were induced by the bacterially derived LPS).

TNF fails to induce death of all tumour cell types. Although many transformed cells are TNF sensitive, the cytokine exerts, at best, a cytostatic effect on others and has no effect on yet others. The cytotoxic activity is invariably enhanced by the presence of IFN-γ. The concurrent presence of this interferon increases the range of transformed cell types sensitive to TNF-α, and can upgrade its cytostatic effects to cytotoxic effects. It can also render many untransformed cells, in particular epithelial and endothelial cells, susceptible to the cytotoxic effects of TNF-α.

TNF-α can mediate death of sensitive cells via apoptosis or necrosis (necrotic death is characterized by clumping of the nuclear chromatin, cellular swelling, disintegration of intracellular organelles and cell lysis; apoptotic death is characterized by cellular shrinking, formation of dense 'apoptotic' masses and DNA fragmentation).

In addition to its cytotoxic effects, TNF-α appears to regulate the growth of some (non-transformed) cell types. It is capable of stimulating the growth of macrophages and fibroblasts, while suppressing division of haematopoietic stem cells. The systemic effects of this cytokine on cellular growth *in vitro* are thus complex and, as yet, only poorly appreciated.

9.5.4   Tumour necrosis factor receptors

Two TNF receptor types have been characterized. The type I receptor displays a molecular mass of 55 kDa, and is known as TNF-R55 (also P-55, TNFR1 or CD120α). The type II receptor is larger (75 kDa) and is known as TNF-R75 (also P-75, TNFR2 or CD120β). Both receptors bind both TNF-α and TNF-β. A TNF receptor is present on the plasma membrane of almost all nucleated cell types, generally in numbers varying from 100 to 10 000. Although virtually all cells express TNF-R55, the TNF-R75 cell surface receptor distribution is more restricted, this being most prominent on haematopoietic cells. Differential regulation of expression of these two receptor types is also apparent. Generally, low constitutive expression of TNF-R55 is observed, with TNF-R75 expression being inducible.

Both receptor types are members of the NGF receptor superfamily and exhibit the characteristic four (cysteine-rich) repeat units in their extracellular domain. The extracellular domains of TNF-R55 and TNF-R75 exhibit only 28 per cent homology and their intracellular domains are devoid of any homology, indicating the likely existence of distinct signalling mechanisms.

It appears that TNF-R55 is capable of mediating most TNF activities, whereas the biological activities induced via the TNF-R75 receptor are more limited. For example, TNF's cytotoxic activity, as well as its ability to induce synthesis of various cytokines and prostaglandins, is all mediated mainly/exclusively by TNF-R55. TNF-R75 appears to play a more prominent role in the induction of synthesis of T-lymphocytes. All of the biological activities mediated by TNF-R75 can also be triggered via TNF-R55, and usually at much lower densities of receptors. TNF-R75 thus appears to play more of an accessory role, mainly to enhance effects mediated via TNF-R55.

Binding of TNF to either receptor type results in oligomerization of the receptor (Figure 9.5). Indeed, antibodies raised against the extracellular domains of the receptors can induce TNF activity, indicating that the major/sole function of the TNF ligand is to promote such clustering of receptors. In most cases, binding of TNF to TNF-R55 results in rapid internalization of the ligand—receptor complex followed by lysosomal degradation. In contrast, binding of TNF to TNF-R75 does not induce such receptor internalization. In some cases, ligand binding appears to activate selective cleavage of the extracellular domain, resulting in the release of soluble TNF-R75. Soluble forms of both receptor types have been found in both the blood and urine.

The exact molecular mechanisms by which TNF-induced signal transduction are mediated remain to be characterized in detail. Oligomerization of the receptors is often followed by their phosphorylation, most likely by accessory kinases that associate with the intracellular domain of the receptor (neither receptor type displays intrinsic protein kinase activity). The existence of several phosphoproteins capable of associating with (the intracellular domain of) TNF-R55 and TNF-R75 have also been established. Following clustering of the TNF receptors, these

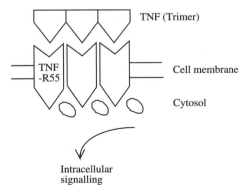

Figure 9.5 TNF binding to its receptor (TNF-R55), with resulting clustering of the receptor and generation of intracellular signals. Binding of TNF to its other receptor (TNF-R75) also induces receptor clustering. (See text for details)

associated proteins are likely to become activated, thus mediating additional downstream events that eventually trigger characteristic TNF molecular responses. The downstream events are complex and varied. Experimental evidence from various studies implicates a variety of mechanisms, including phosphorylation events, as well as activation of various phospholipases, resulting in the generation of messengers such as diacylglycerol, inositol phosphates and arachidonic acid.

9.5.5   Tumour necrosis factor: therapeutic aspects

The initial interest in utilizing TNF as a general anti-cancer agent has diminished, largely due to the realization that:

- many tumours are not susceptible to destruction mediated by TNF (indeed, some tumours produce TNF as an autocrine growth factor);

- tumour cell necrosis is not TNF's major biological activity;

- severe side effects usually accompany systemic administration of therapeutically relevant doses of this cytokine.

One such product (tradename Beromun) has been approved for general medical use (Box 9.2). Most clinical interest in TNF, however, now centres around neutralizing its biological effects in situations where overexpression of TNF induces negative clinical effects. TNF has been firmly implicated in mediating many of the adverse effects associated with dozens of diseases (Table 9.7). Administration of anti-TNF monoclonal antibodies or soluble forms of the TNF receptor should help reduce the severity of many of the symptoms of these diseases.

Enbrel is a product now approved for medical use that is based upon this strategy. The product is an engineered hybrid protein consisting of the extracellular domain of the TNF p75 receptor fused directly to the F_c (constant) region of human IgG (see Box 13.2 for a discussion of antibody structure) The product is expressed in a CHO cell line from which it is excreted as a dimeric soluble protein of approximately 150 kDa. After purification and excipient addition (mannitol, sucrose and trometamol), the product is freeze-dried. It is indicated for the treatment of rheumatoid arthritis and is usually administered as a twice-weekly s.c. injection of 25 mg product reconstituted in WFI. Enbrel functions as a competitive inhibitor of TNF, a major pro-inflammatory cytokine. Binding of TNF to Enbrel prevents it from binding to its true cell surface receptors. The antibody F_c component of the hybrid protein confers an extended serum half-life on the product, increasing it by fivefold relative to the soluble TNF receptor portion alone.

More recently, an additional approach to preventing TNF toxicity has been proposed. Several metalloprotease inhibitors (most notably hydroxamic acid) prevent proteolytic processing (i.e. release) of TNF-α from producer cell surfaces. Such inhibitors may also prove useful in preventing TNF-induced illness. The extent to which TNF (and inhibitors of TNF) will serve as future therapeutic agents remains to be determined by future clinical trials.

Box 9.2

Product case study: Beromun

Beromun (tradename, also known as tasonermin) is human TNF-α-1a produced by recombinant means in *E. coli*. The recombinant product is identical to the native human protein, with three 17.3 kDa, 157 amino acid polypeptides forming the biologically active homotrimer. It gained approval for use within the EU in 1999. Beromun is indicated for the treatment of soft-tissue sarcoma (STS) in the limbs, mainly so as to prevent or delay amputation. It is usually used in combination with the chemotherapeutic drug melphalan. STS is a relatively rare malignancy (0.6 per cent of newly diagnosed cancers) and only some 10 per cent of sufferers require amputation or severely debilitating surgery.

After initial cell fermentation and product extraction from the producer cells, the crude preparation is subject to multiple chromatographic steps, including ion-exchange, hydrophobic interaction chromatography and gel-filtration chromatography. The purified product is presented in lyophilized form in vials (1 mg active/vial) and excipients include a phosphate buffer, sodium chloride and serum albumin.

The product (reconstituted in sterile physiological saline to a concentration of 0.2 mg ml^{21}) is administered by isolated limb perfusion in which 3–4 mg of product is circulated through the limb over 90 min via a perfusion circuit divorced from general body circulation. The exact mechanisms by which the product induces haemorrhagic necrosis in tumours remain to be elucidated. Pharmacokinetic studies indicate that TNF administered systemically has a relatively short terminal half-life (20–30 min). Studies indicate some systemic leakage will occur during the procedure and/or that some TNF will enter systemic circulation upon restoration of circulation to the isolated limb subsequent to the procedure. Common (systemic) side effects include nausea, cardiac arrhythmias and liver toxicity, whereas regional side effects (in the isolated limb) can include oedema, nerve injury and infection. Clinical trials verified that beromun delays/prevents the need for whole limb amputation in a majority of patients treated.

Table 9.7 Some diseases in which TNF is known to mediate many of the symptoms

Disease	Symptoms induced by TNF
Cancer	Cachexia
	Stimulation of growth of some tumours
Septic shock	Vascular leakage
	Tissue necrosis
	Hypotension
	Activates blood clotting
Rheumatoid arthritis	Tissue inflammation
	Possible role in destruction of joints
MS	Inflammation
Diabetes	Death of pancreatic islet cells induces insulin resistance

Further reading

Books

Balkwill, F. 2000. *The Cytokine Network*. Oxford University Press.

Fitzgerald, K.A., O'Neill, L.A.J., Gearing, A.J.H., and Callard, RE. 2001. *The Cytokine Factsbook*. Academic Press.

Mantovani, A. 2000. *Pharmacology of Cytokines*. Oxford University Press.

Moreland, L. and Emery, P. (eds). 2003. *TNF-α Inhibition in the Treatment of Rheumatoid Arthritis*. Taylor and Francis, London.

Mire-Sluis, A. 1998. *Cytokines*. Academic Press.

Thompson, A. (ed.). 2003. *The Cytokine Handbook*. Academic Press, London.

Articles

Interleukins

Atkins, M. 2002. Interleukin 2: clinical applications. *Seminars in Oncology* **29**(3), 12–17.

Aulitzky, W.E., Schuler, M., Peschel, C., and Huber, C. 1994. Interleukins. Clinical pharmacology and therapeutic use. *Drugs* **48**(5), 667–677.

Brown, M. and Hural, J. 1997. Function of IL-4 and control of its expression. *Critical Reviews in Immunology* **17**(1), 1–32.

Devos, R., Plaetinck, G., Cornelis, S., Guisez, Y., Van der Heyden, J., and Tavernier, J. 1995. Interleukin-5 and its receptor: a drug target for eosinophilia associated with chronic allergic disease. *Journal of Leukocyte Biology* **57**, 813–818.

Fehniger, T.A., Cooper, M.A., and Caligiuri, M.A. 2002. Interleukin-2 and interleukin-15: immunotherapy for cancer. *Cytokine and Growth Factor Reviews* **13**(2), 169–183.

Fickenscher, H., Hor, S., Kupers, H., Knappe, A., Wittmann, S., and Sticht, H. 2002. The interleukin-10 family of cytokines. *Trends in Immunology* **23**(2), 89–96.

Fry, J. and Mackall, C. 2002. Interleukin-7: from bench to clinic. *Blood* **99**(11), 3892–3904.

Hawrylowich, C. and O'Garra, A. 2005. Potential role of interleukin-10 secreting regulatory T-cells in allergy and asthma. *Nature Reviews Immunology* **5**(4), 271–283.

Hohlfeld, R. 1997. Biotechnological agents for the immunotherapy of multiple sclerosis – principles, problems and perspectives. *Brain* **120**, 865–916.

Jeal, W. and Goa, K. 1997. Aldesleukin (recombinant interleukin-2): a review of its pharmacological properties, clinical efficacy and tolerability in patients with renal cell carcinoma. *Biodrugs* **7**(4), 285–317.

Komschlies, K.L., Grzegorzewski, K.J., and Wiltrout, R.H. 1995. Diverse immunological and haematological effects of interleukin-7: implications for clinical application. *Journal of Leukocyte Biology* **58**, 623–631.

Krawczenko, A., Kieda, C., and Dus, D. 2005. The biological role and potential therapeutic applications of interleukin 7. *Archivum Immunologiae et Therapiae Experimentalis* **53**(6), 518–525.

Lee, Y. and Hirani, A. 2006. Interleukin-4 in atherosclerosis. *Archives of Pharmacal Research* **29**(1), 1–15.

Leonard, W. and Spolski, R. 2005. Interleukin 21: a modulator of lymphoid proliferation, apoptosis and differentiation. *Nature Reviews Immunology* **5**(9), 688–698.

Martin, M. and Falk, W. 1997. The interleukin-1 receptor complex and interleukin-1 signal transduction. *European Cytokine Network* **8**(1), 5–17.

McInnes, I. and Gracie, J. 2004. Interleukin-15: a new cytokine target for the treatment of inflammatory diseases. *Current Opinion in Pharmacology* **4**(4), 392–397.

Noble, S. and Goa, K. 1997. Aldesleukin (recombinant interleukin-2): a review of its pharmacological properties, clinical efficacy and tolerability in patients with metastatic melanoma. *Biodrugs* **7**(5), 394–422.

Renauld, J.C., Kermouni, A., Vink, A., Louahed, J., and Van Snick, J. 1995. Interleukin-9 and its receptor: involvement in mast cell differentiation and T-cell oncogenesis. *Journal of Leukocyte Biology* **57**, 353–359.

Ryan, J. 1997. Interleukin-4 and its receptor – essential mediators of the allergic response. *Journal of Allergy and Clinical Immunology* **99**(1), 1–5.

Simpson, R.J., Hammacher, A., Smith, D.K., Matthews, J.M., and Ward, L.D. 1997. Interleukin-6 structure—function relationships. *Protein Science*, **6**(5), 929–955.

Scott, P. 1993. IL-12: initiation cytokine for cell mediated immunity. *Science* **260**, 496–497.

Wigginton, J. and Wiltrout, R. 2002 IL-12/IL-2 combination cytokine therapy for solid tumors: translation from benchside to bedside. *Expert Opinion on Biological Therapy* **2**(5), 513–524.

Tumour necrosis factor

Argiles, J. and Lepezsoriano, F. 1997. Cancer cachexia – a key role for TNF. *International Journal of Oncology* **10**(3), 565–572.

Bodmer, J.-L., Schneider, P., and Tschopp, J. 2002. The molecular architecture of the TNF superfamily. *Trends in Biochemical Sciences* **27**(1), 19–26.

Bondeson, J., Feldmann, M., and Maini, R.N. 2001. Tumour necrosis factor as a therapeutic target: from the laboratory to the clinic. *Immunologist* **8**(6), 136–140.

Brandt, J. and Braun, J. 2006. Anti-TNF-alpha agents in the treatment of psoriatic arthritis. *Expert Opinion on Biological Therapy* **6**(2), 99–107.

Crowe, P.D., VanArsdale, T.L., Walter, B.N., Ware, C.F., Hession, C., Ehrenfels, B., Browning, J.L., Din, W.S., Goodwin, R.G., and Smith, C.A. 1994. A lymphotoxin-β-specific receptor. *Science* **264**, 707–709.

Darnay, B. and Aggarwal, B. 1997. Early events in TNF signalling – a story of associations and disassociations. *Journal of Leukocyte Biology* **61**(5), 559–566.

Dellinger, R.P., Opal, S.M., Rotrosen, D., Suffredini, A.F., and Zimmerman, J.L. 1997. From the bench to the bedside – the future of sepsis research. *Chest* **111**(3), 744–753.

Feldmann, M., Elliott, M.J., Woody, J.N., Maini, R.N. 1997. Anti-tumor necrosis factor alpha therapy of rheumatoid arthritis. *Advances in Immunology* **64**, 283–350.

Liu, Z. 2005. Molecular mechanism of TNF signaling and beyond. *Cell Research* **15**(1), 24–27.

MacEwan, D. 2002. TNF receptor subtype signalling: differences and cellular consequences. *Cellular Signalling* **14**(6), 477–492.

McDermot, M. 2001. TNF and TNFR biology in health and disease. *Cellular and Molecular Biology* **47**(4), 619–635.

Moreland, L.W., Heck, L.W., and Koopman, W.J. 1997. Biologic agents for treating rheumatod arthritis: concepts and progress. *Arthritis and Rheumatism* **40**(3), 397–409.

McGeer, E. and McGeer, P. 1997. Inflammatory cytokines in the CNS – possible role in the pathogenesis of neurodegenerative disorders and therapeutic implications. *CNS Drugs* **7**(3), 214–228.

Nestorov, I. 2005. Clinical pharmacokinetics of TNF antagonists: how do they differ? *Seminars in Arthritis and Rheumatism* **34**(5)(Suppl. 1), 12–19.

Pestka, S., Krause, C.D., Sarkar, D., Walter, M.R., Shi, Y., and Fisher, P.B. 2004 Interleukin-10 and related cytokines and receptors. *Annual Review of Immunology* **22**, 929–979.

Terranova, P. 1997. Potential roles of tumor necrosis factor-alpha in follicular development, ovulation and the life-span of the corpus luteum. *Domestic Animal Endocrinology* **14**(1), 1–15.

Tracey, K. and Cerami, A. 1994. Tumor necrosis factor: a pleiotropic cytokine and therapeutic target. *Annual Review of Medicine* **45**, 491–503.

Vandenabeele, P., Declercq, W., Beyaert, R., and Fiers, W. 1995. Two tumour necrosis factor receptors: structure and function. *Trends in Cell Biology* **5**, 392–399.

10
Growth factors

10.1 Introduction

The differentiation, growth and division of eukaryotic cells is modulated by various influences, of which growth factors are amongst the most important for many cell types. A wide range of polypeptide growth factors have been identified (Table 10.1) and more, undoubtedly, remain to be characterized. Factors that inhibit cell growth also exist. For example, interferons and TNF inhibit proliferation of various cell types.

Some growth factors may be classified as cytokines (e.g. interleukins, TGF-β and CSFs). Others (e.g. IGFs) are not members of this family. Each growth factor has a mitogenic (promotes cell division) effect on a characteristic range of cells. Whereas some such factors affect only a few cell types, most stimulate growth of a wide range of cells.

The ability of growth factors to promote accelerated cellular growth, differentiation and/or division has predictably attracted the attention of the pharmaceutical industry. Several such products, most notably a range of haematopoietic growth factors, have now gained approval for general medical use (Table 10.2), and such haematopoietic growth factors are considered directly below. A number of additional polypeptide growth factors are considered subsequently (Table 10.3).

10.2 Haematopoietic growth factors

Blood consists of red and white cells which, along with platelets, are all suspended in plasma. All peripheral blood cells are derived from a single cell type, i.e. the stem cell (also known as a pluripotential, pluripotent or haemopoietic stem cell; see also Chapter 14). These stem cells reside in the bone marrow, alongside additional cell types, including (marrow) stromal cells. Pluripotential stem cells have the capacity to undergo prolonged or indefinite self-renewal. They also have the potential to differentiate, thereby yielding the range of cells normally found in blood (Table 10.4). This process, by which a fraction of stem cells is continually 'deciding' to differentiate (thus continually producing new blood cells and platelets to replace aged cells), is known as haemopoiesis.

Pharmaceutical biotechnology: concepts and applications   Gary Walsh
© 2007 John Wiley & Sons, Ltd   ISBN 978 0 470 01244 4 (HB)   978 0 470 01245 1 (PB)

Table 10.1 Overview of some polypeptide growth factors. Many can be grouped into families on the basis of amino acid sequence homology, or the cell types affected. Most growth factors are produced by more than one cell type and display endocrine, paracrine or autocrine effects on target cells by interacting with specific cell surface receptors

Growth factor	Major target cells
Interleukins	Various, mainly cells mediating immunity and inflammation
IFN-γ	Mainly lymphocytes and additional cells mediating immunity (and inflammation)
CSFs	Mainly haemopoietic cells
EPO	Erythroid precursor cells
TPO	Mainly megakaryocytes
Neurotrophic factors	Several, but mainly neuronal cell populations
Insulin	Various
IGFs	A very wide range of cells found in various tissue types
EGF	Various, including epithelial and endothelial cells and fibroblasts
PDGF	Various, including fibroblasts, glial cells and smooth muscle cells
FGFs	Various, including fibroblasts, osteoblasts and vascular endothelial cells
TGFs-α	Various
LIF	Mainly various haemopoietic cells

Table 10.2 Growth factors approved for general medical use

Product[a]	Indication	Company
Neupogen (filgrastim; G-CSF)	Neutropenia caused by chemotherapy Bone marrow transplants	Amgen Inc.
Leukine (sargramostim, GM-CSF)	Autologous bone marrow transplantation Neutrophil recovery after bone marrow transplantation	Berlex Labs
Neulasta (PEGylated filgrastim, see above)	Neutropenia	Amgen
Epogen (epoetin alfa, rEPO)	Anaemia associated with various medical conditions	Amgen
Procrit (epoetin alfa, rEPO)	Anaemia associated with various medical conditions	Ortho Biotech
Neorecormon (epoetin beta, rEPO)	Anaemia associated with various medical conditions	Boehringer-Mannheim
Aranesp (darbepoetin alfa, a rEPO analogue)	Anaemia associated with various medical conditions	Amgen
Nespo (darbepoetin alfa, a rEPO analogue	Anaemia associated with various medical conditions	Dompe Biotec
Regranex (becaplermin, rPDGF)	Neuropathic diabetic ulcers	Janssen & Ortho-McNeil
Kepivance (palifermin, rKGF)	Severe oral mucositis	Amgen
Increlex (mecasermin, rh IGF-1)	Growth failure in children	Tercica/Baxter
Iplex (mecasermin rinfabate, complex of rhIGF-1 and IGFBP-3)	Growth failure in children	Insmed
GEM 21S (implantable product containing rhPDGF-BB)	Periodontally related defects	Luitpold Pharmaceuticals and Biomimetic Pharmaceuticals

[a]IGFBP: insulin like growth factor binding protein.

Table 10.3 Some growth factors that may have significant future therapeutic application, and the conditions they aim to treat

Growth factor	Possible medical indication
TPO	Thrombocytopenia
EGF	Wound healing, skin ulcers
TGF-β	Bone healing, skin ulcers, detached retinas
FGFs	Soft tissue ulcers, wound healing
Neurotrophic factors	Mainly conditions caused by/associated with neurodegeneration, including peripheral neuropathies, amyotrophic lateral sclerosis and neurodegenerative diseases of the brain

The study of the process of haemopoiesis is rendered difficult by the fact that it is extremely difficult to distinguish or separate individual stem cells from their products during the earlier stages of differentiation. However, a picture of the process of differentiation is now beginning to emerge (Figure 10.1). During the haemopoietic process, the stem cells differentiate, producing cells that become progressively more restricted in their choice of developmental options.

The production of many mature blood cells begins when a fraction of the stem cells differentiates, forming a specific cell type termed CFU-S (where CFU refers to colony-forming unit). These cells, in turn, differentiate to yield CFU-GEMM cells, a mixed CFU that has the potential to differentiate into a range of mature blood cell types, including granulocytes, monocytes, erythrocytes, platelets, eosinophils and basophils. Note that lymphocytes are not derived from the CFU-GEMM pathway, but differentiate via an alternative pathway from stem cells (Figure 10.1).

The details of haemopoiesis presented thus far prompt two very important questions: (1) How is the correct balance between stem cell self-renewal and differentiation maintained? (2) What forces exist that regulate the process of differentiation? The answer to both questions, in particular the latter, is beginning to emerge in the form of a group of cytokines termed 'haemopoietic growth factors'. This group includes:

- several interleukins, which primarily affect production and differentiation of lymphocytes;

- CSFs, which play a major role in the differentiation of stem-derived cells into neutrophils, macrophages, megakaryocytes (from which platelets are derived), eosinophils and basophils;

- EPO, which is essential in the production of red blood cells;

- TPO, which is essential in the production of platelets.

Table 10.4 The range of blood cells that are ultimately produced upon the differentiation of pluripotential stem cells (see text for details). (Note that osteoclasts are multinucleated cells often associated with small depressions on the surface of bone. They function to reabsorb calcified bone)

Neutrophils	T- and B-lymphocytes
Eosinophils	Erythrocytes
Basophils	Monocytes
Megakaryocytes	Osteoclasts

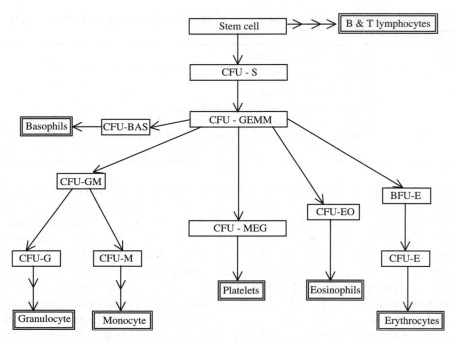

Figure 10.1   A simplified overview of the haematopoietic process as currently understood. Refer to text for details (BFU: burst forming unit)

Most of these haemopoietic growth factors are glycoproteins, displaying a molecular mass in the region of 14–24 kDa. Most are produced by more than one cell type, and several such regulators can stimulate proliferation of any one haemopoietic cell lineage. This is due to the presence of receptors for several such factors on their surface. Receptor numbers for any one growth factor are low (less than 500 per cell), and proliferation can be stimulated even when only a small proportion of these are occupied.

During normal haemopoiesis, only a small fraction of stem cells undergo differentiation at any given time. The remainder continue to self-renew. The molecular detail underpinning self-renewal is poorly understood. However, it has been shown that certain transformed stem cells can be induced to undergo continuous proliferation *in vitro* under the influence of IL-3. The concentration of IL-3 is critical, with differentiation occurring below certain threshold concentrations of this cytokine. The delicate balance between stem cell renewal and differentiation is likely affected not only by the range of growth factors experienced, but also by the concentration of each growth factor.

10.2.1   The interleukins as haemopoietic growth factors

The interleukin family of cytokines has been overviewed in Chapter 9, and a number of cytokines are known to influence haemopoiesis. The IL-3 receptor, for example, is found on a wide variety of progenitor haemopoietic cells, and appears to stimulate not only CFU-GEMM, but also the precursor cells of basophils, eosinophils and platelets. The role of IL-11, which also plays a role, was also discussed in Chapter 9.

Table 10.5   Summary of some of the properties of G-CSF, M-CSF and GM-CSF

	G-CSF	M-CSF	GM-CSF
Molecular mass (kDa)	21	45–90	22
Main producer cells	Bone marrow stromal cell Macrophages Fibroblasts	Lymphocytes Myoblasts Osteoblasts Monocytes Fibroblasts Endothelial cells	Macrophages T-lymphocytes Fibroblasts Endothelial cells
Main target cells	Neutrophils. Also other haemopoietic progenitors and endothelial cells	Macrophages and their precursor cells	Haematopoietic progenitor cells Granulocytes Monocytes Endothelial cells Megakaryocytes T-lymphocytes Erythroid cells

10.2.2   Granulocyte colony-stimulating factor

G-CSF is also known as pluripoietin and CSF-β. Two slight variants are known, one consisting of 174 amino acids and the other of 177. The smaller polypeptide predominates and also displays significantly greater biological activity than the larger variant.

G-CSF is a glycoprotein. It displays a single O-linked glycosylation site and an apparent molecular mass in the region of 21 kDa, and is synthesized by various cell types (Table 10.5). It functions as a growth and differentiation factor for neutrophils and their precursor cells. It also appears to activate mature neutrophils (which are leukocytes capable of ingesting and killing bacteria). G-CSF also appears to act in synergy with additional haemopoietic growth factors to stimulate growth/differentiation of various other haemopoietic progenitor cells. In addition, this cytokine promotes the proliferation and migration of endothelial cells.

The G-CSF receptor has been well characterized. It is a single transmembrane polypeptide found on the surface of neutrophils, as well as in various haemopoietic precursor cells, platelets, endothelial cells and, notably, various myeloid leukaemias. (Myeloid means derived from bone marrow; leukaemia refers to a cancerous condition in which there is uncontrolled overproduction of white blood cells in the bone marrow or other blood-forming organs. The white cells produced are generally immature/abnormal and result in the suppression of production of healthy white blood cells.)

10.2.3   Macrophage colony-stimulating factor

M-CSF serves as a growth, differentiation and activation factor for macrophages and their precursor cells. It is also known as CSF-1. This cytokine is produced by various cell types (Table 10.5). The mature form is a glycoprotein containing three potential N-linked glycosylation sites. Three related forms of human M-CSF have been characterized. All are ultimately derived from the same gene and share common C and N termini. The largest consists of 522 amino acids, with the 406 and 224 amino acid forms lacking different lengths of the internal sequence of the 522 form. The molecular masses of these mature M-CSFs range from 45 to 90 kDa.

The biologically active form of M-CSF is a homodimer (two identical subunits). These homodimers can exist as integral cell surface proteins, or may be released from their producer cell by proteolytic cleavage, thus yielding the soluble cytokine. The M-CSF receptor is a single-chain, heavily glycosylated, polypeptide of molecular mass 150 kDa.

10.2.4   Granulocyte macrophage colony-stimulating factor

GM-CSF is also known as CSF-α or pluripoietin-α. It is a 127 amino acid, single-chain, glycosylated polypeptide, exhibiting a molecular mass in the region of 22 kDa. It is produced by various cells (Table 10.5), and studies have indicated that its biological activities include:

- Proliferation/differentiation factor of haemopoietic progenitor cells, particularly those yielding neutrophils (a variety of granulocyte) and macrophages, but also eosinophils, erythrocytes and megakarycytes. *In vivo* studies also demonstrate this cytokine's ability to promote haemopoiesis.

- Activation of mature haemopoietic cells, resulting in:

  - enhanced phagocytic activity;

  - enhanced microbiocidal activity;

  - augmented anti-tumour activity;

  - enhanced leukocyte chemotaxis.

The intact GM-CSF receptor is a heterodimer, consisting of a low-affinity α-chain and a β-chain, which also forms part of the IL-3 and IL-5 receptors. (The β-chain alone does not bind GM-CSF.) The α-chain is an 80 kDa glycoprotein and exhibits only a short intracellular domain. The larger β-chain (130 kDa) displays a significant intracellular domain. Signal transduction involves the (tyrosine) phosphorylation of a number of cytoplasmic proteins (Figure 10.2).

10.2.5   Clinical application of colony-stimulating factors

Several CSF preparations have gained regulatory approval (Table 10.2). G-CSF and GM-CSF have proven useful in the treatment of neutropenia. All three CSF types are (or are likely to be) useful also in the treatment of infectious diseases, some forms of cancer and the management of bone marrow transplants, as they stimulate the differentiation/activation of white blood cell types most affected by such conditions.

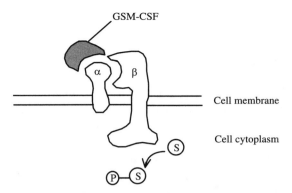

Figure 10.2 The GM-CSF receptor. Ligand binding appears to promote the phosphorylation of various cytoplasmic polypeptide substrates (at least in part via an associated JAK2), leading to signal transduction

Neutropenia is a condition characterized by a decrease in blood neutrophil count below 1.5×10^9 cells per litre; a normal blood count is $(2.0–7.5) \times 10^9$ cells per litre. Its clinical symptoms include the occurrence of frequent and usually serious infections, often requiring hospitalization. Neutropenia may be caused by a number of factors (Table 10.6), at least some of which are responsive to CSF treatment. Particularly noteworthy is neutropenia triggered by administration of chemotherapeutic drugs to cancer patients. Chemotherapeutic agents (e.g. cyclophosphamide, doxorubicin and methotrexate), when administered at therapeutically effective doses, often induce the destruction of stem cells and/or compromise stem cell differentiation.

Filgrastim is a recombinant human G-CSF (produced in *E. coli*), approved for chemotherapy-induced neutropenia (Table 10.2 and Box 10.1). Neulasta is the tradename given to a PEGylated form of filgrastim approved for general medical use in the USA in 2002 (Table 10.2). Manufacture of this product entails covalent attachment of an activated monomethoxypolyethylene glycol molecule to the N-terminal methionyl residue of filgrastim. The product is formulated in the presence of acetate buffer, sorbitol and polysorbate and is presented in pre-filled syringes for s.c. injection. As in the case of PEGylated interferons (Chapter 8), the rationale for PEGylation is to increase the drug's plasma half-life, thereby reducing the frequency of injections required.

Table 10.6 Some causes of neutropenia

Genetic (particularly in black populations)
Severe bacterial infection
Severe sepsis
Severe viral infection
Aplastic anaemia[a]
Acute leukaemia
Hodgkin's/non-Hodgkin's lymphoma
Various drugs, especially anti-cancer drugs
Autoimmune neutropenia

[a]Aplastic anaemia describes bone marrow failure, characterized by serious reduction in the number of stem cells present.

Box 10.1

Product case study: Neupogen

Neupogen (tradename, also known as filgrastim) is a recombinant G-CSF first approved for medical use in the USA in 1991. It is indicated for the treatment of neutropenia associated with various medical conditions, often in cancer patients undergoing chemotherapy. The 175 amino acid, 18.8 kDa protein is produced in engineered *E. coli* and is identical to native human G-CSF except for (a) the presence of an additional methionine residue at the molecule's N terminal end and (b) the absence of glycosylation. Both differences are a consequence of expression in *E. coli*. Production entails cell growth and harvest, followed by product extraction and purification using several chromatographic and ultrafiltration procedures. It is presented as a sterile solution in either single-use vials or pre-filled syringes and contains acetate buffer, sorbitol and Tween 80 as excipients.

Details of administration vary according to exact indication, but it is administered by s.c. or i.v. infusion daily, usually for several days at dosage levels of 5–10 µg kg^{-1} day^{-1}. Peak plasma concentrations are usually observed within 2–8 h after s.c. administration and the drug's elimination half-life ranges between 2 and 5 h. The product is used in conjunction with cancer chemotherapy in one of two ways: prophylactically, to prevent neutropenia onset, or therapeutically, to reverse established neutropenia. Numerous clinical trials have been undertaken with various cancer patient types, and the product has generally been shown to be safe and effective in accelerating neutrophil count recovery following various chemotherapeutic regimes. The product was found to reduce the median number of days of severe neutropenia observed in patients with severe chronic neutropenia, acute myeloid leukaemia and non-myeloid malignancies who receive myeloblative chemotherapy followed by bone marrow transplant.

Commonly reported side effects include bone and muscle pain. Serious (sometimes life threatening) but rare side effects have included splenic rupture and adult respiratory distress syndrome. Neupogen is manufactured and marketed by Amgen Inc.

G-CSF and GM-CSF have also found application after allogenic or autologous bone marrow transplantation, to accelerate neutrophil recovery. (Allogenic means that donor and recipient are different individuals, and autologous means that donor and recipient are the same.)

10.2.6 Erythropoietin

EPO is an additional haemopoietic growth factor. It is primarily responsible for stimulating and regulating erythropoiesis (i.e. erythrogenesis, the production of red blood cells) in mammals.

The erythron is a collective term given to mature erythrocytes, along with all stem-cell-derived progeny that have committed to developing into erythrocytes. It can thus be viewed as a disperse organ whose primary function relates to transport of oxygen and carbon dioxide (haemoglobin constitutes up to one-third of the erythrocyte cytoplasm), as well as maintaining blood pH. An average adult contains in the region of 2.3×10^{13} erythrocytes (weighing up to 3 kg). They are

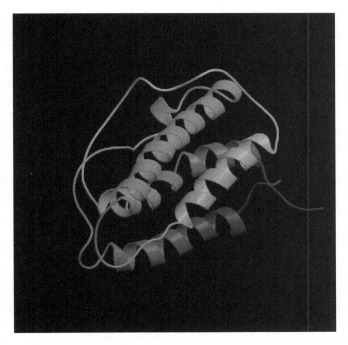

Figure 10.3   Three-dimensional structure of EPO. Structural details courtesy of the Protein Data Bank, http://www.rcsb.org/pdb/

synthesized at a rate of about 2.3 million cells per second and have a circulatory life of approximately 120 days, during which they travel almost 200 miles.

EPO is an atypical cytokine, in that it acts as a true (endocrine) hormone and is not synthesized by any type of white blood cell. It is encoded by a single copy gene, located on (human) chromosome 7. The gene consists of four introns and five exons. The mature EPO gene product contains 165 amino acids and exhibits a molecular mass in the region of 36 kDa (Figure 10.3). EPO is a glycoprotein, almost 40 per cent of which is carbohydrate. Three N-linked and one O-linked glycosylation sites are evident. The O-linked carbohydrate moiety appears to play no essential role in the (*in vitro* or *in vivo*) biological activity of EPO. Interestingly, removal of the N-linked sugars, although having little effect on EPO's *in vitro* activity, all but destroys its *in vivo* activity. The sugar components of EPO are likely to contribute to the molecule's solubility, cellular processing and secretion, as well as its *in vivo* metabolism.

Incomplete (N-linked) glycosylation prompts decreased *in vivo* activity due to more rapid hepatic clearance of the EPO molecule. Enzymatic removal of terminal sialic acid sugar residues from oligosaccharides exposes otherwise hidden galactose residues. These residues are then free to bind specific hepatic lectins, which promote EPO removal from the plasma. The reported plasma $t_{1/2}$ value for native EPO is 4–6 h. The $t_{1/2}$ for desialated EPO is 2 min. Comparison of native human EPO with its recombinant form produced in CHO cells reveals very similar glycosylation patterns.

EPO in the human adult is synthesized almost exclusively by specialized kidney cells (peritubular interstitial cells of the kidney cortex and upper medulla). Minor quantities are also synthesized in the liver, which represents the primary EPO-producing organ of the foetus.

EPO is present in serum and (at very low concentrations) in urine, particularly of anaemic individuals. This cytokine/hormone was first purified in 1971 from the plasma of anaemic sheep, and small quantities of human EPO were later purified (in 1977) from over 2500 l of urine collected from anaemic patients. Large-scale purification from native sources was thus impractical. The isolation (in 1985) of the human EPO gene from a genomic DNA library facilitated its transfection into CHO cells. This now facilitates large-scale commercial production of the recombinant human product (rhEPO), which has found widespread medical application.

EPO stimulates erythropoiesis by:

- increasing the number of committed cells capable of differentiating into erythrocytes;

- accelerating the rate of differentiation of such precursors;

- increasing the rate of haemoglobin synthesis in developing cells.

An overview of the best-characterized stages in the process of erythropoiesis is given in Figure 10.4. The erythroid precursor cells, BFU-E (burst forming unit-erythroid), display EPO receptors on their surface. The growth and differentiation of these cells into CFU-Es (where E stands for erythroid) require the presence of not only EPO, but also IL-3 and/or GM-CSF. CFU-E cells display the greatest density of EPO cell surface receptors. These cells, not surprisingly, also display the greatest biological response to EPO. Progressively more mature erythrocyte precursors display progressively less EPO receptors on their cell surfaces. Erythrocytes themselves are devoid of EPO receptors. EPO binding to its receptor on CFU-E cells promote their differentiation into proerythroblasts and the rate at which this differentiation occurs appears to determine the rate of erythropoiesis. CFU-E cells also are responsive to IGF-1.

Although the major physiological role of EPO is certainly to promote red blood cell production, EPO mRNA has also been detected in bone marrow macrophages, as well as some multipotential haemopoietic stem cells. Although the physiological relevance is unclear, it is possible that EPO produced by such sources may play a localized paracrine (or autocrine) role in promoting erythroid differentiation. The level of EPO production in the kidneys (or liver) is primarily regulated by the oxygen demand of the producer cells, relative to their oxygen supply.

The EPO receptor is a member of the haemopoietic cytokine receptor superfamily. Its intracellular domain displays no known catalytic activity, but it appears to couple directly to the JAK2 kinase (Chapter 8) that likely promotes the early events of EPO signal transduction. Other studies have implicated additional possible signalling mechanisms, including the involvement of G proteins, protein kinase C and Ca^{2+}. The exact molecular events underlining EPO signal transduction remain to be elucidated in detail.

10.2.6.1 Therapeutic applications of erythropoietin

A number of clinical circumstances have been identified which are characterized by an often profoundly depressed rate of erythropoiesis (Table 10.7). Many, if not all, such conditions could

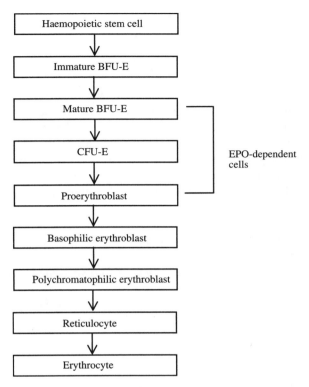

Figure 10.4 Stages in the differentiation of haemopoietic stem cells, yielding mature erythrocytes. The EPO-sensitive cells are indicated. Each cell undergoes proliferation as well as differentiation; thus, greater numbers of the more highly differentiated daughter cells are produced. The proliferation phase ends at the reticulocyte stage; each reticulocyte matures over a 2-day period, yielding a single mature erythrocyte

be/are responsive to administration of exogenous EPO. The prevalence of anaemia, and the medical complications that ensue, prompts tremendous therapeutic interest in this haemopoietic growth factor. EPO has been approved for use to treat various forms of anaemia (Table 10.2). It was the first therapeutic protein produced by genetic engineering whose annual sales value topped US$1 billion. Current combined annual sales value of commercialized recombinant EPO products is now close to US$10 billion.

Table 10.7 Diseases (and other medical conditions) for which anaemia is one frequently observed symptom

Renal failure
Rheumatoid arthritis
Cancer
AIDS
Infections
Bone marrow transplantation

Box 10.2

Product case study: Neorecormon

Neorecormon (tradename, also known as epoetin beta) is a recombinant human EPO first approved for medical use in the EU in 1997. It is indicated for the treatment of anaemia associated with various medical conditions, most commonly chronic renal failure and cancer patients receiving chemotherapy. Neorecormon is produced by recombinant DNA technology in a CHO cell line and is manufactured as outlined in Figure 10.5. It is presented in lyophilized format at various strengths (500–10 000 IU/vial) and contains phosphate buffer, sodium chloride, calcium chloride, urea, polysorbate and various amino acids as excipients.

The product displays a terminal half-life of 4–12 h and 8–22 h after i.v. and s.c. administration respectively. Dosage regime is dependent upon the exact disease condition, but generally involves administration once/several times weekly. Various clinical trials investigating various indications proved product efficacy in the treatment and prevention of anaemia, with increased haemocrit values observed.

Common side effects include increased blood pressure, increased respiratory infections and increased platelet counts. Serious (rare) side effects were most often related to cardiovascular complications. Neorecormon is marketed by Roche.

Neorecormon is one such product (Box 10.2), an overview of whose production is provided in Figure 10.5. More recently, an engineered form of EPO has gained marketing approval. Darbepoetin-alfa is its international non-proprietary name and it is marketed under the tradenames Aranesp (Amgen) and Nespo (Dompé Biotec, Italy). The 165 amino acid protein is altered in amino acid sequence when compared with the native human product. The alteration entails introducing two new N-glycosylation sites so that the recombinant product, produced in an engineered CHO cell line, displays five glycosylation sites as opposed to the normal three. The presence of two additional carbohydrate chains confers a prolonged serum half-life on the molecule (up to 21 h, compared with 4–6 h for the native molecule).

EPO was first used therapeutically in 1989 for the treatment of anaemia associated with chronic kidney failure. This anaemia is largely caused by insufficient endogenous EPO production by the diseased kidneys. Prior to EPO approval this condition could only be treated by direct blood transfusion. It responds well, and in a dose-dependent manner, to the administration of recombinant human EPO (rhEPO). The administration of EPO is effective, both in the case of patients receiving dialysis and those who have not yet received this treatment.

Administration of EPO doses ranging from 50 to 150 IU kg^{-1} three times weekly is normally sufficient to elevate the patient's haematocrit values to a desired 32–35 per cent. (Haematocrit refers to 'packed cell volume', i.e. the percentage of the total volume of whole blood that is composed of erythrocytes.) Plasma EPO concentrations generally vary between 5 and 25 IU l^{-1} in healthy individuals. One IU (international unit) of EPO activity is defined as the activity that promotes the same level of stimulation of erythropoiesis as 5 mmol of cobalt.

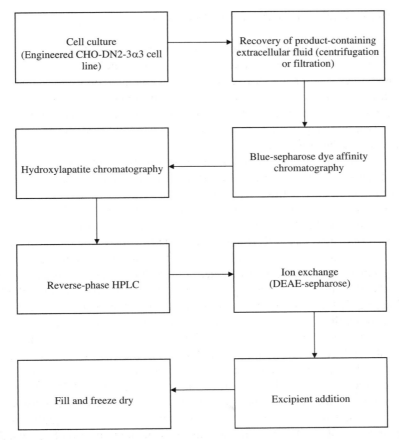

Figure 10.5 Schematic overview of the production of the EPO-based product 'Neorecormon'. Refer to text for further details

In addition to enhancing erythropoiesis, EPO treatment also improves tolerance to exercise, as well as a patient's sense of well-being. Furthermore, reducing/eliminating the necessity for blood transfusions also reduces/eliminates the associated risk of accidental transmission of blood-borne infectious agents, as well as the risk of precipitating adverse transfusion reactions in recipients. The therapeutic spotlight upon EPO has now shifted to additional (non-renal) applications (Table 10.8).

Table 10.8 Some non-renal applications of EPO (refer to text for details)

Treatment of anaemia associated with chronic disease
Treatment of anaemia associated with cancer/chemotherapy
Treatment of anaemia associated with prematurity
To facilitate autologous blood donations before surgery
To reduce transfusion requirements after surgery
To prevent anaemia after bone marrow transplantation

10.2.6.2 *Chronic disease and cancer chemotherapy*

Anaemia often becomes a characteristic feature of several chronic diseases, such as rheumatoid arthritis. In most instances this can be linked to lower than normal endogenous serum EPO levels (although in some cases a deficiency of iron or folic acid can also represent a contributory factor). Several small clinical trials have confirmed that administration of EPO increases haematocrit and serum haemoglobin levels in patients suffering from rheumatoid arthritis. A satisfactory response in some patients, however, required a high-dose therapy that could render this therapeutic approach unattractive from a cost:benefit perspective.

Severe, and in particular chronic, infection can also sometimes induce anaemia, which is often made worse by drugs used to combat the infection. For example, anaemia is evident in 8 per cent of patients with asymptomatic HIV infection. This incidence increases to 20 per cent for those with AIDS-related complex, and is greater than 60 per cent for patients who have developed Kaposi's sarcoma. Up to a third of AIDS patients treated with zidovudine also develop anaemia. Again, several trials have confirmed that EPO treatment of AIDS sufferers (be they receiving zidovudine or not) can increase haematocrit values and decrease transfusion requirements.

Various malignancies can also induce an anaemic state. This is often associated with decreased serum EPO levels, although iron deficiency, blood loss or tumour infiltration of the bone marrow can be complicating factors. In addition, chemotherapeutic agents administered to this patient group often adversely affect stem cell populations, thus rendering the anaemia even more severe.

Administration of EPO to patients suffering from various cancers/receiving various chemo-therapeutic agents yielded encouraging results, with significant improvements in haematocrit levels being recorded in approximately 50 per cent of cases. In one large US study (2000 patients; most receiving chemotherapy) s.c. EPO administration of an average of 150 IU kg^{-1}, three times weekly, for 4 months, reduced the number of patients requiring blood transfusions from 22 per cent to 10 per cent. Improvement in the sense of well-being and overall quality of life was also noted. The success rate of EPO in alleviating cancer-associated anaemia has varied in different trials, ranging from 32 per cent to 85 per cent.

On a more cautionary note, the EPO receptor is expressed not only by specific erythrocyte precursor cells, but also by endothelial, neural, and myeloma cells. Concern has been expressed that EPO, therefore, might actually stimulate growth of some tumour types, particularly those derived from such cells. To date, no evidence (*in vitro* or *in vivo*) has been obtained to support this hypothesis.

10.2.7 Thrombopoietin

Human TPO is a 332 amino acid, 60 kDa glycoprotein, containing six potential N-linked glycosylation sites. These are all localized towards the C-terminus of the molecule. The N-terminal half exhibits a high degree of amino acid homology with EPO and represents the biologically active domain of the molecule.

TPO is the haemopoietic growth factor now shown to be the primary physiological regulator of platelet production. This molecule may, therefore, represent an important future therapeutic agent in combating thrombocytopenia, a condition characterized by reduced blood platelet levels. The most likely initial TPO therapeutic target is thrombocytopenia induced by cancer chemo- or

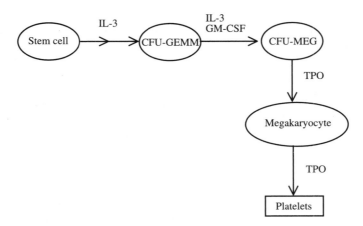

Figure 10.6 Simplified representation of the production of platelets from stem cells. CFU-megakaryocytes and in particular, mature megakaryocytes, are most sensitive to the stimulatory actions of TPO. These two cell types also display a limited response to IL-6, IL-11 and LIF

radio-therapy. This indication generally accounts for up to 80 per cent of all platelet transfusions undertaken. In the USA alone, close to 2 million people receive platelet transfusions annually.

Platelets (thrombocytes) carry out several functions in the body, all of which relate to the arrest of bleeding. They are disc-shaped structures 1–2 μm in diameter, and are present in the blood of healthy individuals at levels of approximately $250 \times 10^9 \, l^{-1}$. They are formed by a lineage-specific stem cell differentiation process, as depicted in Figure 10.6. The terminal stages of this process entail the maturation of large progenitor cells termed 'megakaryocytes'. Platelets represent small vesicles that bud off from the megakaryocyte cell surface and enter the circulation.

10.3   Growth factors and wound healing

The wound healing process is complex and not yet fully understood. The area of tissue damage becomes the focus of various events, often beginning with immunological and inflammatory reactions. The various cells involved in such processes, as well as additional cells at the site of the wound, also secrete various growth factors. These mitogens stimulate the growth and activation of various cell types, including fibroblasts (which produce collagen and elastin precursors, and ground substance), epithelial cells (e.g. skin cells) and vascular endothelial cells. Such cells advance healing by promoting processes such as granulation (growth of connective tissue and small blood vessels at the healing surface) and subsequent epithelialization. The growth factors that appear most significant to this process include FGFs, TGFs, PDGF, IGF-1 and EGF.

Wounds can be categorized as acute (healing quickly on their own) or chronic (healing slowly, and often requiring medication). Chronic wounds, such as ulcers (Table 10.9), occur if some influence disrupts the normal healing process. Such influences can include diabetes, malnutrition, rheumatoid arthritis and ischaemia (inadequate flow of blood to any part of the body). Elderly people are particularly susceptible to developing chronic wounds, often resulting in the necessity for hospitalization. Ulceration (particularly of the limbs or extremities) associated with old age,

Table 10.9   Various types of ulcers along with their underlying cause. An ulcer may simply be described as a break or cut in the skin or membrane lining the digestive tract which fails to heal. The damaged area may then become inflammed

Ulcer category	Description
Decubitus ulcer (e.g. bed sores, pressure sores)	Ulcer due to continuous pressure exerted on a particular area of skin. Often associated with bed-ridden patients
Diabetic ulcers	Ulcers (e.g. 'diabetic leg') caused by complications of diabetes
Varicose ulcers	Due to defective circulation, sometimes associated with varicose veins
Rodent ulcers	An ulcerous cancer (basal cell carcinoma), usually affecting the face
Peptic ulcers	Ulcer of the digestive tract, caused by digestion of the mucosa by acid and pepsin. May occur in, for example, the duodenum (duodenal ulcer) or the stomach (gastric ulcer)

diabetes, etc., remains the underlining cause of up to 50 per cent of all amputations carried out in the USA.

The fluid exuded from a fresh or acute wound generally exhibits high levels of various growth factors (as determined by bioassay or immunoassay analysis). In contrast, the concentration of such mitogens present in chronic wounds is usually several-fold lower. Direct (topical) application of exogenous growth factors results in accelerated wound healing in animals. Although some encouraging results have been observed in human trials, the overall results obtained thus far have been disappointing. Having said this, one such factor (rPDGF, tradename Regranex; Table 10.2) has been approved for topical administration on diabetic ulcers. Future studies may well also focus on application of a cocktail of growth factors instead of a single such factor to a wound surface.

A greater understanding of wound physiology/biochemistry may also facilitate greater success in future trials. It has been established, for example, that the fluid exuded from chronic wounds harbours high levels of proteolytic activity (almost 200-fold higher than associated with acute wounds). Failure of mitogens to stimulate wound healing may thus be due, in part, to their rapid proteolytic degradation (and/or the degradation of growth factor receptors present on the surface of susceptible cells). Identification of suitable protease inhibitors, and their application in conjunction with exogenous growth factor therapy, may improve clinical results recorded in the future.

Overall, therefore, a range of growth factors may demonstrate potential in wound management/healing, or for other therapeutic indications. A summary of these factors and their biological activities/potential, therefore, constitutes the remainder of this chapter.

10.3.1   Insulin-like growth factors

The IGFs (also termed 'somatomedins'), constitute a family of two closely related (small) polypeptides: IGF-I and IGF-II. As the names suggest, these growth factors bear a strong structural resemblance to insulin (or, more accurately, proinsulin). Infusion of IGF-I decreases circulating

Table 10.10 Overview of some of the effects of the IGFs

Promotes cell cycle progression in most cell types
Foetal development: promotes growth and differentiation of foetal cells and organogenesis
Promotes longitudinal body growth and increased body weight
Promotes enhanced functioning of the male and female reproductive tissue
Promotes growth and differentiation of neuronal tissue

levels of insulin and glucagon, increases tissue glucose uptake and inhibits hepatic glucose export. IGFs display pluripotent activities, regulating the growth, activation, differentiation (and maintenance of the differentiated state) of a wide variety of cell and tissue types (Table 10.10). The full complexity and variety of their biological activities are only now beginning to be appreciated.

The liver represents the major site of synthesis of the IGFs, from where they enter the blood stream, thereby acting in a classical endocrine fashion. A wide variety of body cells express IGF receptors, of which there are two types. Furthermore, IGFs are also synthesized in smaller quantities at numerous sites in the body and function in an autocrine or paracrine manner at these specific locations. IGF activity is also modulated by a family of IGFBPs, of which there are at least six.

10.3.2 Insulin-like growth factor biological effects

IGFs exhibit a wide range of gross physiological effects (Table 10.10), all of which are explained primarily by the ability of these growth factors to stimulate cellular growth and differentiation. Virtually all mammalian cell types display surface IGF receptors. IGFs play a major stimulatory role in promoting the cell cycle (specifically, it is the sole mitogen required to promote the G1b phase, i.e. the progression phase; various other phases of the cycle can be stimulated by additional growth factors). IGF activity can also contribute to sustaining the uncontrolled cell growth characteristic of cancer cells. Many transformed cells exhibit very high levels of IGF receptors, and growth of these cells can be inhibited *in vitro* by the addition of antibodies capable of blocking IGF-receptor binding.

Most of the growth-promoting effects of GH are actually mediated by IGF-I. Direct injection of IGF-I into hypophysectionized animals (animals whose pituitary, i.e. source of GH, is surgically removed) stimulates longitudinal bone growth, as well as growth of several organs/glands (e.g. kidney, spleen, thymus).

Such effects render IGFs likely therapeutic candidates in treating the various forms of dwarfism caused by a dysfunction in some element of the GH–IGF growth axis. Initial trials show that s.c. administration of recombinant human IGF-I over a 12-month period significantly increases the growth rate of Laron-type dwarfs.

IGFs also play a core role in tissue renewal and repair (e.g. wound healing) during adulthood. For example, these growth factors play a central role in bone remodelling (i.e. reabsorption and rebuilding, which helps keep bones strong and contributes to whole-body calcium homeostasis). Reabsorption of calcified bone is undertaken by osteoclasts, cells of haemopoietic origin whose formation is stimulated by IGFs. These mitogens may, therefore, influence the development of osteoporosis, a prevalent condition (especially amongst the elderly), which is characterized by brittle, uncalcified bone.

IGFs are often localized within various areas of the kidney. Direct infusion of IGF-I influences (usually enhances) renal function by a number of means, including promoting increased:

- glomerular filtration rate;

- renal plasma flow;

- kidney size and weight.

These responses are obviously mediated by multiple effects on the growth and activity of several renal cell types, suggesting that IGFs play a physiological role in regulating renal function. Not surprisingly, IGF-I is currently being assessed as a potential therapeutic agent in the treatment of various forms of kidney disease.

IGF-I is widely expressed in the central nervous system. IGF-II is also present, being produced mainly by tissues at vascular interfaces with the brain. Both growth factors, along with insulin, play a number of important roles in the nervous system. They stimulate the growth and development of various neuronal populations and promote neurotrophic effects (discussed later) and may, therefore, be of potential use in the treatment of various neurodegenerative diseases.

10.3.3   Epidermal growth factor

EGF, a 6 kDa, 53 amino acid unglycosylated peptide, was one of the first growth factors discovered. Its existence was initially noted in the 1960s, as a factor present in saliva which could promote premature tooth eruption and eyelid opening in neonatal mice. EGF has subsequently proven to exert a powerful mitogenic effect on many cell types, and its receptor (Figure 10.7) is expressed by most cells. Its influence on endothelial cells, epithelial cells and fibroblasts is particularly noteworthy, and the skin appears to be its major physiological target. It stimulates growth

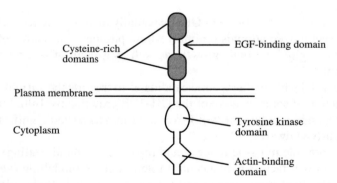

Figure 10.7   The EGF receptor. The N-terminal, extracellular region of the receptor contains 622 amino acids. It displays two cysteine-rich regions, between which the ligand-binding domain is located. A 23 amino acid hydrophobic domain spans the plasma membrane. The receptor cytoplasmic region contains some 542 amino acids. It displays a tyrosine kinase domain, which includes several tyrosine autophosphorylation sites, and an actin-binding domain that may facilitate interaction with the cell cytoskeleton

Table 10.11 Range of cells producing PDGF, and its major biological activities

Synthesized by		
	Platelets	Macrophages
	Fibroblasts	Endothelial cells
	Astrocytes	Megakaryocytes
	Myoblasts	Kidney epithelial cells
	Vascular smooth muscle cells	Many transformed cell types
Biological activities		
mitogen for:		
	Fibroblasts	Variety of transformed cells
	Smooth muscle cells	Glial cells
chemoattractant for:		
	Fibroblasts	Monocytes
	Neutrophils	Smooth muscle cells

of the epidermal layer. Along with several other growth factors, EGF plays a role in the wound healing process, rendering a potential medical application obvious.

EGF may also find a novel agricultural application in the defleecing of sheep. Administration of EGF to sheep has a transient effect on the wool follicle bulb cell, which results in a weakening of the root that holds the wool in place. Although novel, this approach to defleecing is unlikely to be economically attractive.

10.3.4 Platelet-derived growth factor

PDGF is a polypeptide growth factor that is sometimes termed osteosarcoma-derived growth factor or glioma-derived growth factor. It was first identified over 20 years ago as being the major growth factor synthesized by platelets. It is also produced by a variety of cell types. PDGF exhibits a mitogenic effect on fibroblasts, smooth muscle cells and glial cells, and exerts various additional biological activities (Table 10.11).

PDGF plays an important role in the wound healing process. It is released at the site of damage by activated platelets, and acts as a mitogen/chemoattractant for many of the cells responsible for initiation of tissue repair. It thus tends to act primarily in a paracrine manner. It also represents an autocrine/paracrine growth factor for a variety of malignant cells.

Active PDGF is a dimer. Two constituent polypeptides, A and B, have been identified, and three active PDGF isoforms are possible: AA, BB and AB. Two slightly different isoforms of the human PDGF A polypeptide (generated by differential mRNA splicing) have been identified. The short A form contains 110 amino acids and the long form contains 125 amino acids. Both exhibit one potential glycosylation site and three intrachain disulfide bonds. Two PDGF receptor subunits have been identified. Both are transmembrane glycoproteins whose cytoplasmic domains display tyrosine kinase activity upon activation.

In vitro and *in vivo* studies support the thesis that PDGF is of value in wound management, particularly with regard to chronic wounds. All three isoforms of PDGF are available from a range of recombinant systems. *In vitro* studies, using various cell lines, suggest that PDGF AB or BB dimeric isoforms are the most potent.

Normal skin appears to be devoid of PDGF receptors. Animal studies illustrate that rapid expression of both α and β receptor subunits is induced upon generation of an experimental wound (e.g. a surgical incision). Receptor expression is again switched off following re-epithelialization and complete healing of the wound.

Initial human trials have found that daily topical application of PDGF (BB isoform) stimulated higher healing rates of chronic pressure wounds, although the improvement recorded fell just short of being statistically significant. A second trial found that daily topical application of PDGF (BB) did promote statistically significant accelerated healing rates of chronic diabetic ulcers. The product (tradename Regranex) was approved for general medical use in the late 1990s. Its active ingredient is manufactured by Chiron Corporation, in an engineered strain of *Saccharomycies cerevisiae* harbouring the PDGF B chain gene. Regranex is notable in that it is formulated as a non-sterile (low bioburden) gel, destined for topical administration. The final formulation contains methylparaben, propylparaben and *m*-cresol as preservatives. In addition, as is the case with EGF, PDGF antagonists may also prove valuable in the treatment of some cancer types in which inappropriately high generation of PDGF-like mitogenic signals leads to the transformed state.

10.3.5 Fibroblast growth factors

FGFs constitute a family of about 20 proteins (numbered consecutively FGF-1 to FGF-20). Typically, they display a molecular mass in the region of 18–28 kDa and induce a range of mitogenic, chemotactic and angiogenic responses. Classification as an FGF is based upon structural similarity. All display a 140 amino acid central core that is highly homologous between all family members. All FGFs also tightly bind heparin and heparin-like glycosaminoglycans found in the extracellular matrix. This property has been used to purify several such FGFs via heparin affinity chromatography. Although many of the original members of this family stimulate the growth/development of fibroblasts (hence the name), several newer members have little/no effect upon fibroblasts. Keratinocyte growth factor (FGF-7) is thus far the only member of the FGF family to have gained approval for general medical use (Box 10.3).

10.3.6 Transforming growth factors

TGFs represent yet another family of polypeptide mitogens. The members of this family include TGF-α, as well as several species of TGF-β. TGF-α is initially synthesized as an integral membrane protein. Proteolytic cleavage releases the soluble growth factor, which is a 50 amino acid polypeptide. This growth factor exhibits a high amino acid homology with EGF, and it induces its biological effects by binding to the EGF receptor. It is synthesized by various body tissues, as well as by monocytes and keratinocytes. It is also manufactured by many tumour cell types, for which it can act as an autocrine growth factor.

TGF-β was first described as a growth factor capable of inducing transformation of several fibroblast cell lines (hence the name TGFs). It is now recognized that 'TGF-β' actually

Box 10.3

Product case study: Kepivance

Kepivance (tradename, also known as palifermin) is a keratinocyte growth factor produced by recombinant means in *E. coli*. It was first approved in the USA in 2004 and is indicated to decrease the incidence and duration of severe oral mucositis in patients with haematologic malignancies receiving myelotoxic therapy requiring haematopoietic stem cell support. Oral mucositis, characterized by ulcerative lesions of the mouth, is a common and debilitating side effect of high-dose chemotherapy or radiotherapy. It is particularly common (incidence of 70–80 per cent) and severe in the indication patient grouping.

Keratinocyte growth factor is a 140 amino acid, 16.3 kDa member of the FGF family. It differs from native keratinocyte growth factor in that the first 23 N-terminal amino acids have been deleted, which improves its stability. After cell growth the product is recovered and purified by a multistep chromatographic protocol. It is presented in lyophilized format in single-use vials and containing mannitol, sucrose, polysorbate 20 and histidine as excipients. It is administered by daily i.v. injection, usually for several days.

The cell surface keratinocyte growth factor receptor is present in a wide variety of tissues, including the tongue, buccal mucosa and oesophagus, but also additional tissues including the liver, kidneys and bladder. Keratinocyte growth factor receptor binding triggers proliferation, differentiation and migration of epithelial cells.

After single-dose i.v. administration, total body clearance appears two- to four-fold higher in cancer patients than in healthy subjects, although elimination half-life (3.3–5.7 h) was similar in both groups.

The initial trial establishing safety and efficacy was a randomized, placebo-controlled study involving 212 patients. The treated group received 60 μg kg^{-1} day^{-1} of the product for 6 days. The primary end-point measured was the number of days during which the patients experienced severe oral mucositis, which treatment reduced from 9 to 3 days. The incidence of mucositis was also reduced from 98 per cent to 63 per cent.

Side effects noted included rash, erythema, tongue thickening and discoloration, as well as alteration of taste. Kepivance is manufactured and marketed by Amgen Inc.

---

consists of three separate growth factors: TGF-β1, -β2 and -β3. Although the product of distinct genes, all exhibit close homology. In the mature form, they exist as homodimers, with each subunit containing 112 amino acid residues. Most body cells synthesize TGF-β, singly or in combination.

TGF-βs are pleiotrophic cytokines. They are capable of inhibiting the cell cycle and, hence, cell growth of several cell types, most notably epithelial and haematopoietic cells. These factors, however, stimulate the growth of other cell types, most notably cells that give rise to connective tissue, cartilage and bone. They induce the synthesis of extracellular matrix proteins and modulate the expression of matrix proteases. They also serve as a powerful chemoattractant for monocytes and fibroblasts. Given such activities, it is not surprising that the major physiological impact of TGF-βs appears to relate to:

- tissue remodelling;

- wound repair;

- haemopoiesis.

Such activities render them potentially useful therapeutic agents, and several are being assessed medically.

10.3.7   Neurotrophic factors

Neurotrophic factors constitute a group of cytokines that regulate the development, maintenance and survival of neurons, both in the central and peripheral nervous systems (Table 10.12). Although the first member of this family (NGF) was discovered more than 50 years ago, it is only in the last decade that the other members have been identified and characterized. The major subfamily of neurotrophic factors is the neurotrophins.

The original understanding of the term 'neurotrophic factor' was that of a soluble agent found in limiting quantities in the environment of sensitive neurons, and being generally manufactured by the neuronal target cells. It specifically promoted the growth and maintenance of those neurons. This description is now considered to be oversimplistic. Many neurotrophic factors are also synthesized by non-nerve target cells, and influence cells other than neurons (e.g. NGF is synthesized by mast cells and influences various cells of the immune system). Furthermore, various cytokines (including several growth factors; Table 10.12), discovered because of their ability to stimulate the growth of non-neuronal cells, are now also known to influence neuronal cells.

Each neurotrophic factor influences the growth and development of a specific group of neuronal types, with some cells being sensitive to several such factors. Many sustain specific neuronal populations whose death underlines various neurodegenerative diseases. This raises the possibility that these regulatory molecules may be of benefit in treating such diseases. Results from early clinical trials have been at best mixed, but many remain optimistic that neurotrophic factors may provide future effective treatments for some currently incurable neurodegenerative conditions.

Table 10.12   Molecules displaying neurotrophic activity *in vivo* and/or with neurons in culture

NGF	FGFs
BDNF	PDGF
Neurotrophin-3 (NT-3)	IGF-I and -II
Neurotrophin-4/5 (NT 4/5)	TGF-β1
Neurotrophin-6 (NT-6)	GM-CSF
CNTF	EPO
GDNF	LIF

Further reading

Books

Dallman, M. 2000. *Haemopoietic and Lymphoid Cell Culture.* Cambridge University Press.

Hoke, A. (ed.). 2006. *Erythropoietin and the Nervous System.* Springer Verlag.

Korholz, D. 2003. *Cytokines and Colony Stimulating Factors.* Humana Press.

Kuter, D., Hunt, P., Sheridan, W.P., and Zucker-Franklin, D. (eds). 1997. *Thrombopoiesis and Thrombopoietins.* Humana Press.

McKay, I. 1998. *Growth factors and receptors.* Oxford University Press.

Medkalf, D. 1995. *Haemopoietic Colony-Stimulating Factors: From Biology to Clinical Applications.* Cambridge University Press.

Morstyn, G. 1998. *Filgrastim in Clinical Practice.* Marcel Dekker.

Orlic, D. 1999. *Haematopoietic Stem Cells.* New York Academy of Science.

Patel, T. 2006. *Epidermal Growth Factor.* Humana Press.

Sytkowski, A. 2004. *Erythropoietin.* Wiley.

Yee, D. (ed.). 2004. *Insulin-like Growth Factors.* IOS Press.

Articles

Erythropoietin and thrombopoietin

Basser, R. 2002. The impact of thrombopoietin on clinical practice. *Current Pharmaceutical Design* **8**(5), 369–377.

Bottomley, A., Thomas, R., van Steen, K., Flechtner, H., Djulbegovic, B. 2002. Human recombinant erythropoietin and quality of life: a wonder drug or something to wonder about? *Lancet Oncology* **3**(3), 145–153.

Buemi, M., Aloisi, C., Cavallaro, E., Corica, F., Floccari, F., Grasso, G., Lasco, A., Pettinato, G, Ruello, A., Sturiale, A., and Frisina, N. 2002. Recombinant human erythropoietin: more than just the correction of uremic anemia. *Journal of Nephrology* **15**(2), 97–103.

Corwin, H. 2006. The role of erythropoietin therapy in the critically ill. *Transfusion Medicine Reviews* **20**(1), 27–33.

Engert, A. 2005. Recombinant human erythropoietin in oncology: current status and further developments. *Annals of Oncology* **16**(10), 1584–1595.

Fried, W. 1995. Erythropoietin. *Annual Review of Nutrition* **15**, 353–377.

Geddis, A.E., Linden, H.M., and Kaushansky, K. 2002. Thrombopoietin: a pan-hematopoietic cytokine. *Cytokine and Growth Factor Reviews* **13**(1), 61–73.

Heuser, M. and Ganser, A. 2006. Recombinant human erythropoietin in the treatment of nonrenal anemia. *Annals of Hematology* **85**(2), 69–78.

Kaushansky, K. 1997. Thrombopoietin – understanding and manipulating platelet production. *Annual Review of Medicine* **48**, 1–11.

Kaushansky, K. and Drachman, J. 2002. The molecular and cellular biology of thrombopoietin: the primary regulator of platelet production. *Oncogene* **21**(21), 3359–3367.

Kaushansky, K. 2005. The molecular mechanisms that control thrombopoiesis. *Journal of Clinical Investigation* **115**(12), 3339–3347.

Kendall, R. 2001. Erythropoietin. *Clinical and Laboratory Haematology* **23**(2), 71–80.

Lok, S., Kaushansky, K., Holly, R.D., Kuijper, J.L., Lofton-Day, C.E., Oort, P.J., Grant, F.J., Heipel, M.D., Burkhead, S.K., Kramer, J.M., Bell, L.A., Sprecher, C.A., Blumberg, H., Johnson, R., Prunkard, D., Ching, A.F.T., Mathewes, S.L., Bailey, M.C., Forstrom, J.W., Buddle, M.M., Osborn, S.G., Evans, S.J., Sheppard, P.O.,

Presnell, S.R., O'Hara, P.J., Hagen, F.S., Roth, G.J., and Foster, D.C. 1994. Cloning and expression of murine thrombopoietin cDNA and stimulation of platelet production *in vivo*. *Nature* **369**, 565–568.

Markham, A. and Bryson, H. 1995. Epoetin alfa, a review of its pharmacodynamic and pharmacokinetic properties and therapeutic use in non-renal applications. *Drugs* **49**(2), 232–254.

Miyazaki, H. and Kato, T. 1999. Thrombopoietin: biology and clinical potential. *International Journal of Haematology* **70**(4), 216–225.

Ogden, J. 1994. Thrombopoietin – the erythropoietin of platelets? *Trends in Biotechnology* **12**, 389–390.

Strauss, R. 2006. Controversies in the management of the anemia of prematurity using single-donor red blood cell transfusions and/or recombinant human erythropoietin. *Transfusion Medicine Review* **20**(1), 34–44.

Colony-stimulating factors

Duarte, R. and Frank, D. 2002. The synergy between stem cell factor (SCF) and granulocyte colony stimulating factor (G-CSF): molecular basis and clinical relevance. *Leukaemia and Lymphoma* **43**(6), 1179–1187.

Esser, M. and Brunner, H. 2003. Economic evaluations of granulocyte colony stimulating factor in the prevention and treatment of chemotherapy-induced neutropenia. *Pharmacoeconomics* **21**(18), 1295–1313.

Fleetwood, A.J., Cook, A.D., and Hamilton, J.A. 2005. Functions of granulocyte-macrophage colony stimulating factor. *Critical Reviews in Immunology* **25**(5), 405–428.

Frampton, J.E., Lee, C.R., and Faulds, D. 1994. Filgrastim. A review of its pharmacological properties and therapeutic efficacy in neutropenia. *Drugs* **48**(5), 731–760.

Frampton, J.E., Yarker, Y.E., and Goa, K.L. 1995. Lenograstim. A review of its pharmacological properties and therapeutic efficacy in neutropenia and related clinical settings. *Drugs* **49**(5), 767–793.

Hamilton, J. 1997. CSF-1 signal transduction. *Journal of Leukocyte Biology* **62**(2), 145–155.

Harousseau, J. 1997. The role of colony-stimulating factors in the treatment of acute leukaemia. *Biodrugs* **7**(6), 448–460.

Heuser, M. and Ganser, A. 2005. Colony stimulating factors in the management of neutropenia and its complications. *Annals of Hematology* **84**(11), 697–708.

Lyman, G. and Kuderer, N. 2004. The economics of the colony stimulating factors in the prevention and treatment of febrile neutropenia. *Critical Review in Oncology and Hematology* **50**(2), 129–146.

Tabbara, I.A., Ghazal, C.D., and Ghazal, H.H. 1996. The clinical applications of granulocyte-colony-stimulating factor in haematopoietic stem cell transplantation: a review. *Anticancer Research* **16**(6B), 3901–3905.

Watowich, S.S., Wu, H., Socolovsky, M., Klingmuller, U., Constantinescu, S.N., and Lodish, H.F. 1996. Cytokine receptor signal transduction and the control of haematopoietic cell development. *Annual Review of Cell and Developmental Biology* **12**, 91–128.

Insulin-like growth factors

Bach, L. 1999. The insulin-like growth factor system: basic and clinical aspects. *Australian and New Zealand Journal of Medicine*. **29**(3), 355–361.

Furstenberger, G. and Senn, H. 2002. Insulin like growth factors and cancer. *Lancet Oncology*, **3**(5), 298–302.

Kostecka, Z. and Blahovec, J. 2002. Animal insulin-like growth factor binding proteins and their biological functions. *Veterinarni Medicina* **47**(2–3), 75–84.

LeRoith, D. 1997. Insulin-like growth factors. *New England Journal of Medicine* **336**(9), 633–640.

Schmid, C. 1995. Insulin-like growth factors. *Cell Biology International* **19**(5), 445–457.

Yee, D. 2006. Targeting insulin-like growth factor pathways. *British Journal of Cancer* **94**(4), 465–468.

Zumkeller, W. 2002. The insulin-like growth factor system in hematopoietic cells. *Leukaemia and Lymphoma* **43**(3), 487–491.

Epidermal growth factor and platelet-derived growth factor

Board, R. and Jayson, G. 2005. Platelet-derived growth factor receptor (PDGFR): a target for anti cancer therapeutics. *Drug Resistance Updates* **8**(1–2), 75–83.

Boonstra, J., Rijken, P., Humbel, B., Cremers, F., Verkleij, A., and van Bergen en Henegouwen, P. 1995. The epidermal growth factor. *Cell Biology International* **19**(5), 413–430.

Johnstone, P. 2002. The epidermal growth factor receptor: a new target for anticancer therapy – introduction. *Current Problems in Cancer* **26**(3), 114–164.

Kane, S. 2006. Cancer therapies targeted to the epidermal growth factor receptor and its family members. *Expert Opinion on Therapeutic Patents* **16**(2), 147–164.

Leserer, M., Gschwind, A., and Ullrich, A. 2000. Epidermal growth factor receptor signal transactivation. *IUBMB Life* **49**(5), 405–430.

Shih, A. and Holland, E. 2006. Platelet-derived growth factor (PDGF) and glial tumorigenesis. *Cancer Letters* **232**(2), 139–147.

Wakeling, A. 2002. Epidermal growth factor receptor tyrosine kinase inhibitors. *Current Opinion in Pharmacology* **2**(4), 382–387.

Fibroblast growth factor/transforming growth factor

Dennler, S., Goumans, M.J., and ten Dijke, P. 2002. Transforming growth factor beta signal transduction. *Journal of Leukocyte Biology* **71**(5), 731–740.

Hu, X. and Zuckerman, K. 2001. Transforming growth factor: signal transduction pathways, cell cycle mediation, and effects on hematopoiesis. *Journal of Hematotherapy and Stem Cell Research* **10**(1), 67–74.

Imel, E. and Econs, M. 2005. Fibroblast growth factor 23: roles in health and disease. *Journal of the American Society of Nephrology* **16**(9), 2565–2575.

Kim, I.Y., Kim, M.M., and Kim, S.J. 2005. Transforming growth factor beta: biology and clinical relevance. *Journal of Biochemistry and Molecular Biology* **38**(1), 1–8.

Le, Y., Yu, X., Ruan, L., Wang, O., Qi, D., Zhu, J., Lu, X., Kong, Y., Cai, K., Pang, S., Shi, X., and Wang, J.M. 2005. The immunopharmacological properties of transforming growth factor beta. *International Immunopharmacology* **9**(13–14), 1771–1782.

Narayan, S., Thangasamy, T., and Balusu, R. 2005. Transforming growth factor beta receptor signaling in cancer. *Frontiers in Bioscience* **10**, 1135–1145.

Powers, C.J., McLeskey, S.W., and Wellstein, A. 2000. Fibroblast growth factors, their receptors and signalling. *Endocrine Related Cancer* **7**(3), 165–197.

Neurotrophic and related factors

Allen, S. and Dawbarn, D. 2006. Clinical relevance of the neurotropins and their receptors. *Clinical Science* **110**(2), 175–191.

Butte, M. 2001. Neurotrophic factor structure reveals clues to evolution, binding, specificity and receptor activation. *Cellular and Molecular Life Sciences* **58**(8), 1003–1013.

Chao, M.V., Rajagopal, R., and Lee, F.S. 2006. Neurotrophin signaling in health and disease. *Clinical Science* **110**(2), 167–173.

Fu, S. and Gordon, T. 1997. The cellular and molecular basis of peripheral nerve regeneration. *Molecular Neurobiology* **14**, 67–116.

Haque, N.S.K., Borghesani, P., and Isacson, O. 1997. Therapeutic strategies for Huntington's disease, based on a molecular understanding of the disorder. *Molecular Medicine Today* **3**(4), 175–183.

Hefti, F. 1997. Pharmacology of neurotrophic factors. *Annual Review of Pharmacology and Toxicology* **37**, 239–267.

Kwon, Y. 2002. Effect of neurotrophic factors on neuronal stem cell death. *Journal of Biochemistry and Molecular Biology* **35**(1), 87–93.

Sofroniew, M.V., Howe, C.L., and Mobley, W.C. 2001. Nerve growth factor signalling, neuroprotection and neural repair. *Annual Review of Neuroscience* **24**, 1217–1281.

Yamada, K., Mizuno, M., and Nabeshima, T. 2002. Role for brain-derived neurotrophic factor in learning and memory. *Life Sciences* **70**(7), 735–744.

Thorne, R. and Frey, W. 2001. Delivery of neurotrophic factors to the central nervous system – pharmacokinetic considerations. *Clinical Pharmacokinetics* **40**(12), 907–946.

Tatagiba, M., Brosamle, C., and Schwab, M.E. 1997. Regeneration of injured axons in the adult mammalian central nervous system. *Neurosurgery* **40**(3), 541–546.

Walsh, G. 1995. Nervous excitement over neurotrophic factors. *Bio/Technology* **13**, 1167–1171.

Yamada, M., Ikeuchi, T., and Hatanaka, H. 1997. The neurotrophic action and signalling of epidermal growth factor. *Progress in Neurobiology* **51**(1), 19–37.

11

Therapeutic hormones

11.1 Introduction

Hormones are amongst the most important group of regulatory molecules produced by the body. Originally, the term hormone was defined as a substance synthesized and released from a specific gland in the body that, by interacting with a receptor present in/on a distant sensitive cell, brought about a change in that target cell. Hormones travel to the target cell via the circulatory system. This describes what is now termed a true endocrine hormone.

At its loosest definition, some now consider a hormone to be any regulatory substance that carries a signal to generate some alteration at a cellular level. This embraces the concept of paracrine regulators (i.e. produced in the immediate vicinity of their target cells) and autocrine regulators (i.e. producer cell is also the target cell). Under such a broad definition, all cytokines, for example, could be considered hormones. The delineation between a cytokine and a hormone is already quite fuzzy using any definition.

True endocrine hormones, however, remain a fairly well defined group. Virtually all of the hormones used therapeutically (discussed below) fit into this grouping. Examples include insulin, glucagon, GH and the gonadotrophins.

11.2 Insulin

Insulin is a polypeptide hormone produced by the beta cells of the pancreatic islets of Langerhans. It plays a central role in regulating blood glucose levels, generally keeping it within narrow defined limits (3.5–8.0 mmol l^{-1}), irrespective of the nutritional status of the animal. It also has a profound effect on the metabolism of proteins and lipids and displays some mitogenic activity. The latter is particularly evident in *in vitro* studies and at high insulin concentrations. Some of these mitogenic effects are likely mediated via the IGF-1 receptor, and their physiological relevance is questionable.

Although many cells in the body express the insulin receptor, its most important targets are skeletal muscle fibres, hepatocytes and adipocytes, where it often antagonizes the effects of

Pharmaceutical biotechnology: concepts and applications   Gary Walsh
© 2007 John Wiley & Sons, Ltd   ISBN 978 0 470 01244 4 (HB)   978 0 470 01245 1 (PB)

Table 11.1   Some metabolic effects of insulin. These effects are generally countered by other hormones (glucagon and, in some cases, adrenaline). Hence, the overall effect noted often reflects the relative rates of these hormones present in the plasma

Metabolic pathway	Target tissue	Effect of insulin	Effect of glucagon
Glycogen synthesis	Liver	↑	↓
Glycogen degradation	Liver	↓	↑
Gluconeogenesis	Liver	↓	↑
Glycogen synthesis	Muscle	↑	–
Glycogen degradation	Muscle	–	–
Fatty acid synthesis	Adipose	↑	↓
Fatty acid degradation	Adipose	↓	↑

glucagon (Table 11.1). The most potent known stimulus of pancreatic insulin release is an increase in blood glucose levels, often occurring after meal times. Insulin orchestrates a suitable metabolic response to the absorption of glucose and other nutrients in a number of ways:

- it stimulates glucose transport (and transport of amino acids, K^+ ions and other nutrients) into cells, thus reducing their blood concentration;

- it stimulates (or helps to stimulate) intracellular biosynthetic (anabolic) pathways, such as glycogen synthesis (Table 11.1), which helps to convert the nutrients into a storage form in the cells;

- it inhibits (or helps to inhibit) catabolic pathways, such as glycogenolysis;

- it stimulates protein and DNA synthesis (which underlines insulin's growth-promoting activity).

In general, insulin achieves such metabolic control by inducing the dephosphorylation of several key regulatory enzymes in mainline catabolic or anabolic pathways. This inhibits the former and stimulates the latter pathway types. These effects are often opposed by other hormones, notably glucagon and adrenaline. Thus, when blood glucose concentrations decrease (e.g. during fasting), insulin levels decrease and the (largely catabolic) effects of glucagon become more prominent. Insulin also induces its characteristic effects by altering the level of transcription of various genes, many of which code for metabolic enzymes. Another gene upregulated by insulin is that of the integral membrane glucose transporter.

11.2.1   Diabetes mellitus

Failure of the body to synthesize sufficient insulin results in the development of insulin-dependent diabetes mellitus (IDDM). This is also known as type-1 diabetes or juvenile-onset diabetes.

IDDM is caused by T-cell-mediated autoimmune destruction of the insulin-producing β-pancreatic islet cells in genetically predisposed individuals. This is probably due to the expression of a 'super antigen' on the surface of the β-cells in such individuals, although the molecular detail of what extent factors trigger onset of the β-cell destruction remain to be elucidated. IDDM may, however, be controlled by parenteral administration of exogenous insulin preparations, usually by regular s.c. injection.

11.2.2 The insulin molecule

Insulin was first identified as an anti-diabetic factor in 1921, and was introduced clinically the following year. Its complete amino acid sequence was determined in 1951. Although mature insulin is a dimeric structure, it is synthesized as a single polypeptide precursor, i.e. preproinsulin. This 108 amino acid polypeptide contains a 23 amino acid signal sequence at its amino terminal end. This guides it through the endoplasmic reticulum membrane, where the signal sequence is removed by a specific peptidase.

Proinsulin-containing vesicles bud off from the endoplasmic reticulum and fuse with the golgi apparatus. Subsequently, proinsulin-containing vesicles (clathrin-coated secretory vesicles), in turn, bud off from the golgi. As they move away from the golgi, they lose their clathrin coat, becoming non-coated secretory vesicles. These vesicles serve as a storage form of insulin in the β-cell. Elevated levels of blood glucose, or other appropriate signals, cause the vesicles to fuse with the plasma membrane, thereby releasing their contents into the blood via the process of exocytosis.

Proinsulin is proteolytically processed in the coated secretory granules, yielding mature insulin and a 34-amino acid connecting peptide (C peptide, Figure 11.1). The C peptide is further proteolytically modified by removal of a dipeptide from each of its ends. The secretory granules thus contain low levels of proinsulin, C peptide and proteases, in addition to insulin itself. The insulin is stored in the form of a characteristic zinc–insulin hexamer, consisting of six molecules of insulin stabilized by two zinc atoms.

Mature insulin consists of two polypeptide chains connected by two interchain disulfide linkages. The A-chain contains 21 amino acids, whereas the larger B-chain is composed of 30 residues. Insulins from various species conform to this basic structure, while varying slightly in their amino acid sequence. Porcine insulin (5777 Da) varies from the human form (5807 Da) by a single amino acid, whereas bovine insulin (5733 Da) differs by three residues.

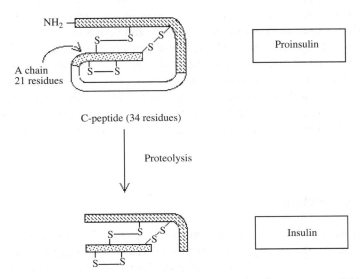

Figure 11.1 Proteolytic processing of proinsulin, yielding mature insulin, as occurs within the coated secretory granules

Although a high degree of homology is evident between insulins from various species, the same is not true for proinsulins, as the C peptide sequence can vary considerably. This has therapeutic implications, as the presence of proinsulin in animal-derived insulin preparations can potentially elicit an immune response in humans.

11.2.3 The insulin receptor and signal transduction

The insulin receptor is a tetrameric integral membrane glycoprotein consisting of two 735 amino acid α-chains and two 620 amino acid β-chains. These are held together by disulfide linkages (Figure 11.2). The α-chain resides entirely on the extracellular side of the plasma membrane and contains the cysteine-rich insulin-binding domain.

Each β-subunit is composed of three regions: the extracellular domain, the transmembrane domain and a large cytoplasmic domain that displays tyrosine kinase activity. In the absence of bovine insulin, tyrosine kinase activity is very weak. Proteolytic digestion of the α-subunit results in activation of this kinase activity. It is believed that the intact α-subunit exerts a negative influence on the endogenous kinase of the β-subunit and that binding of insulin, by causing a conformational shift in α-subunits, relieves this negative influence.

The cytoplasmic domain of the β-subunit displays three distinct sub-domains: (a) the 'juxtamembrane domain', implicated in recognition/binding of intracellular substrate molecules; (b) the tyrosine kinase domain, which (upon receptor activation) displays tyrosine kinase activity; (c) the C-terminal domain, whose exact function is less clear, although site-directed mutagenesis studies implicate it promoting insulin's mitogenic effects.

The molecular mechanisms central to insulin signal transduction are complex and have yet to be fully elucidated. However, considerable progress in this regard has been made over the last decade. Binding of insulin to its receptor promotes the autophosphorylation of three specific tyrosine residues in the tyrosine kinase domain (Figure 11.2b). This, in turn, promotes an alteration in the conformational state of the entire β-subunit, unmasking adenosine triphosphate (the phosphate donor) binding sites and substrate docking sites and activating its tyrosine kinase activity. Depending upon which specific intracellular substrates are then phosphorylated, at least two different signal transduction pathways are initiated (Figure 11.2c). Activation of the 'mitogen-activated protein kinase' pathway is ultimately responsible for triggering insulin's mitogenic effects, whereas activation of the PI-3 kinase pathway apparently mediates the majority of insulin's metabolic effects. Many of these effects, in particular the mitogenic effects, are promoted via transcriptional regulation of insulin-sensitive genes, of which there are probably in excess of 100 (Table 11.2).

11.2.4 Insulin production

Traditionally, commercial insulin preparations were produced by direct extraction from pancreatic tissue of slaughterhouse pigs and cattle, followed by multistep chromatographic purification. However, the use of animal-derived product had a number of potential disadvantages, including:

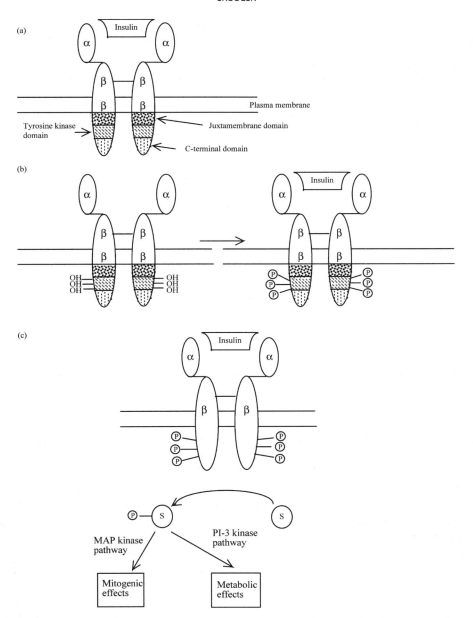

Figure 11.2 Structure of the insulin receptor (a). Binding of insulin promotes autophosphorylation of the β-subunits, where each β-subunit phosphorylates the other β-subunit. Phosphate groups are attached to three specific tyrosine residues (tyrosines 1158, 1162 and 1163), as indicated in (b). Activation of the β-subunit's tyrosine kinase activity in turn results in the phosphorylation of various intracellular (protein) substrates which trigger the mitogen-activated protein kinase and/or the phosphoinositide (PI-3) kinase pathway responsible for inducing insulin's mitogenic and metabolic effects. The underlying molecular events occurring in these pathways are complex (e.g. refer to Combettes-Souverain, M. and Issad, T. 1998. Molecular basis of insulin action. *Diabetes and Metabolism*, **24**, 477–489)

Table 11.2   Selected genes whose rate of transcription is altered by binding of insulin to its receptor. In virtually all instances, the ultimate effect is to promote anabolic events characteristic of insulin action. Two-dimensional gel electrophoresis has also pinpointed dozens of proteins of unknown function whose cellular level is altered by insulin

Protein class	Gene product	Insulin effect ($\uparrow$ or $\downarrow$ in transcription rate)
Integral membrane proteins	Insulin receptor	$\uparrow\downarrow$
	GH receptor	$\uparrow$
	Glucose transporters	$\uparrow$
Enzymes	Fatty acid synthetase	$\uparrow$
	Glutamine synthetase	$\uparrow\downarrow$
	Pyruvate kinase	$\uparrow$
	Fructose 1,6-bis-phosphatase	$\downarrow$
	Phosphoenolpyruvate carboxykinase	$\downarrow$
	Glucokinase	$\uparrow$
Hormones	IGF-1	$\uparrow$
	Glucagon	$\downarrow$
	GH	$\downarrow$
Transcription factors	C-Myc	$\uparrow$
	C-Fos	$\uparrow$
	egr-1	$\uparrow$

- *Immunogenicity.* Bovine insulin differs from human insulin by three amino acids and it elicits an immunological response in humans. This can trigger long-term complications, including insulin resistance (as anti-insulin antibodies neutralize some of the administered products). The presence of these antibodies can also affect the pharmacokinetic profile of the drug, as antibody-bound insulin molecules are largely resistant to the normal insulin degradative process. Porcine insulin, differs from human insulin by only a single amino acid (residue 30 of the B-chain; threonine in humans, alanine in pigs) and is essentially non-immunogenic in humans. However, many of the porcine insulin contaminants (including porcine proinsulin) are immunogenic in humans.

- *Availability.* Some 170 million people suffer from diabetes worldwide, a figure projected to double by 2030. Insulin administration is essential to the survival of those with type-1 (insulin-dependent) diabetes, and is required to control the progression of a minority of those with (the more common) insulin-independent type-2 diabetes. The annual insulin requirement has surpassed 5000 kg and continues to grow, prompting concern of an insulin shortfall from slaughterhouse sources.

Such issues and concerns underpinned the development of recombinant human insulin products, now routinely used in the management of diabetes.

11.2.5 Production of human insulin by recombinant DNA technology

Human insulin produced by recombinant DNA technology was first approved for general medical use in 1982, initially in the USA, West Germany, the UK and The Netherlands. As such, it was the first product of recombinant DNA technology to be approved for therapeutic use in humans. From the 1990s on, several engineered insulin products (discussed later) also gained approval (Table 11.3).

The initial approach to recombinant insulin production taken entailed inserting the nucleotide sequence coding for the insulin A- and B-chains into two different *E. coli* cells (both strain K12). These cells were then cultured separately in large-scale fermentation vessels, with subsequent chromatographic purification of the insulin chains produced. The A- and B-chains are then incubated together under appropriate oxidizing conditions in order to promote interchain disulfide bond formation, forming 'human insulin *crb*'

An alternative method (developed in the Eli Lilly research laboratories), entails inserting a nucleotide sequence coding for human proinsulin into recombinant *E. coli*. This is followed by purification of the expressed proinsulin and subsequent proteolytic excision of the C peptide *in vitro*. This approach has become more popular, largely due to the requirement for a single fermentation and subsequent purification scheme. Such preparations have been termed 'human insulin *prb*'.

Although recombinant product produced by either method is identical in sequence to native insulin, any impurities present will be host microbial-cell-derived and, hence, potentially highly immunogenic. Stringent purification of the recombinant product must thus be undertaken. This entails several chromatographic steps (often gel filtration and ion exchange, along with additional steps that exploit differences in molecular hydrophobicity, e.g. hydrophobic interaction chromatography or reverse-phase chromatography) (Figure 11.3).

A 'clean-up' process-scale RP-HPLC step has been introduced into production of human insulin *prb*. The C8 or C18 RP-HPLC column used displays an internal volume of 80 l or more, and up to 1200 g of insulin may be loaded during a single purification run (Figure 11.4). Separation is achieved using an acidic (often acetic-acid-based) mobile phase (i.e. set at a pH value sufficiently below the insulin pI value of 5.3 in order to keep it fully in solution). The insulin is usually loaded in the water-rich acidic mobile phase, followed by gradient elution using acetonitrile (insulin typically elutes at 15–30 per cent acetonitrile).

The starting material loaded onto the column is fairly pure (~92 per cent), and this step yields a final product of approximately 99 per cent purity. Over 95 per cent of the insulin activity loaded onto the column can be recovered. A single column run takes in the order of 1 h.

The RP-HPLC 'polishing' step not only removes *E. coli*-derived impurities, but also effectively separates modified insulin derivatives from the native insulin product. The resultant extremely low levels of impurities remaining in these insulin preparations fail to elicit any significant immunological response in diabetic recipients.

11.2.6 Formulation of insulin products

Insulin, whatever its source, may be formulated in a number of ways, generally in order to alter its pharmacokinetic profile. Fast (short)-acting insulins are those preparations that yield an elevated blood insulin concentration relatively quickly after their administration (which is usually by s.c. or, less commonly, by i.m. injection). Slow-acting insulins, on the other hand, enter the circulation

Table 11.3 Native and engineered human insulin preparations that have gained approval for general medical use. Reproduced in updated form with permission from: Walsh, G. 2005. Therapeutic insulins and their large-scale manufacture. *Applied Microbiology and Biotechnology*, **67**, 151–159

Product	Description	Structure	Company	Approved
Recombinant products of native human insulin sequence				
Humulin	Recombinant human insulin produced in *E. coli*	Identical to native human insulin	Eli Lilly	1982 (USA)
Novolin	Recombinant human insulin produced in *S. cerevisiae*	Identical to native human insulin	Novo Nordisk	1991 (USA)
Insuman	Recombinant human insulin produced in *E. coli*	Identical to native human insulin	Hoechst AG	1997 (EU)
Actrapid/Velosulin/ Monotard/ Insulatard/ Protaphane/ Mixtard/ Actraphane/ Ultratard	All contain recombinant human insulin produced in *S. cerevisiae* formulated as short/ intermediate/long acting product)	Identical to native human insulin	Novo Nordisk	2002 (EU)
Exubera	Recombinant human insulin produced in *E. coli* but administered via the pulmonary route	Identical to native human insulin	Pfizer	2006 (USA)
Engineered insulins				
Humalog (Insulin lispro)	Recombinant short-acting human insulin analogue produced in *E. coli*	Engineered: inversion of native B28–B29 proline–lysine sequence	Eli Lilly	1996 (USA and EU)
Liprolog (Insulin lispro)	Recombinant short-acting human insulin analogue produced in *E. coli*	Engineered: inversion of native B28–B29 proline–lysine sequence	Eli Lilly	1997 (EU)
NovoRapid (Insulin Aspart)	Recombinant short-acting human insulin analogue produced in *S. cerevisiae*	Engineered: B28 proline replaced by aspartic acid	Novo Nordisk	1999 (EU)
Novolog (Insulin Aspart)	Recombinant short-acting human insulin analogue produced in *S. cerevisiae*	Engineered: B28 proline replaced by aspartic acid	Novo Nordisk	2001 (USA)
Levemir (Insulin detemir)	Recombinant long-acting human insulin analogue produced in *S. cerevisiae*	Engineered: devoid of B30 threonine and a C14 fatty acid is covalently attached to B29 lysine	Novo Nordisk	2004 (EU)
Apidra (Insulin Glulisine)	Recombinant rapid-acting insulin analogue produced in *E. coli*	Engineered: B3 asparagine is replaced by a lysine and B29 lysine is replaced by glutamic acid.	Aventis pharmaceuticals	2004 (USA)
Lantus (Insulin glargine; optisulin)	Recombinant long-acting human insulin analogue produced in *E. coli*	Engineered: A 21 asparagine replaced by glycine and B chain elongated by two arginines	Aventis pharmaceuticals	2000 (USA and EU)

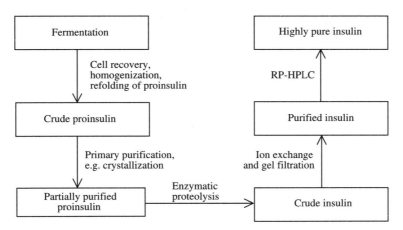

Figure 11.3 A likely purification scheme for human insulin *prb*. A final RP-HPLC polishing step yields a highly pure product. Refer to text for details

Figure 11.4 Process-scale HPLC column. Photograph courtesy of NovaSep Ltd, http://www.novasep.com

Table 11.4 Some pharmacokinetic characteristics of short, intermediate and long-acting insulin preparations

Category	Onset (hours after administration)	Peak activity (hours after administration)	Duration (h)
Short-acting	0.5–1	2–5	6–8
Intermediate-action	2	4–12	up to 24
Long-acting	4	10–20	up to 36

much more slowly from the depot (injection) site. This is characterized by a slower onset of action, but one of longer duration (Table 11.4).

In healthy individuals, insulin is typically secreted continuously into the bloodstream at low basal levels, with rapid increases evident in response to elevated blood sugar levels. Insulin secretion usually peaks approximately 1 h after a meal, falling off to base levels once again within the following 2 h.

The blood insulin level is continuously up- or down-regulated as appropriate for the blood glucose levels at any given instant. Conventional insulin therapy does not accurately reproduce such precise endogenous control. Therapy consists of injections of slow- and fast-acting insulins, as appropriate, or a mixture of both. No slow-acting insulin preparation, however, accurately reproduces normal serum insulin baseline levels. An injection of fast-acting insulin will not produce a plasma hormone peak for 1.5–2 h post injection, and levels then remain elevated for up to 5 h. Hence, if fast-acting insulin is administered at mealtime, diabetics will still experience hyperglycaemia for the first hour, and hypoglycaemia after 4–5 h. Such traditional animal or human insulin preparations must thus be administered 30 min or so before eating, and the patient must not subsequently alter their planned mealtime.

Insulin, at typical normal plasma concentrations (approximately 1×10^{-9} mol l^{-1}) exists in true solution as a monomer. Any insulin injected directly into the bloodstream exhibits a half-life of only a few minutes.

The concentration of insulin present in soluble insulin preparations (i.e. fast-acting insulins), is much higher (approximately 1×10^{-3} mol l^{-1}). At this concentration, the soluble insulin exists as a mixture of monomer, dimer, tetramer and zinc–insulin hexamer. These insulin complexes have to dissociate in order to be absorbed from the injection site into the blood, which slows down the onset of hormone action.

In order to prolong the duration of insulin action, soluble insulin may be formulated to generate insulin suspensions. This is generally achieved in one of two ways:

1. Addition of zinc in order to promote Zn–insulin crystal growth (which take longer to disassociate and, hence, longer to leak into the bloodstream from the injection depot site).

2. Addition of a protein to which the insulin will complex, and from which the insulin will only be slowly released. The proteins normally used are protamines, which are basic polypeptides naturally found in association with nucleic acid in the sperm of several species of fish. Depending on the relative molar ratios of insulin:protamine used, the resulting long-acting insulins generated are termed protamine–Zn–insulin or isophane insulin. Biphasic insulins include mixtures of short- and long-acting insulins, which attempt to mimic normal insulin rhythms in the body.

11.2.7 Engineered insulins

Recombinant DNA technology facilitates not only production of human insulin in microbial systems, but also facilitates generation of insulins of modified amino acid sequences. The major aims of generating such engineered insulin analogues include:

- Identification of insulins with altered pharmacokinetic properties, such as faster-acting or slower-acting insulins.

- Identification of super-potent insulin forms (insulins with higher receptor affinities). This is due to commercial considerations, namely the economic benefits that would accrue from utilizing smaller quantities of insulin per therapeutic dose.

The insulin amino acid residues that interact with the insulin receptor have been identified (A1, A5, A19, A21, B10, B16, B23-25), and a number of analogues containing amino acid substitutions at several of these points have been manufactured. Conversion of histidine to glutamate at the B10 position, for example, yields an analogue displaying fivefold higher activity *in vitro*. Other substitutions have generated analogues with even higher specific activities. However, increased *in vitro* activity does not always translate to increased *in vivo* activity.

Attempts to generate faster-acting insulins have centred upon developing analogues that do not dimerize or form higher polymers at therapeutic dose concentrations. The contact points between individual insulin molecules in insulin dimers/oligomers include amino acids at positions B8, B9, B12–13, B16 and B23–28. Thus, analogues with various substitutions at these positions have been generated. The approach adopted generally entails insertion of charged or bulky amino acids, in order to promote charge repulsion or steric hindrance between individual insulin monomers. Several are absorbed from the site of injection into the bloodstream far more quickly than native soluble (fast-acting) insulin. Such modified insulins could thus be injected at mealtimes rather than 1 h before, and several such fast-acting engineered insulins have now been approved for medical use (Table 11.3). 'Insulin lispro' (tradename 'Humalog') was the first such engineered short-acting insulin to come to market (Box 11.1 and Figure 11.5).

'Insulin Aspart' is a second fast-acting engineered human insulin analogue now approved for general medical use. It differs from native human insulin in that the proline[B28] residue has been replaced by aspartic acid. This single amino acid substitution also decreases the propensity of individual molecules to self-associate, ensuring that they begin to enter the bloodstream from the site of injection immediately upon administration.

A number of studies have also focused upon the generation of longer-acting insulin analogues. The currently used Zn–insulin suspensions, or protamine–Zn–insulin suspensions, generally display a plasma half-life of 20–25 h. Selected amino acid substitutions have generated insulins which, even in soluble form, exhibit plasma half-lives of up to 35 h.

Optisulin or Lantus are the tradenames given to one such analogue that gained general marketing approval in 2000 (Table 11.3). The international non-proprietary name for this engineered molecule is 'insulin glargine'. It differs from native human insulin in that the C-terminal aspargine residue of the A-chain has been replaced by a glycine residue and the β-chain has been elongated (again from its C-terminus) by two arginine residues. The overall effect is to increase the molecule's pI (the pH at which the molecule displays a net overall zero charge and, consequently, at

Box 11.1

Product case study: Humalog

Humalog (tradename, also known as insulin lispro) was the first recombinant fast-acting insulin analogue to gain marketing approval (in 1996). It is indicated for the treatment of diabetes mellitus, for the control of hyperglycaemia and is used in conjunction with long-acting insulins (see main text). It is administered subcutaneously. The product displays an amino acid sequence identical to native human insulin with one alteration, i.e. an inversion of the natural proline–lysine sequence found at positions 28 and 29 of the insulin B-chain. This simple alteration significantly decreased the propensity of individual insulin molecules to self-associate when stored at therapeutic dose concentrations. The dimerization constant for insulin lispro is 300 times lower than that exhibited by unmodified human insulin. Structurally, this appears to occur because the change in sequence disrupts the formation of interchain hydrophobic interactions critical to self-association.

The rationale underlining the sequence alteration was rooted in studies not of insulin, but of IGF-1 (Chapter 10). The latter displays a strong structural resemblance to proinsulin, with up to 50 per cent of amino acid residues within the IGF-1 A- and B-domains being identical to those found in comparative positions in the insulin A- and B-chains. When compared with insulin, IGF-1 molecules display a significantly decreased propensity to self-associate. Sequencing studies earlier revealed that the proline^{B28}–lysine^{B29} sequence characteristic of insulin is reversed in IGF-1. It was suggested that if this sequence difference was responsible for the differences in self-association propensity, then inversion of the proline^{B28}–lysine^{B29} sequence in insulin would result in its decreased self-association. Direct experimentation proved this hypothesis accurate.

Insulin lispro is manufactured commercially in a manner quite similar to the 'proinsulin' route used to manufacture native recombinant human insulin. A synthetic gene coding for Lys^{B28}–Pro^{B29} proinsulin is expressed in *E. coli*. Following fermentation and recovery, the proinsulin is treated with trypsin and carboxypeptidase B, resulting in the proteolytic excision of the engineered insulin molecule. It is then purified to homogeneity by a number of high-resolution chromatographic steps. The final product formulation also contains *m*-cresol (preservative and stabilizer), zinc oxide (stabilizer), glycerol (tonicity modifier) and a phosphate-based buffer and is presented in vial, cartridge and injector pen formats.

Humalog has proven equipotent to regular human insulin, but its effect is more rapid and of shorter duration. After s.c. administration it displays similar bioavailability to regular insulin (typically 55–77 per cent). Peak serum levels are usually recorded 30–90 min after administration, as opposed to 50–120 min in the case of regular insulin. Its serum $t_{1/2}$ is also shorter than that of regular insulin (60 min as opposed to 90 min). Product safety and efficacy have been established in several trials, an earlier major trial being an open label, crossover study of 1008 patients with type-1 diabetes and 722 patients with type-2 diabetes. Predictably, the major potential negative adverse effect is hypoglycaemia. The product was developed and is marketed by Eli Lilly.

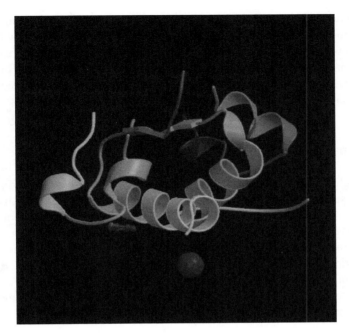

Figure 11.5 Three-dimensional structure of the engineered fast-acting insulin, 'Insulin lispro'. Structural details courtesy of the Protein Data Bank, http://www.rcsb.org/pdb/

which it is least soluble) from 5.4 to a value approaching 7.0. The engineered insulin is expressed in a recombinant *E. coli* K12 host strain and is produced via the 'proinsulin route' as described previously. The purified product is formulated at pH 4.0, a pH value at which it is fully soluble. Upon s.c. injection, the insulin experiences an increase in pH towards more neutral values and, consequently, appears to precipitate in the subcutaneous tissue. It resolubilizes very slowly and, hence, a greatly prolonged duration of release into the bloodstream is noted. Consequently, a single daily injection supports the maintenance of acceptable basal blood insulin levels, and insulin molecules are still detected at the site of injection in excess of 24 h after administration.

Levemir (tradename) is an alternative engineered long-acting insulin product that gained approval for general medical use in 2004 (Table 11.3). This differs from native insulin in that it is devoid of the threonine B30 residue and (more importantly from a pharmacokinetic perspective) contains a 14-carbon fatty acid residue covalently attached to the side chain of lysine residue B29. This allows the insulin to bind reversibly to albumin, both at the site of injection and in plasma (albumin contains three high-affinity fatty acid binding sites). This, in turn, ensures constant and prolonged release of free insulin, bestowing upon it a prolonged duration of action of up to 24 h. Product manufacture entails initial expression of insulin in engineered *Saccharomyces cerevisiae*, purification and acylation (attachment of the fatty acid group).

The generation of engineered insulin analogues raises several important issues relating to product safety and efficacy. Alteration of a native protein's amino acid sequence could render the engineered product immunogenic. Such an effect would be particularly significant in the case of insulin, as the product is generally administered daily for life. In addition, alteration in structure could have unintended influences upon pharmacokinetic and/or pharmacodynamic characteristics of the drug. Preclinical and, in particular, clinical evaluations undertaken upon the analogues thus far approved, however, have confirmed their safety and efficacy. The sequence changes introduced are relatively minor and do not seem to elicit an immunological response. Fortuitously, neither have the alterations made affected the ability of the insulin molecule to interact with the insulin receptor, and trigger the resultant characteristic biological responses.

11.2.8 Additional means of insulin administration

Issues surrounding protein delivery via means alternative to parenteral routes have been outlined in Chapter 4. One such product (inhalable insulin, tradename Exubera) has gained marketing approval (Box 11.2). An additional approach that may mimic more closely the normal changes in blood insulin levels entails the use of infusion systems that constantly deliver insulin to the patient. The simplest design in this regard is termed an 'open-loop system'. This consists of an infusion pump that automatically infuses soluble insulin subcutaneously, via a catheter. Blood glucose levels are monitored manually and the infusion rate is programmed accordingly.

Although the potential of such systems is obvious, they have not as yet become popular in practice, mainly due to complications that can potentially arise, including:

Box 11.2

Product case study: Exubera

Exubera (tradename) is a recombinant human insulin first approved for medical use in the USA in 2006. It is indicated for the treatment of diabetes mellitus for the control of hyperglycaemia and is particularly noteworthy in that it is the first such product delivered by inhalation technology. The recombinant insulin is produced in *E. coli* and, after purification, is formulated as a powder also containing citrate buffer components, mannitol and glycine as excipients. It is sold as blisters containing 1 mg or 3 mg unit doses that are administered using a specially designed inhaler. A fraction of the total dose is emitted as fine particles capable of reaching the deep lung, from where the insulin is absorbed. Pharmacokinetic studies (in both healthy and diabetic subjects) show that the insulin is absorbed as quickly as s.c. administered rapid-acting insulin analogues and, therefore, should be administered within 10 min of mealtime. Glucose-lowering activity usually commences within 10–20 min of administration, with a maximum effect observed after approximately 2 h and an activity duration of approximately 6 h. Actual serum insulin concentrations typically peak 50 min post administration, compared with 100 min or so for s.c. administered regular insulin.

Pre-approval safety and efficacy clinical studies involved product administration to 2500 adults with either type-1 or -2 diabetes. The primary efficacy parameter measured was glycaemic control (as measured by the reduction from baseline in haemoglobin A1c). Hypoglycaemia was the most commonly reported adverse effect. Trials also showed a greater decline in pulmonary function in the Exubera group, and product should not be administered to patients with underlying lung disease, or to smokers. Exubera was developed by Nektar Inc. and is marketed under licence by Pfizer.

- abscess formation or development of cellulitis at the site of injection;

- possible pump malfunction;

- catheter obstruction;

- hypersensitivity reactions to components of the system;

- requirement for manual blood glucose monitoring.

The closed-loop system (often termed the 'artificial pancreas') is essentially a more sophisticated version of the system described above. It consists not only of a pump and infusion device, but also of an integral glucose sensor and computer that analyses the blood glucose data obtained and adjusts the flow rate accordingly. The true potential of such systems remains to be assessed.

Although infusion pumps can go some way towards mimicking normal control of blood insulin levels, transplantation of insulin-producing pancreatic cells should effectively 'cure' the diabetic patient, and research aimed at underpinning this approach continues.

11.3 Glucagon

Glucagon is a single-chain polypeptide of 29 amino acid residues and a molecular mass of 3500 Da. It is synthesized by the A-cells of the islets of Langerhans, and also by related cells found in the digestive tract. Like insulin, it is synthesized as a high molecular mass from which the mature hormone is releases by selective proteolysis.

The major biological actions of glucagon tend to oppose those of insulin, particularly with regard to regulation of metabolism. Glucagon has an overall catabolic effect, stimulating the breakdown of glycogen, lipid and protein. A prominent metabolic effect is to increase blood glucose levels (i.e. it is a hyperglycaemic hormone). Indeed, the major physiological function of glucagon is to prevent hypoglycaemia. The hormone is stored in secretory vesicles, after synthesis in the pancreatic A-cells, and released by exocytosis upon experiencing a drop in blood glucose concentration.

Glucagon initiates its metabolic (and other) effects by binding to a specific cell surface receptor, thus activating a membrane-bound adenylate kinase. This, in turn, promotes activation of a cyclic adenosine monophosphate (cAMP)-dependent protein kinase (Figure 11.6). The kinase phosphorylates key regulatory enzymes in carbohydrate metabolism, thereby modulating their activity. Hepatic glycogen phosphorylase, for example, is activated via phosphorylation, while glycogen synthetase is inactivated, thus promoting glycogen breakdown. The rate of gluconeogenesis is also stimulated by the inactivation of pyruvate kinase and simultaneous activation of fructose 1,6-bisphosphatase.

Hypoglycaemia remains the most frequent complication of insulin administration to diabetics. It usually occurs due to (a) administration of an excessive amount of insulin; (b) administration of insulin prior to a mealtime, but with subsequent omission of the meal; or (c) due to increased physical activity. In severe cases this can lead to loss of consciousness, and even death. Although it may be treated by oral or i.v. administration of glucose, insulin-induced hypoglycaemia is sometimes treated by administration of glucagon.

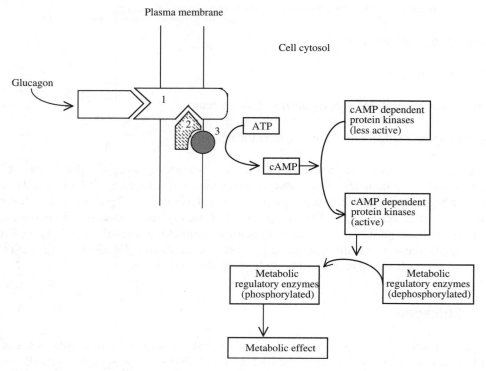

Figure 11.6 Initiation of a metabolic response to the binding of glucagon to its receptor. (1) glucagon cell surface receptor; (2) G protein; (3) adenylate cyclase. (See text for further detail)

Glucagon is also used medically as a diagnostic aid during certain radiological examinations of the stomach and small and large intestine where decreased intestinal motility is advantageous (the hormone has an inhibitory effect on the motility of the smooth muscle lining the walls of the gastrointestinal tract).

Traditionally, glucagon preparations utilized therapeutically are chromatographically purified from bovine or porcine pancreatic tissue. (The structure of bovine, porcine and human glucagon is identical, thus eliminating the possibility of direct immunological complications). Such commercial preparations are generally formulated with lactose and sodium chloride and sold in freeze-dried form. Glucagon, 0.5–1.0 units (approximately 0.5–1.0 mg freeze-dried hormone), is administered to the patient by s.c. or i.m. injection.

More recently, glucagon preparations produced via recombinant means have also become available. 'GlucaGen' is the tradename given to one such product, produced by Novo Nordisk using an engineered *S. cerevisiae* strain. Upstream processing (aerobic batch-fed fermentation) is followed by an upward adjustment of media pH in order to dissolve precipitated product (glucagon is insoluble in aqueous-based media in the pH range 3–9.5). This facilitates subsequent removal of the yeast by centrifugation. Glucagon is then recovered and purified from the media by a series of further precipitation and high-resolution chromatographic steps. Eli Lilly also produces a recombinant glucagon product using an engineered *E. coli* strain.

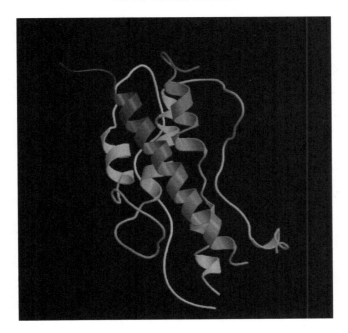

Figure 11.7   Three-dimensional structure of hGH. Structural details courtesy of the Protein Data Bank, http://www.rcsb.org/pdb/

11.4   Human growth hormone

hGH (somatotrophin; Figure 11.7) is a polypeptide hormone synthesized in the anterior pituitary. It promotes normal body growth and lactation and influences various aspects of cellular metabolism.

Mature hGH contains 191 amino acid residues and displays a molecular mass of 22 kDa. It also contains two characteristic intrachain disulfide linkages. hGH mRNA can also undergo alternate splicing, yielding a shortened GH molecule (20 kDa), which appears to display biological activities indistinguishable from the 22 kDa species.

hGH displays significant, although not absolute, species specificity. GHs isolated from other primates are the only preparations biologically active in humans. (This precluded the earlier use of bovine/porcine preparations for medical use in humans.)

GH synthesis and release from the pituitary is regulated by two peptide hypothalamic factors: GH-releasing hormone (GHRH, also known as GH-releasing factor (GHRF) or somatorelin) and GH release inhibiting hormone (GHRIH) or somatostatin. Furthermore, whereas GH directly mediates some of its biological actions, its major influence on body growth is mediated indirectly via IGF-1, as discussed below. GHRH, GHRIH, GH and IGF-1 thus form a hormonal axis, as depicted in Figure 11.8.

11.4.1   The growth hormone receptor

GH induces its characteristic biological effects by binding to a specific cell surface receptor. The human receptor is a single-chain 620 amino acid transmembrane polypeptide. Sequence analysis

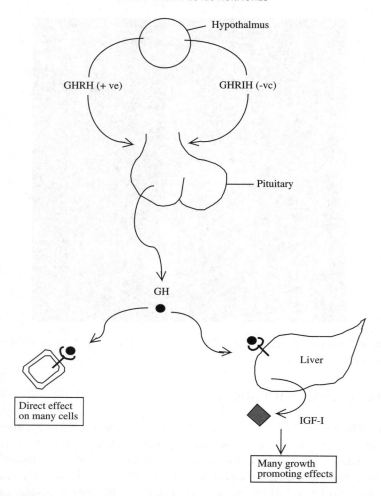

Figure 11.8   Overview of the mechanisms by which GH induces its biological effects and how its secretion from the pituitary is regulated

indicates it is a member of the haemopoietic receptor superfamily (which includes receptors for several interleukins, GM-CSF and EPO).

Soluble GH-binding proteins (GHBPs) are also found in the circulation. In humans, these GHBPs are generated by enzymatic cleavage of the integral membrane receptor, releasing the GH-binding extracellular domain. The exact physiological role of these binding proteins remains to be elucidated. In serum, GH binds to two such GHBPs, an action that prolongs the hormone's plasma half-life.

11.4.2   Biological effects of growth hormone

GH primarily displays an anabolic activity. It partially stimulates the growth of bone, muscle and cartilage cells directly. Binding of GH to its hepatic receptor results in the synthesis and release of

Table 11.5 Some of the major biological effects promoted by growth hormone. Although many of these are direct, other effects are mediated via IGF-1 (Chapter 10)

Increased body growth (particularly bone and skeletal muscle)
Stimulation of protein synthesis in many tissues
Mobilization of depot lipids from adipose tissue (lipolytic effect)
Elevation of blood glucose levels (anti-insulin effect)
Increase of muscle and cardiac glycogen stores
Increased kidney size and enhanced renal function
Reticulocytosis (increased reticulocyte production in the bone marrow)

IGF-1, which mediates most of GH's growth-promoting activity on, for example, bone and skeletal muscle (Chapter 10). The major effects mediated by hGH are summarized in Table 11.5.

A deficiency in the secretion of hGH during the years of active body growth results in pituitary dwarfism (a condition responsive to exogenous hGH administration). On the other hand, overproduction of hGH during active body growth results in gigantism. hGH overproduction after primary body growth has occurred results in acromegaly, a condition characterized by enlarged hands and feet, as well as coarse features.

11.4.3 Therapeutic uses of growth hormone

GH has a potentially wide range of therapeutic uses (Table 11.6). To date, however, its major application has been for the treatment of short stature. hGH extracted from human pituitary glands was first used to treat pituitary dwarfism (i.e. caused by suboptimal pituitary GH secretion) in 1958. It has subsequently proven effective in the treatment of short stature caused by a variety of other conditions, including:

- Turner's syndrome;

- idiopathic short stature;

- chronic renal failure.

The use of hGH extracted from the pituitaries of deceased human donors came to an abrupt end in 1985, when a link between treatment and Creutzfeld–Jacob disease (CJD, a rare, but fatal,

Table 11.6 Some actual or likely therapeutic uses for hGH. Refer to text for details

Treatment of short stature caused by GH deficiency
Treatment of defective growth caused by various diseases/medical conditions
Induction of lactation
Counteracting ageing
Treatment of obesity
Body building
Induction of ovulation

neurological disorder) was discovered. In this year, a young man, who had received hGH therapy some 15 years previously, died from CJD, which, investigators concluded, he had contracted from infected pituitary extract (CJD appears to be caused by a prion). At least an additional 12 CJD cases suspected of being caused in the same way have subsequently been documented. Fortunately, several recombinant hGH (rhGH) preparations were coming on-stream at that time (Table 11.7), and now all hGH preparations used clinically are derived from recombinant sources. Currently, in excess of 20 000 people are in receipt of rhGH therapy.

rhGH was first produced in *E. coli* in the early 1980s. The initial recombinant preparations differed from the native human hormone only in that they contained an extra methionine residue (due to the AUG start codon inserted at the beginning of the gene). Subsequently, a different cloning strategy allowed production in *E. coli* of products devoid of this terminal methionine.

In vitro analysis, including tryptic peptide mapping, amino acid analysis and comparative immunoassays, shows the native and recombinant forms of the molecule to be identical. Clinical trials in humans have also confirmed that the recombinant version promotes identical biological responses to the native hormone. rhGH was first purified (on a laboratory scale) by Genentech scientists using the strategy outlined in Figure 11.9. A somewhat similar strategy is likely used in its process-scale purification.

11.5   The gonadotrophins

The gonadotrophins are a family of hormones for which the gonads represent their primary target (Table 11.8). They directly and indirectly regulate reproductive function and, in some cases, the development of secondary sexual characteristics. Insufficient endogenous production of any

Table 11.7   rhGH preparations approved for general medical use

Product (tradename)	Company	Indication
Humatrope	Eli Lilly	hGH deficiency in children
Nutropin	Genentech	hGH deficiency in children
Nutropin AQ	Schwartz Pharma AG	Growth failure, Turner's syndrome
BioTropin	Biotechnology General	hGH deficiency in children
Genotropin	Pharmacia & Upjohn	hGH deficiency in children
Saizen	Serono Laboratories	hGH deficiency in children
Serostim	Serono Laboratories	Treatment of AIDS-associated catabolism/wasting
Norditropin	Novo Nordisk	Treatment of growth failure in children due to inadequate growth hormone secretion
Omnitrop	Sandoz	Treatment of children and adults with certain forms of growth disturbance
Valtropin	Biopartners	Treatment of children and adults with certain forms of growth disturbance
Somavert (hGH analogue, i.e. antagonist)	Pharmacia	Treatment of selected patients suffering from acromegaly

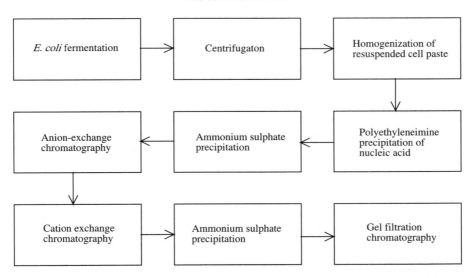

Figure 11.9 Production of rhGH in *E. coli* (as an intracellular protein). Subsequent to fermentation, the cells are collected by centrifugation or filtration. After homogenization, nucleic acids and some membrane constituents are precipitated by the addition of polyethyleneimine. Ammonium sulfate precipitation of the supernatant concentrates the crude rhGH preparation. Chromatographic purification follows, as illustrated

member of this family will adversely affect reproductive function, which generally can be treated by administration of an exogenous preparation of the hormone in question. Most gonadotrophins are synthesized by the pituitary, although some are made by reproductive and associated tissues.

11.5.1 Follicle-stimulating hormone, luteinizing hormone and human chorionic gonadotrophin

FSH and LH play critical roles in the development and maintenance of male and, particularly, female reproductive function (Box 11.3). hCG, produced by pregnant women, plays a central role in maintaining support systems for the developing embryo during early pregnancy. All three are

Table 11.8 The gonadotrophins, their site of synthesis and their major biological effects

Gonadotrophin	Site of synthesis	Major effects
FSH	Pituitary	Stimulates follicular growth (female)
		Enhanced spermatogenesis (male)
LH	Pituitary	Induction of ovulation (female)
		Synthesis of testosterone (male)
CG		Maintenance of the corpus luteum in pregnant females
PMSG (horses only)	Endometrial cups	Maintenance of pregnancy in equids
Inhibin	Gonads	Inhibition of FSH synthesis
		Probably tumour repressor
Activin	Gonads	Stimulation of FSH synthesis

Box 11.3

An overview of the female reproductive cycle

The human female reproductive (ovarian) cycle is initiated and regulated by gonadotrophic hormones. Day 1 of the cycle is characterized by commencement of menstruation: the discharge of fragments of the endometrium (wall of the womb) from the body, signifying fertilization has not occurred in the last cycle. At this stage, plasma levels of FSH and LH are low, but these begin to increase slowly over the subsequent 10–14 days.

During the first phase of the cycle, a group of follicles (each of which houses an egg) begins to develop and grow, largely under the influence of FSH. Shortly thereafter, a single dominant follicle normally emerges and the remainder regress. The growing follicle begins to synthesize oestrogens, which, in turn, trigger a surge in LH secretion at the cycle midpoint (day 14). A combination of elevated FSH and LH levels (along with additional factors such as prostaglandin F_{2a}) promotes follicular rupture. This releases the egg cell (ovulation) and converts the follicle into the progesterone-secreting follicular remnant: the corpus luteum. Release of the egg marks the end of the first half (follicular phase) of the cycle and the commencement of the second (luteal) phase.

In the absence of fertilization subsequent to ovulation, the maximum life span of the corpus luteum is 14 days, during which time it steadily regresses. This, in turn, promotes slowly decreasing levels of corpus luteum hormones, i.e. oestrogen and progesterone. Progesterone normally serves to prepare (thicken) the lining of the womb for implantation of an embryo should fertilization occur. Withdrawal of hormonal support as the corpus luteum regresses results in the shedding of the endometrial tissue, which marks commencement of the next cycle. However, if ovulation is followed by fertilization, then the corpus luteum does not regress but is maintained by hCG, synthesized in the placenta of pregnant females (Figure 11.B1).

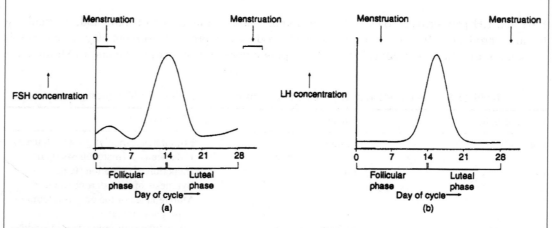

Figure 11.B1   Changes in plasma FSH (a) and LH (b) levels during the reproductive cycle of a healthy human female

heterodimeric hormones containing an identical α-polypeptide subunit and a unique β-polypeptide subunit that confers biological specificity to each gonadotrophin. In each case, both subunits of the mature proteins are glycosylated. Human FSH displays four N-linked (asparagines or Asn-linked) glycosylation sites, located at positions Asn 52 and 78 of the α-subunit and Asn 7 and 24 of the β-subunit. Some 30 per cent of the hormone's overall molecular mass is accounted for by its carbohydrate component. Structurally, the attached oligosaccharides are heterogeneous in nature, varying in particular in terms of the content of sialic acid residues and sulfate groups present. This represents the structural basis of the charge heterogeneity characteristic of this (and other) gonadotrophins.

The oligosaccharide components play a direct and central role in the biosynthesis, secretion, serum half-life and potency of the gonadotrophins. The sugar components attached to the α-subunits play an important role in dimer assembly and stability, as well as hormone secretion and possibly signal transduction. The sugars associated with the β-subunit, while contributing to dimer assembly and secretion, appear to play a more prominent role in clearance of the hormone from circulation.

The functional effects of glycosylation take on added significance in the context of producing gonadotrophins by recombinant means. As discussed subsequently, several are now produced for clinical application in recombinant (animal cell line) systems. Although the glycosylation patterns observed on the recombinant molecules can vary somewhat in composition from those associated with the native hormone, these slight differences bear no negative influence upon their clinical applicability.

The synthesis and release of both FSH and LH from the pituitary is stimulated by a hypothalamic peptide, gonadotrophin-releasing hormone (also known as gonadorelin, LH-releasing hormone, or LH/FSH-releasing factor).

FSH exhibits a molecular mass of 34 kDa. The α-subunit gene (containing four exons) is present on chromosome 6, and the β gene (three exons) is found on chromosome 11. mRNA coding for both sununits is translated separately on the rough endoplasmic recticulum, followed by removal of their signal peptides upon entry into the endoplasmic recticulum. N-linked glycosylation also takes place, as does intrachain disulfide bond formation. The α- and β-subunits combine non-covalently and appear to be stored in secretory vesicles separately to those containing LH. Although free α-subunits are also found within the pituitary, few β-subunits are present in unassociated form. Such free β-subunits are rapidly degraded.

The major FSH target in the male is the Sertoli cells, found in the walls of the seminiferous tubules of the testis. They function to anchor and nourish the spermatids, which subsequently are transformed into spermatozoa during the process of spermatogenesis. Sertoli cells also produce inhibin (discussed later), which functions as a negative feedback regulator of FSH. The major physiological effect of FSH in the male is thus sperm cell production.

In the female, FSH mainly targets the granulosa cells of the ovarian follicle (Box 11.4). FSH exhibits a mitogenic effect upon these cells, stimulating their division and, hence, follicular growth and development. This activity is enhanced by the paracrine action of locally produced growth factors. FSH also triggers enzymatic production of glycosaminoglycans, as well as expression of aromatase and other enzymes involved in oestrogen synthesis. Glycosaminoglycans form an essential component of the follicular fluid, and granulosa-cell-derived oestrogens play multiple roles in sustaining and regulating female reproductive function.

Box 11.4

Female follicular structure

The major female reproductive organs are a pair of ovaries, situated in the lower abdomen. At birth, each ovary houses approximately 1 million immature follicles. Each follicle is composed of an egg cell (ovum) surrounded by two layers of cells: an inner layer of granulosa cells and an outer layer of theca cells. During the follicular phase of the female reproductive cycle (Box 11.3), a group of follicles (~20), approximately 5 mm in diameter, are recruited by FSH (i.e. they begin to grow). FSH targets the granulosa cells, prompting them to synthesize oestrogen. The dominant follicle continues to grow to a diameter of 20–25 mm (Figure 11.B2). At this stage, it contains a fluid-filled cavity with the ovum attached to one side. Ovulation is characterized by bursting of the follicle and release of the ovum.

Typically, 400 follicles will mature and fully ovulate during an average woman's reproductive lifetime. The remaining 99.98 per cent of her follicles begin to develop, but regress due to inadequate FSH stimulation. The molecular detail of how FSH (and LH) promotes follicular growth is described in the main body of the text.

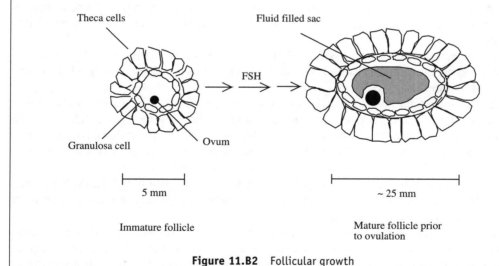

Figure 11.B2   Follicular growth

Prior to puberty, serum FSH levels are insufficient to promote follicular recruitment and development. Subsequent to puberty, as a group of follicles begin to develop at the beginning of a cycle, the one that is most responsive to FSH (i.e. displays the lowest FSH threshold) becomes the first to secrete oestrogen. As one effect of oestrogen is to suppress FSH release from the pituitary, blood FSH levels then plateau or decline slightly. This slightly lower FSH concentration is insufficient to sustain growth of follicles of higher FSH thresholds, so they die, leaving only the single oestrogen-producing dominant follicle (Boxes 11.3 and 11.4) to mature and ovulate.

FSH exerts its molecular effects via a specific receptor on the surface of sensitive cells. This receptor contains a characteristic seven transmembrane-spanning regions and is functionally coupled (via membrane-associated G-proteins) to adenylate cyclase. This generates the second messenger cAMP. FSH itself can promote increased expression of its own receptor in the short term, although longer-term exposure to elevated FSH levels down-regulates receptor numbers. Cloning and analysis of gonadotrophin receptors from several species indicate a high level of homology between the FSH, LH and CG receptors.

LH exhibits a molecular mass of 28.5 kDa. The gene coding for the β-subunit of this glycoprotein hormone is present on human chromosome 19. This subunit exhibits significant amino acid homology to placental CG. Both promote identical biological effects and act via the same 93 kDa cell surface receptor. The LH receptor is present on testicular Leydig cells in males and on female ovarian theca, as well as granulosa, luteal and interstitial cells.

LH promotes synthesis of testosterone, the major male androgen (Box 11.5) by the testicular Leydig cells. FSH sensitizes these cells to the activities of LH, probably by increasing LH receptor numbers on the cell surface. Leydig cells have a limited storage capacity for testosterone (~25 μg), but secrete 5–10 mg of the hormone into the bloodstream daily in young healthy males.

The primary cellular targets of LH in the females are the follicular theca cells, which constitutively express the LH receptor. Under the influence of LH, these cells produce androgens. The androgens (principally testosterone) are then taken up by granulosa cells and converted into oestrogens (Box 11.5) by the already-mentioned aromatase complex. Thus, the follicle represents the major female gonadal endocrine unit, in which granulosa and theca cells cooperate in the synthesis of oestrogens. Physiologically, LH in the female plays a major role in maturation of the dominant follicle and appears central to triggering ovulation.

11.5.2 Pregnant mare serum gonadotrophin

Pregnant mare serum gonadotrophin (PMSG) is a unique member of the gonadotrophin family of hormones. It is synthesized only by pregnant mares (i.e. is not found in other species). Furthermore, it displays both FSH-like and LH-like biological activities.

This glycoprotein hormone is a heterodimer, composed of an α and a β subunit and approximately 45 per cent of its molecular mass is carbohydrate. Reported molecular masses range from 52 to 68 kDa, a reflection of the potential variability of the hormone's carbohydrate content.

PMSG is secreted by cup-shaped outgrowths found in the horn of the uterus of pregnant horses. These equine-specific endometrial cups are of foetal, rather than maternal, origin. They first become visible around day 40 of gestation, and reach maximum size at about day 70, after which they steadily regress. They synthesize high levels of PMSG and secrete it into the blood, where it is detectable between days 40 and 130 of gestation.

11.5.3 The inhibins and activins

The inhibins and activins are a family of dimeric growth factors synthesized in the gonads. They exert direct effects both on gonadal and extra-gonadal tissue, and are members of the TGF-β

Box 11.5

The androgens and oestrogens

The androgens and oestrogens represent the major male and female sex hormones respectively (Figure 11.B3). The testicular Leydig cells represent the primary source of androgens in the male, of which testosterone is the major one. Testosterone, in turn, serves as a precursor for two additional steroids, i.e. dihydrotestosterone and the oestrogen called oestradiol. These mediate many of its biological effects.

Females, too, produce androgens, principally in the follicular theca cells. Androgens are also produced in the adrenals in both male and females.

The biological activities of androgens (only some of which are specific to males) may be summarized as:

- promoting and regulating development of the male phenotype during embryonic development;

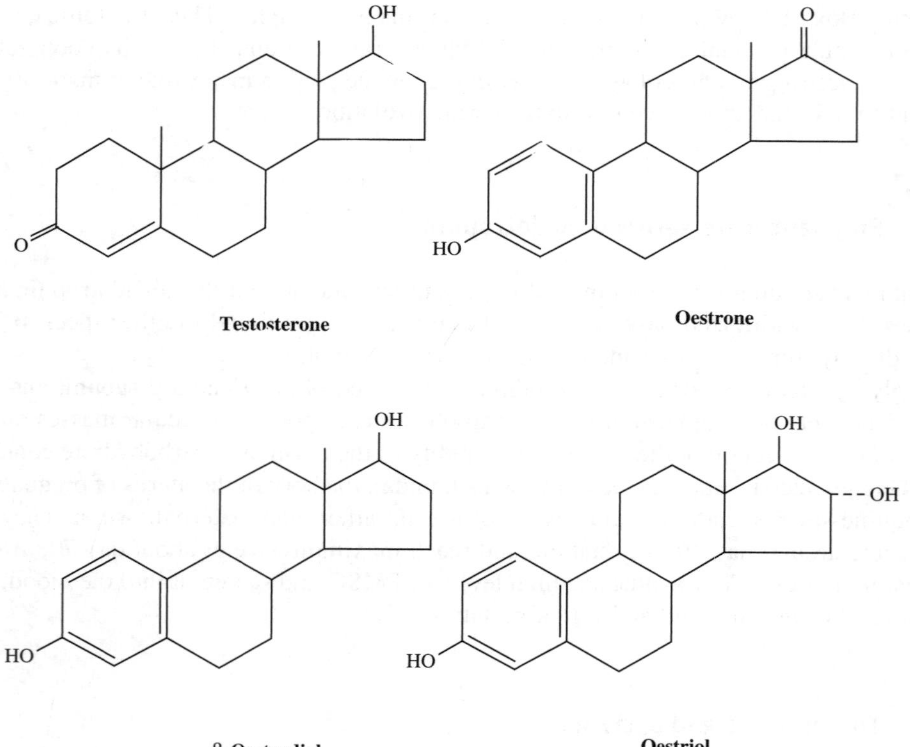

Figure 11.B3 Androgen and oestrogen structures

- promoting sperm cell synthesis;

- promoting development and maintenance of male secondary sexual characteristics at/after puberty;

- general growth-promoting effects;

- behavioural effects (e.g. male aggressiveness, etc.);

- regulation of serum gonadotrophin levels.

The follicular granulosa cells are the major site of synthesis of female steroid sex hormones: the oestrogens. β-Oestradiol represents the principal female follicular oestrogen. Oestriol is produced by the placenta of pregnant females. Oestriol and oestrone are also produced in small quantities as products of β-oestradiol metabolism.

Testosterone represents the immediate precursor of the oestrogens, the conversion being catalysed by the aromatase complex, i.e. a microsomal enzyme system. The biological actions of oestrogens may be summarized as:

- growth and maturation of the female reproductive system;

- maintenance of reproductive capacity;

- development and maintenance of female secondary sexual characteristics;

- female behavioural effects;

- complex effects upon lipid metabolism and distribution of body fat;

- regulation of bone metabolism (oestrogen deficiency promotes bone decalcification, as seen in postmenopausal osteoporosis).

family of proteins. The inhibins are heterodimers consisting of α- and β-polypeptide subunits. Activins are ββ dimers. The mature form of the α-subunit is termed α_c, and it consists of 134 amino acid residues. Two closely related (but structurally distinct) β-subunit forms have been characterized: β_A and β_B. These exhibit in excess of 70 per cent amino acid homology and differ in size by only a single amino acid. The naming and polypeptide composition of the inhibin/activin family may be summarized as follows:

- inhibin A = $\alpha_c\beta_A$

- inhibin B = $\alpha_c\beta_B$

- activin A $= \beta_A\beta_A$

- activin AB $= \beta_A\beta_B$

- activin B $= \beta_B\beta_B$.

Inhibins and activins were initially identified as gonadal-derived proteins capable of inhibiting (inhibin) or stimulating (activin) pituitary FSH production (Figure 11.10). The major gonadal sites of inhibin synthesis are the Sertoli cells (male) and granulosa cells (female). In addition to

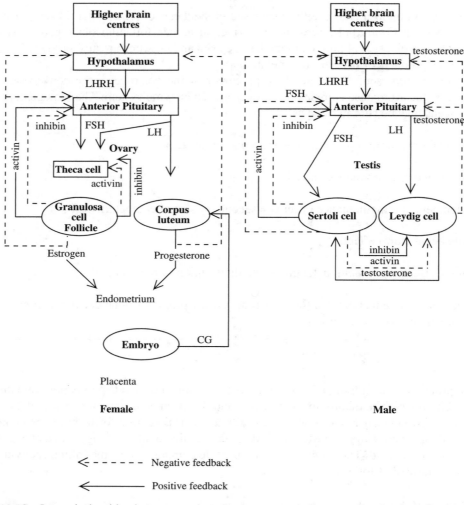

Figure 11.10   Interrelationships between various hormones regulating reproductive function in the male and female. Particular emphasis is placed upon the regulatory effect many have on the production levels of additional reproductive hormones

targeting the pituitary, the inhibins/activins likely play a direct (mutually antagonistic) role as paracrine/autocrine regulators of gonadal function.

They also likely induce responses in tissues other than the pituitary and gonads. In adults, for example, inhibin is also synthesized by the adrenal glands, spleen and nervous system. Recent studies involving inhibin-deficient transgenic mice reveal a novel role for inhibin as a gonadal-specific tumour suppressor. These mice, in which the α inhibin gene was missing, all developed normally, but all ultimately developed gonadal stromal tumours.

11.6   Medical and veterinary applications of gonadotrophins

Because of their central role in maintaining reproductive function, the therapeutic potential of gonadotrophins in treating subfertility/some forms of infertility was obvious. Gonadotrophins are also used to induce a superovulatory response in various animal species, as outlined later. The market for these hormones, though modest by pharmaceutical standards, is, none the less, substantial. By the late 1990s the annual human market stood at about US$250 million, of which the USA accounts for ~US$110 million, Europe ~US$90 million and Japan US$50 million.

11.6.1   Sources and medical uses of follicle-stimulating hormone, luteinizing hormone and human chorionic gonadotrophin

Although the human pituitary is the obvious source of human gonadotrophins, it also constitutes an impractical source of medically useful quantities of these hormones. However, the urine of post-menopausal women does contain both FSH and LH activity. Up until relatively recently, this has served as the major source used medically, particularly of FSH.

Menotrophin (human menopausal gonadotrophin) is the name given to FSH-enriched extracts from human urine. Such preparations contain variable levels of LH activity, as well as various other proteins normally present in urine. As much as 2.5 l of urine may be required to produce one dose (75 IU, ~7.5 mg) of human FSH (hFSH).

As mentioned previously, hCG exhibits similar biological activities to hLH and is excreted in the urine of pregnant women. Traditionally, hCG from this source has found medical application in humans (as an alternative to hLH; Figure 11.11).

In females, menotrophins and hCG are used for the treatment of anovulatory infertility. This condition is due to insufficient endogenous gonadotrophin production. Menotrophin is administered to stimulate follicular maturation, with subsequent administration of hCG to promote ovulation and corpus luteum formation. Mating at this point should lead to fertilization.

Dosage regimes attempt to mimic as closely as possible normal serum gonadotrophin profiles as occur during the reproductive cycle of fertile females. This is achieved by monitoring the resultant oestrogen production, or by using ultrasonic equipment to monitor follicular response. Depending upon the basal hormonal status of the female, calculation of the optimal dosage levels can be tricky. (Treatments are tailored to meet the physiological requirements of individual patients). Overdosage can, and does, result in multiple follicular development with consequent risk of multiple pregnancy.

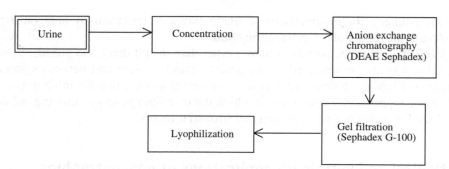

Figure 11.11 Overview of the procedure by which hCG may be purified from the urine of pregnant females at laboratory scale. Production-scale systems would be at least partially based upon such a purification strategy. Although initial concentration steps could involve precipitation, the use of ultrafiltration would now be more common

Treatment typically entails daily i.m. administration of gonadotrophin, often for 12 days or more, followed by a single dose of hCG. Alternatively, three equal larger doses of menotrophin may be administered on alternate days, followed by hCG administration 1–2 days after the final menotrophin dose.

Gonadotrophins are also used in assisted reproduction procedures. Here the aim is to administer therapeutic doses of FSH that exceed individual follicular FSH threshold requirements, thus stimulating multiple follicular growth. This, in turn, facilitates harvest of multiple eggs, which are then available for *in vitro* fertilization. This technique is often employed when a woman has a blocked fallopian tube or some other impediment to normal fertilization of the egg by a sperm cell. After treatment, the resultant eggs are collected, incubated *in vitro* with her partner's sperm, incubated in culture media until the embryonic blastocyst is formed, and then implanted into the mother's uterus.

FSH and hCG also find application in the treatment of male subfertility or related conditions. Both are administered to males exhibiting hypogonadotrophic hypogonadism to stimulate sperm synthesis and normal sexual function. hCG has found limited application in the treatment of pre-pubertal cryptorchidism (a condition characterized by failure of the testes to descend fully into the scrotum from the abdomen). The ability of this hormone to stimulate testosterone production also caught the attention of some athletes, and, as a result, the International Olympic Committee has banned its use.

The LH/hCG cell surface receptor is found in a number of non-gonadal tissues, indicating that these hormones may exert physiologically relevant non-gonadal functions (Table 11.9). In addition, whereas liver, kidney and muscle cells are devoid of such a receptor, it is expressed by a number of these tissues before birth, hinting at a potential developmental role. Receptor levels in non-gonadal tissues are generally much lower than in gonadal tissue. hCG, therefore, probably has a number of pregnancy/non-pregnancy-related non-gonadal functions that may give rise to future additional clinical applications.

11.6.2 Recombinant gonadotrophins

Gonadotrophins are now also produced by recombinant DNA technology. The genes, or cDNAs coding for gonadotrophins from several species, have been identified and expressed in various

Table 11.9 Some notable non-gonadal tissues that express functional LH/hCG receptors

Pregnancy/fertility related tissue	Other tissue
Uterus	Skin
Cervix	Blood vessels
Placenta	Adrenal cortex
Oviduct	Brain tissue
Foetal membranes	Prostate gland
Seminal vesicles	Bladder
Sperm cells	Monocytes
Breast	Macrophages

recombinant host systems, particularly mammalian cell lines. rhFSH produced in CHO cells has proven clinically effective. Although exhibiting an amino acid sequence identical to the human molecule, its carbohydrate composition differs slightly. When administered to humans, the preparation is well tolerated and yields no unexpected side effects. It does not elicit an immunological response, and its plasma half-life is similar to the native hormone. rhFSH has proven efficacious in stimulating follicular growth in females suffering from hypogonadotrophic hypogonadism and is effective in the treatment of males suffering from similar conditions. rhFSH was amongst the first biopharmaceutical substances to be approved for general medical use in Europe by the European Medicines Agency via the centralized application procedure (Chapter 4). Recombinant gonadotrophins approved for general medical use are listed in Table 11.10 and additional details of one representative product (Ovitrelle) are provided in Box 11.6.

11.6.3 Veterinary uses of gonadotrophins

Gonadotrophins may be utilized to treat subfertility in animals and are routinely used to induce a superovulatory response in valuable animals, most notably valuable horses and cattle.

The theory and practice of superovulation is quite similar to the use of gonadotrophins to assist *in vitro* fertilization procedures in humans. Exogenous FSH is administered to the animal such that multiple follicles develop simultaneously. After administration of LH to help promote ovulation, the animal is mated, thus fertilizing the released egg cells. Depending upon the specific animal and the superovulatory regime employed, anything between 0 and 50 viable embryos may be produced, although, more typically, the number is between 4 and 10. The embryos are then

Table 11.10 Recombinant gonadotrophins now approved for general medical use in the EU and/or the USA

Product (tradename)	Company	Indication
Gonal F (rhFSH)	Serono	Anovulation and Superovulation
Puregon (rhFSH)	N.V. Organon	Anovulation and Superovulation
Follistim (rhFSH)	Organon	Some forms of infertility
Luveris (rhLH)	Ares-Serono	Some forms of infertility
Ovitrelle (rhCG)	Serono	Used in selected assisted reproductive techniques

Box 11.6

Product case study: Ovitrelle

Ovitrelle (tradename in EU, sold as Ovidrel in the USA and also known as choriogonadotropin alfa) is a recombinant hCG approved for general medical use in the EU and USA in 2001. It is indicated for the treatment of female infertility due to anovulation and for patients undergoing assisted reproductive technology. It is used to trigger final follicle maturation and luteinization after follicle stimulation.

The glycoprotein hormone, consisting of an α- and a β-subunit, displays an identical amino acid sequence and very similar glycosylation detail when compared with native hCG. It is produced in an engineered CHO cell line and, after cell culture, the product is purified from the media by multistep chromatography, ultrafiltration and nanofiltration. The final product also contains mannitol, methionine, poloxamer188 sodium hydroxide and ortho-phosphoric acid and is presented in either vials or pre-filled syringes. The product is administered subcutaneously. Following s.c. administration, maximum serum concentration is usually witnessed after 12–24 h and the product is eliminated with a mean terminal half-life of approximately 29 h. Absolute bioavailability is of the order of 40 per cent.

Pivotal pre-approval safety and efficacy was assessed in a randomized, open label, multicentre study of infertile females undergoing *in vitro* fertilization and embryo transfer. The primary efficacy parameter was the mean number of oocytes retrieved, which (at 13.6) was similar to the number retrieved when urinary-derived hCG was used. Serious potential side effects can include ovarian over/hyper-stimulation, sometimes with pulmonary or vascular complications. The product is marketed by Serono Inc.

recovered from the animal (either surgically or, more usually, non-surgically), and are often maintained in cell culture for a short period of time. A single embryo is then usually reimplanted into the donor female, while the remaining embryos are implanted into other recipient animals, who act as surrogate mothers, carrying the offspring to term.

This technology is most often applied to valuable animals (e.g. prize winning horses, or high milk-yield dairy cattle) in order to boost their effective reproductive capacity several-fold. All the offspring will inherit its genetic complement from the biological mother (and father), irrespective of what recipient animal carries it to term.

Gonadotrophins usually utilized to induce a superovulatory response include porcine FSH (p-FSH), porcine LH (p-LH) and PMSG. p-FSH is extracted from the pituitary glands of slaughterhouse pigs. The crude pituitary extract is usually subject to a precipitation step, using either ethanol or salts. The FSH-containing precipitate is normally subjected to at least one subsequent chromatographic step. The final product often contains some LH and low levels of additional pituitary-derived proteins.

p-LH is obtained, again, by its partial purification from the pituitary glands of slaughterhouse pigs. Although the target recipients are almost always cattle or horses, in both cases a porcine source is utilized. The use of a product derived from a species other than the intended recipient species is encouraged as it helps minimize the danger of accidental transmission of disease via infected source material (many pathogens are species specific).

Most superovulatory regimes utilizing p-FSH entail its administration to the recipient animal twice daily for 4–5 days. Regular injections are required due to the relatively short half-life of FSH in serum; s.c. administration helps prolong the duration of effectiveness of each injection. The 4 or 5 days of treatment with FSH is followed by a single dose of LH, promoting final follicular maturation and ovulation.

The causes of variability of superovulatory responses are complex and not fully understood. The general health of the animal, as well as its characteristic reproductive physiology, is important. The exact composition of the gonadotrophin preparations administered and the exact administration protocol also influence the outcome. The variability of FSH:LH ratios in many p-FSH preparations can affect the results obtained, with the most consistent superovulatory responses being observed when FSH preparations exhibiting low LH activity are used. The availability of recombinant FSH and LH will overcome these difficulties at least.

An alternative superovulatory regime entails administration of PMSG, which, as described earlier, exhibits both FSH and LH activity. The major rationale for utilizing PMSG is its relatively long circulatory half-life. In cattle, clearance of PMSG may take up to 5 days. The slow clearance rate appears to be due to the molecule's high content of *N*-acetyl-neuramic acid. This extended serum half-life means that a single dose of PMSG is sufficient to induce a superovulatory response. Paradoxically, however, its extended half-life limits its use in practice. Post-ovulatory stimulation of follicular growth can occur, resulting in the recovery of a reduce number of viable embryos. Attempts to negate this biological effect have centred around administration of anti-PMSG antibodies several days after PMSG administration. However, this gonadotrophin is still not widely used.

11.7 Additional recombinant hormones now approved

Three additional recombinant hormones have recently gained marketing approval: thyroid-stimulating hormone, parathyroid hormone and calcitonin.

Structurally, thyroid-stimulating hormone (TSH or thyrotrophin) is classified as a member of the gonadotrophin family, although functionally it targets the thyroid gland as opposed to the gonads. As with other gonadotrophins, it is a heterodimeric glycoprotein displaying a common α-subunit and a unique β-subunit. The β-subunit shows less homology to that of other members of the group. It consists of 118 amino acids, is particularly rich in cysteine residues and contains one N-linked glycosylation site (Asn 23).

TSH is synthesized by a distinct pituitary cell type: the thyrotroph. Its synthesis and release are promoted by thyrotrophin-releasing hormone (a hypothalamic tripeptide hormone). TSH exerts its characteristic effects by binding specific receptors found primarily, but not exclusively, on the surface of thyroid gland cells. Binding to the receptor activates adenylate cyclase, leading to increased intracellular cAMP levels. Ultimately, this triggers TSH's characteristic effects on thyroid function, including promoting iodine uptake from the blood, synthesis of the iodine-containing thyroid hormones thyroxine (T_4) and triiodothyronine (T_3) and the release of these hormones into the blood, from where they regulate many aspects of general tissue metabolic activity. Elevated plasma levels of T_4 and T_3 also promote decreased TSH synthesis and release by a negative feedback mechanism.

TSH is approved for medical use as a diagnostic aid in the detection of thyroid cancer/thyroid remnants in post-thyroidectomy patients. Thyroid cancer is relatively rare, exhibiting highest incidence in adults, particularly females. First-line treatment is surgical removal of all/most of the thyroid gland (thyroidectomy). This is followed by thyroid hormone suppression therapy, which entails administration of T_3 or T_4 at levels sufficient to maintain low serum TSH levels through the negative feedback mechanism mentioned earlier. TSH suppression is required in order to prevent TSH-mediated stimulation of remnant thyroid cancer cells. The reoccurrence of active thyroid cancer can be detected by administration of TSH along with radioactive iodine. TSH promotes uptake of radioactivity, which can then be detected by appropriate radioimaging techniques.

The commercial recombinant TSH product (tradename, thyrogen; international non-proprietary name; thyrotrophin alfa) is produced in a CHO cell line co-transfected with plasmids harbouring the DNA sequences coding for the α- and β-TSH subunits. The cells are grown in batch harvest animal cell culture bioreactors. Following recovery and concentration (ultrafiltration), the TSH is chromatographically purified and formulated to a concentration of 0.9 mg ml^{-1} in phosphate buffer containing mannitol and sodium chloride as excipients. After sterile filtration and aseptic filling into glass vials, the product is freeze-dried. Finished product has been assigned a shelf life of 3 years when stored at 2–8 °C.

Human parathyroid hormone (hPTH) is an 84 amino acid polypeptide that functions as a primary regulator of calcium and phosphate metabolism in bones. It stimulates bone formation by osteoblasts, which display high-affinity cell surface receptors for the hormone. PTH also increases intestinal absorption of calcium.

A truncated version of PTH (tradenames Forsteo and Forteo) has been approved for the treatment of osteoporosis in postmenopausal women. This 4 kDa polypeptide is identical in sequence to the N-terminal residues 1–34 of endogenous hPTH, it binds to the native PTH receptor and triggers the same effects. The product is produced in *E. coli*, purified and presented as a sterile solution containing 250 μg active substance per millilitre.

Osteoporosis affects some 75 million people in Europe, Japan and the USA combined. The condition is characterized by progressive thinning of the bones, leading to bone fragility and increased risk of fracture. Treatment with Forsteo increases bone mineral density and generally entails daily s.c. injection for several months at dosage levels of 20 μg active/dose.

Calcitonin is a polypeptide hormone that (along with PTH and the vitamin D derivative, 1,25-dihydroxycholecalciferol) plays a central role in regulating serum ionized calcium (Ca^{2+}) and inorganic phosphate (Pi) levels. The adult human body contains up to 2 kg of calcium, of which 98 per cent is present in the skeleton (i.e. bone). Up to 85 per cent of the 1 kg of phosphorus present in the body is also found in the skeleton (the so-called mineral fraction of bone is largely composed of $Ca_3(PO_4)_2$, which acts as a body reservoir for both calcium and phosphorus). Calcium concentrations in human serum approximate to 0.1 mg ml^{-1} and are regulated very tightly (serum phosphate levels are more variable).

Calcitonin lowers serum Ca^{2+} and Pi levels, primarily by inhibiting the process of bone resorption, but also by decreasing resorption of Pi and Ca^{2+} in the kidney. Calcitonin receptors are predictably found primarily on bone cells (osteoclasts) and renal cells, and generation of cAMP via adenylate cyclase activation plays a prominent role in hormone signal transduction.

Calcitonin is used clinically to treat hypercalcaemia associated with some forms of malignancy and Paget's disease. The latter condition is a chronic disorder of the skeleton in which bone grows abnormally in some regions. It is characterized by substantially increased bone turnover rates, which reflects overstimulation of both osteoclasts (promote bone resorption, i.e. degradation of old bone) and osteoblasts (promotes synthesis of new bone).

In most mammals, calcitonin is synthesized by specialized parafollicular cells in the thyroid. In sub-mammalian species, it is synthesized by specialized anatomical structures known as ultimobranchial bodies.

Calcitonin produced by virtually all species is a single-chain 32 amino acid residue polypeptide, displaying a molecular mass in the region of 3500 Da. Salmon calcitonin differs in sequence from the human hormone by nine amino acid residues. It is noteworthy, however, as it is approximately 100-fold more potent than the native hormone in humans. The higher potency appears due to both a greater affinity for the receptor and a greater resistance to degradation *in vivo*. As such, salmon, as opposed to human calcitonin, is used clinically. Traditional clinical preparations were manufactured by direct chemical synthesis, although a recombinant form of the molecule has now gained marketing approval. The recombinant calcitonin is produced in an engineered *E. coli* strain. Structurally, salmon calcitonin displays C-terminal amidation. A C-terminal amide group ($—CONH_2$) replacing the usual carboxyl group is a characteristic feature of many polypeptide hormones. If present, it is usually required for full biological activity/stability. As *E. coli* cannot carry out post-translational modifications, the amidation of the recombinant calcitonin is carried out *in vitro* using an α-amidating enzyme, which is itself produced by recombinant means in an engineered CHO cell line. The purified, amidated finished product is formulated in an acetate buffer and filled into glass ampoules. The (liquid) product exhibits a shelf life of 2 years when stored at 2–8 °C.

11.8 Conclusion

Several hormone preparations have a long history of use as therapeutic agents. In virtually all instances they are administered simply to compensate for lower than normal endogenous production of the hormone in question. Since it first became medically available, insulin has saved or prolonged the lives of millions of diabetics. Gonadotrophins have allowed tens, if not hundreds, of thousands of sub-fertile individuals to conceive. GH has improved the quality of life of thousands of people of short stature. Most such hormones were in medical use prior to the advent of genetic engineering. Recombinant hormonal preparations, however, are now gaining greater favour, mainly on safety grounds. Hormone therapy will remain a central therapeutic tool for clinicians for many years to come.

Further reading

Books

Anonymous. 2004. *Follicle Stimulating Hormone*. Icon Health Publications, CA, USA.
Ashcroft, F. and Ashcroft, S. (eds). 2006. *Insulin*. IRL Press, NY, USA.
Bercu, B. 1998. *Growth Hormone Secretagogues in Clinical Practice*. Marcel Dekker.
Fauser, B. 1997. *FSH Action and Intraovarian Regulation*. Parthenon.
Hakin, N. 2002. *Pancreas and Islet Transplantation*. Oxford University Press.

Jorgensen, J. and Christiansen, J. (eds). 2005. *Growth Hormone Deficiency in Adults*. S. Karger AG.
Juul, A. 2000. *Growth Hormone in Adults*. Cambridge University Press.
Mac Hadley, E. 1999. *Endocrinology*, Prentice Hall.
Norman, A. 1997. *Hormones*. Academic Press.
O'Malley, B. 1997. *Hormones and Signalling*. Academic Press.

Articles

Insulin/diabetes/glucagon

Bristow, A. 1993. Recombinant DNA derived insulin analogues as potentially useful therapeutic agents. *Trends in Biotechnology* **11**, 301–305.
Cao, Y. and Lam, L. 2002. Projections for insulin treatment for diabetics. *Drugs of Today* **38**(6), 419–427.
Docherty, K. 1997. Gene therapy for diabetes mellitus. *Clinical Science* **92**(4), 321–330.
Drucker, D. 2002. Biological actions and therapeutic potential of the glucagon-like peptides. *Gastroenterology* **122**(2), 531–544.
Goa, K.L., Haria, M., and Wilde, M.I. 1997. Lisinopril: a review of its pharmacology and use in the management of the complications of diabetes mellitus. *Drugs* **53**(6), 1081–1105.
Hinds, K. and Kim, S. 2002. Effects of PEG conjugation on insulin properties. *Advanced Drug Delivery Reviews* **54**(4), 505–530.
Home, P. and Kurtzhals, P. 2006. Insulin detemir: from concept to clinical experience. *Expert Opinion on Pharmacotherapy* **7**(3), 325–343.
Johnson, I. 1983. Human insulin from recombinant DNA technology. *Science* **219**, 632–637.
Kurukulasuriya, R. and Link, J. 2005. Progress towards glucacon receptor antagonist therapy for type 2 diabetes. *Expert Opinion on Therapeutic Patents* **15**(12), 1739–1749.
Patton, J.S., Bukar, J., and Nagarajan, S. 1999. Inhaled insulin. *Advanced Drug Delivery Reviews* **35**, 235–247.
Rhodes, C. and White, M. 2002. Molecular insights into insulin action and secretion. *European Journal of Clinical Investigation* **32**, 3–13.
Scherbaum, W. 2005. Inhaled insulin: clinical efficacy. *Diabetes Obesity and Metabolism* **7**(Suppl. 1), S1–S13.
Secchi, A., Di Carlo, V., and Pozza, G. 1997. Pancreas and islet transplantation – current progress, problems and perspectives. *Hormone and Metabolic Research* **29**(1), 1–8.
Selam, J. 1997. Management of diabetes with glucose sensors and implantable insulin pumps – from the dream of the 60s to the realities of the 90s. *ASAIO Journal* **43**(3), 137–142.
Shaikh, I.M., Jadhav, K.R., Ganga, S., Kadam, V.J., and Pisal, S.S. 2005. Advanced approaches in insulin delivery. *Current Pharmaceutical Biotechnology* **6**(5), 387–395.
Sloop, K.W., Michael, M.D., and Moyers, J.S. 2005. Glucagon as a target for the treatment of type 2 diabetes. *Expert Opinion on Therapeutic Targets* **9**(3), 593–600.
Vajo, Z., Fawcett, J., and Duckworth, W.C. 2001. Recombinant DNA technology in the treatment of diabetes: insulin analogues. *Endocrine Reviews* **22**(5), 706–717.
Walsh, G. 2005. Therapeutic insulins and their large scale manufacture. *Applied Microbiology and Biotechnology* **67**, 151–159.

Growth hormone

Ayuk, J. and Sheppard, M. 2006. Growth hormone and its disorders. Postgraduate Medical Journal **82**(963), 24–30.
Carter, C.S., Ramsey, M.M., and Sonntag, W.E. 2002. A critical analysis of the role of growth hormone and IGF-1 in aging and lifespan. *Trends in Genetics* **18**(6), 295–301.

Gibson, F. and Hinds, C. 1997. Growth hormone and insulin-like growth factors in critical illness. *Intensive Care Medicine* **23**(4), 369–378.

Hathaway, D. 2002. Growth hormone: challenges and opportunities for the biotechnology sector. *Journal of Anti-Aging Medicine* **5**(1), 57–62.

Piwien-Pilipuk, G., Huo, J.S., and Schwartz, J. 2002. Growth hormone signal transduction. *Journal of Pediatric Endocrinology and Metabolism* **15**(6), 771–786.

Simpson, H., Savine, R., Sonksen, P., Bengtsson, B.A., Carlsson, L., Christiansen, J.S., Clemmons, D., Cohen, P., Hintz, R., Ho, K., Mullis, P., Robinson, I., Strasburger, C., Tanaka, T., Thorner, M., and GRS Council. 2002. Growth hormone replacement therapy for adults: into the new millennium. *Growth Hormone and IGF Research* **12**(1), 1–33.

Waters, M.J., Hoang, H.N., Fairlie, D.P., Pelekanos, R.A., and Brown, R.J. 2006. New insights into growth hormone action. *Journal of Molecular Endocrinology* **36**(1), 1–7.

Gonadotrophins and thyroid-stimulating hormone

Al-Inany, H., Aboulghar, M.A., Mansour, R.T., and Proctor, M. 2005. Recombinant versus urinary gonadotrophins for triggering ovulation in assisted conception. *Human Reproduction* **20**(8), 2061–2073.

Angelopoulos, N., Goula, A., and Tolis, G. 2005. The role of luteinizing hormone activity in controlled ovarian stimulation. *Journal of Endocrinological Investigation* **28**(1), 79–88.

Bernard, D.J. Chapman, S.C., and Woodruff, T.K. 2002. Minireview: inhibin binding protein (InhBP/p120), betaglycan and the continuing search for the inhibin receptor. *Molecular Endocrinology* **16**(2), 207–212.

Cho, B. 2002. Clinical applications of TSH receptor antibodies in thyroid diseases. *Journal of Korean Medical Science* **17**(3), 293–301.

Dardenne, M. and Savino, W. 1996. Interdependence of the endocrine and immune systems. *Advances in Neuroimmunology* **6**(4), 297–307.

Depaolo, L. 1997. Inhibins, activins and follistatins – the saga continues. *Proceedings of the Society for Experimental Biology and Medicine* **214**(4), 328–339.

Gregory, S. and Kaiser, U. 2004. Regulation of gonadotrophins by inhibin and activin. *Seminars in Reproductive Medicine* **22**(3), 253–267.

Hayden, C.J., Balen, A.H., and Rutherford, A.J. 1999. Recombinant gonadotrophins. *British Journal of Obstetrics and Gynaecology.* **106**(3), 188–196.

Hillier, S. 1994. Current concepts of the role of follicle stimulating hormone and luteinizing hormone in folliculogenesis. *Human Reproduction* **9**(2), 188–191.

Lunenfeld, B. 2004. Historical perspectives in gonadotrophin therapy. *Human Reproduction Update* **10**(6), 453–467.

Macklon, N. and Fauser, B. 2001. Follicle stimulating hormone and advanced follicle development in the human. *Archives of Medical Research* **32**(6), 595–600.

Magner, J. 2004. Thyroid stimulating hormone mediated hyperthyroidism. *Endocrinologst* **14**(4), 201–211.

McCann, S.M., Karanth, S., Mastronardi, C.A., Dees, W.L., Childs, G., Miller, B., Sower, S., and Yu, W.H. 2001. Control of gonadotrophin secretion by follicle stimulating hormone-releasing factor, luteinizing hormone-releasing hormone and leptin. *Archives of Medical Research* **32**(6), 476–485.

Menon, K.M.J., Clouser, C.L., and Nair, A.K. 2005. Gonadotropin receptors: role of post-translational modifications and post-transcriptional regulation. *Endocrine* **26**(3), 249–257.

Risbridger, G.P., Schmitt, J.F., and Robertson, D.M. 2001. Activins and inhibins in endocrine and other tumors. *Endocrine Reviews* **22**(6), 836–858.

Schwartz, N. 1995. The 1994 Stevenson Award lecture. Follicle-stimulating hormone and luteinizing hormone: a tale of two gonadotrophins. *Canadian Journal of Physiology & Pharmacology* **73**, 675–684.

Stewart, E. 2001. Gonadotrophins and the uterus: is there a gonad-independent pathway? *Journal of the Society for Gynaecologic Investigation* **8**(6), 319–326.

Siarim, M. and Krishnamurthy, H. 2001. The role of follicle stimulating hormone in spermatogenesis: lessons from knockout animal models. *Archives of Medical Research* **32**(6), 601–608.

Sturgeon, C. and McAllister, E. 1998. Analysis of hCG: clinical applications and assay requirements. *Annals of Clinical Biochemistry* **35**, 460–491.

Ulloa-Aguirre, A., Timossi, C., Damian-Matsumura, P., and Dias, J.A. 1999. Role of glycosylation in function of follicle stimulating hormone. *Endocrine* **11**(3), 205–215.

Rao, C. and Sanfilippo, J. 1997. New understanding in the biochemistry of implantation – potential direct roles of luteinizing hormone and human chorionic gonadotropin. *Endocrinologist* **7**(2), 107–111.

Fauser, B. and Vanheusden, A. 1997. Manipulation of human ovarian function – physiological concepts and clinical consequences. *Endocrine Reviews* **18**(1), 71–106.

Wonerow, P., Neumann, S., Gudermann, T., and Paschke, R. 2001. Thyrotropin receptor mutations as a tool to understand thyrotropin receptor action. *Journal of Molecular Medicine* **79**(12), 707–721.

Calcitonin and parathyroid hormone

Compston, J. 2005. Recombinant parathyroid hormone in the management of osteoporosis. *Calcified Tissue International* **77**(2), 65–71.

Finday, D. and Sexton, P. 2004. Calcitonin. *Growth Factors* **22**(4), 217–224.

Inzerillo, A.M., Zaidi, M., and Huang, C.L. 2004. Calcitonin: physiological actions and clinical applications. *Journal of Pediatric Endocrinology and Metabolism* **17**(7), 931–940.

Lane, N. and Morris, S. 2005. New perspectives on parathyroid hormone therapy. *Current Opinion in Rheumatology* **17**(4), 467–474.

Yallampalli, C., Chauhan, M., Thota, C.S., Kondapaka, S., and Wimalawansa, S.J. 2002. Calcitonin gene-related peptide in pregnancy and its emerging receptor heterogeneity. *Trends in Endocrinology and Metabolism* **13**(6), 263–269.

12
Recombinant blood products and therapeutic enzymes

12.1 Introduction

Blood and blood products constitute a major group of traditional biologics. The main components of blood are the red and white blood cells, along with platelets and the plasma in which these cellular elements are suspended. Whole blood remains in routine therapeutic use, as do red blood cell and platelet concentrates. A variety of therapeutically important blood proteins also continue to be purified from plasma. These include various clotting factors and immunoglobulins. However, in keeping with the scope of this book, we focus in this chapter upon blood proteins/blood-related proteins produced by genetic engineering. These include recombinant coagulation factors, anticoagulants (such as hirudin) and thrombolytics (such as tPA). Also considered towards the end of the chapter are a number of recombinant enzymes that have found therapeutic application.

12.2 Haemostasis

Blood plays various vital roles within the body and it is not surprising that a number of processes have evolved capable of effectively maintaining haemostasis (the rapid arrest of blood loss upon vascular damage, in order to maintain a relatively constant blood volume). In humans, three main mechanisms underline the haemostatic process:

- The congregation and clumping of blood platelets at the site of vascular injury, thus effectively plugging the site of blood leakage.

- Localized constriction of the blood vessel, which minimizes further blood flow through the area.

- Induction of the blood coagulation cascade. This culminates in the conversion of a soluble serum protein, fibrinogen, into insoluble fibrin. Fibrin monomers then aggregate at the site of

Pharmaceutical biotechnology: concepts and applications   Gary Walsh
© 2007 John Wiley & Sons, Ltd   ISBN 978 0 470 01244 4 (HB)   978 0 470 01245 1 (PB)

damage, thus forming a clot (thrombus) to seal it off. These mechanisms are effective in dealing with small vessel injuries (e.g. capillaries and arterioles), although they are ineffective when the damage relates to large veins/arteries.

12.2.1   The coagulation pathway

The process of blood coagulation is dependent upon a large number of blood clotting factors, which act in a sequential manner. At least 12 distinct factors participate in the coagulation cascade, along with several macromolecular cofactors. The clotting factors are all designated by Roman numerals (Table 12.1) and, with the exception of factor IV, all are proteins. Most factors are proteolytic zymogens, which become sequentially activated. An activated factor is indicated by inclusion of a subscript 'a' (e.g. factor XIIa is activated factor XII).

Although the final steps of the blood clotting cascade are identical, the initial steps can occur via two distinct pathways: extrinsic and intrinsic. Both pathways are initiated when specific clotting proteins make contact with specific surface molecules exposed only upon damage to a blood vessel. Clotting occurs much more rapidly when initiated via the extrinsic pathway.

Two coagulation factors function uniquely in the extrinsic pathway: factor III (tissue factor) and factor VII. Tissue factor is an integral membrane protein present in a wide variety of tissue types (particularly lung and brain). This protein is exposed to blood constituents only upon rupture of

Table 12.1   The coagulation factors that promote the blood clotting process. Note that the factor originally designated as VI was later shown to be factor Va

Factor number	Common name	Pathway in which it functions	Function
I	Fibrinogen	Both	Forms structural basis of clot after its conversion to fibrin
II	Prothrombin	Both	Precursor of thrombin, which activates factors I, V, VII, VIII and XIII
III	Tissue factor (thromboplastin)	Extrinsic	Accessory tissue protein which initiates extrinsic pathway
IV	Calcium ions	Both	Required for activation of factor XIII and stabilizes some factors
V	Proaccelerin	Both	Accessory protein, enhances rate of activation of X
VII	Proconvertin	Extrinsic	Precursor of convertin (VIIa) which activates X (extrinsic system)
VIII	Antihaemophilic factor	Intrinsic	Accessory protein, enhances activation of X (intrinsic system)
IX	Christmas factor	Intrinsic	Activated IX directly activates X (intrinsic system)
X	Stuart factor	Both	Activated form (Xa) converts prothrombin to thrombin
XI	Plasma thromboplastin antecedent	Intrinsic	Activated form (XIa) serves to activate IX
XII	Hageman factor	Intrinsic	Activated by surface contact or the kallikrenin system. XIIa helps initiate intrinsic system
XIII	Fibrin-stabilizing factor	Both	Activated form cross-links fibrin, forming a hard clot

a blood vessel, and it initiates the extrinsic coagulation cascade at the site of damage as described below.

Factor VII contains a number of γ-carboxyglutamate residues (as do factors II, IX and X), which play an essential role in facilitating their binding of Ca^{2+} ions. The initial events initiating the extrinsic pathway entail the interaction of factor VII with Ca^{2+} and tissue factor. In this associated form, factor VII becomes proteolytically active. It displays both binding affinity for, and catalytic activity against, factor X. It thus activates factor X by proteolytic processing, and factor Xa, which initiates the terminal stages of clot formation, remains attached to the tissue factor–Ca^{2+} complex at the site of damage. This ensures that clot formation only occurs at the point where it is needed (Figure 12.1).

The initial steps of the intrinsic pathway are somewhat more complicated. This system requires the presence of clotting factors VIII, IX, XI and XII, all of which, except for factor VIII, are endo-acting proteases. As in the case of the extrinsic pathway, the intrinsic pathway is triggered upon exposure of the clotting factors to proteins present on the surface of body tissue exposed by vascular injury. These protein binding/activation sites probably include collagen.

Additional protein constituents of the intrinsic cascade include prekallikrein, an 88 kDa protein zymogen of the protease kallikrein, and high molecular mass kininogen, a 150 kDa plasma glycoprotein that serves as an accessory factor.

The intrinsic pathway appears to be initiated when factor XII is activated by contact with surface proteins exposed at the site of damage. High molecular mass kininogen also appears to form part of this initial activating complex (Figure 12.2).

Factor XIIa can proteolytically cleave and, hence, activate two substrates:

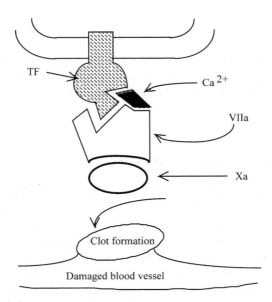

Figure 12.1 Schematic diagram of the initial steps of the extrinsic blood coagulation pathway. See text for details (TF: tissue factor)

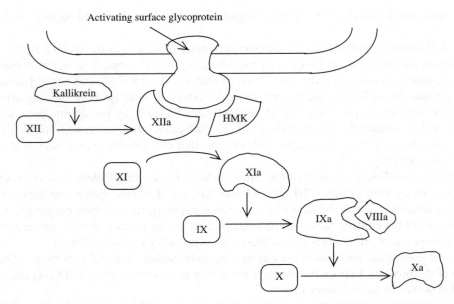

Figure 12.2   The steps unique to the intrinsic coagulation pathway. Factor XIIa can also convert prekallikrein to kallikrein by proteolysis, but this is omitted for the sake of clarity. Full details are given in the main text. The final steps of the coagulation cascade, which are shared by both extrinsic and intrinsic pathways, are outlined in Figure 12.3

- prekallikrein, yielding kallikrein (which, in turn, can directly activate more XII to XIIa).

- factor XI, forming XIa.

Factor XIa, in turn, activates factor IX. Factor IXa then promotes the activation of factor X, but only when it (i.e. IXa) is associated with factor VIIIa. Factor VIIIa is formed by the direct action of thrombin on factor VIII. The thrombin will be present at this stage because of prior activation of the intrinsic pathway.

12.2.2   Terminal steps of coagulation pathway

Both intrinsic and extrinsic pathways generate activated factor X. This protease, in turn, catalyses the proteolytic conversion of prothrombin (factor II) into thrombin (IIa). Thrombin, in turn, catalyses the proteolytic conversion of fibrinogen (I) into fibrin (Ia). Individual fibrin molecules aggregate to form a soft clot. Factor XIIIa catalyses the formation of covalent crosslinks between individual fibrin molecules, forming a hard clot (Figures 12.3 and 12.4).

Prothrombin (factor II) is a 582 amino acid, 72.5 kDa glycoprotein, which represents the circulating zymogen of thrombin (IIa). It contains up to six γ-carboxyglutamate residues towards its N-terminal end, via which it binds several Ca^{2+} ions. Binding of Ca^{2+} facilitates prothrombin binding to factor Xa at the site of vascular injury. The factor Xa complex then proteolytically

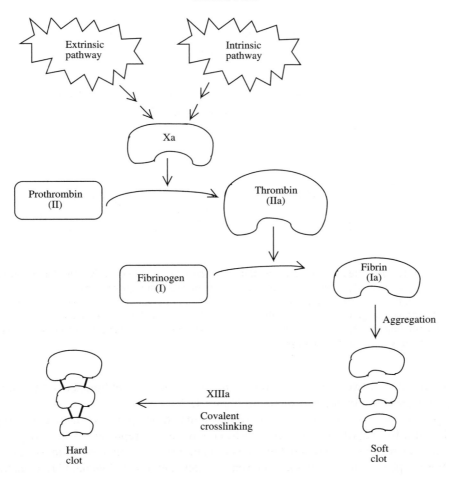

Figure 12.3 Overview of the blood coagulation cascade, with emphasis upon the molecular detail of its terminal stages. Refer to text for specific detail

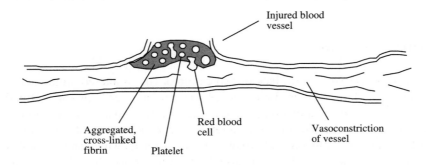

Figure 12.4 Terminal steps of a blood clot (haemostatic plug): cross-linked fibrin molecules bind together platelets and red blood cells congregated at the site of damage, thus preventing loss of any more blood

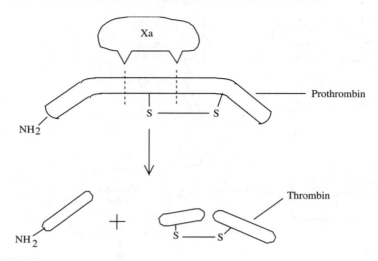

Figure 12.5 Proteolytic cleavage of prothrombin by factor Xa, yielding active thrombin. Although pro-thrombin is a single-chain glycoprotein, thrombin consists of two polypeptides linked by what was originally the prothrombin intrachain disulfide bond. The smaller thrombin polypeptide fragment consists of 49 amino acid residues, and the large polypeptide chain contains 259 amino acids. The N-terminal fragment released from prothrombin contains 274 amino acid residues. Activation of prothrombin by Xa does not occur in free solution, but at the site of vascular damage

cleaves prothrombin at two sites (arg^{274}–thr^{275} and arg^{323}–ile^{324}), yielding active thrombin and an inactive polypeptide fragment, as depicted in Figure 12.5.

Fibrinogen (factor I) is a large (340 kDa) glycoprotein consisting of two identical tri-polypeptide units, α, β and γ. Its overall structural composition may thus be represented as $(\alpha\,\beta\,\gamma)_2$.

The N-terminal regions of the α and β fibrinogen chains are rich in charged amino acids, which, via charge repulsion, play an important role in preventing aggregation of individual fibrinogen molecules. Thrombin, which catalyses the proteolytic activation of fibrinogen, hydrolyses these N-terminal peptides. This renders individual fibrin molecules more conducive to aggregation, therefore promoting soft clot formation. The soft clot is stabilized by the subsequent introduction of covalent cross-linkages between individual participating fibrin molecules. This reaction is catalysed by factor XIIIa.

12.2.3   Clotting disorders

Genetic defects characterized by (a) lack of expression or (b) an altered amino acid sequence of any clotting factor can have serious clinical consequences. In order to promote effective clotting, both intrinsic and extrinsic coagulation pathways must be functional, and the inhibition of even one of these pathways will result in severely retarded coagulation ability. The result is usually occurrence of spontaneous bruising and prolonged haemorrhage, which can be fatal. With the ex-ception of tissue factor and Ca^{2+}, defects in all other clotting factors have been characterized. Up to 90 per cent of these, however, relate to a deficiency in factor VIII, and much of the remainder is due to a deficiency in factor IX.

Table 12.2 Recombinant blood coagulation factors that have been approved for the management of coagulation disorders

Product (tradename)	Company	Indication
Advate (rhFactor VIII)	Baxter	Haemophilia A
Bioclate (rhFactor VIII)	Aventis	Haemophilia A
Benefix (rhFactor IX)	Genetics Institute	Haemophilia B
Kogenate (rhFactor VIII)	Bayer	Haemophilia A
Helixate NexGen (rhFactor VIII)	Bayer	Haemophilia A
NovoSeven (rhFactor VIIa)	Novo-Nordisk	Some forms of haemophilia
Recombinate (rhFactor VIII)	Baxter Healthcare/Genetics Institute	Haemophilia A
ReFacto (B-domain deleted rhFactor VIII)	Genetics Institute	Haemophilia A

Such clotting disorders are generally treated by ongoing administration of whole blood or, more usually, concentrates of the relevant coagulation factor purified from whole blood. This entails significant risk of accidental transmission of blood-borne disease, particularly hepatitis and AIDS. In turn, this has hastened the development of blood coagulation factors produced by genetic engineering, several of which are now approved for general medical use (Table 12.2).

12.2.4 Factor VIII and haemophilia

Haemophilia A (classical haemophilia, often simply termed haemophilia) is an X-linked recessive disorder caused by a deficiency of factor VIII. Von Willebrand disease is a related disorder, also caused by a defect in the factor VIII complex, as discussed below.

Intact factor VIII, as usually purified from the blood, consists of two distinct gene products: factor VIII and (multiple copies of) von Willebrand's factor (vWF; Figure 12.6). This complex displays a molecular mass ranging from 1 to 2 MDa, of which up to 15 per cent is carbohydrate. The fully intact factor VIII complex is required to enhance the rate of activation of factor IX of the intrinsic system.

The factor VIII polypeptide portion of the factor VIII complex is coded for by an unusually long gene (289 kb). Transcription and processing of the mRNA generates a shorter, mature, mRNA that codes for a 300 kDa protein. Upon its synthesis, this polypeptide precursor is subsequently proteolytically processed, with removal of a significant portion of its mid region. This yields two fragments: an amino terminal 90 kDa polypeptide and an 80 kDa carboxyl terminal polypeptide. These associate non-covalently (a process requiring Ca^{2+} ions) to produce mature factor VIII (sometimes called factor VIII:C). This mature factor VIII is then released into the plasma where it associates with multiple copies of vWF forming the biologically active factor VIII complex. vWF stabilizes factor VIII in plasma (particularly against proteolytic degradation). It also can associate with platelets at the site of vascular damage and, hence, presumably plays a role in docking the factor VIII complex in an appropriate position where it can participate in the coagulation cascade.

Persons suffering from haemophilia A exhibit markedly reduced levels (or the complete absence) of factor VIII complex in their blood. This is due to the lack of production of factor VIII:C.

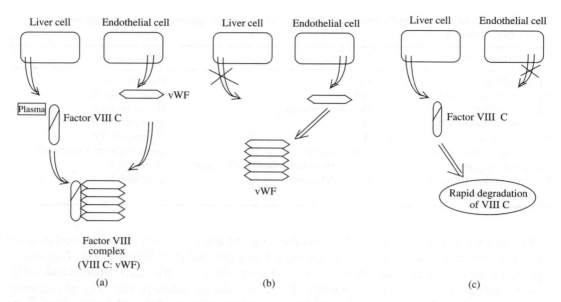

Figure 12.6 (a) Synthesis of factor VIII complex as occurs in healthy individuals. (b) In the case of persons suffering from haemophilia A, synthesis of factor VIII:C is blocked, thus preventing constitution of an active factor VIII complex in plasma. (c) Persons suffering from von Willebrand's disease fail to synthesize vWF. Although they can synthesize VIII:C, this is rapidly degraded upon entering the blood due to lack of its vWF stabilizing factor

Persons suffering from (the rarer) von Willebrand's disease lack both components of mature factor VIII complex (Figure 12.6). The severity of the resultant disease is somewhat dependent upon the level of intact factor VIII complex produced. Persons completely devoid of it (or expressing levels below 1 per cent of normal values) will experience frequent, severe and often spontaneous bouts of bleeding.

Persons expressing 5 per cent or above of the normal complex levels experience less severe clinical symptoms. Treatment normally entails administration of factor VIII complex purified from donated blood. More recently, recombinant forms of the product have also become available. Therapeutic regimens can require product administration on a weekly basis, for life. About 1 in 10 000 males are born with a defect in the factor VIII complex and there are approximately 25 000 haemophiliacs currently resident in the USA.

12.2.5 Production of factor VIII

Native factor VIII is traditionally purified from blood donations first screened for evidence of the presence of viruses such as hepatitis B and HIV. A variety of fractionation procedures (initially mainly precipitation procedures) have been used to produce a factor VIII product. The final product is filter-sterilized and filled into its finished product containers. The product is then freeze-dried and the containers are subsequently sealed under vacuum, or are flushed with an inert gas (e.g. N_2) before sealing. No preservative is added. The freeze-dried product is then stored below 8 °C until shortly before its use.

Although earlier factor VIII preparations were relatively crude (i.e. contained lower levels of various other plasma proteins), many of the modern preparations are chromatographically purified to a high degree. The use of immunoaffinity chromatography has become widespread in this regard since 1988 (Figure 12.7). The extreme bioselectivity of this method can yield a single-step purification factor of several thousand-fold.

Although fractionation can reduce very significantly the likelihood of pathogen transmission, it cannot entirely eliminate this possibility. Blood-derived factor VIII products, including those prepared by immunoaffinity chromatography, generally undergo further processing steps designed to remove/inactivate any virus present. The raw material is often heated for up to 10 h at 60 °C or treated with solvent or dilute detergent prior to chromatography. Recombinant factor VIII is also often treated with dilute detergent in an effort to inactivate any viral particles potentially present.

Production of recombinant factor VIII (Table 12.2) has ended dependence on blood as the only source of this product, and eliminated the possibility of transmitting blood-borne diseases specifically derived from infected blood. In the past, over 60 per cent of haemophiliacs were likely to be accidentally infected via contaminated products at some stage of their life.

Several companies have expressed the cDNA coding for human factor VIII:C in a variety of eukaryotic production systems (human VIII:C contains 25 potential glycosylation sites). CHO cells and BHK cell lines have been most commonly used, in addition to other cell lines, such as various mouse carcinoma cell lines. The recombinant factor VIII product generally contains only VIII:C (i.e. is devoid of vWF). However, both clinical and preclinical studies have shown that administration of this product to patients suffering from haemophilia A is equally as effective as administering blood-derived factor VIII complex. The recombinant VIII:C product appears to bind plasma

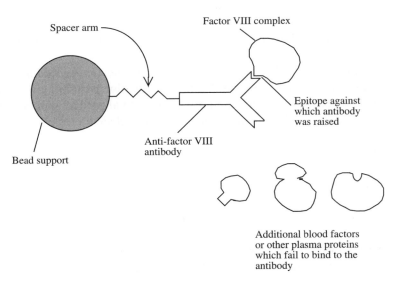

Figure 12.7 Purification of factor VIII complex using immunoaffinity chromatography. The immobilized anti-factor VIII antibody is of mouse origin. Antibodies raised against specific epitopes on both the VIII:C and vWF components have both been successfully used. Industrial-scale columns would often exhibit a bed volume of several litres. Note that the different elements in this diagram are not drawn to the correct scale relative to each other

vWF with equal affinity to native VIII:C upon its injection into the patient's circulatory system. Animal and human pharmacokinetic data reveal no significant difference between the properties of recombinant and native products.

Some patients, particularly those suffering from severe haemophilia A (i.e. those naturally producing little or no VIII:C), will mount an immune response against injected factor VIII:C whatever its source.

The production of anti-factor VIII:C antibodies renders necessary administration of higher therapeutic doses of the product. In severe cases, the product may even become ineffective. Several approaches may be adopted in order to circumvent this problem. These include:

- Exchange transfusion of whole blood. This will transiently decrease circulating anti-factor VIII: C antibodies.

- Direct administration of factor Xa, thus bypassing the non-functional step in the coagulation cascade (Figures 12.2 and 12.3).

- Administration of high levels of a mixture of clotting factors II, VII, IX and X, which works effectively in 50 per cent of treated patients.

- Administration of factor VIIa, as discussed subsequently.

- Administration of porcine factor VIII, which may or may not cross-react with the antibodies raised against human factor VIIIa. (However, the immune system will soon begin to produce antibodies against the porcine clotting factor.)

- Administration of factor VIII, with concurrent administration of immunosuppressive agents.

Owing to the frequency of product administration, the purification procedure for recombinant factor VIII:C must be particularly stringent. Unlike the situation pertaining when the product is purified from human blood, any contaminant present in the final product will be non-human and, hence, immunogenic. Sources of such contaminants would include:

- host cell proteins;

- animal cell culture medium;

- antibody leaked from the immunoaffinity column.

Emphasis is placed not only upon ensuring the absence of contaminant proteins, but also other potential contaminants, particularly DNA. (Host cell-line-derived DNA could harbour oncogenes; Chapter 7.)

Recombinant factor VIII is gaining an increasing market share of the factor VIII market. Researchers are also attempting to develop modified forms of VIII:C (by site-directed mutagenesis) that display additional desirable characteristics. Particularly attractive in this regard would be the development of a product exhibiting an extended circulatory half-life. This could reduce the frequency of injections required by haemophilia A sufferers. However, any alteration of the primary sequence of the molecule carries with it the strong possibility of rendering the resultant mutant immunogenic.

12.2.6 Factors IX, IIVa and XIII

Individuals who display a deficiency of factor IX develop haemophilia B, also known as Christmas disease. Although its clinical consequences are very similar to that of a deficiency of factor VIII, its general incidence in the population is far lower. Persons suffering from haemophilia B are treated by i.v. administration of a concentrate of factor IX. This was traditionally obtained by fractionation of human blood. Recombinant factor IX is now also produced in genetically engineered CHO cells (Table 12.2 and Box 12.1).

Factor IX obtained from blood donations is normally only partially pure. In addition to factor IX, the product contains lower levels of factors II, VII and X and has also been used to treat bleeding disorders caused by a lack of these factors.

Factor IX may also be purified by immunoaffinity chromatography, using immobilized anti-IX murine monoclonals. Purification to homogeneity is particularly important in the case of

---

Box 12.1

Product case study: BeneFix

BeneFix (tradename, also known as nonacog alfa) is a recombinant human blood factor IX approved for general medical use in the USA and EU since 1997. It is indicated for the control and prevention of haemorrhagic episodes in patients with haemophilia B (Christmas disease), including use in a surgical setting. The 55 kDa, 415 amino acid, single-chain protein displays identical amino acid sequence and a relatively similar post-translational modification profile to that of native, serum-derived factor IX. Post-translational modifications present include both N- and O-linked glycosylation, γ-carboxylation, β-hydroxylation, sulfation and phosphorylation. The recombinant blood factor IX is produced in an engineered CHO cell line, and downstream processing entails multiple chromatographic steps, ultrafiltration and diafiltration. The final product is presented in lyophilized form in single-use vials and contains histidine, sucrose, glycine and polysorbate 80 as excipients. Vials contain 250, 500 or 1000 IU of factor IX (as determined by a one-stage clotting assay against a World Health Organization international standard). The product is intended for i.v. administration and exact dosage/administration regimes depend upon the severity of the factor IX deficiency, the location and extent of bleeding, and the patient's clinical condition.

The product displays a mean serum half-life of 18.8 h in humans. It has been evaluated in four clinical trials involving a total of 128 subjects and in the context of both spontaneous bleeding and surgery. Some 88 per cent of the total infusions administered for bleeding were rated as providing a 'good' or 'excellent' response. Reported side effects, although uncommon, included hypersensitivity, as well as headache, fever and nausea. BeneFix is marketed by Wyeth.

recombinant products. At least one monoclonal antibody has been raised which specifically binds only to factor IX which contains pre-bound Ca^{2+} (i.e. the Ca^{2+}-dependent conformation of factor IX). Immobilization of this antibody allowed the development of an immunoaffinity system in which factor IX binds to the column in the presence of a Ca^{2+}-containing buffer. Subsequent elution is promoted simply by inclusion of a chelating agent (e.g. EDTA) in the elution buffer.

Some 5–25 per cent of individuals suffering from haemophilia A develop anti-factor VIII antibodies, and 3–6 per cent of haemophilia B sufferers develop anti-factor IX antibodies. This complicates treatment of these conditions and, as mentioned previously, one approach to their treatment is direct administration of factor VIIa. The therapeutic rationale is that factor VIIa could directly activate the final common steps of the coagulation cascade, independently of either factor VIII or IX (Figure 12.1). Factor VIIa forms a complex with tissue factor that, in the presence of phospholipids and Ca^{2+}, activates factor X.

A recombinant form of factor VIIa (called 'NovoSeven' or 'eptacog alfa-activated') is marketed by Novo-Nordisk (Table 12.2). The recombinant molecule is produced in a BHK cell line, and the final product differs only slightly (in its glycosylation pattern only) from the native molecule.

Purification entails use of an immunoaffinity column containing immobilized murine anti-factor VII antibody. It is initially produced as an unactivated, single-chain 406 amino acid polypeptide, which is subsequently proteolytically converted into the two-chain active factor VIIa complex. After sterilization by filtration, the final product is aseptically filled into its final product containers, and freeze-dried.

A (very rare) genetic deficiency in the production of factor XIII also results in impaired clotting efficacy in affected persons. In this case, covalent links that normally characterize transformation of a soft clot into a hard clot are not formed. Factor XIII preparations, partially purified from human blood, are used to treat individuals with this condition; to date, no recombinant version of the product has been commercialized.

12.3   Anticoagulants

Although blood clot formation is essential to maintaining haemostasis, inappropriate clotting can give rise to serious, sometimes fatal medical conditions. The formation of a blood clot (a thrombus) often occurs inappropriately within diseased blood vessels. This partially or completely obstructs the flow of blood (and hence oxygen) to the tissues normally served by that blood vessel.

Thrombus formation in a coronary artery (the arteries that supply the heart muscle itself with oxygen and nutrients) is termed coronary thrombosis. This results in a heart attack, characterized by the death (infarction) of oxygen-deprived heart muscle; hence the term myocardial infarction. The development of a thrombus in a vessel supplying blood to the brain can result in development of a stroke. In addition, a thrombus (or part thereof) that has formed at a particular site in the vascular system may become detached. After travelling through the blood, this may lodge in another blood vessel, obstructing blood flow at that point. This process, which can also give rise to heart attacks or strokes, is termed embolism.

Anticoagulants are substances that can prevent blood from clotting and, hence, are of therapeutic use in cases where a high risk of coagulation is diagnosed. They are often administered to patients with coronary heart disease and to patients who have experienced a heart attack or

Table 12.3 Anticoagulants that are used therapeutically or display therapeutic potential

Anticoagulant[a]	Structure	Source	Molecular mass (Da)
Heparin	Glycosaminoglycan	Beef lung, pig gastric mucosa	3 000–40 000
Dicoumarol	Coumarin-based	Chemical manufacture	336.3
Warfarin	Coumarin-based	Chemical manufacture	308.4
Hirudin	Polypeptide	Leech saliva, genetic engineering	7 000
Ancrod	Polypeptide	Snake venom, genetic engineering	35 000
Protein C	Glycoprotein	Human plasma	62 000

[a]Dicoumarol and related molecules are generally used over prolonged periods, whereas heparin is used over shorter periods. Hirudin has recently been approved for general medical use, while ancrod remains under clinical investigation.

stroke (in an effort to prevent recurrent episodes). The major anticoagulants used for therapeutic purposes are listed in Table 12.3.

Heparin is a carbohydrate-based (glycosaminoglycan) anticoagulant associated with many tissues, but mainly found stored intracellularly as granules in mast cells that line the endothelium of blood vessels. Upon release into the bloodstream, heparin binds to and thereby activates an additional plasma protein, namely antithrombin. The heparin–antithrombin complex then binds a number of activated clotting factors (including IIa, IXa, Xa, XIa and XIIa), thereby inactivating them. The heparin now disassociates from the complex and combines with another antithrombin molecule, thereby initiating another turn of this inhibitory cycle.

Heparin was originally extracted from liver (hence its name), but commercial preparations are now obtained by extraction from beef lung or porcine gastric mucosa.

Although the product has proven to be an effective (and relatively inexpensive) anticoagulant, it does suffer from a number of clinical disadvantages, including the need for a cofactor (antithrombin III) and poorly predictable dose responses. Despite such disadvantages, however, heparin still enjoys widespread clinical use.

The vitamin K antimetabolites dicoumarol and warfarin are related coumarin-based anticoagulants which, unlike heparin, may be administered orally. These compounds induce their anticoagulant effect by preventing the vitamin K-dependent γ-carboxylation of certain blood factors, specifically factors II, VII, IX and X. Upon initial hepatic synthesis of these coagulation factors, a specific carboxylase catalyses the γ-carboxylation of several of their glutamate residues (for example, 10 of the first 33 residues present in prothrombin are γ-carboxyglutamate). This post-translational modification is required in order to allow these factors to bind Ca^{2+} ions, which is a prerequisite to their effective functioning. Vitamin K is an essential cofactor for the carboxylase enzyme, and its replacement with the antimetabolite dicoumarol renders this enzyme inactive. As a consequence, defective blood factors are produced that hinder effective functioning of the coagulation cascade. The only major side effect of these oral anticoagulants is prolonged bleeding; thus, the dosage levels are chosen with care. Dicoumarol was first isolated from spoiled sweet clover hay, as the agent that promoted haemorrhage disease in cattle. Both dicoumarol and warfarin have also been utilized (at high doses) as rat poisons.

12.3.1   Hirudin

Hirudin is a leech-derived anticoagulant that functions by directly inhibiting thrombin. A range of blood-sucking animals contain substances in their saliva that specifically inhibit some element of the blood coagulation system (Table 12.4).

A bite from any such parasite is characterized by prolonged host bleeding. This property led to the documented use of leeches as an aid to blood-letting as far back as several hundred years BC. The method was particularly fashionable in Europe at the beginning of the 19th century. Many doctors at that time still believed that most illnesses were related in some way to blood composition, and blood-letting was a common, if ineffective, therapy. The Napoleonic Army surgeons, for example, used leeches to withdraw blood from soldiers suffering from conditions as diverse as infections and mental disease.

With the advent of modern medical principles, the medical usage of leeches waned somewhat. In more recent years, however, they did stage a limited comeback. They were occasionally used to drain blood from inflamed tissue, and in procedures associated with plastic surgery.

The presence of an anticoagulant in the saliva of the leech, *Hirudo medicinalis*, was first described in 1884. However, it was not until 1957 that the major anticoagulant activity present was purified and named hirudin. Hirudin is a short (65 amino acid) polypeptide, of molecular mass 7000 Da. The tyrosine residue at position 63 is unusual in that it contains a sulfate group. The molecule appears to have two domains. The globular N-terminal domain is stabilized by three disulfide linkages, whereas the C-terminal domain is more elongated and exhibits a high content of acidic amino acids.

Hirudin exhibits its anticoagulant effect by tightly binding thrombin, thus inactivating it. In addition to its critical role in the production of a fibrin clot, thrombin displays several other (non-enzymatic) biological activities important in sustaining haemostasis. These include:

- it is a potent inducer of platelet activation and aggregation;

- it functions as a chemoattractant for monocytes and neutrophils;

- it stimulates endothelial transport.

Table 12.4   Some substances isolated from bloodsucking parasites which inhibit their host's haemostatic mechanisms. All are polypeptides of relatively low molecular mass

Polypeptide	Molecular mass (Da)	Producer	Haemostatic effect disrupted
Hirudin	7 000	*Hirudo medicinalis*	Binds to and inhibits thrombin
Rhodniin	11 100	*Rhodnius prolixus*	Binds to and inhibits thrombin
Antistatin	15 000	*Haementeria officinalis*	Inhibits factor Xa
Tick anticoagulant peptide (TAP)	6 800	*Ornithodoros moubata*	Inhibits factor Xa
Calin	55 000	*Hirudo medicinalis*	Inhibits platelet adhesion
Decorsin	4 400	*Macrobdella decora*	Inhibits platelet adhesion

One molecule of hirudin binds a single molecule of thrombin with very high affinity ($K_d \approx 10^{-12}$ mol l^{-1}). Binding and inactivation occur as a two-step mechanism. The C-terminal region of hirudin first binds along a groove on the surface of thrombin, resulting in a small conformational change of the enzyme. This then facilitates binding of the N-terminal region to the active site area (Figure 12.8). Binding of hirudin inhibits all the major functions of thrombin. Fragments of hirudin can also bind thrombin, but will generally only inhibit some of thrombin's range of activities. For example, binding of an N-terminal hirudin fragment to thrombin inhibits only the thrombin catalytic activity.

Hirudin displays several potential therapeutic advantages as an anticoagulant. These include:

- it acts directly upon thrombin;

- it does not require a cofactor;

- it is less likely than many other anticoagulants to induce unintentional haemorrhage;

- it is a weak immunogen.

Although the therapeutic potential of hirudin was appreciated for many years, insufficient material could be purified from the native source to support clinical trials, never mind its widespread medical application. The hirudin gene was cloned in the 1980s, and it has subsequently been expressed in a number of recombinant systems, including *E. coli, Bacillis subtilis* and *Saccharomyces cerevisiae*. A recombinant hirudin (tradename Refludan) was first approved for general medical use in 1997. The recombinant production system was constructed by insertion (via a plasmid) of a synthetic hirudin gene into a strain of *S. cerevisiae*. The yeast cells secrete the product, which is then purified by various fractionation techniques (Figure 12.9). The recombinant molecule displays a slightly altered amino acid sequence compared with the native product. Its first two amino acids, leucine and threonine, replace two valines of native hirudin. It is also devoid of the sulfate

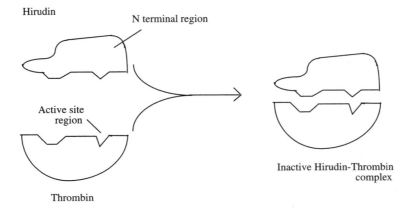

Figure 12.8 Binding of hirudin to thrombin, thus inactivating this activated coagulation factor

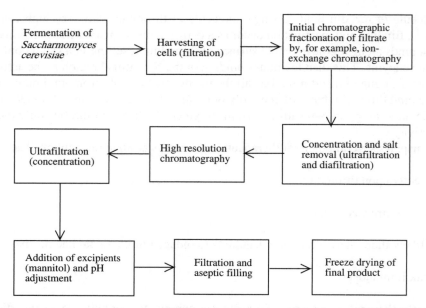

Figure 12.9   Overview of the production of Refludan (recombinant hirudin). The exact details of many steps remain confidential for obvious commercial reasons. A number of QC checks are carried out on the final product to confirm the product's structure. These include amino acid composition, HPLC analysis and peptide mapping

group normally present on tyrosine[63]. Clinical trials, however, have proven this slightly altered product to be both safe and effective. The final product is presented in freeze-dried form with the sugar mannitol representing the major added excipient. The product, which displays a useful shelf-life of 2 years when stored at room temperature, is reconstituted with saline or WFI immediately prior to its i.v. administration. A second recombinant product (tradename revasc, also produced in *S. cerevisiae*) has also been approved.

12.3.2   Antithrombin

Antithrombin, already mentioned in the context of heparin, is the most abundantly occurring natural inhibitor of coagulation. It is a single-chain 432 amino acid glycoprotein displaying four oligosaccharide side chains and an approximate molecular mass of 58 kDa. It is present in plasma at concentrations of 150 μg ml^{-1} and is a potent inhibitor of thrombin (factor IIa), as well as of factors IXa and Xa. It inhibits thrombin by binding directly to it in a 1:1 stoichiometric complex.

Plasma-derived antithrombin concentrates have been used medically since the 1980s for the treatment of hereditary and acquired antithrombin deficiency. Hereditary (genetic) deficiency is characterized by the presence of little/no native antithrombin activity in plasma and results in an increased risk of inappropriate blood clot/emboli formation. Acquired antithrombin deficiency can be induced by drugs (e.g. heparin and oestrogens), liver disease (decreased antithrombin

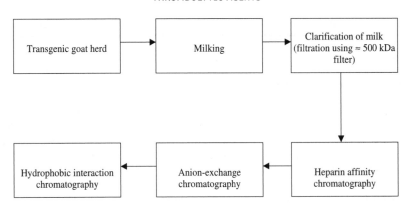

Figure 12.10 Outline of the production and purification of antithrombin from the milk of transgenic goats. Purification achieves an overall product yield in excess of 50 per cent, with a purity greater than 99 per cent

synthesis) or various other medical conditions. Recombinant antithrombin has been expressed in the milk of transgenic goats (Chapter 5), and this product (tradename Atryn) was approved for general medical use in Europe in 2006 (Figure 12.10). The recombinant product displays an identical amino acid sequence to that of native human antithrombin, although its oligosaccharide composition does vary somewhat from the native protein.

A related product (tradename Xigiris, also known as drotrecogin alfa) has also been approved for medical use. Xigiris is a recombinant human activated protein C, a molecule that plays an important role in controlling coagulation *in vivo*. The recombinant product is produced in an engineered mammalian cell line and, like several other blood proteins, is characterized by the presence of several γ-carboxyglutamate and β-hydroxylated residues (Chapter 2). Activated protein C is indicated for the treatment of severe sepsis, largely in order to prevent multiple organ failure that can be triggered by sepsis-associated blood clot formation.

12.4 Thrombolytic agents

The natural process of thrombosis functions to plug a damaged blood vessel, thus maintaining haemostasis until the damaged vessel can be repaired. Subsequent to this repair, the clot is removed via an enzymatic degradative process known as fibrinolysis. Fibrinolysis normally depends upon the serine protease plasmin, which is capable of degrading the fibrin strands present in the clot.

In situations where inappropriate clot formation results in the blockage of a blood vessel, the tissue damage that ensues depends, to a point, upon how long the clot blocks blood flow. Rapid removal of the clot can often minimize the severity of tissue damage. Thus, several thrombolytic (clot-degrading) agents have found medical application (Table 12.5). The market for an effective thrombolytic agent is substantial. In the USA alone, it is estimated that 1.5 million people suffer acute myocardial infarction each year, and there are another 0.5 million suffer strokes.

Table 12.5   Thrombolytic agents approved for general medical use

Product[a]	Company
Activase (rh-tPA)	Genentech
Ecokinase (rtPA; differs from human tPA in that three of its five domains have been deleted)	Galenus Mannheim
Retavase (rtPA; see Ecokinase)	Boehringer Manheim/Centocor
Rapilysin (rtPA; see Ecokinase)	Boehringer Manheim
Tenecteplase (also marketed as Metalyse) (TNK-tPA, modified rtPA)	Boehringer Ingelheim
TNKase (tenecteplase; modified rtPA; see Tenecteplase)	Genentech
Streptokinase (produced by *Streptokinase haemolyticus*)	Various
Urokinase (extracted from human urine)	Various
Staphylokinase (extracted from *Staphylococcus aureus* and produced in various recombinant systems	Various

[a]r: recombinant; rh: recombinant human.

12.4.1   Tissue plasminogen activator

The natural thrombolytic process is illustrated in Figure 12.11. Plasmin is a protease that catalyses the proteolytic degradation of fibrin present in clots, thus effectively dissolving the clot. Plasmin is derived from plasminogen, its circulating zymogen. Plasminogen is synthesized in, and released from, the kidneys. It is a single-chain 90 kDa glycoprotein that is stabilized by several disulfide linkages.

tPA (also known as fibrinokinase) represents the most important physiological activator of plasminogen. tPA is a 527 amino acid serine protease. It is synthesized predominantly in vascular endothelial cells (cells lining the inside of blood vessels) and displays five structural domains, each of which has a specific function (Table 12.6). tPA displays four potential glycosylation sites, three of which are normally glycosylated (residues 117, 184 and 448). The carbohydrate moieties play an important role in mediating hepatic uptake of tPA and, hence, its clearance from plasma. It is normally found in the blood in two forms: a single-chain polypeptide (type I tPA) and a two-chain structure (type II) proteolytically derived from the single-chain structure. The two-chain form is the one predominantly associated with clots undergoing lysis, but both forms display fibrinolytic activity.

Table 12.6   The five domains that constitute human tPA, and the biological function of each domain

tPA domain	Function
Finger domain (F domain)	Promotes tPA binding to fibrin with high affinity
Protease domain (P domain)	Displays plasminogen-specific proteolytic activity
Epidermal growth factor domain (EGF domain)	Binds hepatic receptors, thereby mediating hepatic clearance of tPA from blood
Kringle-1 domain (K_1 domain)	Associated with binding to the hepatic receptor
Kringle-2 domain (K_2 domain)	Facilitates stimulation of tPA's proteolytic activity by fibrin

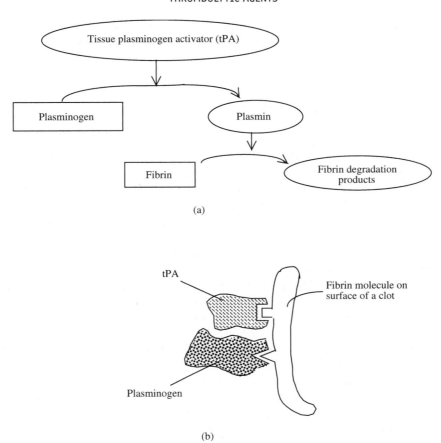

Figure 12.11   (a) The fibrinolytic system, in which tPA proteolytically converts the zymogen plasminogen into active plasmin, which in turn degrades the fibrin strands, thus dissolving the clot. tPA and plasminogen both bind to the surface of fibrin strands (b), thus ensuring rapid and efficient activation of the thrombolytic process

Fibrin contains binding sites for both plasminogen and tPA, thus bringing these into close proximity. This facilitates direct activation of the plasminogen at the clot surface (Figure 12.11). This activation process is potentiated by the fact that binding of tPA to fibrin (a) enhances the subsequent binding of plasminogen and (b) increases tPA's activity towards plasminogen by up to 600-fold.

Overall therefore, activation of the thrombolytic cascade occurs exactly where it is needed, i.e. on the surface of the clot. This is important, as the substrate specificity of plasmin is poor, and circulating plasmin displays the catalytic potential to proteolyse fibrinogen, factor V and factor VIII. Although soluble serum tPA displays a much reduced activity towards plasminogen, some free-circulating plasmin is produced by this reaction. If uncontrolled, this could increase the risk of subsequent haemorrhage. This scenario is usually averted because circulating plasmin is rapidly neutralized by anther plasma protein, α_2-antiplasmin (a 70 kDa single-chain glycoprotein that binds plasmin very tightly in a 1:1 complex). In contrast to free plasmin, plasmin present on a clot surface is very slowly inactivated by α_1-antiplasmin. The thrombolytic system has thus evolved in a self-regulating fashion, which facilitates efficient clot degradation with minimal potential disruption to other elements of the haemostatic mechanism.

12.4.2   First-generation tissue plasminogen activator

Although tPA was first studied in the late 1940s, its extensive characterization was hampered by the low levels at which it is normally synthesized. Detailed studies were facilitated in the 1980s after the discovery that the Bowes melanoma cell line produces and secretes large quantities of this protein. This also facilitated its initial clinical appraisal. The tPA gene was cloned from the melanoma cell line in 1983, and this facilitated subsequent large-scale production in CHO cell lines by recombinant DNA technology. The tPA cDNA contains 2530 nucleotides and encodes a mature protein of 527 amino acids. The glycosylation pattern was similar, though not identical, to the native human molecule. A marketing licence for the product was first issued in the USA to Genentech in 1987 (under the tradename Alteplase). The therapeutic indication was for the treatment of acute myocardial infarction. The production process entails an initial (10 000 l) fermentation step, during which the cultured CHO cells produce and secrete tPA into the fermentation medium. After removal of the cells by sub-micrometre filtration and initial concentration, the product is purified by a combination of several chromatographic steps. The final product has been shown to be greater than 99 per cent pure by several analytical techniques, including HPLC, SDS-PAGE, tryptic mapping and N-terminal sequencing.

   Alteplase has proven effective in the early treatment of patients with acute myocardial infarction (i.e. those treated within 12 h after the first symptoms occur). Significantly increased rates of patient survival (as measured 1 day and 30 days after the initial event) are noted when tPA is administered in favour of streptokinase, a standard therapy (see later). tPA has thus established itself as a first-line option in the management of acute myocardial infarction. A therapeutic dose of 90–100 mg (often administered by infusion over 90 min) results in a steady-state alteplase concentration of 3–4 mg l^{-1} during that period. However, the product is cleared rapidly by the liver, displaying a serum half-life of approximately 3 min. As is the case for most thrombolytic agents, the most significant risk associated with tPA administration is the possible induction of severe haemorrhage.

12.4.3   Engineered tissue plasminogen activator

Modified forms of tPA have also been generated in an effort to develop a product with an improved therapeutic profile (e.g. faster acting or exhibiting a prolonged plasma half-life). Reteplase is the international non-proprietary name given to one such modified human tPA produced in recombinant *E. coli* cells and is sold under the tradenames Ecokinase, Retavase and Rapilysin (Table 12.5). This product's development was based upon the generation of a synthetic nucleotide sequence encoding a shortened (355 amino acid) tPA molecule. This analogue contained only the tPA domains responsible for fibrin selectivity and catalytic activity. The nucleotide sequence was integrated into an expression vector subsequently introduced into *E. coli* (strain K12) by treatment with calcium chloride. The protein is expressed intracellularly, where it accumulates in the form of an inclusion body. Owing to the prokaryotic production system, the product is non-glycosylated. The final sterile freeze-dried product exhibits a 2-year shelf life when stored at temperatures below 25 °C. An overview of the production process is presented in Figure 12.12.

The lack of glycosylation, as well as the absence of the EGF and K_1 domains (Table 12.6), confers an extended serum half-life upon the engineered molecule. Reteplase-based products display a serum half-life of up to 20 min, facilitating its administration as a single bolus injection as opposed to continuous infusion. Absence of the molecule's F_1 domain also reduces the product's fibrin-binding affinity. It is theorized that this may further enhance clot degradation, as it facilitates more extensive diffusion of the thrombolytic agent into the interior of the clot. Tenecteplase (also marketed under the tradename Metalyse) is yet an additional engineered tPA now on the market. Produced in a CHO cell line, this glycosylated variant differs in sequence to native tPA by six amino acids (Thr 103 converted to Asn; Asn 117 converted to Gln and the Lys–His–Arg–Arg sequence at position 296–299 converted to Ala–Ala–Ala–Ala). Collectively, these modifications result in a prolonged plasma half-life (to between 15 and 19 min), as well as an increased resistance to PAI-1 (plasminogen activator inhibitor 1, a natural tPA inhibitor).

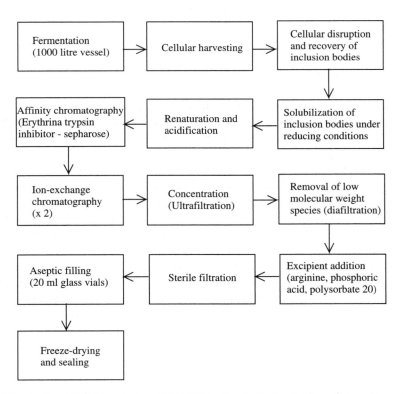

Figure 12.12 Production of Ecokinase, a modified tPA molecule that gained regulatory approval in Europe in 1996. The production cell line is recombinant *E. coli* K12, which harbours a nucleotide sequence coding for the shortened tPA molecule. The product accumulates intracellularly in the form of inclusion bodies

12.4.4   Streptokinase

Streptokinase is a 48 kDa extracellular bacterial protein produced by several strains of *Strepto-coccus haemolyticus* group C. Its ability to induce lysis of blood clots was first demonstrated in 1933. Early therapeutic preparations administered to patients often caused immunological and other complications, usually prompted by impurities present in these products. Chromatographic purification (particularly using gel filtration and ion-exchange columns) overcame many of these initial difficulties. Modern chromatographically pure streptokinase preparations are usually supplied in freeze-dried form. These preparations (still obtained by non-recombinant means) often contain albumin as an excipient. The albumin prevents flocculation of the active ingredient upon its reconstitution.

Streptokinase is a widely employed thrombolytic agent. It is administered to treat a variety of thrombo-embolic disorders, including:

- pulmonary embolism (blockage of the pulmonary artery – which carries blood from the heart to the lungs for oxygenation – by an embolism), which can cause acute heart failure and sudden death;

- deep-vein thrombosis (thrombus formation in deep veins, usually in the legs);

- arterial occlusions (obstruction of an artery);

- acute myocardial infarction.

Streptokinase induces its thrombolytic effect by binding specifically and tightly to plasminogen. This induces a conformational change in the plasminogen molecule that renders it proteolytically active. In this way, the streptokinase–plasminogen complex catalyses the proteolytic conversion of plasminogen to active plasmin.

As a bacterial protein, streptokinase is viewed by the human immune system as an antigenic substance. In some cases, its administration has elicited allergic responses that have ranged from mild rashes to more serious anaphylactic shock (an extreme and generalized allergic response characterized by swelling, constriction of the bronchioles, circulatory collapse and heart failure).

Another disadvantage of streptokinase administration is the associated increased risk of haemorrhage. Streptokinase-activated plasminogen is capable of lysing not only clot-associated fibrin, but also free plasma fibrinogen. This can result in low serum fibrinogen levels and, hence, compromise haemostatic ability. It should not be administered to, for example, patients suffering from coagulation disorders or bleeding conditions such as ulcers. Despite such potential clinical complications, careful administration of streptokinase has saved countless thousands of lives.

12.4.5   Urokinase

The ability of some component of human urine to dissolve fibrin clots was first noted in 1885. It was not until the 1950s, however, that the active substance was isolated and named urokinase.

Urokinase is a serine protease produced by the kidney and is found in both the plasma and urine. It is capable of proteolytically converting plasminogen into plasmin. Two variants of the enzyme have been isolated: a 54 kDa species and a lower molecular mass (33 kDa) variant. The lower molecular mass form appears to be derived from the higher molecular mass moiety by proteolytic processing. Both forms exhibit enzymatic activity against plasminogen.

Urokinase is used clinically under the same circumstances as streptokinase; because of its human origin, adverse immunological responses are less likely. Following acute medical events such as pulmonary embolism, the product is normally administered to the patient at initial high doses (by infusion) for several minutes. This is followed by hourly i.v. injections for up to 12 h.

Urokinase utilized medically is generally purified directly from human urine. It binds to a range of adsorbents, such as silica gel and, especially, kaolin (hydrated aluminium silicate), which can be used initially to concentrate and partially purify the product. It may also be concentrated and partially purified by precipitation using sodium chloride, ammonium sulfate or ethanol as precipitants.

Various chromatographic techniques may be utilized to purify urokinase further. Commonly employed methods include anion-exchange (DEAE-based) chromatography, gel filtration on Sephadex G-100 and chromatography on hydroxyapatite columns. Urokinase is a relatively stable molecule. It remains active subsequent to incubation at 60 °C for several hours, or brief incubation at pHs as low as 1.0 or as high as 10.0.

After its purification, sterile filtration and aseptic filling, human urokinase is normally freeze-dried. Because of its heat stability, the final product may also be heated to 60 °C for up to 10 h in an effort to inactivate any undetected viral particles present. The product utilized clinically contains both molecular mass forms, with the higher molecular mass moiety predominating. Urokinase can also be produced by techniques of animal cell culture utilizing human kidney cells or by recombinant DNA technology.

12.4.6 Staphylokinase

Staphylokinase is a protein produced by a number of strains of *S. aureus* that also displays therapeutic potential as a thrombolytic agent. The protein has been purified from its natural source by a combination of ammonium sulfate precipitation and cation-exchange chromatography on CM cellulose. Affinity chromatography using plasmin or plasminogen immobilized to sepharose beads has also been used. The pure product is a 136 amino acid polypeptide displaying a molecular mass in the region of 16.5 kDa. Lower molecular mass derivatives lacking the first 6 or 10 NH_2-terminal amino acids have also been characterized. All three appear to display similar thrombolytic activity *in vitro* at least.

The staphylokinase gene has been cloned in *E. coli*, as well as various other recombinant systems. The protein is expressed intracellularly in *E. coli* at high levels, representing 10–15 per cent of total cellular protein. It can be purified directly from the clarified cellular homogenate by a combination of ion-exchange and hydrophobic interaction chromatography.

Although staphylokinase shows no significant homology with streptokinase, it induces a thrombolytic effect by a somewhat similar mechanism: it also forms a 1:1 stoichiometric complex with plasminogen. The proposed mechanism by which staphylokinase induces plasminogen activation is outlined in Figure 12.13. Binding of the staphylokinase to plasminogen appears initially to yield an inactive staphylokinase–plasminogen complex. However, complex formation somehow induces subsequent proteolytic cleavage of the bound plasminogen to form plasmin, which remains complexed to the staphylokinase. This complex (via the plasmin) then appears to catalyse the conversion of free plasminogen to plasmin, and it may even accelerate the process of conversion of other staphylokinase–plasminogen complexes into staphylokinase–plasmin complexes. The net effect is generation of active plasmin, which displays a direct thrombolytic effect by degrading clot-based fibrin, as described previously (Figure 12.11).

The serum protein α_2-antiplasmin can inhibit the activated plasmin–staphylokinase complex. It appears that the α_2-antiplasmin can interact with the active plasmin moiety of the complex, resulting in dissociation of staphylokinase and consequent formation of an inactive plasmin–α_2-antiplasmin complex.

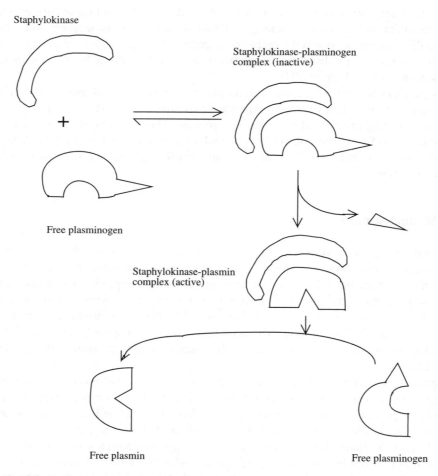

Figure 12.13 Schematic representation of the mechanism by which staphylokinase appears to activate the thrombolytic process via the generation of plasmin. See text for details

The thrombolytic ability of (recombinant) staphylokinase has been evaluated in initial clinical trials with encouraging results. Some 80 per cent of patients suffering from acute myocardial infarction who received staphylokinase responded positively (10 mg staphylokinase was administered by infusion over 30 min). The native molecule displays a relatively short serum half-life (of 6.3 min), although covalent attachment of PEG reduces the rate of serum clearance, hence effectively increasing the molecule's half-life significantly. As with streptokinase, patients administered staphylokinase develop neutralizing antibodies. A number of engineered (domain-deleted) variants have been generated that display significantly reduced immunogenicity.

12.4.7 α_1-Antitrypsin

The respiratory tract is protected by a number of defence mechanisms, including:

- particle removal in the nostril/nasopharynx;

- particle expulsion (e.g. by coughing);

- upward removal of substances via mucociliary transport;

- presence in the lungs of immune cells, such as alveolar macrophages;

- production/presence of soluble protective factors, including α_1-antitrypsin, lysozyme, lactoferrin and interferon.

Failure/ineffective functioning of one or more of these mechanisms can impair normal respiratory function. Emphysema, for example, is a condition in which the alveoli of the lungs are damaged. This compromises the lung's capacity to exchange gases, and breathlessness often results. This condition is often promoted by smoking, respiratory infections or a deficiency in the production of serum α_1-antitrypsin.

α_1-Antitrypsin is a 394 amino acid, 52 kDa serum glycoprotein. It is synthesized in the liver and secreted into the blood, where it is normally present at concentrations of 2–4 g l^{-1}. It constitutes in excess of 90 per cent of the α_1-globulin fraction of blood.

The α_1-antitrypsin gene is located on chromosome 14. A number of α_1-antitrypsin gene variants have been described. Their gene products can be distinguished by their differential mobility upon gel electrophoresis. The normal form is termed M, but point mutations in the gene have generated two major additional forms, i.e. S and Z. These mutations result in a greatly reduced level of synthesis and secretion into the blood of the mature α_1-antitrypsin. In particular, persons inheriting two copies of the Z gene display greatly reduced levels of serum α_1-antitrypsin activity. This is often associated with the development of emphysema (particularly in smokers). The condition may be treated by the administration of purified α_1-antitrypsin. This protein constitutes the major serine protease inhibitor present in blood. It is a potent inhibitor of the protease elastase, which serves to protect the lung from proteolytic damage by inhibiting neutrophil elastase. The product is administered on an ongoing basis to sufferers, who receive up to 200 g of the inhibitor each year. It is normally prepared by limited fractionation of whole human blood, although the large quantities required by patients heightens the risk of accidental transmission of blood-borne pathogens. The α_1-antitrypsin gene has been expressed in a number of recombinant systems, including in the milk of transgenic sheep. Although use of the recombinant product would all but preclude blood pathogen transmission, it appears significantly more costly to produce.

12.4.8 Albumin

HSA is the single most abundant protein in blood (Table 12.7). Its normal concentration is approximately 42 g l^{-1}, representing 60 per cent of total plasma protein. The vascular system of an average adult thus contains in the region of 150 g of albumin. HSA is responsible for over 80 per cent of the colloidal osmotic pressure of human blood. More than any other plasma constituent, HSA is thus responsible for retaining sufficient fluid within blood vessels. It has been aptly described as the protein that makes blood thicker than water.

Albumin molecules also temporarily leave the circulation, entering the lymphatic system, which harbours a large pool of this protein (up to 230 g in an adult). Lower quantities of albumin are also present in the skin.

In addition to its osmoregulatory function, HSA serves a transport function. Various metabolites travel throughout the vascular system predominantly bound to HSA. These include fatty acids, amino acids, steroid hormones and heavy metals (e.g. copper and zinc), as well as many drugs.

HSA is a 585 amino acid, 65.5 kDa polypeptide. It is one of the few plasma proteins that are unglycosylated. A prominent feature is the presence of 17 disulfide bonds, which help stabilize the molecule's three-dimensional structure. HSA is synthesized and secreted from the liver, and its gene is present on human chromosome number 4.

Table 12.7 The major plasma proteins of known function found in human blood

Protein	Normal plasma concentration (g l^{-1})	Molecular mass (kDa)	Function
Albumin	35–45	66.5	Osmoregulation transport
Retinol-binding protein	0.03–0.06	21	Retinol transport
Thyroxine binding globulin	0.01–0.02	58	Binds/transports thyroxine
Transcortin	0.03–0.04	52	Cortisol and corticosterone transport
Ceruloplasmin	0.1–0.6	151	Copper transport
Haptoglobin			Binds and helps conserve haemoglobin
type 1–1	1.0–2.2	100	
type 2–1	1.6–3.0	200	
type 2–2	1.2–2.6	400	
Transferrin	2.0–3.2	76.5	Iron transport
Hemopexin	0.5–1.0	57	Binds haem destined for disposal
β2-Microglobulin	0.002	11.8	Associated with human leukocyte antigen histocompatibility antigen
γ-Globulins	7.0–15.0	150	Antibodies
Transthyretin	0.1–0.4	55	Binds thyroxine

HSA is used therapeutically as an aqueous solution; it is available in concentrated form (15–25 per cent protein) or as an isotonic solution (4–5 per cent protein). In both cases, in excess of 95 per cent of the protein present is albumin. It can be prepared by fractionation from normal plasma or serum, or purified from placentas. The source material must first be screened for the presence of indicator pathogens. After purification, a suitable stabilizer (often sodium caprylate) is added, but no preservative. The solution is then sterilized by filtration and aseptically filled into final sterile containers. The relative heat stability of HSA allows a measure of subsequent heat treatment, which further reduces the risk of accidental transmission of viable pathogens (particularly viruses). This treatment normally entails heating the product to 60 °C for 10 h. It is then normally incubated at 30–32 °C for a further 14 days and subsequently examined for any signs of microbial growth.

HSA is used as a plasma expander in the treatment of haemorrhage, shock, burns and oedema, as well as being administered to some patients after surgery. For adults, an initial infusion containing at least 25 g of albumin is used. The annual world demand for HSA exceeds 300 t, representing a market value of the order of US$1 billion.

Despite screening of raw material and heat treatment of final product, HSA derived from native blood sometimes (though rarely) will harbour pathogens. rDNA technology provides a way of overcoming such concerns, and the HSA gene and cDNA have been expressed in a wide variety of microbial systems, including *E. coli*, *Bacillus subtilis*, *S. cerevisiae*, *Pichia pastoris* and *Aspergillus niger*. (Its lack of glycosylation renders possible production of native HSA in prokaryotic and eukaryotic systems.) However, HSA's relatively large size, as well as the presence of so many disulfide bonds, can complicate recombinant production of high levels of correctly folded products in some production systems. The main stumbling block in replacing native HSA with a recombinant version, however, is an economic one. Unlike most biopharmaceuticals, HSA can be produced in large quantities and inexpensively by direct extraction from its native source. Native HSA currently sells at US$2–3 per gram. Although it can be guaranteed blood pathogen free, recombinant HSA products will find it difficult to compete with this price.

12.5   Enzymes of therapeutic value

Enzymes are used for a variety of therapeutic purposes, the most significant of which are listed in Table 12.8. A number of specific examples have already been discussed in detail within this chapter, including tPA, urokinase, and factor IXa. The additional therapeutic enzymes now become the focus of the remainder of the chapter. Although a limited number of polymer-degrading enzymes (used as digestive aids) are given orally, most enzymes are administered intravenously.

12.5.1   Asparaginase

Asparaginase is an enzyme capable of catalysing the hydrolysis of L-asparagine, yielding aspartic acid and ammonia (Figure 12.14). In the late 1970s, researchers illustrated that serum transferred from healthy guinea pigs into mice suffering from leukaemia contained some agent capable of inhibiting the proliferation of the leukaemic cells. A search revealed the agent to be asparaginase.

Table 12.8 Enzymes used therapeutically

Enzyme	Application
Tissue plasminogen activator	Thrombolytic agent
Urokinase	Thrombolytic agent
Ancrod	Anticoagulant
Factor IXa	Haemophilia B
Asparaginase	Anti-cancer agent
Nuclease (DNase)	Cystic fibrosis
Glucocerebrosidase	Gaucher's disease
α-Galactosidase	Fabry disease
Urate oxidase	Hyperuricaemia
Laronidase	Mucopolysaccharidosis
Superoxide dismutase (SOD)	Oxygen toxicity
Acid α-glucosidase	Pompe disease
α-L-Iduronidase	Mucopolysaccharidosis I (MPS I)
N-Acetylgalactosamine-4-sulfactase	Mucopolysaccharidosis IV
Trypsin/papain/collagenase	Debriding/anti-inflammatory agents
Lactase/pepsin/papain/pancrelipase	Digestive aids

Most healthy (untransformed) mammalian cells are capable of directly synthesizing asparagine from glutamine (Figure 12.14). Hence, asparagine is generally classified as a non-essential amino acid (i.e. we do not require it as an essential component of our diet). However, many transformed cells lose the ability to synthesize asparagine themselves. For these, asparagine becomes an essential amino acid. In the case of the leukaemic mice, the guinea pigs' asparaginase deprived the transformed cells of this amino acid by hydrolysing plasma asparagine. This approach has been successfully applied to treating some forms of human leukaemia. For example, the PEG–L-asparaginase previously mentioned was approved for the treatment of refractory childhood acute lymphoblastic leukaemia.

Generally, the plasma concentration of asparagine is quite low (~40 μmol l^{-1}). Therefore, therapeutically useful asparaginases must display a high substrate affinity (i.e. low K_m values). Asparaginase from *E. coli* and *Erwinia*, as well as from *Pseudomonas* and *Acinetobacter*, has been studied in greatest detail. It has proven effective in inhibiting growth of various leukaemias and other transformed cell lines. PEG-coupled enzymes are often preferred, as they display an extended plasma half-life.

Although asparaginase therapy has proven effective, a number of side effects have been associated with initiation of therapy. These have included severe nausea, vomiting and diarrhoea, as well as compromised liver and kidney function. Side effects are probably due to a transient asparaginase deficiency in various tissues. Under normal circumstances, dietary-derived plasma asparagine levels are sufficient to meet normal tissue demands, and the cellular asparagine biosynthetic pathway remains repressed. Reduced plasma asparagine levels result in the induction of cellular asparagine synthesis. High-dose asparaginase administration will immediately reduce plasma asparagine levels. However, the ensuing initiation of cellular asparagine synthesis may not occur for several hours. Thus, a more suitable therapeutic regimen may entail initial low-dose asparaginase administration, followed by stepwise increasing dosage levels.

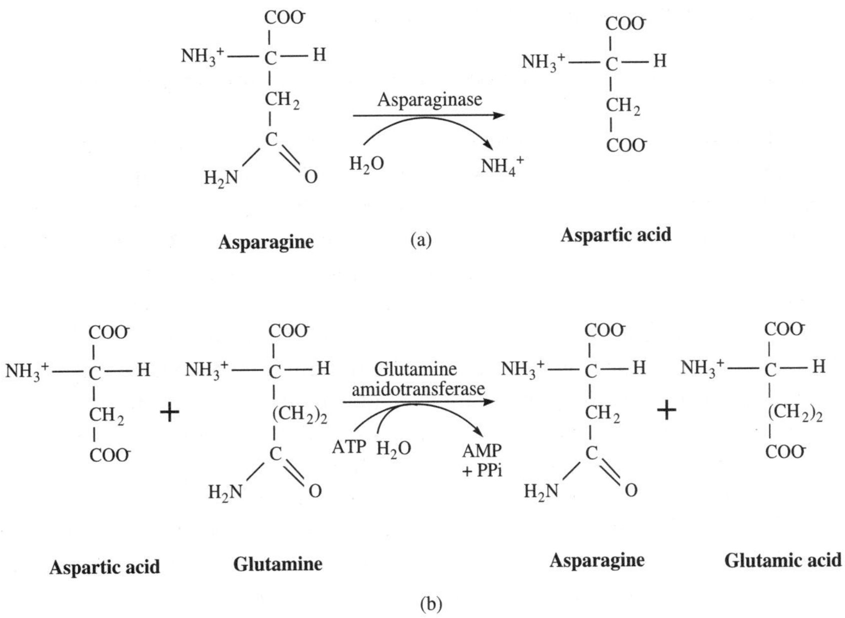

Figure 12.14 (a) Hydrolytic reaction catalysed by L-asparaginase. (b) Reaction by which asparagine is synthesized in most mammalian cells

12.5.2 DNase

Recombinant DNase preparations have been used in the treatment of cystic fibrosis since the end of 1993. This genetic disorder is common, particularly in ethnic groups of northern European extraction, where the frequency of occurrence can be as high as 1 in 2500 live births. A higher than average incidence has also been recorded in southern Europe, as well as in some Jewish and African-American populations.

A number of clinical symptoms characterize cystic fibrosis. Predominant among these is the presence of excess sodium chloride in cystic fibrosis patient sweat. Indeed, measurement of chloride levels in sweat remains the major diagnostic indicator of this disease. Another characteristic is the production of an extremely viscous, custard-like mucus in various body glands/organs that severely compromises their function. Particularly affected are:

- The lungs, in which mucus compromises respiratory function.

- The pancreas, in which the mucus blocks its ducts in 85 per cent of cystic fibrosis patients, causing pancreatic insufficiency. This is chiefly characterized by secretion of greatly reduced levels of digestive enzymes into the small intestine.

- The reproductive tract, in which changes can render males, in particular, subfertile or infertile.

- The liver, in which bile ducts can become clogged.

- The small intestine, which can become obstructed by mucus mixed with digesta.

These clinical features are dominated by those associated with the respiratory tract. The physiological changes induced in the lung of cystic fibrosis sufferers render this tissue susceptible to frequent and recurrent microbial infection, particularly by *Pseudomonas* species. The presence of microorganisms in the lung attracts immune elements, particularly phagocytic neutrophils. These begin to ingest the microorganisms, and large quantities of DNA are released from damaged microbes and neutrophils at the site of infection. High molecular mass DNA is itself extremely viscous and increases substantially the viscosity of the respiratory mucus.

The genetic basis of this disease was underlined by the finding of a putative cystic fibrosis gene in 1989. Specific mutations in this gene, which resides on human chromosome 7, were linked to the development of cystic fibrosis, and the gene is expressed largely by cells present in sweat glands, the lung, pancreas, intestine and reproductive tract.

Some 70 per cent of all cystic fibrosis patients exhibit a specific three-base-pair deletion in the gene, which results in the loss of a single amino acid (phenylalanine 508) from its final polypeptide product. Other cystic fibrosis patients display various other mutations in the same gene.

The gene product is termed cystic fibrosis transmembrane conductance regulator (CFTR), and it codes for a chloride ion channel. It may also carry out additional (as yet undetermined) functions.

Although therapeutic approaches based upon gene therapy (Chapter 14) may well one day cure cystic fibrosis, current therapeutic intervention focuses upon alleviating cystic fibrosis symptoms, particularly those relating to respiratory function. Improved patient care has increased life expectancy of cystic fibrosis patients to well into their 30s. The major elements of cystic fibrosis management include:

- chest percussion (physically pounding on the chest) in order to help dislodge respiratory tract mucus, rendering the patient better able to expel it;

- antibiotic administration, to control respiratory and other infections;

- pancreatic enzyme replacement;

- attention to nutritional status.

The relatively recent innovation in cystic fibrosis therapy is the use of DNase to reduce the viscosity of respiratory mucus. Scientists had been aware since the 1950s that free DNA concentrations in the lung of cystic fibrosis sufferers were extremely high (3–14 mg ml^{-1}). They realized that this could contribute to the mucus viscosity. Pioneering experiments, entailing inhalation of DNase-enriched extracts of bovine pancrease, were undertaken, but both product safety and efficacy were called into question. The observed toxicity was probably due to trypsin, or other contaminants, which were damaging to the underlying lung tissue. The host immune system was also probably neutralizing much of the bovine DNase.

The advent of genetic engineering and improvements in chromatographic methodology facilitated the production of highly purified recombinant human DNase (rhDNase) preparations. Initial *in vitro* studies proved encouraging: incubation of the enzyme with sputum derived from a cystic fibrosis patient resulted in a significant reduction of the sputum's viscosity. Clinical trials also showed the product to be safe and effective, and Genentech received marketing authorization for the product in December 1993, under the tradename Pulmozyme. The annual cost of treatment varies, but often falls between US$10 000 and US$15 000.

Pulmozyme is produced in an engineered CHO cell line harbouring a nucleotide sequence coding for native human DNase. Subsequent to upstream processing, the protein is purified by tangential flow filtration followed by a combination of chromatographic steps. The purified 260 amino acid glycoprotein displays a molecular mass of 37 kDa. It is formulated as an aqueous solution at a concentration of 1.0 mg ml^{-1}, with the addition of calcium chloride and sodium chloride as excipients. The solution, which contains no preservative, displays a final pH of 6.3. It is administered directly into the lungs by inhalation of an aerosol mist generated by a compressed-air-based nebulizer system.

12.5.3 Glucocerebrosidase

Glucocerebrosidase preparations are administered to relieve the symptoms of Gaucher's disease, which affects some 5000 people worldwide. This is a lysosomal storage disease affecting lipid metabolism, specifically the degradation of glucocerebrosides. Glucocerebrosides are a specific class of lipid, consisting of a molecule of sphingosine, a fatty acid and a glucose molecule (Figure 12.15). They are found in many body tissues, particularly in the brain and other neural tissue, in which they are often associated with the myelin sheath of nerves. Glucocerebrosides, however, are not abundant structural components of membranes, but are mostly formed as intermediates in the synthesis and degradation of more complex glycosphingolipids. Their degradation is undertaken by specific lysosomal enzymes, particularly in cells of the reticuloendothelial system (i.e. phagocytes, which are spread throughout the body and which function as (a) a defence against microbial infection and (b) removal of worn-out blood cells from the plasma; these phagocytes are particularly prevalent in the spleen, bone marrow and liver).

Gaucher's disease is an inborn error of metabolism characterized by lack of the enzyme glucocerebrosidase, with consequent accumulation of glucocerebrosides, particularly in tissue-based macrophages. Clinical systems include enlargement and compromised function of these macrophage-containing tissues, particularly the liver and spleen, as well as damage to long bones and, sometimes, mental retardation. Administration of exogenous glucocerebrosidase as enzyme replacement therapy has been shown to reduce the main symptoms of this disease. The enzyme is normally administered by slow i.v. infusion (over a period of 2 h) once every 2 weeks.

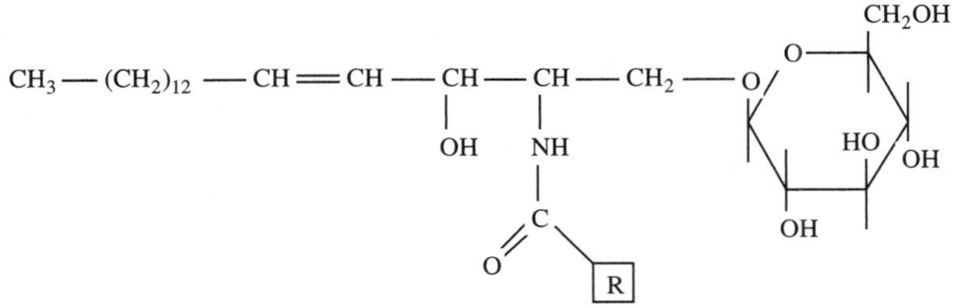

Figure 12.15 Generalized structure of a glucocerebroside

Genzyme Corporation was granted marketing authorization in 1991 for a glucocerebrosidase preparation to be used for the treatment of Gaucher's disease. This Genzyme product (tradename Ceredase) was extracted from placentas (afterbirths) obtained from maternity hospital wards. The enzyme displays a molecular mass of 65 kDa and four of its five potential glycosylation sites are glycosylated. It had been estimated that a 1-year's supply of enzyme for an average patient required extraction of 27 000 placentas, which rendered treatment extremely expensive. Genzyme then gained regulatory approval for a recombinant version of glucocerebrosidase produced in CHO cells. This product (tradename Cerezyme) has been on the market since 1994, and the total world market for glucocerebrosidase is estimated to be in the region of US$200 million.

Cerezyme is produced in a CHO cell line harbouring the cDNA coding for human β-glucocerebrosidase. The purified product is presented as a freeze-dried powder, which also contains mannitol, sodium citrate, citric acid and polysorbate 80 as excipients. It exhibits a shelf life of 2 years when stored at 2–8 °C.

An integral part of the downstream processing process entails the modification of cerezyme's oligosaccharide components. The native enzyme's sugar side-chains are complex and, for the most part, are capped with a terminal sialic acid or galactose residue. Animal studies indicate that in excess of 95 per cent of injected glucocerebrosidase is removed from the circulation by the liver via binding to hepatocyte surface lectins. As such, the intact enzyme is not available for uptake by the affected cell type, i.e. the tissue macrophages. These macrophages display high levels of surface mannose receptors. Treatment of native glucocerebrosidase with exoglycosidases, by removing terminal sugar residues, can expose mannose residues present in their sugar side-chains, resulting in their binding to and uptake by the macrophages. In this way, the 'mannose-engineered' enzyme is selectively targeted to the affected cells.

12.5.4 α-Galactosidase, urate oxidase and laronidase

Recombinant α-galactosidase, urate oxidase and laronidase represent additional biopharmaceuticals recently approved for general medical use. α-Galactosidase is approved for long-term enzyme replacement therapy in patients with Fabry disease. Like Gaucher's disease, Fabry disease is a genetic disease of lipid metabolism. Sufferers display little or no liposomal α-galactosidase-A activity. This results in the progressive accumulation of glycosphingolipids in several body cell types. Resultant clinical manifestations are complex, affecting the nervous system, vascular endothelial cells and major organs. Although the condition is rare (500–1000 patients within the EU), untreated sufferers usually die in their 40s or 50s.

Two recombinant α-galactosidases are now on the market (Fabrazyme, produced by Genzyme and Replagal, produced by TKT Europe). Fabrazyme is produced in an engineered CHO cell line, and downstream processing entails a combination of five chromatographic purification steps followed by concentration and diafiltration. Excipients added include mannitol and sodium phosphate buffering agents, and the final product is freeze-dried after filling into glass vials. Replagal is produced in a continuous human cell line and is also purified by a combination of five chromatographic purification steps, although it is marketed as a liquid solution.

Human α-galactosidase is a 100 kDa homodimeric glycoprotein. Each 398 amino acid monomer displays a molecular mass of 45.3 kDa (excluding the glycocomponent) and is glycosylated at three positions (asparagines 108, 161 and 184). After administration (usually every second week by a 40 min infusion), the enzyme is taken up by various body cell types and directed to the lysosomes. This cellular uptake and delivery process appear to be mediated by mannose-

6-phosphate residues present in the oligosaccharide side-chains of the enzyme. Mannose-6-phosphate receptors are found on the surfaces of various cell types, and also intracellularly, associated with the golgi complex, which then directs the enzyme to the lysosomes.

The enzyme urate oxidase has also found medical application for the treatment of acute hyperuricaemia (elevated plasma uric acid levels), associated with various tumours, particularly during their treatment with chemotherapy.

Uric acid is the end-product of purine metabolism in humans, other primates, birds and reptiles. It is produced in the liver by the oxidation of xanthine and hypoxanthine (Figure 12.16),

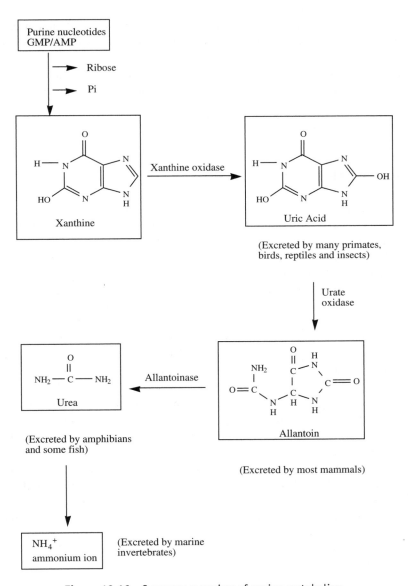

Figure 12.16   Summary overview of purine metabolism

and is excreted via the kidneys. Owing to its relatively low solubility, an increase in serum uric acid levels often triggers the formation and precipitation of uric acid crystals, typically resulting in conditions such as gout or urate stones in the urinary tract. Significantly elevated serum uric acid concentrations can also be associated with rapidly proliferating cancers or, in particular, with onset of chemotherapy. In the former instance, rapid cellular turnover results in increased rates of nucleic acid catabolism and, hence, uric acid production. In the latter case, chemotheraphy-induced cellular lysis results in the release of intracellular contents, including free purines and purine-containing nucleic acids, into the bloodstream. The increased associated purine metabolism then triggers hyperuricaemia. The elevated uric acid concentrations often trigger crystal formation in the renal tubules, and hence renal failure.

Purine metabolism in some mammals is characterized by a further oxidation of uric acid to allantoin by the enzyme urate oxidase. Allantoin is significantly more water soluble than uric acid and is also freely excreted via the renal route.

Administration of urate oxidase to humans suffering from hyperuricaemia results in the reduction of serum uric acid levels through its conversion to allantoin. Urate oxidase purified directly

---

Box 12.2

Product case study: Aldurazyme

Aldurazyme (tradename, also known as laronidase) is a recombinant version of one polymorphic variant of the human enzyme α-L-iduronidase. It was approved for general medical use in the USA in 2003 and is indicated for the treatment of patients with certain forms of the rare inherited disease MPS I. MPS I is caused by a deficiency of a lysosomal α-L-iduronidase, which normally catalyses the hydrolysis of terminal α-L-iduronic acid residues from the glycosaminoglycans dermatan sulfate and heparin sulfate. The deficiency results in accumulation of the glycosaminoglycans throughout the body, causing widespread cell and tissue dysfunction.

The 628 amino acid, 83 kDa monomeric glycosylated enzyme containing six N-linked oligosaccharide side-chains is produced in an engineered CHO cell line. After cell culture it is purified by a combination of dye affinity, metal chelate and hydrophobic interaction chromatography. The final product is formulated as a liquid concentrate containing laronidase, as well as sodium phosphate buffer, sodium chloride and polysorbate 80. It is filled in 5 ml single-use vials and is usually administered intravenously by infusion (0.58 mg per kilogram body weight) over 3–4 h, once weekly. Fortuitously, two of the enzyme's oligosaccharide side-chains terminate in mannose-6-phosphate, facilitating product cellular uptake via the mannose-6-phosphate cell surface receptor.

Clinical evaluation entailed administration to 45 MPS I patients in a randomized, placebo-controlled clinical trial. The primary efficacy outcomes assessed were forced vital capacity and distance walked in 6 min, both of which were statistically higher relative to placebo after 26 weeks of treatment. The most serious adverse reaction noted was that of a severe anaphylactic reaction in one patient. The most common adverse effects reported were respiratory tract infection, rash and injection-site reactions. The product is manufactured by BioMarin Inc. and is distributed by Genzyme Corporation.

from cultures of the fungus *Aspergillus flavus* has been used to treat this condition for a number of years. More recently, a recombinant form of the fungal enzyme (tradename Fasturtec) has gained regulatory approval in the EU. Produced in an engineered strain of *S. cerevisiae*, the enzyme is a tetramer composed of four identical polypeptide subunits. Each subunit contains 301 amino acids, displays a molecular mass of 34 kDa and is N-terminal acetylated.

Laronidase is yet an additional recombinant enzyme now approved for general medical use. The product, used to treat mucopolysaccharidosis, is overviewed in Box 12.2.

12.5.5 Superoxide dismutase

Under normal circumstances in aerobic metabolism, oxygen is reduced by four electrons, forming H_2O. Although this usually occurs uneventfully, incomplete reduction will result in the generation of oxygen radicals and other reactive species. These are: the superoxide radical O_2^-, hydrogen peroxide (H_2O_2) and the hydroxyl radical (OH^-). The superoxide and hydroxyl radicals are particularly reactive and can attack membrane components, nucleic acids and other cellular macromolecules, leading to their destruction/modification. O_2^- and OH^- radicals, for example, are believed to be amongst the most mutagenic substances generated by ionizing radiation.

Oxygen-utilizing organisms have generally evolved specific enzyme-mediated systems that serve to protect the cell from such reactive species. These enzymes include SOD and catalase or glutathione peroxidase (GSH-px), which catalyse the following reactions:

$$O_2^{\bullet-} + O_2^{\bullet-} + 2H^+ \xrightarrow{\text{SOD}} H_2O_2 + O_2$$

$$H_2O_2 + H_2O_2 \xrightarrow{\text{catalase or GSH-px}} 2H_2O + O_2$$

In general, all aerobic organisms harbour these oxygen-defence systems. At least three types of SOD have been identified: a cytosolic eukaryotic dismutase, generally a 31 kDa dimer, containing both copper and zinc; a 75 kDa mithocondrial form and a 40 kDa bacterial form, each of which contains two manganese atoms. There is also an iron-containing form found in some bacteria, blue–green algae and many plants. The metal ions play a direct role in the catalytic conversion, serving as transient acceptors/donors of electrons.

In humans, increased generation of O_2^- and/or reduced SOD levels have been implicated in a wide range of pathological conditions, including ageing, asthma, accelerated tumour growth, neurodegenerative diseases and inflammatory tissue necrosis. Furthermore, administration of SOD has been found to reduce tissue damage due to irradiation, or to other conditions that generate O_2^-. Increased SOD production in *Drosophila melanogaster* leads to increased oxygen tolerance and, interestingly, increased life span.

SOD isolated from bovine liver or erythrocytes has been used medically as an anti-inflammatory agent. Human SOD has also been expressed in several recombinant systems, and is currently being evaluated to assess its ability to prevent tissue damage induced by exposure to excessively oxygen-rich blood.

12.5.6   Debriding agents

Debridement refers to the process of cleaning a wound by removal of foreign material and dead tissue. Cleansing of the wound facilitates rapid healing and minimizes the risk of infection due to the presence of bacteria at the wound surface. The formation of a clot, followed by a scab, on a wound surface can trap bacteria, which then multiply (usually evidenced by the production of pus), slowing the healing process. Although debridement may be undertaken by physical means (e.g. cutting away dead tissue, washing/cleaning the wound), proteolytic enzymes are also often used to facilitate this process.

The value of proteases in cleansing tissue wounds has been appreciated for several hundred years. Wounds were sometimes cleansed in the past by application of protease-containing maggot saliva. Nowadays, this is usually more acceptably achieved by topical application of the enzyme to the wound surface. In some cases, the enzyme is formulated in an aqueous-based cream, and in others it is impregnated into special bandages. Trypsin, papain, collagenase and various microbial enzymes have been used in this regard.

Trypsin is a 24 kDa proteolytic enzyme synthesized by the mammalian pancreas in an inactive zymogen form: trypsinogen. Upon its release into the small intestine, it is proteolytically converted into trypsin by an enteropeptidase. Active trypsin plays a digestive role, hydrolysing peptide bonds in which the carboxyl group has been contributed by an arginine or lysine. Trypsin used medically is generally obtained by the enzymatic activation of trypsinogen, extracted from the pancreatic tissue of slaughterhouse animals.

Papain is a cysteine protease isolated from the latex of the immature fruit and leaves of the plant *Carica papaya*. It consists of a single 23.4 kDa, 212 amino acid polypeptide, and the purified enzyme exhibits broad proteolytic activity. Although it can be used as a debriding agent, it is also used for a variety of other industrial processes, including meat tenderizing and for the clarification of beverages.

Collagenase is a protease that can utilize collagen as a substrate. Although it can be produced by animal cell culture, certain microorganisms also produce this enzyme, most notably certain species of *Clostridia* (the ability of these pathogens to produce collagenase facilitates their rapid spread throughout the body). Collagenase used therapeutically is usually obtained from cell fermentation supernatants of *Clostridium histolyticum*. Such preparations are applied topically to promote debridement of wounds, skin ulcers and burns.

Chymotrypsin has also been utilized to promote debridement, as well as the reduction of soft tissue inflammation. It is also used in some opthalmic procedures, particularly in facilitating cataract extraction. It is prepared by activation of its zymogen, chymotrypsinogen, which is extracted from bovine pancreatic tissue.

Yet another proteolytic preparation used for debridement of wounds and skin ulcers consists of proteolytic enzymes derived from *B. subtilis*. The preparation displays broad proteolytic activity and is usually applied several times daily to the wound surface.

12.5.7   Digestive aids

A number of enzymes may be used as digestive aids (Table 12.9). In some instances, a single enzymatic activity is utilized, whereas other preparations contain multiple enzyme activities. These enzyme preparations may be used to supplement normal digestive activity, or to confer upon an individual a new digestive capability.

Table 12.9 Enzymes that are used as digestive aids

Enzyme	Application
α-Amylase	Aids in digestion of starch
Cellulase	Promotes partial digestion of cellulose
α-Galactosidase	Promotes degradation of flatulence factors
Lactase	Counteracts lactose intolerance
Papain Pepsin Bromelain	Enhanced degradation of dietary protein
Pancreatin	Enhanced degradation of dietary carbohydrate, fat and protein

The use of enzymes as digestive aids is only applied under specific medical circumstances. Some medical conditions (e.g. cystic fibrosis) can result in compromised digestive function due to insufficient production/secretion of endogenous digestive enzymes. Digestive enzyme preparations are often formulated in powder (particularly tablet) form, and are recommended to be taken orally immediately prior, or during, meals. As the product never enters the blood stream, the product purity need not be as stringent as enzymes (or other proteins) administered intravenously. Most digestive enzymes are, at best, semi-pure preparations.

In some instances there is a possibility that the efficacy of these preparations may be compromised by conditions associated with the digestive tract. Most function at pH values approaching neutrality. They would thus display activity possibly in saliva and particularly in the small intestine. However, the acidic conditions of the stomach (where the pH can be below 1.5) may denature some of these enzymes. Furthermore, the ingested enzymes would also be exposed to endogenous proteolytic activities associated with the stomach and small intestine. Some of these difficulties, however, may be at least partially overcome by formulating the product as a tablet coated with an acid-resistant film to protect the enzyme as it passes through the stomach.

Pancreatin is a pancreatic extract usually obtained from the pancrease of slaughterhouse animals. It contains a mixture of enzymes, principally amylase, protease and lipase, and, thus, exhibits a broad digestive capability. It is administered orally mainly for the treatment of pancreatic insufficiency caused by cystic fibrosis or pancreatitis. As it is sensitive to stomach acid, it must be administered in high doses or, more usually, as enteric-coated granules or capsules that may be taken directly or sprinkled upon the food prior to its ingestion. Individual digestive activities, such as papain, pepsin or bromelains (proteases), or α-amylase are sometimes used in place of pancreatin.

Cellulase is not produced in the human digestive system. Cellulolytic enzyme preparations obtained from *A. niger* or other fungal sources are available, and it is thought that their ingestion may improve overall digestion, particularly in relation to high-fibre diets.

α-Galactosides are oligosaccharides present in plant matter, particularly in beans. They are not normally degraded in the human digestive tract due to the absence of an appropriate endogenous digestive enzyme (i.e. an α-galactosidase). However, upon their entry into the large intestine, these oligosaccharides are degraded by microbial α1–6 galactosidases, thus stimulating microbial fermentation. The end-products of fermentation include volatile fatty acids, carbon dioxide, methane and hydrogen, which lead to flatulence. This can be avoided by minimizing dietary intake of food containing α-galactosides. Another approach entails the simultaneous ingestion of tablets containing α-galactosidase activity. If these 'flatulence factors' are degraded before or upon reaching the small intestine, then the monosaccharides released will be absorbed and, hence, will be subsequently unavailable to promote undesirable microbial fermentations in the large intestine.

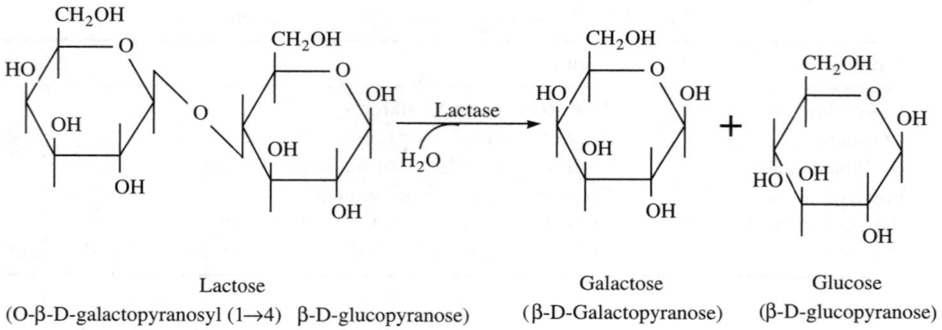

Lactose	Galactose	Glucose
(O-β-D-galactopyranosyl (1→4) β-D-glucopyranose)	(β-D-Galactopyranose)	(β-D-glucopyranose)

Figure 12.17   Hydrolysis of lactose by lactase (β-galactosidase)

Lactose, the major disaccharide present in milk, is composed of a molecule of glucose linked via a glycosidic bond to a molecule of galactose. The digestive tract of young (suckling) animals generally produces significant quantities of the enzyme β-galactosidase (lactase), which catalyses the hydrolysis of lactose, releasing the constituent monosaccharides (Figure 12.17). This is a pre-requisite to their subsequent absorption.

The digestive tracts of many adult human populations, however, produce little or no lactase, rendering these individuals lactose intolerant. This is particularly common in Asia, Africa, Latin America and the Middle East. It severely curtails the ability of these people to drink milk without feeling ill. In the absence of sufficient endogenous digestive lactase activity, milk lactose is not absorbed and, thus, serves as a carbon source for intestinal microorganisms. The resultant production of lactic acid, CO_2 and other gases causes gastrointestinal irritation and diarrhoea. A number of approaches have been adopted in an effort to circumvent this problem. Most involve the application of microbial lactase enzymes. In some instances, the enzyme has been immobilized in a column format, such that passage of milk through the column results in lactose hydrolysis. Free lactase has also been added to milk immediately prior to its bottling, so that lactose hydrolysis can slowly occur prior to its eventual consumption (i.e. during transport and storage).

Fungal and other microbial lactase preparations have also been formulated into tablet form, or sold in powder form. These can be ingested immediately prior to the consumption of milk or lactose-containing milk products, or can be sprinkled over the food before eating it. Such lactose preparations are available in supermarkets in many parts of the world.

Further reading

Books

Becker, R. 2000. *Thrombolytic and Antithrombolytic Therapy*. Oxford University Press.
Goodnight, S. 2001. *Disorders of Hemostasis and Thrombosis*. McGraw Hill.
Kirchmaier, C. 1991. *New Aspects on Hirudin*. Karger.
Lauwers, A. and Scharpe, S. (eds). 1997. *Pharmaceutical Enzymes*. Marcel Dekker.
McGrath, B. and Walsh, G (eds). 2006. *Directory of Therapeutic Enzymes*. Taylor and Francis.
Poller, H. 1996. *Oral Anticoagulants*. Arnold.

Articles

Plasma proteins, including coagulation factors

Bowen, D. 2002. Haemophilia a and haemophilia b: molecular insights. *Journal of Clinical Pathology – Molecular Pathology* **55**(1), 1–18.

Chuang, V.T., Kragh-Hansen, U., and Otagiri, M. 2002. Pharmaceutical strategies utilizing recombinant human serum albumin. *Pharmaceutical Research* **19**(5), 569–577.

Cervenakova, L., Brown, P., Hammond, D.J., Lee, C.A., and Saenko, E.L. 2002. Factor VIII and transmissible spongiform encephalopathy: the case for safety. *Haemophilia* **8**(2), 63–75.

DeLoughrey, T. 2006. Management of bleeding emergencies: when to use recombinant activated factor VII. *Expert Opinion on Pharmacotherapy* **7**(1), 25–34.

Farrugia, A. 2004. Safety and supply of haemophilia products: worldwide perspectives. *Haemophilia* **10**(4), 327–333.

Federici, A. and Mannucci, P. 2002. Advances in the genetics and treatment of von Willebrand disease. *Current Opinion in Pediatrics* **14**(1), 23–33.

Giangrande, P. 2005. Haemophila B: Christmas disease. *Expert Opinion on Pharmacotherapy* **6**(9), 1517–1524.

Graw, J., Brackmann, H.H., Oldenburg, J., Schneppenheim, R., Spannagl, M., and Schwaab, R. 2005. Haemophilia A: from mutation analysis to new therapies. *Nature Reviews Genetics* **6**(6), 488–501.

Kingdon, H. and Lundblad, R. 2002. An adventure in biotechnology: the development of haemophilia A therapeutics – from whole blood transfusion to recombinant DNA technology to gene therapy. *Biotechnology and Applied Biochemistry* **35**, 141–148.

Klinge, J., Ananyeva, N.M., Hauser, C.A., and Saenko, E.L. 2002. Hemophilia A – from basic science to clinical practice. *Seminiars in Thrombosis and Hemostasis* **28**(3), 309–321.

Legaz, M.E., Schmer, G., Counts, R.B., and Davie, E.W. 1973. Isolation and characterization of human factor VIII (antihaemophilic factor). *Journal of Biological Chemistry* **248**, 3946–3955.

Nicholson, J.P., Wolmarans, M.R., and Park, G.R. 2000. The role of albumin in critical illness. *British Journal of Anaesthesia* **85**(4), 599–610.

Ogden, J. 1992. Recombinant haemoglobin in the development of red blood cell substitutes. *Trends in Biotechnology* **10**, 91–95.

Winslow, R. 2000. Blood substitutes: refocusing an elusive goal. *British Journal of Haematology* **111**(2), 387–396.

Anticoagulants and related substances

Bates, S. and Weitz, J. 2005. New anticoagulants: beyond heparin, low molecular weight heparin and warfarin. *British Journal of Pharmacology* **144**(8), 1017–1028.

Dahlback, A. and Villotreix, B. 2005. The anticoagulant protein C pathway. *FEBS Letters* **579**(15), 3310–3316.

De Kort, M., Buijsman, R.C., and van Boeckel, C.A. 2005. Synthetic heparin derivatives as new anticoagulant drugs. *Drug Discovery Today* **10**(11), 769–779.

Dodt, J. 1995. Anti-coagulatory substances of bloodsucking animals: from hirudin to hirudin mimetics. *Angewandte Chemie International Edition in English* **34**, 867–880.

Eldor, A., Orevi, M., and Rigbi, M. 1996. The role of the leech in medical therapeutics. *Blood Reviews* **10**(4), 201–209.

Fischer, K. 2004. The role of recombinant hirudins in the management of thrombotic disorders. *Biodrugs* **18**(4), 235–268.

Linkins, L. and Weitz, J. 2005. New anticoagulant therapy. *Annual Review of Medicine* **56**, 63–77.

O'Brien, L.A., Gupta, A., and Grinnell, B.W. 2006. Activated protein C and sepsis. *Frontiers in Bioscience* **11**, 676–698.

Pineo, G. and Hull, R. 1997. Low molecular weight heparin-prophylaxis and treatment of venous thromboembolism. *Annual Review of Medicine* **48**, 79–91.

Salzet, M. 2002. Leech thrombin inhibitors. *Current Pharmaceutical Design* **8**(7), 493–503.

Sohn, J.H., Kang, H.A., Rao, K.J., Kim, C.H., Choi, E.S., Chung, B.H., and Rhee, S.K. 2001. Current status of the anticoagulant hirudin: its biotechnological production and clinical practice. *Applied Microbiology and Biotechnology* **57**(5–6), 606–613.

Walker, C. and Royston, D. 2002. Thrombin generation and its inhibition: a review of the scientific basis and mechanism of action of anticoagulant therapies. *British Journal of anaesthesia* **88**(6), 848–863.

Thrombolytics

Al-Buhairi, A. and Jan, M. 2002. Recombinant tissue plasminogen activator for acute ischemic stroke. *Saudi Medical Journal* **23**(1), 13–19.

Banerjee, A., Chisti, Y., and Banerjee, U.C. 2004. Streptokinase – a clinically useful thrombolytic agent. *Biotechnology Advances* **22**(4), 287–307.

Bansal, V. and Roychoudhury, P. 2006. Production and purification of urokinase: a comprehensive review. *Protein Expression and Purification* **45**(1), 1–14.

Blasi, F. 1999. The urokinase receptor. A cell surface, regulated chemokine. *APMIS* **107**(1), 96–101.

Castillo, P.A., Palmer, C.S., Halpern, M.T., Hatziandreu, E.J., and Gersh, B.J. 1997. Cost-effectiveness of thrombolytic therapy for acute myocardial infarction. *Annals of Pharmacotherapy* **31**(5), 596–603.

Collen, D. 1998. Staphylokinase: a potent, uniquely fibrin-selective thrombolytic agent. *Nature Medicine* **4**(3), 279–284.

Gillis, J.C., Wagstaff, A.J., and Goa, K.L. 1995. Alteplase. A reappraisal of its pharmacological properties and therapeutic use in acute myocardial infarction. *Drugs* **50**(1), 102–136.

Marder, V. and Stewart, D. 2002. Towards safer thrombolytic therapy. *Seminars in Haematology* **39**(3), 206–216.

Peng, Y., Yang, X., and Zhang, Y. 2005. Microbial fibrinolytic enzymes: an overview of source, production, properties and thrombolytic activity *in vivo*. *Applied Microbiology and Biotechnology* **69**(2), 126–132.

Perler, B. 2005. Thrombolytic therapeutics: the current state of affairs. *Journal of Endovascular Therapy* **12**(2), 224–232.

Preissner, K.T., Kanse, S.M., and May, A.E. 2000. Urokinase receptor: a molecular organizer in cellular communication. *Current Opinion in Cell Biology* **12**(5), 621–628.

Rabasseda, X. 2001. Tenecteplase (TNK tissue plasminogen activator): a new fibrinolytic for the acute treatment of myocardial infarction. *Drugs of Today* **37**(11), 749–760.

Verstraete, M. 2000. Third generation thrombolytic drugs. *American Journal of Medicine* **109**(1), 52–58.

Enzyme therapeutics

Avramis, V. and Panosyan, E. 2005. Pharmacokinetic/pharmacodynamic relationships of asparaginase formulations – the past, the present and recommendations for the future. *Clinical Pharmacokinetics* 44(4), 367–393.

Barranger, J.A., Tomich, J., Weiler, S., Sakallah, S., Sansieri, C., Mifflin, T., Bahnson, A., Wei, F.S., Wei, J.F., Vallor, M., Nimgaonkar, M., Ball, E., Mohney, T., Dunigan., J., Ohashi, T., Bansal, V., Mannion Henderson, J., Liu, C.M., and Rice, E. 1995. Molecular biology of glucocerebrosidase and the treatment of Gaucher disease. *Cytokines and Molecular Therapy* **1**(3), 149–163.

Chakrabarti, R. and Schuster, S. 1997. L-Asparaginase – perspectives on the mechanisms of action and resistance. *International Journal of Pediatric Haematology/Oncology* **4**(6), 597–611.

Chapple, L. 1997. Reactive oxygen species and antioxidants in inflammatory diseases. *Journal of Clinical Periodontology* **24**(5), 287–296.

Conway, S. and Watson, A. 1997. Nebulized bronchodilators, corticosteroids and rhDNase in adult patients with cystic fibrosis. *Thorax* **52**(2), S64–S68.

Grabowski, G. 2005. Recent clinical progress in Gaucher disease. *Current Opinion in Pediatrics* **17**(4), 519–524.

James, E. 1994. Superoxide dismutase. *Parasitology Today* **10**(12), 482–484.

Ronghe, M., Burke, G.A., Lowis, S.P., and Estlin, E.J. 2001. Remission induction therapy for childhood acute lymphoblastic leukaemia: clinical and cellular pharmacology of vincristine, corticosteroids, L-asparaginase and anthracyclines. *Cancer Treatment Reviews* **27**(6), 327–337.

Zhao, H. and Grabowski. G. 2002. Gaucher disease: perspectives on a prototype lysosomal disease. *Cellular and Molecular Life Sciences* **59**(4), 694–707.

13

Antibodies, vaccines and adjuvants

13.1 Introduction

Few substances have had a greater positive impact upon human healthcare management than antibodies, vaccines and adjuvants. For most of this century, these immunological agents have enjoyed widespread medical application, predominantly for the treatment/prevention of infectious diseases. As a group, they are often referred to as 'biologics' (Chapter 1).

Polyclonal antibody preparations have been used to induce passive immunity against a range of foreign (harmful) agents, and vaccines are used efficiently, and safely, to promote active immunization. Adjuvants are usually co-administered with the vaccine preparation, in order to enhance the immune response against the vaccine.

The development of modern biotechnological methodology has had a significant impact upon the therapeutic application of immunological agents, as discussed later. Monoclonal/engineered antibodies find a range of therapeutic uses, and many of the newer vaccine preparations are now produced by recombinant DNA technology. This chapter focuses predominantly upon those modern biotech products. Many currently used products, however, are still produced by more traditional means, and these, too, are also considered, in summary at least.

13.2 Traditional polyclonal antibody preparations

Polyclonal antibody preparations have been used for several decades to induce passive immunization against infectious diseases and other harmful agents, particularly toxins. The antibody preparations are usually administered by direct i.v. injection. While this affords immediate immunological protection, its effect is transitory, usually persisting for only 2–3 weeks (i.e. until the antibodies are excreted). Passive immunization can be used prophylactically (i.e. to prevent a future medical episode) or therapeutically (i.e. to treat a medical condition that is already established). An example of the former would be prior administration of a specific anti-snake toxin antibody preparation to an individual before they travel to a world region in which these snakes are commonly found. An example of the latter would be administration of the anti-venom antibody immediately after the individual has experienced a snake bite.

Pharmaceutical biotechnology: concepts and applications   Gary Walsh
© 2007 John Wiley & Sons, Ltd   ISBN 978 0 470 01244 4 (HB)   978 0 470 01245 1 (PB)

Antibody preparations used to induce passive immunity may be obtained from either animal or human sources. Preparations of animal origin are generally termed 'antisera', and those sourced from humans are called 'immunoglobulin preparations'. In both cases, the predominant antibody type present is IgG.

Antisera are generally produced by immunizing healthy animals (e.g. horses) with appropriate antigen. Small samples of blood are subsequently withdrawn from the animal on a regular basis and quantitatively analysed for the presence of the desired antibodies (often using ELISA-based immunoassays). This facilitates harvesting of the blood at the most appropriate time points. Large animals, such as horses, can withstand withdrawal of 1 or 2 l of blood every 10–14 days, and antibody levels are usually maintained by administration of repeat antigen booster injections.

The blood is collected using an aseptic technique into sterile containers. It can then be allowed to clot with subsequent recovery of the antibody-containing antisera by centrifugation. Alternatively, the blood may be collected in the presence of heparin, or another suitable anticoagulant, with subsequent removal of the suspended cellular elements, again by centrifugation. In this case, the resultant antibody-containing solution is termed 'plasma'.

The antibody fraction is then purified from the serum (or plasma). Traditionally, this entailed precipitation steps, usually using ethanol and/or ammonium sulfate as precipitants. The precipitated antibody preparations, however, are only partially purified and modern preparations are generally subjected to additional high-resolution chromatographic fractionation (Figure 13.1). Ion-exchange chromatography is often employed, as is protein A affinity chromatography. (IgG from many species binds fairly selectively to protein A.)

Following high resolution purification, the antibody titre is determined, usually using an appropriate bioassay, or an immunoassay. Stabilizing agents, such as NaCl (0.9 per cent w/v) or glycine (2–3 per cent w/v) are often added, as are antimicrobial preservatives. Addition of preservative is particularly important if the product is subsequently filled into multi-dose containers. Phenol, at concentrations less than 0.25 per cent, is often used. After adjustment of the potency to fall within specification, the product is sterile filtered and aseptically filled into sterile containers. These are sealed immediately if the product is to be marketed in liquid form. Such antibody solutions are often filled under an oxygen-free nitrogen atmosphere in order to prevent oxidative degradation during subsequent storage. Such a product, if stored between 2 and 8 °C, should exhibit a shelf life of up to 5 years.

Although specific antisera have proven invaluable in the treatment of a variety of medical conditions (Table 13.1), they can also induce unwanted side effects. Particularly noteworthy is their ability to induce hypersensitivity reactions; some such sensitivity reactions (e.g. 'serum sickness') are often not acute, whereas others (e.g. anaphylaxis) can be life threatening. Because of such risks, antibody preparations derived from human donors (i.e. immunoglobulins) are usually preferred as passive immunizing agents.

Immunoglobulins are purified from the serum (or plasma) of human donors by methods similar to those used to purify animal-derived antibodies. In most instances, the immunoglobulin preparations are enriched in antibodies capable of binding to a specific antigen (usually an infectious microorganism/virus). These may be purified from donated blood of individuals who have recently:

- been immunized against the antigen of interest;

- recovered from an infection caused by the antigen of interest.

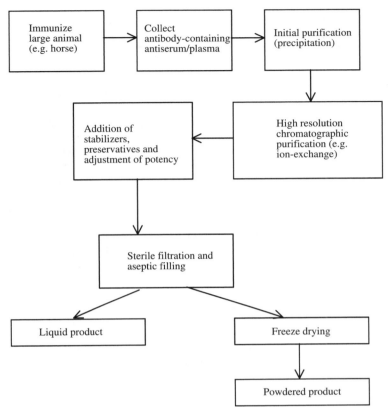

Figure 13.1 Overview of the production of antisera for therapeutic use to induce passive immunization. Refer to text for specific details

Although hypersensitivity reactions can occur upon administration of immunoglobulin preparations, the incidence of such events is far less frequent than is the case upon administration of antibody preparations of animal origin. As with all blood-derived products, the serum from which the immunoglobulins are due to be purified is first assayed for the presence of infectious agents before its use.

The major polyclonal antibody preparations used therapeutically are listed in Table 13.1. These may generally be categorized into one of several groups upon the basis of their target specificities. These groups include antibodies raised against:

- specific microbial or viral pathogens;

- microbial toxins;

- snake/spider venoms (anti-venins).

Table 13.1 Polyclonal antibody preparations of human or animal origin used to induce passive immunity against specific biological agents

Antibody	Source	Specificity
Anti-D immunoglobulin	Human	Specificity against rhesus D antigen.
Botulism antitoxin	Horse	Specificity against toxins of type A, B or E *Clostridium botulinum*
Cytomegalovirus immunoglobulin	Human	Antibodies exhibiting specificity for cytomegalovirus
Diphtheria antitoxin	Horse	Antibodies raised against diphtheria toxoid
Diphtheria immunoglobulin	Human	Antibodies exhibiting specificity for diphtheria toxoid
Endotoxin antibodies	Horse	Antibodies raised against gram negative bacterial LPS
Gas gangrene antitoxins	Horse	Antibodies raised against a-toxin of *Clostridum novyi*, *Clostridum perfringens* and *Clostridum septicum*
H. influenzae immunoglobulins	Human	Antibodies raised against surface capsular polysaccharide of *H. influenzae*
Hepatitis A immunoglobulin	Human	Specificity against hepatitis A surface antigen
Hepatitis B immunoglobulin	Human	Specificity against hepatitis B surface antigen
Leptospira antisera	Animal	Antibodies raised against *Leptospira icterohaemorrhagiae* (used to treat Weil's disease)
Measles immunoglobulin	Human	Specificity against measles virus
Normal immunoglobulin	Human	Specificities against variety of infectious and other biological agents prevalent in general population
Rabies immunoglobulin	Human	Specificity against rabies virus
Scorpion venom antisera	Horse	Specificity against venom of one or more species of scorpion
Snake venom antisera	Horse	Antibodies raised against venom of various poisonous snakes
Spider antivenins	Horse	Antibodies raised against venom of various spiders
Tetanus antitoxin	Horse	Specificity against toxin of *Clostridium tetani*
Tetanus immunoglobulin	Human	Specificity against toxin of *C. tetani*
Tick-borne encephalitis immunoglobulin	Human	Antibodies against tick-borne encephalitis virus
Varicella-zoster immunoglobulin	Human	Specificity for causative agent of chicken pox

13.3 Monoclonal antibodies

In the last 20 years or so, antibody-based therapeutics have mainly focused upon the medical application of monoclonal antibodies. Monoclonal antibody technology was first developed in the mid 1970s, when Kohler and Milstein successfully fused immortal myeloma cells with antibody-producing B-lymphocytes. A proportion of the resultant hybrids were found to be stable, cancerous, antibody-producing cells. These 'hybridoma' cells represented an inexhaustible source of monospecific (monoclonal) antibody. Hybridoma technology facilitates the relatively straightforward production of monospecific antibodies against virtually any desired antigen.

The production process (Box 13.1) entails initial immunization of a mouse with the antigen of interest. The mouse is subsequently sacrificed and its spleen removed. (The spleen is an organ enriched in B-lymphocytes. Because of the immunization process, a significant proportion of these lymphocytes are likely capable of producing antibodies recognizing specific epitopes on the antigen.)

Box 13.1

The basis of monoclonal antibody production by hybridoma technology

When antigen enters the body, it stimulates an immune response. A major element of this response entails activation of selected B-lymphocytes to produce antibodies capable of binding the antigen (the humoral immune response). The binding of antibody can reduce/inactivate the biological activity of the antigen (especially if it is a toxin, for example), and also marks the antigen for destruction by other elements of the immune system. Any given antibody will bind

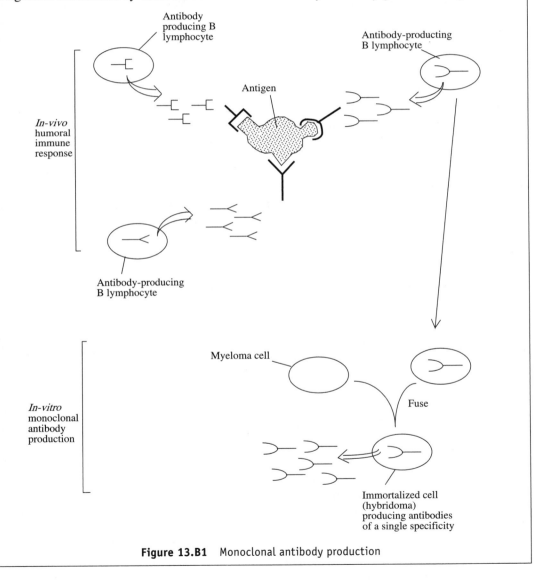

Figure 13.B1 Monoclonal antibody production

only to a specific region of the antigen called an epitope (in the example above, the antigen contains just three epitopes). Most antigens encountered naturally (e.g. proteins, viruses, bacteria, etc.) contain hundreds, if not thousands, of different epitopes. A typical epitope region on a protein surface would comprise five to seven amino acid residues. Each specific antibody, which recognizes a specific epitope, is produced by a specific B-lymphocyte. If one single antibody-producing cell could be isolated and cultured *in vitro*, then it would be a source of monoclonal (monospecific) antibody. However, B-lymphocytes die after a short time when cultured *in vitro* and, hence, are an impractical source of long-term antibody production (Figure 13.B1).

Monoclonal antibody technology entails isolation of such B-lymphocytes, with subsequent fusion of these cells with transformed (myeloma) cells. Many of the resultant hybrid cells retain immortal characteristics, while producing large quantities of the monospecific antibody. These hybridoma cells can be cultured long term to effectively produce an inexhaustible supply of the monoclonal antibody of choice.

Spleen-derived B-lymphocytes are then incubated with mouse myeloma cells in the presence of propylene glycol. This promotes fusion of the cells. The resultant immortalized antibody-producing hybridomas are subsequently selected from unfused cells by culture in a specific selection medium. Individual hybridomas can be separated from each other by simple dilution and subsequently grown in culture, producing a clone. Individual clones can be screened to identify which ones produce murine (monoclonal) antibody that binds the antigen of interest. Appropriate clones are then selected and grown on a larger scale in order to produce biotechnologically useful quantities of antibody. Whereas many of the monoclonal antibodies approved in the 1980s and early 1990s were produced by such means, the majority of more recent approvals are engineered products produced by recombinant means, as described later.

13.3.1 Antibody screening: phage display technology

Phage display technology provides an extremely powerful modern way to generate a library of (protein) ligands and, subsequently, screen these ligands for their ability to bind a selected target molecule. The technique, as the name implies, employs filamentous phage (bacteriophage) that replicate in *E. coli*.

The principle of phage display is presented in Figure 13.2. A library of genes (one of which codes for the protein of interest) is first generated/obtained. These genes are inserted (batch cloned) into a phage library fused to a gene encoding one of the phage coat proteins (pIII, pIV or pVIII). The phage are then incubated with *E. coli*, which facilitates phage replication. Expression of the fusion gene product during replication and the subsequent incorporation of the fusion product into the mature phage coat results in the gene product being 'presented' on the phage surface. The entire phage library can then be screened in order to identify the one(s) coding for the protein of interest. This is usually achieved by affinity selection (biopanning). Biopanning entails passing the library over immobilized target molecules, usually in immobilized column format. Only the phage expressing the protein of desired specificity should be retained in the immobilized column. The bound phage can subsequently be eluted, e.g. by reducing the pH of the elution buffer or inclusion of a competitive ligand (usually free target molecules) in the buffer. Eluted phage can then be repassed over the affinity column in order to isolate those binding the immobilized

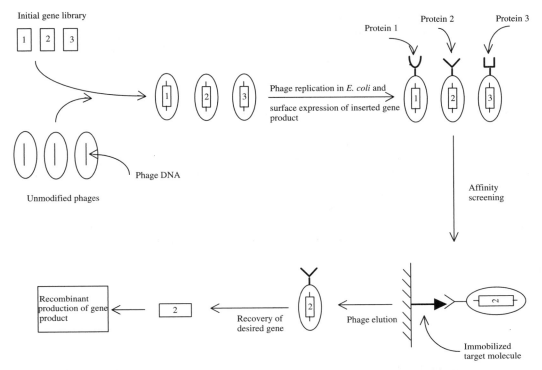

Figure 13.2 Phage display technology and the screening of a library for a protein capable of binding to a desired target molecule. In this simplified example an initial library of three genes is inserted into three phages. One (gene 2) codes for the desired protein. In reality, libraries typically consist of hundreds of thousands/millions of different genes. Refer to text for further details

ligand with the highest specificity/affinity. Once this is achieved, the gene coding for the protein of interest can be excised from the phage genome by standard techniques of molecular biology. It can then be incorporated into an appropriate microbial/animal cell/transgenic expression system (Chapter 3), facilitating large-scale production of the gene product. Variations of the phage display approach have been developed, some of which utilize engineered phage (phagemids), whereas others achieve library expression not on the surface of phage but on the surface of bacteria.

Amongst the first and still most prominent application of phage display technology is the production and screening of antibody libraries in order to isolate/identify an antibody capable of binding to a desired target molecule. As such, this technology has now come to the fore in identifying antibodies suited to clinical application.

Genes/cDNA libraries coding for antibodies/antigen-binding antibody fragments have been obtained from human and animal (e.g. mice, rabbit and chicken) sources. Two types of library can be generated. 'Immune libraries' are obtained by cloning antibody/antibody fragment coding sequences derived from B-lymphocytes (usually from spleens) of donors previously immunized with the target antigen. A high number of hits (positive clones) should be obtained from such libraries.

Non-immune libraries are produced in a similar fashion, but using B-lymphocytes from non-immunized donors as a source of antibody genes. This approach becomes necessary if initial immunization with the antigen of interest is not possible (e.g. due to ethical considerations). Although such libraries will generate a lower number of positive clones, they can be screened against multiple antigens. Such native non-immune libraries can also be the starting point for the generation of so-called synthetic immune libraries. Their generation entails initial *in vitro* engineering of the non-immune library of antibody genes in order to increase still further the level of antibody diversity generated.

13.3.2 Therapeutic application of monoclonal antibodies

The unrivalled specificity of monoclonal antibodies, coupled to their relatively straightforward production and their continuity of supply, renders them attractive biochemical tools. Therapeutically, they represent by far the single largest category of biopharmaceutical substances under investigation. Several hundred such preparations are currently undergoing preclinical and clinical trials. Throughout the 1980s the focus of attention rested upon their use either as *in vivo* imaging (i.e. diagnostic) agents or as direct therapeutic agents. Initial studies centred mainly around cancer, but monoclonal antibody preparations are now used in a variety of other medical circumstances:

- induction of passive immunity;

- diagnostic imagining;

- therapeutically (e.g. treatment of cancer, transplantation and cardiovascular disease).

All *in vivo* diagnostic/therapeutic applications are dependent upon the selective interaction of a monoclonal antibody with a specific target cell type in the body (e.g. a cancer cell). Therefore, a prerequisite to application of monoclonal antibody-based products in this way is the identification of a cell surface antigen unique to the target cell type (Figure 13.3). Once identified and characterized, monoclonal antibodies may be raised against that unique surface antigen (USA or USAg).

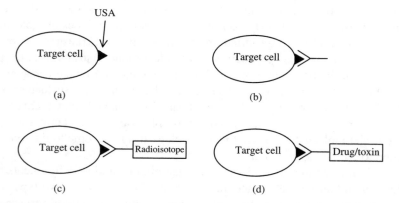

Figure 13.3 Underlying principle/approaches taken during the development and use of antibody-mediated target cell detection/destruction. A prerequisite for adoption of this strategy is the identification and characterization of a surface antigen unique to the target cell type ('unique surface antigen', USA). Antibodies raised against the USA should selectively interact with the target cell (b). In some instances the antibody is chemically coupled to a radioactive tag (c), a drug or a toxin (d)

The specificity of antibody-antigen binding should ensure that, once injected into the body, the antibody will selectively congregate on the surface of only the target cells. Depending upon the specific intended therapeutic/diagnostic application, antibodies employed may have nothing attached to them or may be conjugated to a radioisotope, drug or toxin (Figure 13.3). With the latter approaches, the antibody is used as a 'magic bullet', delivering a radioactive/drug load to specific cells in the body. All of these various strategies have been adopted in practice, and specific examples will be provided in subsequent sections of this chapter.

In the context of antibody-mediated cell targeting, it is also important to appreciate that binding of an antibody to a USA can, by itself, trigger a number of responses. In some instances, the antibody–antigen (Ab–USA) complex is quickly internalized. In other instances, the Ab–USA complex is shed from the cell surface, whereas in yet other cases binding induces neither response. The specific response triggered is generally only determined by direct experimentation. The induction of Ab–USA shedding renders an anti-type-specific antigen (TSA) antibody clinically useless.

By 2006, a total of 29 such antibody-based products had gained marketing approval in some world regions at least (Table 13.2). Over half of these aim to detect/treat various cancers, and cancer represents the single most significant indication for antibody-based products currently in clinical trials. The application of antibodies in the context of cancer is overviewed in the next section. A minority of the products approved/in trials are intact antibodies produced by classical hybridoma technology. The majority are engineered antibodies ('chimaeric' or 'humanized') or antibody fragments. The generation/rationale for use of such engineered products is also discussed subsequently in this chapter.

13.3.3 Tumour immunology

The transformation of a cell to the cancerous state is normally associated with increased surface expression of antigens recognized as foreign by the host immune system. These surface antigens, often termed tumour antigens or tumour surface antigens, are either not expressed at all by the untransformed cell or are expressed at such low levels that they fail to induce immunological tolerance.

The presence of tumour-specific antigens implies that the immune system should be capable of recognizing and destroying transformed cells. This concept, known as immunosurveillance, probably does function to some extent in the body. The immune system does respond to the presence of some tumours, causing their partial or complete regression. The major anti-tumour immune elements include:

- T-lymphocytes, which are capable of recognizing and lysing malignant cells.

- NK cells, which, like some T-cells, can induce lysis of tumour cells. The tumouricidal activity of NK cells is potentiated by various cytokines (e.g. IL-2 and TNF).

- Macrophages can destroy tumour cells, largely by releasing damaging lysosomal enzymes and reactive oxygen metabolites at the tumour cell surface. Macrophages also produce TNF, which can kill tumour cells by (a) binding to high-affinity TNF cell surface receptors (which is directly toxic to the cells) and (b) promoting synthesis of additional cytokines (which can, indirectly, lead to tumour destruction via activation of other elements of immunity).

Table 13.2 Monoclonal/engineered antibodies thus far approved for medical use. The sponsoring companies and therapeutic indications are also listed

Product	Company	Indication
CEA-Scan (Arcitumomab, murine Mab fragment (Fab), directed against human CEA)	Immunomedics	Detection of recurrent/metastatic colorectal cancer
MyoScint (Imiciromab-Pentetate, murine Mab fragment directed against human cardiac myosin)	Centocor	Myocardial infarction imaging agent
OncoScint CR/OV (Satumomab Pendetide, murine Mab directed against TAG-72, a high molecular weight tumour associated glycoprotein)	Cytogen	Detection/staging/follow up of colorectal and ovarian cancers
Orthoclone OKT3 (Muromomab CD3, murine Mab directed against the T-lymphocyte surface antigen CD3)	Ortho Biotech	Reversal of acute kidney transplant rejection
ProstaScint (Capromab Pentetate murine Mab directed against the tumour surface antigen PSMA)	Cytogen	Detection/staging/follow up of prostate adenocarcinoma
ReoPro (Abciximab, Fab fragments derived from a chimaeric Mab, directed against the platelet surface receptor $GPII_b/III_a$)	Centocor	Prevention of blood clots
Rituxan (Rituximab chimaeric Mab directed against CD20 antigen found on the surface of B-lymphocytes)	Genentech/IDEC Pharmaceuticals	Non-Hodgkin's lymphoma
Verluma (Nofetumomab murine Mab fragments (Fab) directed against carcinoma-associated antigen)	Boehringer Ingelheim/ NeoRx	Detection of small cell lung cancer
Zenapax (Daclizumab, humanized Mab directed against the α chain of the IL-2 receptor)	Hoffman La Roche	Prevention of acute kidney transplant rejection
Simulect (Basiliximab, chimaeric Mab directed against the α chain of the IL-2 receptor)	Novartis	Prophylaxis of acute organ rejection in allogeneic renal transplantation
Remicade (Infliximab, chimaeric Mab directed against TNF-α)	Centocor	Treatment of Crohn's disease
Synagis (Palivizumab, humanized Mab directed against an epitope on the surface of respiratory syncytial virus)	MedImmune (USA) Abbott (EU)	Prophylaxis of lower respiratory tract disease caused by respiratory syncytial virus in paediatric patients
Herceptin (Trastuzumab, humanized antibody directed against HER2, i.e. human epidermal growth factor receptor 2)	Genentech (USA) Roche Registration (EU)	Treatment of metastatic breast cancer if tumour overexpresses HER2 protein
Indimacis 125 (Igovomab, Murine Mab fragment (Fab$_2$) directed against the tumour-associated antigen CA 125)	CIS Bio	Diagnosis of ovarian adenocarcinoma
Tecnemab KI (murine Mab fragments (Fab/ Fab$_2$ mix) directed against HMW-MAA, i.e. high molecular weight melanoma-associated antigen)	Sorin	Diagnosis of cutaneous melanoma lesions

Table 13.2   (*Continued*)

Product	Company	Indication
LeukoScan (Sulesomab, murine Mab fragment (Fab) directed against NCA 90, a surface granulocyte non-specific cross-reacting antigen)	Immunomedics	Diagnostic imaging for infection/inflammation in bone of patients with osteomyelitis
Humaspect (Votumumab, human Mab directed against cytokeratin tumour-associated antigen)	Organon Teknika	Detection of carcinoma of the colon or rectum
Mabthera (Rituximab, chimaeric Mab directed against CD 20 surface antigen of B-lymphocytes)	Hoffmann La Roche (see also Rituxan)	Non-Hodgkin's lymphoma
Mabcampath (EU) or Campath (USA); alemtuzumab; a humanized monoclonal antibody directed against CD52 surface antigen of B-lymphocytes)	Millennium & ILEX (EU); Berlex, ILEX Oncology & Millennium Pharmaceuticals (USA)	Chronic lymphocytic leukaemia
Mylotarg (Gemtuzumab zogamicin; a humanized antibody-toxic antibiotic conjugate targeted against CD33 antigen found on leukaemic blast cells)	Wyeth Ayerst	Acute myeloid leukaemia
Zevalin (Ibritumomab Tiuxetan; murine monoclonal antibody, produced in a CHO cell line, targeted against the CD20 antigen)	IDEC Pharmaceuticals	Non-Hodgkin's lymphoma
Humira (EU & USA; also sold as Trudexa in EU) (adalimumab; r (anti-TNF) human monoclonal antibody created using phage display technology	Cambridge Antibody Technologies & Abbott (USA) Abbott (EU)	Rheumatoid arthritis
Bexxar (tositumomab; murine monoclonal raised against CD 20 surface antigen, found on the surface of B-lymphocytes)	Corixa and GlaxoSmithKline	Non-Hodgkin's lymphoma
Erbitux (cetuximab; chimaeric antibody raised against human EGF receptor)	ImClone Systems and Bristol-Myers-Squibb (USA) Merck (EU)	Treatment of EGF receptor-expressing metastatic colorectal cancer
Xolair (omalizumab; humanized monoclonal which binds immunoglobulin E at the site of high affinity IgE receptor binding)	Genentech/Novartis/Tanox/Sankyo	Treatment of adults/adolescents with moderate to severe persistent asthma
Raptiva (efalizumab; humanized antibody expressed in CHO cell line. Binds to the LFA-1, which is expressed on all leukocytes)	Genentech (USA) Serono (EU)	Treatment of adult patients with chronic moderate to severe plaque psoriasis
Avastin (humanized monoclonal raised against vascular endothelial growth factor).	Genentech USA) Roche (EU)	Carcinoma of the colon or rectum
Tysabri (humanized monoclonal raised against selected leukocyte integrins)	Biogen Idec/Elan	Treatment of patients with relapsing forms of MS
NeutroSpec (murine monoclonal antibody raised against CD15 surface antigen of selected leukocytes)	Palatin Technologies/Mallinckrodt Inc.	Imaging of equivocal appendicitis

- Antibodies that, by binding to the cell surface antigen, mark the tumour cell for destruction. NK cells and macrophages express cell surface receptors that bind to the antibody F_c region (Box 13.2). Thus, antibody bound to tumour antigens directs these immune elements directly to tumour surface. Antibodies also activate complement, which is capable of directly lysing tumour cells.

Box 13.2

Antibody architecture

Five major classes of antibodies (immunoglobulins, Igs) have been characterized: IgM, IgG, IgA, IgD and IgE). Immunoglobulins of all classes display a similar basic four-chain structure consisting of two identical light (L) chains and two identical heavy (H) chains (Figure 13.B2). The overall structure is held together by disulfide linkages and non-covalent interactions. Different H chain types are present in immunoglobulins of different classes. In addition, some classes can be further subdivided into subclasses (isotypes) based upon more subtle differences. Thus, human IgG can be subdivided into IgG_1, IgG_2, IgG_3 and IgG_4. Murine IgG can be subdivided into IgG_1, IgG_{2a}, IgG_{2b} and IgG_3.

In their native conformations, each immunoglobulin chain is seen to be composed of discrete domain structures, stabilized by intrachain disulfide linkages (not shown below). Each domain contains approximately 110 amino acid residues. H chains and L chains contain both variable (V) and constant (C) domains. Variable regions house the actual antigen-binding site of the antibody. Variable regions of antibodies displaying different (antigen-binding) specificities differ in amino acid sequence. Constant regions (within any one antibody class/subclass) do not. L chains contain one variable (V_L) and one constant (C_L) domain. H chains contain one variable (V_H) and three constant (C_H1, C_H2 and C_H3) domains. In addition, H chains display a single short sequence joining C_H1 and C_H2. This is the flexible hinge (H) region, which contains several proline residues.

Treatment with certain proteolytic enzymes (e.g. papain) results in cleavage of the immunoglobulin at the hinge region, yielding two separate antigen-binding fragments ($2 \times F_{(ab)}$), and a constant fragment (F_c). The F_c region mediates the various antibody effector functions. F_{ab} fragments, although retaining their antigen-binding properties, are no longer capable of precipitating antigen *in vitro*. However, immunoglobulin incubation with other proteases (e.g. pepsin) results in antibody fragmentation immediately below the hinge region. This leaves intact two interchain disulfide linkages towards the C-terminus of the hinge region. This holds the two antigen-binding fragments together. The products of this fragmentation are donated $F(ab)_2$ and F_c. Because of its bivalent nature, $F(ab)_2$ retains the ability to precipitate antigen *in vitro*.

F_V fragments consist of V_H and V_L domains, and can easily be produced by recombinant DNA technology (as can other antibody fragments). Two F_V domains can be stabilized by the introduction of an interchain covalent linkage (e.g. a disulfide linkage or via direct chemical coupling). 'Single-chain' F_V fragments may also be generated by the introduction of a short peptide linker sequence between the two F_V domains.

Selected regions within the antibody's variable domain display greater variability in amino acid sequence (from one antibody to another) than do other variable regions. These so-called 'hypervariable' regions (complementarity-determining regions (CDRs)) are brought into close proximity upon antibody folding into its native conformation, and represent the antigen binding sites. The remaining areas of the variable domain are termed framework regions.

Immunoglobulins are glycoproteins. The carbohydrate moiety is attached to the heavy chain (usually the C_H2 domain) via an N-linked glycosidic bond. Removal of the carbohydrate group has no effect upon antigen binding, but it does affect various antibody effector functions and alters its serum half-life.

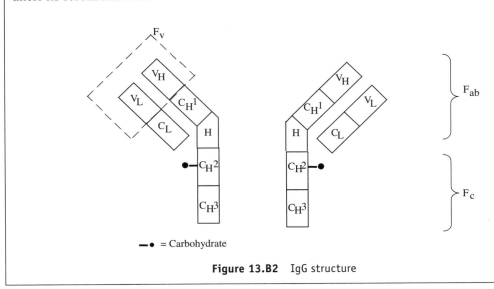

Figure 13.B2 IgG structure

13.3.3.1 Antibody-based strategies for tumour detection/destruction

Clear identification of tumour-associated antigens would facilitate the production of monoclonal antibodies capable of selectively binding to tumour tissue. Such antibodies could be employed to detect and/or destroy the tumour cells.

The antibody preparations could be administered unaltered or (more commonly) after their conjugation to radioisotopes or toxins. Binding of unaltered monoclonal antibodies to a tumour surface alone should facilitate increased destruction of tumour cells (Figure 13.4). This approach, however, has yielded disappointing results, as the monoclonal antibody preparations used to date have been murine in origin. The F_c region of such mouse antibodies is a very poor activator of human immune function. Technical advances, allowing the production of human/humanized monoclonals (see later) may render this therapeutic approach more attractive in the future.

Several clinical trials have evaluated (or continue to evaluate) monoclonal antibodies to which a radioactive tag has been conjugated. These are usually employed as potential anti-cancer agents. The rationale is selective delivery of the radioactivity directly to the tumour site. Most of the radioisotopes being evaluated are β-emitters. These include istopes of iodine (^{125}I, ^{131}I), rhenium (^{186}Re, ^{188}Re) and yttrium (^{90}Y). The medium-energy radioactivity these emit is capable of penetrating a thickness of several cells. Congregation of radioactivity at the tumour surface could thus promote irradiation of several layers of tumour cells, as well as nearby healthy cells. Higher energy α-emitters are also being evaluated. Although their effective path length is only about one cell deep, each emission has a greater likelihood of killing all cells in its path.

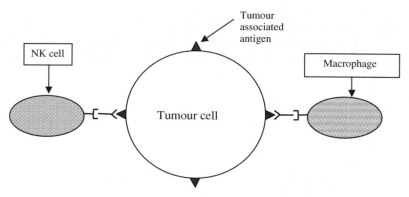

Figure 13.4 Binding of appropriate antibody to tumour-associated antigens marks the tumour cell for destruction. This is largely due to the presence of a domain on the antibody F_c region (see also Box 13.2), which is recognized and bound by macrophages and NK cells. Therefore, congregation of such cells on the surface of the tumour is encouraged. This greatly facilitates their cytocidal activity towards the transformed cells

An allied application of radiolabelled anti-tumour monoclonal antibodies is that of diagnostic imaging (immunoscintigraphy). In this case, the radioisotope employed must be a γ-emitter (such that the radioactivity can penetrate outward through the body for detection purposes). Although various radioisotopes of iodine have been evaluated, technetium (^{99m}Tc) is the one most commonly employed. It has a γ-ray emission energy that is sufficient, but a relatively short half life of 6 h (this minimizes long-term exposure of patient to high-energy γ-rays). It can also be generated at nuclear installations relatively easily and is inexpensive. Equally important, chemical methodologies exist which facilitate its (stable) coupling to antibody molecules. Direct labelling with ^{99m}Tc generally entails initial reduction of antibody disulfide residues, forming free sulfahydryl (—SH) groups. This is achieved by incubation with a suitable reducing agent, such as ascorbic acid or sodium dithionite. A source of ^{99m}Tc (e.g. Na^{99m}TcO$_4$) is then reduced separately. Subsequent mixing under nitrogen gas (to maintain reducing conditions) results in direct linkage of the radioisotope to the antibody.

Upon administration, the anti-tumour ^{99m}Tc conjugate will congregate at the tumour site. The tumour can then be visualized using suitable γ-ray detection equipment, such as a planar gamma camera.

A number of radiolabelled monoclonal antibodies have been approved as tumour diagnostic imaging agents (Table 13.2). Carcinoembryonic antigen (CEA)-SCAN, for example, is an antigen-binding fragment (Fab) of a specific murine monoclonal raised against human CEA. As discussed in detail in Section 13.3.4, CEA is expressed at high levels by some tumours. This is particularly true of tumours of the gastrointestinal tract, such as carcinomas of the colon or rectum. CEA-SCAN (non-proprietary name: Arcitumomab) is used to detect these carcinomas. However, CEA is expressed naturally (all be it at much lower levels) by some non-transformed cells. Therefore, this antibody fragment is used mostly to image recurrence and/or metastases of histologically demonstrated carcinoma of the colon or rectum. It is used as an adjunct to standard imaging techniques, such as a computed tomography scan or ultrasonography. Its industrial method of production is overviewed in Figure 13.5. The product is administered by i.v. injection, and only relatively mild side effects are usually noted. These can include nausea, fever, rash and headaches.

Anti-tumour monoclonal antibodies can also be used to deliver toxins to tumour sites. Toxins conjugated to therapeutic antibodies include ricin, pokeweed toxin, *Pseudomonas* toxin and

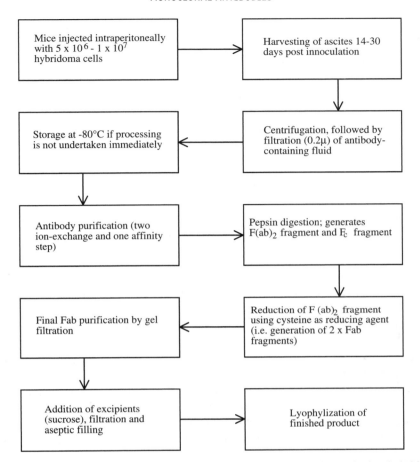

Figure 13.5 Outline of the production strategy of CEA-SCAN. The antibody-producing hybridoma cell line was originally obtained by standard methods of hybridoma generation. Spleen-derived murine B-lymphocytes were fused with murine myeloma calls. The resulting stable hybridomas were screened for the production of anti-CEA monoclonals. The clone chosen produces an IgG anti-CEA antibody. Note that the finished product outlined above is not radiolabelled. The freeze-dried antibody preparation (which has a shelf life of 2 years at 2–8 °C) is reconstituted immediately prior to its medical use. The reconstituting solution contains ^{99m}Tc, and is formulated to facilitate direct conjugation of the radiolabel to the antibody fragment

diphtheria toxin. After binding to the cell surface, the antibody–toxin conjugate is often internalized via endocytosis. It is presumed that, rather than being destroyed, the toxin is subsequently made available inside the cell, such that it can induce its toxic effects. One such antibody-based product now approved for general medical use is Mylotarg (Table 13.2). The product consists of an engineered antibody (a 'humanized' antibody, as described later), conjugated to a cytotoxic anti-tumour antibiotic, calicheamicin (Figure 13.6). The antibody binds specifically to a cell surface antigen, CD33. This is a sialic-acid-dependent adhesion protein found on the surface of leukaemic cells in more than 80 per cent of patients suffering from acute myeloid leukaemia. The product production process entails initial culture of the antibody-producing mammalian cell line with subsequent purification of the antibody by a series of chromatographic steps. Downstream processing

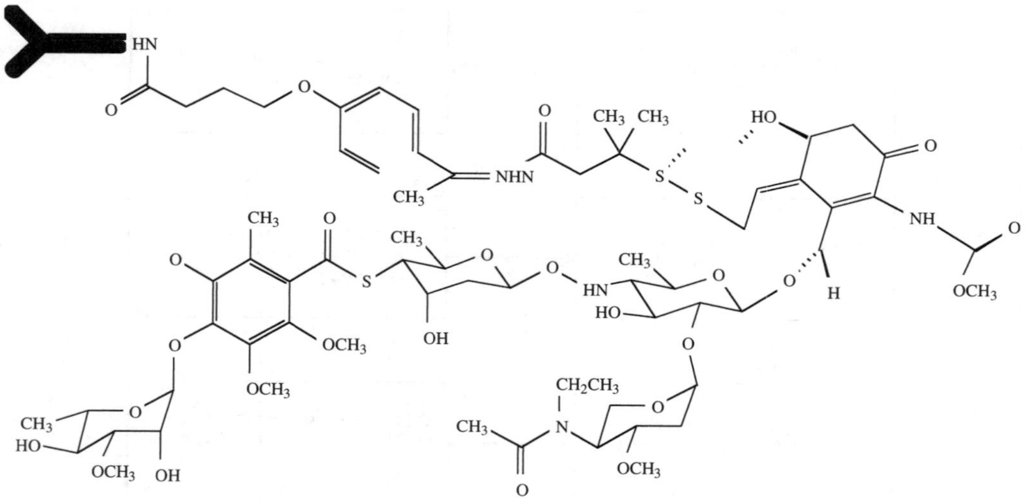

Figure 13.6   Schematic diagram of the antibody based product Mylotarg, with emphasis upon the toxin's chemical structure. In reality three to five molecules of toxin are attached to each antibody molecule

also incorporates ultrafiltration and low-pH incubation steps designed to remove/inactivate any virus potentially present. The cytotoxic antibiotic is obtained separately via fermentation of its producer microrganism, *Micromonospora echinospora* sp. *calichensis*. Direct chemical linkage of antibiotic to antibody is achieved using a bifunctional linker.

Mylotarg administration results in congregation of the antibody–toxin conjugate on the surface of (CD33 positive) leukaemic cells. Binding triggers internalization of the conjugate. Lysosomal degradation ensues, but a significant proportion of the intact antibiotic escapes and induces its cytotoxic affect by binding DNA in its minor groove. This, in turn, induces double-strand breakage.

Mylotarg (like most other drugs) does induce some side effects, the most significant of which is immunosuppression. This is induced because certain additional (non-cancerous) white blood cell precursors also display the CD33 antigen on their surface. The immunosuppressive effect is reversed upon termination of treatment, as pluripotent haematopoietic stem cells (Chapter 10) are unaffected by the product.

13.3.3.2   *Drug-based tumour immunotherapy*

In addition to tumour-selective delivery of toxins and radioisotopes, antibodies may also be used to mediate tumour-targeted drug delivery. At its simplest, this involves conjugation of a chemotherapeutic drug to a tumour-specific antibody. Therapeutic drugs used include adriamycin, aminopterin, methotrexate and vinca alkaloids. This direct approach to tumour drug delivery has met with some success, mainly in animal studies. However, a limited number of drug molecules can be conjugated to each antibody molecule, thus somewhat limiting the drug delivery load.

An alternative approach is the use of a tumour-specific antibody to which a prodrug activating enzyme has been attached. Therapeutically inactive prodrugs could be administered by,

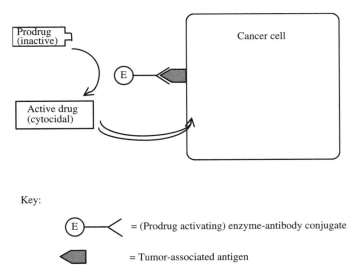

Figure 13.7 Outline of antibody-directed enzyme prodrug therapy (ADEPT). Subsequent to its enzymatic activation, the active drug is taken up by the cell, upon which it exhibits a cytocidal effect. Refer to text for specific detail

for example, i.v. injection. This would subsequently be activated only at the tumour surface (Figure 13.7). This approach has been termed ADEPT or antibody-directed catalysis.

Because of its catalytic nature, a single antibody–enzyme conjugate would activate many molecules of the prodrug in question. Much of the active cytocidal agent released at the tumour surface would be taken up by the tumour cells via simple diffusion or carrier-mediated active transport.

Administration of etoposide in prodrug form exemplifies this approach (Figure 13.8). Etoposide ($C_{29}H_{32}O_{13}$; molecular mass 588.6) is a semi-synthetic derivative of podophyllotoxin, produced naturally by the North American plant *Podophyllum peltatum*. It is used as an anti-cancer agent. Its cellular uptake is diffusion dependent, and once inside the cell it exerts its cytocidal effect. Phosphorylated etoposide is non-diffusable and, hence, represents an inactive prodrug form of etoposide. (Attachment of a charged group to most diffusion-dependent drugs prevents their cellular uptake.) Alkaline phosphatase, however, can cleave the phosphate group, releasing free cytocidal agent. Administration of a tumour-detecting antibody–alkaline phosphatase conjugate thus effectively targets the enzyme to the tumour surface. Subsequent administration of phosphorylated etoposide results in etoposide liberation at the tumour surface, which can then enter tumour cells by diffusion (Figure 13.8). Various other prodrug–enzyme combinations have now been developed, including phenoxyacetamide derivatives of doxorubicin (activated by penicillin amidase) and 5-fluorocytosine (activated by cytosine deaminase).

The prodrugs used should be inexpensive, readily available and should be stable to chemical/enzymatic degradation *in vivo*. Enzymes used should also be stable under physiological conditions, display a reasonable turn over number *in vivo* and not be dependent upon a co-factor for activity. Mammalian enzymes would be likely less immunogenic than microbial enzymes. However, the use of a prodrug capable of being activated by a mammalian enzyme can lead to complications if that enzyme's (human) endogenous counterpart is capable of activating the drug at sites distant from the tumour.

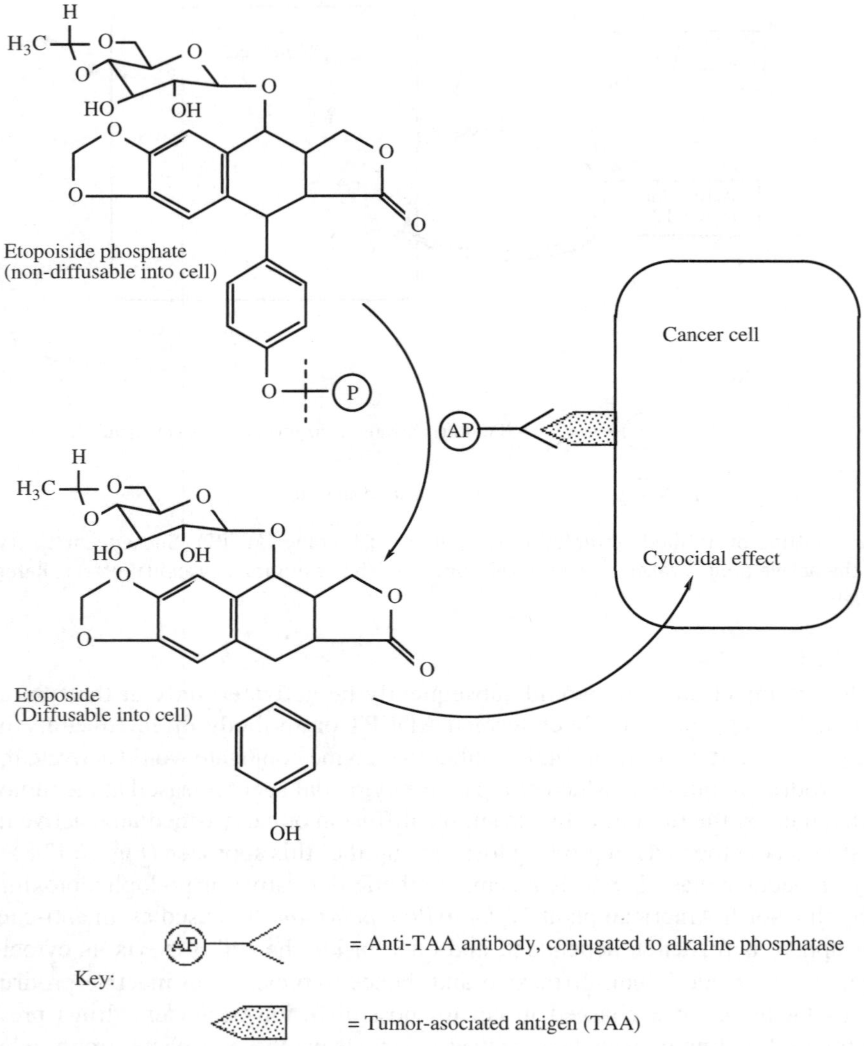

Figure 13.8 The etoposide–alkaline phosphatase ADEPT system. Refer to text for specific details

13.3.3.3 First-generation anti-tumour antibodies: clinical disappointment

Despite the scientific elegance of the antibody-mediated approach to tumour detection/destruction, initial clinical trials proved disappointing. A number of factors contributed to their poor therapeutic performance, particularly against solid tumours. Most such factors relate directly/indirectly to the fact that the first generation of such drugs utilized whole monoclonal antibody preparations of murine origin. These factors include:

- insufficient information exists regarding tumour antigens;

- murine monoclonals prompt an immune response when administered to humans;

- poor penetration of tumour mass by antibody;

- murine monoclonals display a relatively short half-life when administered to humans;

- poor recognition of murine antibody F_c domain by human effector mechanisms.

13.3.4 Tumour-associated antigens

In this text, the term 'tumour-associated antigen' represents any antigen associated with any cancer cell, no matter what factor(s) originally prompted cellular transformation. (In some circles, the term 'tumour-associated antigen' is often applied more specifically: to antigens associated with virally transformed cells.) Identification of tumour-associated antigens forms a core requirement for effective tumour immunotherapy. Identification of such tumour-associated antigens remains a very active area of biomedical research. From the limited data generated to date, tumour-associated antigens can generally be categorized into one of three groups (Table 13.3).

Many tumour types are induced by chemical and/or physical carcinogens we encounter in our living/working environments. Most of these tumours are initiated when the carcinogen induces a point mutation in a nucleotide sequence, thus perhaps altering expression levels of the gene product or altering its functional characteristics. Because such mutations are random, it is not surprising that each resulting tumour displays its own unique tumour-associated antigen(s). This renders an immunological approach to detection/therapy impractical in such instances, as the specific tumour-associated antigens unique to that case would first have to be identified.

In contrast to the above situation, cancers induced by viruses generally exhibit immunological cross-reactivity. Any specific virus will often induce expression of the same tumour antigen no matter what cell type it transforms. Moreover, in some cases, different transforming viruses can induce production of the same tumour antigen(s). Immunodetection/immunotherapy of such cancers is thus rendered attractive. Once a tumour antigen is identified, antibodies raised against it will likely cross-react with several other tumour types.

DNA viruses, such as adenoviruses and papovaviruses (e.g. polyoma and SV40), induce cellular transformation in rodents. Other viruses have been implicated in human cancers. Epstein–Barr virus, for example, has been implicated with nasopharyngeal carcinoma, β-cell lymphomas and Hodgkin's lymphoma. Human papilloma virus is linked to most cervical cancers.

Certain RNA viruses, particularly retroviruses, have also proven capable of inducing cancer. Retroviruses known to induce cancer in animals include Rous sarcoma virus, Kirsten murine

Table 13.3 Characterizing of tumour-associated antigens. Antigens commonly expressed by a number of different tumour types render practical application of tumour immunodetection/immunotherapy in those cases

Cellular-transforming factors	Associated tumour antigen
Tumours induced by chemical carcinogens/irradiation	Each tumour usually displays distinct antigen specificity
Virally induced tumours	Various tumour types display identical tumour-associated antigens (especially if tumours are induced by the same virus)
Various induction factors (often unknown)	The same oncofoetal antigen can be expressed by a number of different tumour types

sarcoma virus, avian myelocytomatosis virus, and various murine leukaemia viruses. Thus far, the only well-characterized human RNA transforming virus is that of human T-cell lymphotropic virus-1 (HTLV-1), which can induce adult T-cell leukaemia/lymphoma. The identification of antigens uniquely associated with various tumour types and the identification of additional cancer-causing viruses remain areas of very active research.

Another group of antigens associated with some tumour types are oncofoetal antigens. These antigens are proteins that are normally expressed during certain stages of foetal development. Subsequent repression of their structural genes, however, prevents their expression at later stages of development and/or into adulthood. Characteristic of some cancers is the re-expression of oncofoetal antigens. Some such antigens remain attached to the cancer cell surface, whereas others are secreted in soluble form. Although these oncofoetal proteins are not recognized as foreign by the host's own immune system, they do represent important potential diagnostic markers. Although some such markers have been identified, efforts continue to identify additional members of this family.

CEA and α-fetoprotein (AFP) represent the most extensively characterized oncofoetal antigens thus far. We have already encountered CEA in the guise of its use as a marker for cancers of the colon and rectum. CEA is a 180 kDa integral membrane glycoprotein. It also may be secreted into the blood in soluble form. It is expressed mainly in the gut, liver and pancreas, during the first 6 months of foetal development. However, it is now known to be expressed (although at greatly reduced levels) by adult colonic mucosal cells and in the lactating breast. Elevated levels of either soluble (serum) or cell-bound CEA are normally indicative of cancers of the gastrointestinal tract.

AFP is a 70 kDa glycoprotein found in the circulatory system of the developing foetus. It is synthesized primarily by the yolk sac and (foetal) liver. AFP is present only in vanishing low quantities in the serum of adults (where it is replaced by serum albumin). Elevated adult serum levels of this marker are often associated with various cancers of the liver, as well as germ cell tumours. It is also sometimes expressed by gastric and pancreatic cancer cells. Although a useful tumour marker, increased serum AFP levels also often accompany cirrhosis and some other noncancerous liver diseases.

CA125 represents an oncofoetal protein that is expressed by up to 90 per cent of ovarian adenocarcinomas. Some of the protein is released from the tumour site into the general circulation. Elevated serum CA125 levels, therefore, have some diagnostic value. Imaging of actual tumour sites can also be undertaken using radiolabelled antibody coupled to immunoscintigraphy. Indimacis-125 is the trade name given to such a product approved in 1996 for use in the EU. The product is an ^{111}In-labelled F(ab)$_2$ fragment derived from a murine hybridoma cell line. Although some of the product is likely absorbed by the circulatory form of the oncofoetal protein, the product has proven effective in imaging relapsing ovarian adenocarcinoma.

In summary, TSA-based complications in the context of developing antibody-based cancer therapies include:

- Limited numbers of TSAs currently characterized.

- Some cancer types, depending upon their cause, may display unique TSAs in different patients.

- TSAs are often expressed, albeit at lower levels, by one or more additional (non-transformed) body cell types.

- In some instances, binding of antibody results in immediate shedding of Ab–TSA from the cell surface.

- TSA expression can be transitory. For many tumours, only a proportion (albeit a large one) of the tumour cells express TSAs at any given time.

13.3.5   Antigenicity of murine monoclonals

Antibody immunogenicity remains one of the inherent therapeutic limitations associated with administration of murine monoclonals to human subjects. In most instances, a single injection of the murine monoclonal will elicit an immune response in 50–80 per cent of patients. Human anti-mouse antibodies (HAMA) will generally be detected within 14 days of antibody administration. Repeated administration of the monoclonal (usually required if the monoclonal is used for therapeutic purposes) will increase the HAMA response significantly. It will also induce an HAMA response in the majority of individuals who display no such response after the initial injection. The HAMA response will effectively and immediately destroy subsequent doses of monoclonal administered. In practice, therefore, therapeutic efficacy of murine monoclonals is limited to the first and, at most, the second dose administered.

An obvious strategy for overcoming the immunogenicity problem would be the generation and use of monoclonal antibodies of human origin. This is possible but difficult. Human antibody-producing lymphocytes can potentially be rendered immortal by:

- transformation by Epstein–Barr virus infection;

- fusion with murine monoclonals;

- fusion with human lymphoblastoid cell lines.

However, a number of technical hurdles remain that prevent routine production of human monoclonal preparations. These include:

- source of antibody-producing cell;

- reliable methods for lymphocyte immortalization;

- stability and antibody-producing capacity of resulting immortalized cells.

Initial stages in the production of murine monoclonals entail administration of the antigen of interest to a mouse. This is followed by sacrifice and recovery of activated B-lymphocytes from the spleen. A similar approach to the production of human monoclonals would be unethical. Administration of some antigens to humans could endanger their health. Although B-lymphocytes could be obtained from the peripheral circulation, the majority of these are unstimulated, and recovery of (stimulated) B-lymphocytes from the spleen is impractical.

Although Epstein–Barr virus is capable of inducing cellular transformation, few antibody-producing B-lymphocytes display the viral cell surface receptor. Most, therefore, are immune to Epstein–Barr virus infection. Even upon successful transformation, most produce low-affinity IgM antibodies, and the cells are often unstable. Having said that, one monoclonal antibody approved for medical use (Humaspect, Table 13.2) is produced by a human lymphoblastoid cell line originally transformed by Epstein–Barr virus.

Fusion of human lymphocytes with human lymphoblastoid cell lines is a very inefficient process. Fusion of human lymphocytes with murine myeloma cells lead to very unstable hybrids. Upon fusion, preferential loss of human genetic elements is often observed. Unfortunately, particularly common is the loss of chromosomes 2, 14 and 22, which encode antibody light and heavy chain loci. The production yields of human monoclonals upon immortalization of the human B-lymphocyte (by whatever means) are also low.

13.3.6 Chimaeric and humanized antibodies

Recombinant DNA technology has provided an alternative (and successful) route of reducing the innate immunity of murine monoclonals. The genes for all human immunoglobulin sub-types have been cloned, and this has allowed generation of various hybrid antibody structures of reduced immunogenicity.

The first strategy entails production of 'chimaeric' antibodies, consisting of mouse variable regions and human constant regions (Figure 13.9). The chimaeric antibody would display the specificity of the original murine antibody, but would largely be human in sequence.

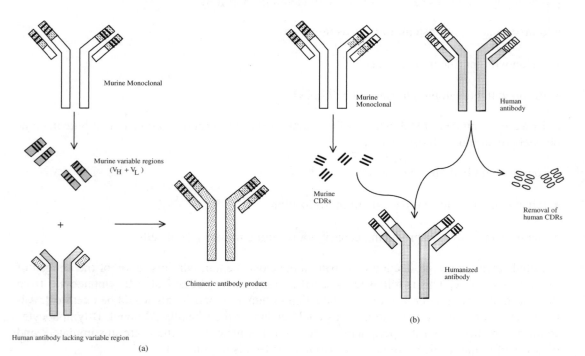

Figure 13.9 Production of chimaeric (a) and humanized (b) antibodies (via recombinant DNA technology). Chimaeric antibodies consist of murine monoclonal V_H and V_L domains grafted onto the F_c region of a human antibody. Humanized antibody consists of murine CDR regions grafted into a human antibody

Table 13.4 The serum half-life values of some IgG antibody preparations when administered to humans

Antibody type	Serum half-life
Intact human monoclonal	14–21 days
Intact murine monoclonal	30–40 h
Chimaeric antibody	200–250 h
Murine F(ab)$_2$ fragment	20 h
Murine Fab fragment	2 h

It was hoped that such chimaeric antibodies, when compared with murine antibodies, would be:

- significantly less immunogenic;

- display a prolonged serum half-life;

- allow activation of various F_c-mediated functions.

Reduced immunogenicity was expected, as only a minor part of the chimaeric antibodies is murine in origin. Furthermore, the HAMA response is normally directed largely at epitopes on the antibody's F_c domains. The variable region appears inherently less immunogenic. In practice, the expected reduced immunogenicity was observed. Early clinical trials with chimaerics have shown them to be generally safe and non-toxic. The rate of immune responses observed after single dose administration dropped from almost 80 per cent (murine) to in the region of 5 per cent (chimaeric). However, repeated administration of chimaerics did eventually raise an immune response in most recipients.

When compared with human monoclonals (half-life 14–21 days), murine monoclonals administered to humans display a relatively short half-life (30–40 h). Chimaerization increased serum half-life by fivefold, with typical values of 230 h being recorded (Table 13.4). A prolonged half-life is desirable if the antibody is to be used therapeutically, as it decreases the required frequency of product administration. Chimaeric antibodies also allow activation of F_c-mediated functions (e.g. activation of complement, etc.), as this domain displays human sequence.

Although chimaeric antibodies contain an entire murine-derived variable region, it is only the CDRs within this variable domain that actually dictate antigen specificity (Box 13.2). A method of reducing still further the antigenicity of murine antibodies is to 'humanize' them. This entails transferring the nucleotide sequences coding for the six CDR regions of the murine antibody of the desired specificity into a human antibody gene. The resulting hybrid antibody will, obviously, be entirely human in nature except for the CDRs (Figure 13.9).

Transfer of murine-derived CDR sequences into human antibody framework regions (Box 13.2), sometimes generates an antibody with greatly reduced antigen binding affinity. Selected murine framework sequences are often also included in the humanized antibody. This process (known as reshaping the human antibody) facilitates folding of the CDRs into their true native conformations. This, in turn, normally restores antibody–antigen-binding affinity. Over 95 per cent of antibodies are human in sequence. Clinical trials indicate that such proteins do indeed behave similarly to native human antibodies.

Humanization has overcome many of the major factors that limited the therapeutic effectiveness of first-generation (murine) monoclonals as therapeutic agents. Several such humanized products have now gained marketing authorization (Table 13.2), and one such product is featured in Box 13.3.

Box 13.3

Product case study: Avastin

Avastin (tradename, also known as bevacizumab) is a 149 kDa recombinant humanized mono-clonal IgG1 antibody first approved for medical use in the USA in 2004, and subsequently in the EU in 2005. It is indicated for first-line treatment of patients with metastatic colorectal cancer, in combination with specified (5-fluorouricil-based) small molecule chemotherapeutic drugs.

The antibody brings about its effect by inhibiting angiogenesis (the formation of new blood vessels), a process required to support tumour growth. Specifically, the antibody binds human vascular endothelial growth factor. This prevents the latter from binding to its cell surface re-ceptor, a process central to triggering new blood vessel formation in both normal and diseased tissue.

The engineered antibody is produced in a CHO cell line using 12 000 l bioreactors. Purifica-tion is achieved mainly by protein A affinity chromatography, followed by sequential anion and cation ion-exchange chromatography. It is formulated as a sterile solution in single-use vials containing either 100 or 400 mg active substance. Excipients include phosphate buffer com-ponents, trehalose and polysorbate 20. The product is generally administered by i.v. infusion (5 mg kg^{-1}) once every 2 weeks.

Pharmacokinetic studies in patients yielded an estimated product half-life of approximately 20 days (11–50 days range) and the product clearance was found to be variable according to body weight, gender and tumour burden. Safety and efficacy were established by three rand-omized, controlled trials. The first study was a randomized double-blind trial involving 813 patients. The primary end-point measured was overall survival, which was extended from a median of 15.6 months to 20.3 months.

Serious, sometimes fatal, side effects reported included gastrointestinal perforation, wound healing complications and haemorrhage, and the product should not be administered for at least 28 days following major surgery. Avastin was developed by Genentech.

13.3.7 Antibody fragments

One limitation to antibody-mediated treatment of solid tumours relates to their poor penetration of tumour mass. Antibody-based tumour therapy (e.g. using radiolabelled murine monoclonals) has proved more effective in treating disseminated cancers (e.g. leukaemias and lymphomas) as opposed to solid tumours. The physical size of the intact antibody likely hinders tumour penetration. As a result, recent interest has focused upon using antibody fragments that retain their antigen-binding capabilities. Fragments, such as F(ab), F(ab)$_2$ and F$_v$ (Box 13.2), can be easily generated, mainly via recombinant DNA technology. These have been labelled with, for example, radioactive tags with the intention of using them as diagnostic and therapeutic agents. Although their lower mo-lecular mass may aid more effective tumour penetration, whole chimaeric/humanized antibodies may be more effective, particularly if used for therapeutic purposes. Fragments generally display greatly reduced serum half-lives (Table 13.4) and cannot initiate effector functions. Radiolabelled fragments may be better suited to diagnostic imaging purposes.

13.3.8 Additional therapeutic applications of monoclonal antibodies

Thus far, the discussion relating to the medical uses of monoclonals has focused exclusively upon cancer. Monoclonal antibodies (and their derivatives), however, have a far broader potential therapeutic application. Actual/potential additional uses include detection and treatment of cardio-vascular disease, infectious agents, and various additional medical conditions (Table 13.2).

Various antibody preparations have been developed that facilitate imaging of vascular-related conditions, including myocardial infarction, deep vein thrombosis and atherosclerosis. Anti-myosin monoclonal antibody fragments (Fab) labelled with ^{111}In, for example, have been used for imaging purposes in conjunction with a planar gamma camera. The antibody displays specificity for intracellular cardiac myosin, which is exposed only upon death of heart muscle tissue induced by a myocardial infarction (heart attack).

Imaging monoclonals could be of use in visualizing the sites/extent of focused bacterial infec-tions. This could be achieved by using radiolabelled antibodies displaying binding affinity for specific bacterial surface antigens. A related, but indirect, approach may entail use of imaging antibodies capable of detecting granulocytes and various other leukocytes that congregate at the sites of infection.

It has been estimated that 1–2 per cent of the US population suffer from autoimmune con-ditions, including rheumatoid arthritis, MS and some forms of diabetes. In many instances, an autoimmune response results from the inappropriate activation of a specific subset of B- and/or T-lymphocytes. The most common immunotherapeutic approach to potentially treat such diseases is to induce depletion of the individual's T- and B-cell populations. This could be achieved by ad-ministration of an antibody raised against a surface antigen present on such cells. Initial trials, for example, have shown that injection of an (unconjugated) anti-CD4 antibody (cell surface glyco-protein present on many T-lymphocytes) over 7 days significantly reduced the clinical symptoms of rheumatoid arthritis for several months.

Antibodies have and likely will find additional use in transplantation-related medicine. In gen-eral, cell-mediated immunological mechanisms are responsible for mediating rejection of trans-planted organs. In many instances, transplant patients must be maintained on immunosuppressive drugs (e.g. some steroids and, often, the fungal metabolite cyclosporine). However, complications may arise if a rejection episode is encountered that proves unresponsive to standard immuno-suppressive therapy. Orthoclone OKT-3 was the first monoclonal antibody-based product to find application in this regard.

This antibody is raised against the protein-based CD3 antigen, present on the cell surface of most T-lymphocytes. I.v. administration of (unconjugated) antibody appears to block normal func-tioning of such T-cells and promote their clearance from the blood. However, upon cessation of antibody administration, CD3 positive cell numbers rapidly revert to normal values. Therefore, maintenance immunosuppressives (e.g. with cyclosporine) must subsequently be restored.

A number of additional antibody-based products aimed at preventing transplant rejection have now gained general marketing authorization. Simulect (chimaeric antibody) and Zenapax (human-ized antibody) were approved in the late 1990s (Table 13.2). Their engineered nature has greatly reduced the HAMA response upon their administration to humans. Both products target the IL-2 receptor and, hence, bind fairly selectively to activated immune system cells, especially activated T-lymphocytes, monocytes and macrophages. Binding prevents further cellular proliferation and, hence, dampens the immune system's attempts to destroy the transplanted tissue.

13.4   Vaccine technology

The application of vaccine technology forms a core element of modern medicinal endeavour. It plays a central role in both human and veterinary medicine and represents the only commonly employed prophylactic (i.e. preventative) approach undertaken to control many infectious diseases. The current (annual) global vaccine market stands at in excess of US$3 billion. Immunization programmes, particularly those undertaken on a multinational scale, have served to reduce dramatically the incidence of many killer/disabling diseases, such as smallpox, polio and tuberculosis.

Continued/increased emphasis upon the implementation of such immunization programmes is likely. This is true not only of poorer world regions, but also amongst the most affluent nations. An estimated 500 000 adults die annually in the USA from conditions that could have been prevented by vaccination. These include pneumococcal pneumonia, influenza and hepatitis B.

Vaccination seeks to exploit the natural defence mechanisms conferred upon us by our immune system. A vaccine contains a preparation of antigenic components consisting of, derived from or related to a pathogen. In most instances upon vaccine administration, both the humoral and cell-mediated arms of the immune system are activated. The long-term immunological protection induced will normally prevent subsequent establishment of an infection by the same or antigenically related pathogens. Although some vaccines are active when administered orally, more are administered parenterally. Normally, an initial dose administration is followed by subsequent administration of one or more repeat doses over an appropriate time-scale. Such booster doses serve to maximize the immunological response.

Traditional vaccine preparations have largely been targeted against viral and bacterial pathogens, as well as some bacterial toxins and, to a lesser extent, parasitic agents, such as malaria. However, an increased understanding of the molecular mechanisms underlying additional human diseases suggests several novel applications of vaccines to treat/prevent autoimmune conditions and cancer (discussed later). Despite such potentially exciting future applications, recent scientific surveys indicate that the most urgently required vaccines are those that protect against more mundane pathogens (Table 13.5). Although the needs of the developing world are somewhat different to those of developed regions, an effective AIDS vaccine is equally important to both. Approaches to development of such AIDS vaccines are discussed later in this chapter. Of particular consequence to developing world regions is the current lack of a truly effective malaria vaccine. With an estimated annual incidence of 300–500 million clinical cases (with up to 2.7 million resulting deaths), development of an effective vaccine in this instance is a priority.

13.4.1   Traditional vaccine preparations

For the purposes of this discussion, the term 'traditional' refers to those vaccines whose development predated the advent of recombinant DNA technology. Approximately 30 such vaccines

Table 13.5   Some diseases against which effective/more effective vaccines are urgently required. Diseases more prevalent in developing world regions differ from those that are most common in developed countries

Developing world regions	Developed world regions
AIDS	AIDS
Malaria	Respiratory syncytial virus
Tuberculosis	Pneumococcal disease

remain in medical use (Table 13.6). These can largely be categorized into one of several groups, including:

- live, attenuated bacteria (e.g. bacillus Calmette–Guérin (BCG), used to immunize against tuberculosis);

- dead or inactivated bacteria (e.g. cholera and pertussis vaccines);

- live attenuated viruses (e.g. measles, mumps and yellow fever viral vaccines);

- inactivated viruses (hepatitis A and polio (Salk) viral vaccines);

- toxoids (e.g. diphtheria and tetanus vaccines);

- pathogen-derived antigens (e.g. hepatitis B, meningococcal, pneumococcal and *Haemophilus influenzae* vaccines).

Table 13.6 Some traditional vaccine preparations that find medical application. In addition to being marketed individually, a number of such products are also marketed as combination vaccines. Examples include diphtheria, tetanus and pertussis vaccines and measles, mumps and rubella vaccines

Product	Description	Application
Anthrax vaccines	*Bacillus anthracis*-derived antigens found in a sterile filtrate of cultures of this microorganism.	Active immunization against anthrax
BCG (bacillus Calmette–Guérin) vaccine	Live attenuated strain of *Mycobacterium tuberculosis*	Active immunization against tuberculosis
Brucellosis vaccine	Antigenic extract of *Brucella abortus*	Active immunization against brucellosis
Cholera vaccine	Dead strain(s) of *Vibrio cholerae*	Active immunization against cholera
Cytomegalovirus vaccines	Live attenuated strain of human cytomegalovirus	Active immunization against cytomegalovirus
Diphtheria vaccine	Diphtheria toxoid formed by treating diphtheria toxin with formaldehyde	Active immunization against diphtheria
Japanese encephalitis vaccine	Inactivated Japanese encephalitis virus	Active immunization against viral agents causing Japanese encephalitis
H. influenzae vaccine	Purified capsular polysaccharide of *H. influenzae* type b (usually linked to a protein carrier, forming a conjugated vaccine)	Active immunization against *H. influenzae* type b infections (major causative agent of meningitis in young children)
Hepatitis A vaccine	(Formaldehyde)-inactivated hepatitis A virus	Active immunization against hepatitis A
Hepatitis B vaccine	Suspension of hepatitis B surface antigen (HBsAg) purified from the plasma of hepatitis B sufferers	Active immunization against hepatitis B (note: this preparation has largely been superseded by HBsAg preparations produced by genetic engineering)
Influenza vaccines	Mixture of inactivated strains of influenza virus	Active immunization against influenza

(*Continued*)

Table 13.6 (*Continued*)

Product	Description	Application
Leptospira vaccines	Killed strain of *Leptospira interogans*	Active immunization against leptospirosis icterohaemorrhagica (Weil's disease)
Measles vaccines	Live attenuated strains of measles virus	Active immunization against measles
Meningococcal vaccines	Purified surface polysaccharide antigens of one or more strains of *Neisseria meningitidis*	Active immunization against *N. meningitidis* (can cause meningitis and septicaemia)
Mumps vaccine	Live attenuated strain of the mumps virus (*Paramyxovirus parotitidus*)	Active immunization against mumps
Pertussis vaccines	Killed strain(s) of *B. pertussis*	Active immunization against whooping cough
Plague vaccine	Formaldehyde-killed *Yersinia pestis*	Active immunization against plague
Pneumococcal vaccines	Mixture of purified surface polysaccharide antigens obtained from differing serotypes of *Streptococcus pneumoniae*	Active immunization against *Streptococcus pneumoniae*
Poliomyelitis vaccine (Sabin vaccine: oral)	Live attenuated strains of poliomyelitis virus	Active immunization against polio
Poliomyelitis vaccine (Salk vaccine: parenteral)	Inactivated poliomyelitis virus	Active immunization against polio
Rabies vaccines	Inactivated rabies virus	Active immunization against rabies
Rotavirus vaccines	Live attenuated strains of rotavirus	Active immunization against rotavirus (causes severe childhood diarrhoea)
Rubella vacines	Live attenuated strain of Rubella virus	Active immunization against rubella (German measles)
Tetanus vaccines	Toxoid formed by formaldehyde treatment of toxin produced by *C. tetani*	Active immunization against tetanus
Typhoid vaccines	Killed *Salmonella typhi*	Active immunization against typhoid fever
Typhus vaccines	Killed epidemic *Rickettsia prowazekii*	Active immunization against louse-borne typhus
Varicella zoster vaccines	Live attenuated strain of Herpes virus varicellae	Active immunization against chickenpox
Yellow fever vaccines	Live attenuated strain of yellow fever virus	Active immunization against yellow fever

13.4.1.1 Attenuated, dead or inactivated bacteria

Attenuation (bacterial or viral) represents the process of elimination or greatly reducing the virulence of a pathogen. This is traditionally achieved by, for example, chemical treatment or heat, growing under adverse conditions or propagation in an unnatural host. The attenuated product should still immunologically cross-react with the wild-type pathogen. Although rarely occurring in practice, a theoretical danger exists in some cases that the attenuated pathogen might revert to its pathogenic state. An attenuated bacterial vaccine is represented by BCG. BCG is a strain of tuberculae bacillus that fails to cause tuberculosis, but retains much of the antigenicity of the pathogen. Killing or inactivation of pathogenic bacteria usually renders them suitable as vaccines. This is usually achieved by:

- heat treatment;

- treatment with formaldehyde or acetone;

- treatment with phenol or phenol and heat;

- treatment with propiolactone.

13.4.1.2   Attenuated and inactivated viral vaccines

Viral particles destined for use as vaccines are generally propagated in a suitable animal cell culture system. Although true cell culture systems are sometimes employed, many viral particles are grown in fertilized eggs or cultures of chick embryo tissue (Table 13.7).

Many of the more prominent vaccine preparations in current medical use consist of attenuated viral particles (Table 13.6). Mumps vaccines consist of live attenuated strains of *Paramyxovirus parotitidis*. In many world regions, it is used routinely to vaccinate children, often a part of a combined measles, mumps and rubella vaccine.

Several attenuated strains have been developed for use in vaccine preparations. The most commonly used is the Jeryl Linn strain, which is propagated in chick embryo cell culture. This vaccine has been administered to well over 50 million people worldwide and, typically, results in seroconversion rates of over 97 per cent. The Sabin (oral poliomyelitis) vaccine consists of an aqueous suspension of poliomyelitis virus, usually grown in cultures of monkey kidney tissue. It contains approximately 1 million particles of poliomyelitis strains 1, 2 or 3 or a combination of all three strains.

Hepatitis A vaccine exemplifies vaccine preparations containing inactivated viral particles. It consists of a formaldehyde-inactivated preparation of the HM 175 strain of hepatitis A virus. Viral particles are normally propagated initially in human fibroblasts.

13.4.1.3   Toxoids and antigen-based vaccines

Diphtheria and tetanus vaccines are two commonly used toxoid-based vaccine preparations. The initial stages of diphtheria vaccine production entail the growth of *Corynebacterium diphtheriae*.

Table 13.7   Some cell culture systems in which viral particles destined for use as viral vaccines are propagated

Viral particle/vaccine	Typical cell culture system
Yellow fever virus	Chick egg embryos
Measles virus (attenuated)	Chick egg embryo cells
Mumps virus (attenuated)	Chick egg embryo cells
Polio virus (live, oral, i.e. Sabin, and inactivated injectable, i.e. Salk)	Monkey kidney tissue culture
Rubella vaccine	Duck embryo tissue culture, human tissue culture
Hepatitis A viral vaccine	Human diploid fibroblasts
V. zoster vaccines (chickenpox vaccine)	Human diploid cells

Table 13.8 Some vaccine preparations that consist not of intact attenuated/inactivated pathogen, but of surface antigens derived from such pathogens

Vaccine	Specific antigen used
Anthrax vaccines	Antigen found in the sterile filtrate of *B. anthracis*
H. influenzae vaccines	Purified capsular polysaccharide of *H. influenzae* type B
Hepatitis B vaccines	Hepatitis B surface antigen (HBsAg) purified from plasma of hepatitis B carriers
Meningococcal vaccines	Purified (surface) polysaccharides from *N. meningitidis* (groups A or C)
Pneumococcal vaccine	Purified polysaccharide capsular antigen from up to 23 serotypes of *S. pneumoniae*

The toxoid is then prepared by treating the active toxin produced with formaldehyde. The product is normally sold as a sterile aqueous preparation. Tetanus vaccine production follows a similar approach. *Clostridium tetani* is cultured in appropriate media. The toxin is recovered and inactivated by formaldehyde treatment. Again, it is usually marketed as a sterile aqueous-based product.

Traditional antigen-based vaccine preparations consist of appropriate antigenic portions of the pathogen (usually surface-derived antigens; Table 13.8). In most cases, the antigenic substances are surface polysaccharides. Many carbohydrate-based substances are inherently less immunogenic than protein-based material. Poor immunological responses are thus often associated with administration of carbohydrate polymers to humans, particularly to infants. The antigenicity of these substances can be improved by chemically coupling (conjugating) them to a protein-based antigen. Several conjugated *H. influenzae* vaccine variants are available. In these cases, the *Haemophilus* capsular polysaccharide is conjugated variously to diphtheria toxoid, tetanus toxoid or an outer membrane protein of *Neisseria meningitidis* (group B).

13.4.2 The impact of genetic engineering on vaccine technology

The advent of recombinant DNA technology has rendered possible the large-scale production of polypeptides normally present on the surface of virtually any pathogen. These polypeptides, when purified from the producer organism (e.g. *E. coli, Saccharomyces cerevisiae*) can then be used as 'subunit' vaccines. This method of vaccine production exhibits several advantages over conventional vaccine production methodologies. These include:

- Production of a clinically safe product; the pathogen-derived polypeptide now being expressed in a non-pathogenic recombinant host. This all but precludes the possibility that the final product could harbour undetected pathogen.

- Production of subunit vaccine in an unlimited supply. Previously, production of some vaccines was limited by supply of raw material (e.g. hepatitis B surface antigen; see below).

- Consistent production of a defined product that would thus be less likely to cause unexpected side effects.

A number of such recombinant (subunit) vaccines have now been approved for general medical use (Table 13.9). The first such product was that of hepatitis B surface antigen (rHBsAg), which gained marketing approval from the FDA in 1986. Two billion people are infected with hepatitis B worldwide, 350 million individuals suffer from life-long chronic infection, and more

Table 13.9 Recombinant subunit vaccines approved for human use

Product	Company	Indication
Recombivax (rHBsAg produced in *S. cerevisiae*)	Merck	Hepatitis B prevention
Comvax (combination vaccine, containing rHBsAg produced in *S. cerevisiae*, as one component)	Merck	Vaccination of infants against *H. influenzae* type b and hepatitis B
Engerix B (rHBsAg produced in *S. cerevisiae*)	Smithkline Beecham	Vaccination against hepatitis B
Tritanrix–HB (combination vaccine, containing rHBsAg produced in *S. cerevisiae* as one component)	Smithkline Beecham	Vaccination against hepatitis B, diphtheria, tetanus and pertussis
Lymerix (rOspA, a lipoprotein found on the surface of *Borrelia burgdorferi*, the major causative agent of Lyme's disease. Produced in *E. coli*)	Smithkline Beecham	Lyme disease vaccine
Infanrix–Hep B (combination vaccine, containing rHBsAg produced in *S. cerevisiae* as one component)	Smithkline Beecham	Immunization against diphtheria, tetanus, pertussis and hepatitis B
Infanrix–Hexa (combination vaccine, containing rHBsAg produced in S. cerevisiae as one component)	Smithkline Beecham	Immunization against diphtheria, tetanus, pertussis, polio, *H. influenzae* b and hepatitis B
Infanrix–Penta (combination vaccine, containing rHBsAg produced in *S. cerevisiae* as one component)	Smithkline Beecham	Immunization against diphtheria, tetanus, pertussis, polio, and hepatitis B
Ambirix (combination vaccine, containing rHBsAg produced in *S. cerevisiae* as one component)	Glaxo SmithKline	Immunization against hepatitis A and B
Twinrix, adult and pediatric forms in EU (combination vaccine containing rHBsAg produced in *S. cerevisiae* as one component)	Smithkline Beecham (EU) Glaxo Smithkline (USA)	Immunization against hepatitis A and B
Primavax (combination vaccine, containing rHBsAg produced in *S. cerevisiae* as one component)	Pasteur Merieux MSD	Immunization against diphtheria, tetanus and hepatitis B
Procomvax (combination vaccine, containing rHBsAg as one component)	Pasteur Merieux MSD	Immunization against *H. influenzae* type b and hepatitis B
Hexavac (combination vaccine, containing rHBsAg produced in *S. cerevisiae* as one component)	Aventis Pasteur	Immunization against diphtheria, tetanus, pertussis, hepatitis B, polio and *H. influenzae* type b
Triacelluvax (combination vaccine containing r(modified) pertussis toxin	Chiron SpA	Immunization against diphtheria, tetanus and pertussis
Hepacare (r S, pre-S and pre-S2 hepatitis B surface antigens, produced in a mammalian (murine) cell line)	Medeva Pharma	Immunization against hepatitis B
HBVAXPRO (rHBsAg produced in *S. cerevisiae*)	Aventis Pharma	Immunization of children & adolescents against hepatitis B
Dukoral (combination vaccine containing rCholera toxin B subunit as one component	SBL Vaccin AB	Active immunization against diseases caused by *Vibrio cholerae*

than 1 million infected patients die each year from the associated complications of liver cirrhosis and/or liver cancer. Prior to its approval, hepatitis B vaccines consisted of HBsAg purified directly from the blood of hepatitis B sufferers. When present in blood, HBsAg exists not in monomeric form, but in characteristic polymeric structures that display a diameter of 22 μm. Production of hepatitis B vaccine by direct extraction from blood suffered from two major disadvantages:

- The supply of finished vaccine was restricted by the availability of infected human plasma.

- The starting material will likely be contaminated by intact, viable hepatitis B viral particles (and perhaps additional viruses, such as HIV). This necessitates introduction of stringent purification procedures to ensure complete removal of any intact viral particles from the product stream. A final product QC test to confirm this entails a 6-month safety test on chimpanzees.

The HBsAg gene has been cloned and expressed in a variety of expression systems, including *E. coli*, *S. cerevisiae* and a number of mammalian cell lines. The product used commercially is produced in *S. cerevisiae*. The yeast cells are not only capable of expressing the gene, but also assembling the resultant polypeptide product into particles quite similar to those found in the blood of infected individuals. This product proved safe and effective when administered to both animals and humans. An overview of its manufacturing process is presented in Figure 13.10.

Various other companies have also produced recombinant HBsAg-based vaccines. SmithKline Beecham secured FDA approval for such a product (tradename Engerix-B) in 1989 (Figure 13.11 and Box 13.4). Subsequently, SmithKline Beecham have also generated various combination vaccines in which recombinant HBsAg is a component. Twinrix (tradename), for example, contains a mixture of inactivated hepatitis A virus and recombinant HBsAg. Tritanrix, on the other hand, contains diphtheria and tetanus toxoids (produced by traditional means), along with recombinant HBsAg. Dukoral is the tradename given to an additional recombinant protein-containing vaccine now on the market. Indicated for active immunization against disease caused by *Vibrio cholerae* (serogroup 01), the product contains recombinant cholera toxin subunit B and four whole (heat- or formalin-inactivated) *V. cholerae* strains.

It seems likely that many such (recombinant) subunit vaccines will gain future regulatory approval. One such example is that of *Bordetella pertussis* subunit vaccine. *B. pertussis* is a Gram-negative coccobacillus. It is transmitted by droplet infection and is the causative agent of the upper respiratory tract infection commonly termed 'whooping cough'.

13.4.3 Peptide vaccines

An alternative approach to the production of subunit vaccines entails their direct chemical synthesis. Peptides identical in sequence to short stretches of pathogen-derived polypeptide antigens can be easily and economically synthesized. The feasibility of this approach was first verified in the 1960s, when a hexapeptide purified from the enzymatic digest of tobacco mosaic virus was found to confer limited immunological protection against subsequent administration of the intact virus. (The hexapeptide hapten was initially coupled to bovine serum albumin, used as a carrier to ensure an immunological response.)

Similar synthetic vaccines have also been constructed that confer immunological protection against bacterial toxins, including diphtheria and cholera toxins. Although coupling to a carrier is

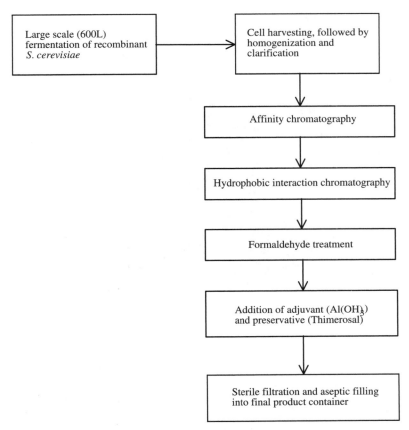

Figure 13.10 Overview of the production of recombinant HBsAg vaccine (Recombivax HB; Merck). A single dose of the product generally contains 10 μg of the antigen

generally required to elicit and immunological response, some carriers are inappropriate due to their ability to elicit a hypersensitive reaction, particularly when repeat injections are undertaken. Such difficulties can be avoided by judicious choice of carrier. Often, a carrier normally used for vaccination itself is used. For example, tetanus toxoid has been used as a carrier for peptides derived from influenza haemaglutanin and *Plasmodium falciparum*.

13.4.4 Vaccine vectors

An alternative approach to the development of novel vaccine products entails the use of live vaccine vectors. The strategy followed involves incorporation of a gene/cDNA coding for a pathogen-derived antigen into a non-pathogenic species. If the resultant recombinant vector expresses the gene product on its surface, then it may be used to immunize against the pathogen of interest (Figure 13.12).

Most vaccine vectors developed to date are viral based, with poxviruses (as well as picorna viruses and adenoviruses) being used most. In general, such recombinant viral vectors elicit both

Figure 13.11 Photographs illustrating some cleanroom-based processing equipment utilized in the manufacture of SmithKline Beecham's hepatitis B surface antigen product. (a) A chromatographic fractionation system, consisting of (from left to right) fraction collector, control tower and chromatographic columns (stacked formation). (b) Some of the equipment used to formulate the vaccine finished product. Photograph courtesy of SmithKline Beecham Biologicals s.a., Belgium

Box 13.4

Product case study: Engerix B

Engerix B (tradename) is a subunit vaccine containing purified recombinant hepatitis B surface antigen (HBsAg) that gained approval in the USA in 1998. It is indicated for active immunization against infection caused by all known serotypes of hepatitis B virus.

The rHBsAg is produced in an engineered *S. cerevisiae* strain and is likely purified subsequent to fermentation by a procedure somewhat similar to that presented in Figure 13.10. The final product is presented as a sterile suspension of the antigen absorbed onto aluminium hydroxide (adjuvant), in either single-use vials or pre-filled syringes. It also contains NaCl and phosphate buffer components as excipients. It is intended for i.m. injection, usually as 10 µg in a volume of 0.5 ml for infants/children or 20 µg (in 1.0 ml) for adults. The normal dosage schedule entails initial administration followed by boosters after 1 and 6 months.

The protective efficacy of Engerix B has been demonstrated in a number of trials, in the context of infants, children and adults. Seroprotection rates (measured as serum anti-hepatitis B antibody titres above a value of 10 mIU ml^{-1}) of over 95 per cent were usually recorded. The product was found to be generally well tolerated. The most frequently reported adverse effects were local reactions at the injection sites, fever, headache and dizziness. Special consideration to risk:benefit ratio should be given to MS patients, as exacerbations of this condition have been (rarely) reported following administration of hepatitis B and other vaccines. Engerix B is manufactured and marketed by GlaxoSmithKline.

strong humoral and, in particular, cell-mediated immunity. The immunological response (especially the cell-mediated response) to subunit vaccines is usually less pronounced.

Poxviruses and, more specifically, the vaccinia virus remain the most thoroughly characterized vector systems developed. These are large, enveloped double-stranded DNA viruses. They are the only DNA-containing viruses that replicate in the cytoplasm of infected cells. The most studied members of this family are variola and vaccinia. The former represents the causative agent of smallpox, and the latter (being antigenically related to variola but non-pathogenic) was used to immunize against smallpox. Vaccinia-based vaccination programmes led to the global eradication of smallpox, finally achieved by the early 1980s.

A number of factors render vaccinia virus a particularly attractive vector system. These include:

- capacity to assimilate large quantities of DNA in its genome successfully;
- prior history of widespread and successful use as a vaccination agent;
- ability to elicit long-lasting immunity;
- ease of production and low production costs;
- stability of freeze-dried finished vaccine product.

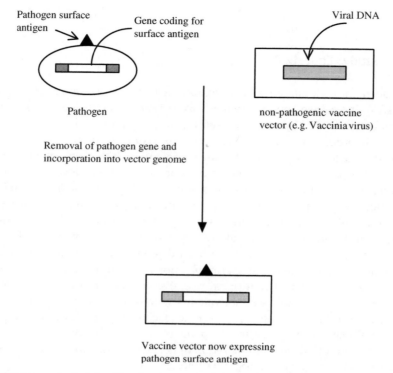

Figure 13.12 Strategy adopted for the development of an engineered vaccine vector. Refer to text for additional detail

Adenoviruses also display potential as vaccine vectors. These double-stranded DNA viruses display a genome consisting of in the region of 36 000 bp, encoding approximately 50 viral genes. Several antigenically distinct human adenovirus serotypes have been characterized, and these viral species are endemic throughout the world. They can prompt respiratory tract infections and, to a lesser extent, gastrointestinal and genitourinary tract infections.

Live adenovirus strains that cause asymptomatic infection and which have proven to be very safe and effective adenovirus vaccines have been isolated. Unlike vaccinia, few sites exist in the adenoviral genome into which foreign DNA can be integrated without comprising viral function. Furthermore, packing limitations curb the quantity of foreign DNA that can be accommodated in the viral genome. However, an approximately 3000 bp region can be removed from a section of the genome termed the E3 region. This facilitates incorporation of pathogen-derived or other DNA at this point.

Recombinant adenoviruses containing the hepatitis B surface antigen gene, the HIV P160 gene, the respiratory syncytial virus F gene, and the herpes simplex virus glycoprotein B gene have all been generated using this approach. Many have been tested in animal models and have been found to elicit humoral and cell-mediated immunity against the pathogen of interest.

The use of recombinant viral vectors as vaccination tools displays considerable clinical promise. One potential complicating factor, however, centres around the possibility that previous recipient exposure to the virus being used as a vector would negate the therapeutic efficacy of the product. Such prior exposure would likely indicate the presence of circulating immune memory

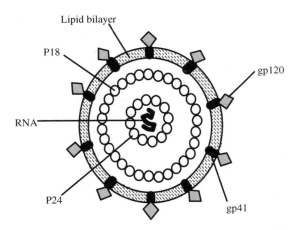

Figure 13.13 Simplified schematic representation of a cross-section of HIV. The central core contains the viral RNA, consisting of two identical single-strand subunits (approximately 9.2 kb long). Associated with the RNA are two (RNA-binding) proteins, P7 and P9, as well as the viral reverse transcriptase complex (not shown above). Surrounding this is the protein P24, which forms the shell of the nuclear caspid. Covering this, in turn, is a lipid bilayer derived from the host cell, still carrying some host cell antigens. The viral protein, P18, is associated with the inner membrane leaflet. Viral gp41 represents a transmembrane protein, and viral gp120, residing on the outside of the lipid bilayer, is attached to gp41 via disulfide bonds

cells that could initiate an immediate immunological response upon re-entry of the virus into the host. Studies involving repeat administration of vaccinia virus have, to some extent, confirmed this possibility. However, the degree to which such an effect limits the applicability of this approach in a clinical setting remains to be elucidated.

13.4.5 Development of an AIDS vaccine

AIDS was initially described in the U.S. in 1981, although sporadic cases probably occurred for at least two decades prior to this. By 1983, the causative agent, now termed HIV, was identified. HIV is a member of the lentivirus subfamily of retroviruses. It is a spherical, enveloped particle, 100–150 nm in diameter, and contains RNA as its genetic material (Figure 13.13).

The viral surface protein, gp 120, is capable of binding to a specific site on the CD4 molecule, found on the surface of susceptible cells (Table 13.10). Some CD4 negative cells may (rarely) also become infected, indicating the existence of an entry mechanism independent of CD4.

Table 13.10 Some cell types whose susceptibility to infection by HIV is believed to be due to the presence of the CD4 antigen on their surface

T-helper lymphocytes
Blood monocytes
Tissue macrophages
Dendritic cells of skin and lymph nodes
Brain microglia

Infection of CD4$^+$ cells commences via interaction between gp 120 and the CD4 glycoprotein, which effectively acts as the viral receptor. Entry of the virus into the cell, which appears to require some additional cellular components, occurs via endocytosis and/or fusion of the viral and cellular membranes. The gp 41 transmembrane protein plays an essential role in this process.

Once released into the cell, the viral RNA is transcribed (by the associated viral reverse transcriptase) into double-stranded DNA. The retroviral DNA can then integrate into the host cell genome (or, in some instances, remain unintegrated). In resting cells, transcription of viral genes usually does not occur to any significant extent. However, commencement of active cellular growth/differentiation usually also triggers expression of proviral genes and, hence, synthesis of new viral particles. Aggressive expression of viral genes usually leads to cell death. Some cells, however (particularly macrophages), often permit chronic low-level viral synthesis and release without cell death.

Entry of the virus into the human subject is generally accompanied by initial viral replication, lasting a few weeks. High-level viraemia (presence of viral particles in the blood) is noted and p24 antigen can be detected in the blood. Clinical symptoms associated with the initial infection include an influenza-like illness, joint pains and general enlargement of the lymph nodes. This primary viraemia is brought under control within 3–4 weeks. This appears to be mediated largely by HIV-specific cytotoxic T-lymphocytes, indicating the likely importance of cell-mediated immunity in bringing the initial infection under control. Although HIV-specific antibodies are also produced at this stage, effective neutralizing antibodies are detected mainly after this initial stage of infection.

After this initial phase of infection subsides, the free viral load in the blood declines, often to almost undetectable levels. This latent phase may last for anything up to 10 years or more. During this phase, however, there does seem to be continuous synthesis and destruction of viral particles. This is accompanied by a high turnover rate of (CD4$^+$) T-helper lymphocytes. The levels of these T-lymphocytes decline with time, as does antibody levels specific for viral proteins. The circulating viral load often increases as a result, and the depletion of T-helper cells compromises general immune function. As the immune system fails, classical symptoms of AIDS-related complex (ARC) and, finally, full-blown AIDS begin to develop.

In excess of 40 million individuals are now thought to be infected by HIV. In 2001 alone, it was estimated that 3 million people died from AIDS and a further 5 million became infected with the virus. Over 20 million people in total are now thought to have died from AIDS. The worst affected geographical region is the southern half of Africa (Table 13.11). Some 90 per cent of sufferers live in poorer world regions. So far, no effective therapy has been discovered, and the main hope of eradicating this disease lies with the development of safe, effective vaccines. The first such putative vaccine entered clinical trials in 1987; but, thus far, no effective vaccine has been developed.

Table 13.11 WHO estimated numbers of individuals infected with HIV by the end of 2005. Almost 75 per cent of these live in the southern half of Africa

World region	Numbers infected
Sub-Saharan Africa	25.8 million
South and South East Asia	7.4 million
Latin America	1.8 million
North America	1.2 million
Eastern Europe and central Asia	1.6 million
North Africa and Middle East	0.51 million
Western and central Europe	0.72 million

13.4.6   Difficulties associated with vaccine development

A number of attributes of HIV and its mode of infection conspire to render development of an effective vaccine less than straightforward. These factors include:

- HIV displays extensive genetic variation even within a single individual. Such genetic variation is particularly prominent in the viral *env* gene whose product, gp 160, is subsequently proteolytically processed to yield gp 120 and gp 41.

- HIV infects and destroys T-helper lymphocytes, i.e. it directly attacks an essential component of the immune system itself.

- Although infected individuals display a wide range of antiviral immunological responses, these ultimately fail to destroy the virus. A greater understanding of what elements of immunity are most effective in combating HIV infection is required.

- After initial virulence subsides, large numbers of cells harbour unexpressed proviral DNA. The immune system has no way of identifying such cells. An effective vaccine must thus induce the immune system to (a) bring the viral infection under control before cellular infection occurs or (b) destroy cells once they begin to produce viral particles and destroy the viral particles released.

- The infection may often be spread not via transmission of free viral particles, but via direct transmission of infected cells harbouring the proviral DNA.

13.4.7   AIDS vaccines in clinical trials

A number of approaches are being assessed with regard to developing an effective AIDS vaccine. No safe attenuated form of the virus has been recognized to date, nor is one likely to be developed in the foreseeable future. The high level of mutation associated with HIV would, in any case, heighten fears that spontaneous reversion of any such product to virulence would be possible.

The potential of inactivated viral particles as effective vaccines has gained some attention, but again fears of accidental transmission of disease if inactivation methods are not consistently 100 per cent effective have dampened enthusiasm for such an approach. In addition, the stringent containment conditions required to produce large quantities of the virus render such production processes expensive.

Not withstanding the possible value of such inactivated viral vaccines, the bulk of products assessed to date are subunit vaccines. Live vector vaccines expressing HIV genes have also been developed and are now coming to the fore (Table 13.12).

Much of the preclinical data generated with regard to these vaccines entailed the use of one of two animal model systems: simian immunodeficiency virus infection of macaque monkeys and HIV infection of chimpanzees. Most of the positive results observed in such systems have been in association with the chimp–HIV model. However, no such system can replace actual testing in humans.

Most of the recombinant subunit vaccines tested in the first half of this decade employed gp 120 or gp 160 expressed in yeast, insect or mammalian (mainly CHO) cell lines. Eukaryotic systems facilitate glycosylation of the protein products. Like all subunit vaccines, these stimulate a humoral-based immune response but fail to elicit a strong T-cell response. The failure to elicit a cell-based

Table 13.12 Some putative HIV vaccines which have made it to clinical trials

Vaccine preparation	Developing company
Inactivated viral particles	Immune Response
rgp 120 subunit vaccines	Genentech/Vaxgen, Biocine, Chiron/Ciba Geigy
rgp 160 subunit vaccines	MicroGenes Sys. Inc., Immuno-Ag.
rp 24 subunit vaccines	MicroGenes Sys. Inc.
Live vaccines based on viral vectors	Biocine, Merck, Sanofi Pasteur, Targeted Genetics
Octameric V3 peptide	UBI

response, in particular a cytotoxic T-cell response (now seen as critical to mounting an effective immune response), explains at least in part why subunit vaccines were a clinical disappointment.

Several HIV vaccine systems based upon live vectors have also been developed in an attempt to stimulate a significant T-cell, as well as B-cell, immune response. Both envelope and core antigens have been expressed in a number of recombinant viral systems. The clinical efficacy of these remains to be established. Expression in engineered vaccinia has been undertaken, but its use as an HIV vaccine is likely precluded by the fact that the virus can apparently disseminate and cause fatal encephalitis in immunosuppressed infected individuals. Modified vaccinia Ankara, canarypox and fowlpox viruses have come to the fore as vectors. These likely can produce sufficient protein to initiate both a humoral and cellular immune response during an abortive replication cycle in humans. A vaccination schedule variation of potential interest entails the use of a vector-based primary dose (to induce a cellular response in particular) followed by a subunit-based booster (to induce mainly a humoral response). Whatever the schedule however, the induction of effective (i.e. broadly neutralizing) antibodies remains a challenge, as the regions of HIV envelope proteins that are most highly conserved seem to be shielded from antibody access by loop structures and sugar side-chains on these surface proteins.

Large-scale clinical trials are likely to be the only way by which any HIV vaccine may be properly assessed. In addition, a greater understanding of the molecular interplay between the virus and immune system may provide clues as to the development of novel vaccine and/or therapeutic products. For example, a small proportion of infected individuals remain clinically asymptomatic for periods considerably greater than the average 10–15 years. An understanding of the immunological or other factors that delay onset of ARC/full-blown AIDS in these individuals may aid in the design of more effective vaccines.

13.4.8 Cancer vaccines

The identification of tumour-associated antigens could pave the way for the development of a range of cancer vaccines. A number of tumour-associated antigens have already been characterized, as described previously. Theoretically, administration of tumour-associated antigens may effectively immunize an individual against any cancer type characterized by expression of the tumour-associated antigen in question. Co-administration of a strong adjuvant (see Section 13.5) would be advantageous, as it would stimulate an enhanced immune response. This is important, as many tumour-associated antigens appear to be weak immunogens. Administration of subunit-based tumour-associated antigen vaccines would primarily stimulate a humoral immune response.

The use of viral vectors may ultimately prove more effective, as a T-cell response appears to be central to the immunological destruction of cancer cells.

The latter approach has been adopted in experimental studies involving malignant melanoma. These transformed cells express significantly elevated levels of a surface glycoprotein, p97. p97 is also expressed (but at far lower levels) on the surface of many normal cell types. Initial animal studies have indicated that administration of a recombinant vaccinia vector expressing p97 has a protective effect against challenge with melanoma cells. However, protracted safety studies would be required in this, or similar, instances to prove that such vaccines would not, for example, induce an autoimmune response if the antigen was not wholly tumour specific. The development of truly effective cancer vaccines probably requires a more comprehensive understanding of the transformed phenotype and how these cells normally evade immune surveillance in the first place. Not withstanding this, limited clinical studies in this field have already begun.

13.4.9 Recombinant veterinary vaccines

Amongst the limited number of biopharmaceuticals approved for animal use, recombinant vaccines represent the single largest subgroup. Several such products target pigs, including 'Porcilis pesti' and 'Bayovac CSF E2'. Porcilis pesti, for example, contains a recombinant form of the classical swine fever virus E2 antigen, the immunodominant surface antigen associated with this viral pathogen. It is used to immunize young pigs. An overview of its manufacture is presented in Figure 13.14. The process is initiated by growth of *Spodoptera frugiperda* cells, typically in a 500 l fermenter. The cells are then infected with the recombinant baculovirus vector, resulting in high-level expression of the recombinant E2 antigen. The antigen is harvested from the production medium by low-speed centrifugation and membrane filtration steps, which serve to remove intact

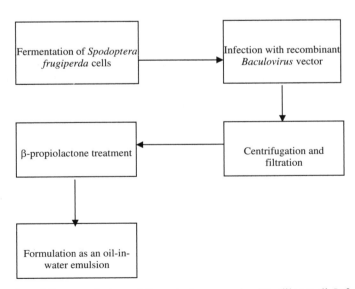

Figure 13.14 Overview of the manufacture of the veterinary vaccine 'Porcilis pesti'. Refer to text for specific details

cells/cellular debris. The antigen-containing supernatant is then treated with β-propiolacetone in order to inactivate viral particles present. The antigen is not subjected to subsequent high-resolution chromatographic purification steps, and hence is not purified to homogeneity. The product is then formulated as an oil-in-water emulsion.

13.5   Adjuvant technology

Administration of many vaccines on their own stimulates a poor host immunological response. This is particularly true of the more recently developed subunit vaccines. An adjuvant is defined as any material that enhances the cellular and/or humoral immune response to an antigen. Adjuvants thus generally elicit an earlier, more potent and longer-lasting, immunological reaction against co-administered antigen. In addition, the use of adjuvants can often facilitate administration of reduced quantities of antigen to achieve an adequate immunological response. This implies consequent economic savings, as vaccines (particularly subunit and vector vaccines) are far more expensive to produce than the adjuvant.

A number of different adjuvant preparations have been developed (Table 13.13). Most preparations also display some associated toxicity and, as a general rule, the greater the product's adjuvanticity, the more toxic it is likely to be. A few different adjuvants may be used in veterinary medicine; however (for safety reasons), aluminium-based products are the only adjuvants routinely used in human medicine. Application of many of the aggressive adjuvant materials is reserved for selected experimentation purposes in animals.

An ideal adjuvant should display several specific characteristics. These include:

- safety (no unacceptable local/systemic responses);

- elicit protective immunity, even against weak immunogens;

Table 13.13 Overview of the adjuvant preparations that have been developed to date, or are under investigation. Of these, aluminium-based substances are the only adjuvants used to any significant degree in humans. Calcium phosphate and oil emulsions find very limited application in human medicine

Mineral compounds	Aluminium phosphate ($AlPO_4$)
	Aluminium hydroxide ($Al(OH)_3$)
	Alum ($AlK(SO_4)_2 \cdot 12H_2O$)
	Calcium phosphate ($CaPO_4$)
Bacterial products	Mycobacterial species
	Mycobacterial components, e.g. trehalose dimycolate (TDM), muramyl dipeptide (MDP)
	Corynebacterium species
	B. pertussis
	LPS
Oil-based emulsions	Freund's complete/incomplete adjuvant (FCA/FIA)
	Starch oil
Saponins	Quil A
Liposomes	
Immunostimulatory complexes (ISCOMs)	
Some cytokines	IL-1 and -2

- be non-pyrogenic;

- be chemically defined (facilitates consistent manufacture and QC testing);

- be effective in infants/young children;

- yield stable formulation with antigen;

- be biodegradable;

- be non-immunogenic itself.

13.5.1 Adjuvant mode of action

Adjuvants are a heterogeneous family of substances, both in terms of their chemical structure and their mode of action. The observed adjuvanticity of any such substance may be due to one or more of the following factors:

- Depot formation of antigen. This results in the subsequent slow release of the antigen from the site of injection, which, in turn, ensures its prolonged exposure to the immune system.

- Enhanced antigen presentation to the cells of the immune system.

- The direct induction of immunostimulatory substances, most notably interleukins and other cytokines.

13.5.2 Mineral-based adjuvants

A number of mineral-based substances display an adjuvant effect. Although calcium phosphate, calcium chloride and salts of various metals (e.g. zinc sulfate and cerium nitrate) display some effect, aluminium-based substances are by far the most potent. Most commonly employed are aluminium hydroxide and aluminium phosphate (Table 13.13). Their adjuvanticity, coupled to their proven safety, render them particularly valuable in the preparation of vaccines for young children. They have been incorporated into millions of doses of such vaccine products so far.

The principal method by which aluminium adjuvanted vaccines are prepared entails mixing the antigen in solution with a preformed aluminium phosphate (or hydroxide) precipitate under chemically defined conditions (e.g. of pH). Adsorption of the antigen to the aluminium-based gel ensues, with such preparations being generally termed 'aluminium-adsorbed vaccines'; 1 mg of aluminium hydroxide will usually adsorb in the order of 50–200 µg of protein.

The major mode of action of such products appears to be depot formation at the site of injection. The antigen is only slowly released from the gel, ensuring its sustained exposure to immune surveillance. The aluminium compounds are also capable of activating complement. This can lead to a local inflammatory response, with consequent attraction of immunocompetent cells to the site of action.

Despite their popularity, aluminium-based adjuvants suffer from several drawbacks. They tend to stimulate only the humoral arm of the immune response effectively. They cannot be frozen or lyophylized, as either process promotes destruction of their gel-based structure. In addition, aluminium-based products display poor/no adjuvanticity when combined with some antigen (e.g. typhoid or *Haemophilus influenzae* type b capsular polysaccharides).

13.5.3 Oil-based emulsion adjuvants

The adjuvanticity of oil emulsions was first recognized in the early 1900s. However, the first such product to gain widespread attention was Freund's complete adjuvant (FCA), developed in 1937. This product essentially contained a mixture of paraffin (i.e. mineral) oil with dead mycobacteria, formulated to form a water-in-oil emulsion. Arlacel A (mannide mono-oleate) is usually added as an emulsifier.

Freund's incomplete adjuvant (FIA) is a similar product. It differs from FCA in that it lacks the mycobacterial component and, consequently, displays somewhat lesser adjuvanticity. The mode of action of FIA is largely attributed to depot formation. The mycobacterial components in FCA have additional direct immunostimulatory activities. Although it is one of the most potent adjuvant substances known, FCA is too toxic for human use.

Latterly, some oil-in-water adjuvants have been developed. Many are squalene-in-water emulsions. Emulsifiers most commonly used include polyalcohols, such as Tween and Span. In some cases, immunostimulatory molecules (including MDP and TDM; see Section 13.5.4) have also been incorporated in order to enhance adjuvanticity. These continue to be carefully assessed and may well form a future family of useful adjuvant preparations.

13.5.4 Bacteria/bacterial products as adjuvants

Selected microorganisms have been identified that can trigger particularly potent immunological responses. The immunostimulatory properties of these cells have generated interest in their potential application as adjuvants. Examples include various mycobacteria, *Corynebacterium parvum*, *Corynebacterium granulosum* and *B. pertussis*. Although some such microorganisms are used as antigens in vaccines, they are considered too toxic to be used solely in the role of adjuvant. Researchers have thus sought to identify the specific microbial biomolecules responsible for the observed immunostimulatory activity. It was hoped that these substances, when purified, might display lesser/no toxic side effects while retaining their immunostimulatory capacity.

Fractionation of mycobacteria resulted in the identification of two cellular immunostimulatory components, namely TDM and MDPs. Both are normally found in association with the mycobacterial cell wall. TDM is composed of a molecule of trehalose (a disaccharide consisting of two molecules of α-D-glucose linked via an α 1–1 glycosidic bond), linked to two molecules of mycolic acid (a long-chain aliphatic hydrocarbon-based acid) found almost exclusively in association with mycobacteria. TDM, although retaining its adjuvanticity, is relatively non-toxic.

The structure of the native immunostimulatory MDPs was found to be *N*-acetyl muramyl-L-alanyl-D-isoglutamine. (*N*-Acetyl muramic acid is a base component of bacterial peptidoglycan.) Native TDM is a potent pyrogen and is too toxic for general use as an adjuvant. The molecular basis underlying MDP's adjuvanticity remains to be fully elucidated. Administration of MDP, however, is known to activate a number of cell types that play direct/indirect roles in immune function, and induces the secretion of various immunomodulatory cytokines (Table 13.14).

Table 13.14 Some cell types activated upon administration of MDP. Activation induces synthesis of a range of immunomodulatory cytokines by these (and other) cells

Cell types activated	Macrophages
	Mast cells
	Polymorphonuclear leukocytes
	Endothelial cells
	Fibroblasts Platelets
Cytokines and other molecules induced	IL-1
	CSFs
	Fibroblast activating factor
	B-cell growth factor
	Prostaglandins

A number of derivatives were synthesized in the hope of identifying a modified form that retained its adjuvanticity but displayed lesser toxicity. Some such derivatives, most notably threonyl-MDP, muramyl tripeptide and murabutide, display some clinical promise in this regard.

Threonlyn-MDP, for example, has been included in the formulation known as Syntex adjuvant formulation-1 (SAF-1). Animal studies suggest that this adjuvant is non-toxic and elicits a good B- and T-cell response.

An additional bacterial component displaying appreciable adjuvanticity is the *C. granulosum*-derived p40 particulate fraction. p40 is composed of fragments of cell wall peptidoglycan and associated glycoproteins.

13.5.5 Additional adjuvants

In addition to the immunostimulatory substances discussed above, the adjuvanticity of a variety of other substances is also being appraised. These include saponins, liposomes and ISCOMs.

Saponins are a family of glycosides (sugar derivatives) widely distributed in plants. Each saponin consists of a sugar moiety bound to a 'sapogenin' (either a steroid or a triterpene). The immunostimulatory properties of the saponin fraction isolated from the bark of *Quillaja* (a tree) has been long recognized. Quil A (which consists of a mixture of related saponins) is used as an adjuvant in selected veterinary vaccines. However, its haemolytic potential precludes its use in human vaccines. Research efforts continue in an attempt to identify individual saponins (or derivatives thereof) that would make safe and effective adjuvants for use in human medicine.

Liposomes are membrane-based supramolecular particles that consist of a number of concentric lipid membrane bilayers separated by aqueous compartments (Figure 13.15). They were developed initially as carriers for therapeutic drugs. Initially, the bilayers were almost exclusively phospholipid based. More recently, non-phospholipid-based liposomes have been developed.

The adjuvanticity of liposomes depends upon their composition, number of layers and charge characteristics. They act as effective adjuvants for both protein- and carbohydrate-based antigens and help stimulate both B- and T-cell responses. Their likely mode of action includes depot formation, but they also possibly increase/enhance antigen presentation to macrophages. The exact molecular mechanism(s) by which they stimulate a T-cell response remains to be elucidated,

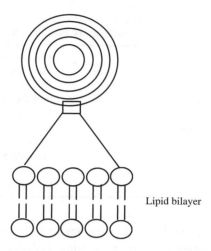

Lipid bilayer

Figure 13.15   Generalized liposome structure. Refer to text for details

but it appears to be associated with their hydrophobicity. Liposomes are likely to gain more widespread use as adjuvants when technical difficulties, associated with their stability and consistent/reproducible production, are resolved.

ISCOMs are stable (non-covalent) complexes composed of a mixture of Quil A, cholesterol and (an amphipathic) antigen. ISCOMs stimulate both humoral and cellular immune responses and have been used in the production of some veterinary vaccines. Their use in humans, however, has not been licensed so far, mainly due to safety concerns relating to the Quil A component.

In summary, therefore, a whole range of adjuvants have thus far been identified/developed. Problems of toxicity have precluded the use of many of these adjuvants (particularly in humans). However, research efforts continue in an attempt to develop the next generation of safe and, hopefully, even more effective vaccine adjuvants.

Further reading

Books

Amyes, S. 2002. *Tumor Immunology*. Taylor and Francis.
Eisenstein, T. (ed.). 2003. *Hepatitis B, the Virus, the Disease and the Vaccine*. Kluwer.
Grossbard, M. 1998. *Monoclonal Antibody Based Therapy of Cancer*. Dekker.
Hackett, C. and Harn, D. 2003. *Vaccine Adjuvants*. Humana Press.
Harris, W. 1997. *Antibody Therapeutics*. CRC Press.
Kontermann, R. 2001. *Antibody Engineering*. Springer Verlag.
Lo, B. 2003. *Antibody Engineering*. Humana Press.
Plotkin, S. 1999. *Vaccines*. W.B. Saunders.
Powell, M. 1995. *Vaccine Design: The Subunit and Adjuvant Approach*. Plenum.
Stern, P. 2000. *Cancer vaccines and immunotherapy*. Cambridge University Press.
Subramanian, G. (ed.). 2004. *Antibodies*. Kluwer.
Woodrow, G. 1997. *New Generation Vaccines*. Marcel Dekker.

Articles

Antibody technology

Benhar, I. 2001. Biotechnological applications of phage and cell display. *Biotechnology Advances* **19**, 1–33.

Berger, M., Shankar, V., and Vafai, A. 2002. Therapeutic applications of monoclonal antibodies. *American Journal of the Medical Sciences* **324**(1), 14–30.

Chapman, P. 2002. PEGylated antibodies and antibody fragments for improved therapy: a review. *Advanced Drug Delivery Reviews* **54**(4), 531–545.

Dinnis, D. and James, D. 2005. Engineering mammalian cell factories for improved recombinant monoclonal antibody production: lessons from nature? *Biotechnology and Bioengineering* **91**(2), 180–189.

Goldenberg, D. 2002. Targeted therapy of cancer with radiolabeled antibodies. *Journal of Nuclear Medicine* **43**(5), 693–713.

He, M. and Khan, F. 2005. Ribosome display: next generation display technologies for production of antibodies *in vitro*. *Expert Review of Proteomics* **2**(3), 421–430.

Hoogenboom, H.R., de Bruine, A.P., Hufton, S.E., Hoet, R.M., Arends, J.W., and Roovers, R.C. 1998. Antibody phage display technology and its applications. *Immunotechnology* **4**, 1–20.

Joshi, A., Bauer, R., Kuebler, P., White, M., Leddy, C., Compton, P., Garovoy, M., Kwon, P., Walicke, P., and Dedrick, R. 2006. An overview of the pharmacokinetics and pharmacodynamics of efalizumab: a monoclonal antibody approved for use in psoriasis. *Journal of Clinical Pharmacology* **46**(1), 10–20.

Keating, G. and Perry, C., 2002. Infliximab – an updated review of its use in Crohn's disease and rheumatoid arthritis. *Biodrugs* **16**(2), 111–148.

Kohler, G. and Milstein, C. 1975. Continuous culture of fused cells secreting antibody of pre-defined specificity. *Nature* **256**, 495–497.

Liossis, S. and Tsokos, G. 2005. Monoclonal antibodies and fusion proteins in medicine. *Journal of Allergy and Clinical Immunology* **116**(4), 721–729.

Longberg, N. 2005. Human antibodies from transgenic animals. *Nature Biotechnology* **23**(9), 1117–1125.

McCarron, P.A., Olwill, S.A., Marouf, W.M.Y., Buick, R.J., Walker, B., and Scott, C.J. 2005. Antibody conjugates and therapeutic strategies. *Molecular Interventions* **5**(6), 368–380.

O'Mahony, D. and Bishop, M. 2006. Monoclonal antibody therapy. *Frontiers in Bioscience* **11**, 1620–1635.

Schrama, D., Reisfeld, R.A., and Becker, J.C. 2006. Antibody targeted drugs as cancer therapeutics. *Nature Reviews Drug Discovery* **5**(2), 147–159.

Umemura, S., Sekido, Y., Itoh, H., and Osamura, R.Y. 2002. Pathological evaluation of HER2 overexpression for the treatment of metastatic breast cancers by humanized anti-HER2 monoclonal antibody (trastuzumab). *Acta Histochemica et Cytochemica, Kyoto* **35**(2), 77–81.

Walsh, G. 2004. Modern antibody-based therapeutics. *Biopharm International* **17**, 18–25.

Wang, R. 1999. Human tumor antigens: implications for cancer vaccine development. *Journal of Molecular Medicine* **77**(9), 640–655.

Weiner, L. 2006. Fully human therapeutic monoclonal antibodies. *Journal of Immunotherapy* **29**(1), 1–9.

Winter, G. and Milstein, C. 1991. Man-made antibodies. *Nature* **349**, 293–299.

Vaccine technology

Andino, R., Silvera, D., Suggett, S.D., Achacoso, P.L., Miller, C.J., Baltimore, D., and Feinberg, M.B. 1994. Engineering poliovirus as a vaccine vector for the expression of diverse antigens. *Science* **265**, 1448–1451.

Bramwell, V. and Perrie, Y. 2005. The rational design of vaccines. *Drug Discovery Today* **10**(22), 1527–1534.

Brewer, J. 2006. How do aluminium adjuvants work? *Immunology Letters* **102**(1), 10–15.

Cox, J. and Coutter, A. 1997. Adjuvants – a classification and review of their modes of action. *Vaccine* **15**(3), 248–256.

Edelman, R. 2002. The development and use of vaccine adjuvants. *Molecular Biotechnology* **21**(2), 129–148.

Francis, J. and Larche, M. 2005. Peptide-based vaccination: where do we stand? *Current Opinion in Allergy and Clinical Immunology* **5**(6), 537–543.

Hilton, L.S., Bean, A.G., and Lowenthal, J.W. 2002. The emerging role of avian cytokines as immunotherapeutics and vaccine adjuvants. *Veterinary Immunology and Immunopathology* **85**(3–4), 119–128.

Jackson, M. and Myers, A. 2005. Vaccines in the pipeline. *Pediatric Emergency Care* **21**(11), 777–783.

Lemon, S. and Thomas, D. 1997. Drug therapy – vaccines to prevent viral hepatitis. *New England Journal of Medicine* **336**(3), 196–204.

Lollini P.L., Cavallo, F., Nanni, P., and Forni, G. 2006. Vaccines for tumour prevention. *Nature Reviews Cancer* **6**(3), 204–216.

Mason, H.S., Warzecha, H., Mor, T., and Arntzen, C.J. 2002. Edible plant vaccines: applications for prophylactic and therapeutic molecular medicine. *Trends in Molecular Medicine* **8**(7), 324–329.

Petrovsky, N. and Aguilar, J. 2004. Vaccine adjuvants: current state and future trends. *Immunology and Cell Biology* **82**(5), 488–496.

Plotkin, S. 2002. Vaccines in the 21st century. *Hybridoma and Hybridomics* **21**(2), 135–145.

Poland, G.A., Murray, D., and Bonilla-Guerrero, R. 2002. Science, medicine and the future: new vaccine development. *British Medical Journal* **324**(7349), 1315–1319.

Puls, R. and Emery, S. 2006. Therapeutic vaccination against HIV: current progress and future possibilities. *Clinical Science* **110**(1), 59–71.

Sela, M., Arnon, R., and Schechter, B. 2002. Therapeutic vaccines: realities of today and hopes for the future. *Drug Discovery Today* **7**(12), 664–673.

Shams, H. 2005. Recent developments in veterinary vaccinology. *Veterinary Journal* **170**(3), 289–299.

Singh, M. and O'Hagan, D. 2002. Recent advances in vaccine adjuvants. *Pharmaceutical Research* **19**(6), 715–728.

Spearman, P. 2006. Current progress in the development of HIV vaccines. *Current Pharmaceutical Design* **12**(9), 1147–1167.

Stevceva, L. and Ferrari, M 2005. Mucosal adjuvants. *Current Pharmaceutical Design* **11**(6), 801–811.

Taylor, D. 2006. Obstacles and advances in SARS vaccine development. *Vaccine* **24**(7), 863–871.

Vajdy, M. and Singh, M. 2005. The role of adjuvants in the development of mucosal vaccines. *Expert Opinion on Biological Therapy* **5**(7), 953–965.

Yokoyama, N., Maeda, K., and Mikami, T. 1997. Recombinant viral vector vaccines for veterinary use. *Journal of Veterinary Medical Science* **59**(5), 311–322.

14

Nucleic-acid- and cell-based therapeutics

14.1 Introduction

Throughout the 1980s and early 1990s, the term 'biopharmaceutical' had become virtually synonymous with 'proteins of therapeutic use' (Chapter 1). Nucleic-acid-based biopharmaceuticals, too, harbour great potential. Current developments in nucleic-acid-based therapeutics centre around gene therapy, as well as antisense technology (including RNAi) and aptamer technology, all of which are discussed later in this chapter. These technologies have the potential to revolutionize medical practice. Despite all the hype, however, it is important to note that by early 2007 at least, only three nucleic-acid-based products had gained approval worldwide: one antisense-based product (tradename Vitravene), one aptamer (tradename Macugen) and one gene therapy product (tradename Gendicine, approved only in China). In contrast, some 165 protein-based biopharmaceuticals had been approved by early 2007. The full benefit of nucleic-acid-based medicines will accrue only after the satisfactory resolution of several technical difficulties currently impeding their routine medical application.

Cell-based medicines also harbour tremendous potential. Although a small number of such products have gained approval, none is a stem-cell-derived product. Cell-based biopharmaceuticals are discussed towards the end of this chapter.

14.2 Gene therapy

The fundamental principle underpinning gene therapy is theoretically straightforward, but difficult to achieve in practice satisfactorily. The principle entails the stable introduction of a gene into the genetic complement of a cell, such that subsequent expression of the gene achieves a therapeutic goal. The potential of gene therapy as a curative approach for inborn errors of metabolism and other conditions induced by the presence of a defective copy of a specific gene (or genes) is obvious.

Pharmaceutical biotechnology: concepts and applications   Gary Walsh
© 2007 John Wiley & Sons, Ltd   ISBN 978 0 470 01244 4 (HB)   978 0 470 01245 1 (PB)

Table 14.1   Some diseases for which gene-based therapeutic approaches are currently being appraised in clinical trials. Many of these examples are discussed in more detail later in this chapter.

Cancer, various forms	AIDS
Cystic fibrosis	Haemophilia
Familial hyper-cholesterolaemia	Severe combined immunodeficiency diseases (SCID)
Gaucher's disease	α_1-antitrypsin deficiency
Purine nucleoside phosphorylase deficiency	CGD
Rheumatoid arthritis	Peripheral vascular disease

An increased understanding of the molecular basis of various other diseases, including cancer, some infectious diseases (e.g. AIDS) and some neurological conditions, also suggests a role for gene therapy in combating these. Indeed two-thirds of all gene therapy trials conducted to date aim to treat cancer. Table 14.1 lists the major disease types for which a gene therapy treatment is currently being assessed in clinical trials. The first such trial was initiated in the USA in 1989. Thus far, some 1200 different clinical studies have/are being undertaken worldwide. The majority (estimated at 67 per cent) have/are being undertaken in the USA, with most of the remaining trials being undertaken in Europe (mainly in the UK and in Germany). The majority of trials (62 per cent) are in early stage (phase I) and only some 2.2 per cent of all trials have reached phase III. Despite the initial enthusiasm, only a handful of such studies have revealed a therapeutic benefit to the patient.

Moreover, gene therapy, like all other medical interventions, is not without associated risk. A US patient died in 1999 as a result of participating in one such trial. Even more disturbingly, the ensuing FDA investigation unearthed allegations that at least six other deaths attributed to clinical trial treatments had gone unreported to the regulatory agency; and further, that only a fraction of serious adverse effects had been reported. As a result, regulatory regulation and monitoring of gene therapy trials has been increased. Further serious adverse effects, including some fatalities, have been reported subsequently, as discussed later.

Such disappointing results do not reflect any flaw in the concept of gene therapy. They instead reflect the need to develop more effective technical means of accomplishing gene therapy in practice. These initial studies have highlighted the technical innovations required to achieve successful gene transfer and expression. These, in turn, should render future ('second-generation') gene therapy protocols more successful.

14.2.1   Basic approach to gene therapy

The basic approach to gene therapy is outlined in Figure 14.1. The desired gene must usually be packaged into a vector system capable of delivering it safely inside the intended recipient cells. A variety of vectors can be used to effect gene transfer. These include both viruses (particularly retroviruses and adenoviruses) and non-viral carriers, such as plasmid-containing liposomes/lipoplexes (Table 14.2 and Figure 14.2). Each such vector has its own unique set of advantages and disadvantages, as discussed subsequently in this chapter.

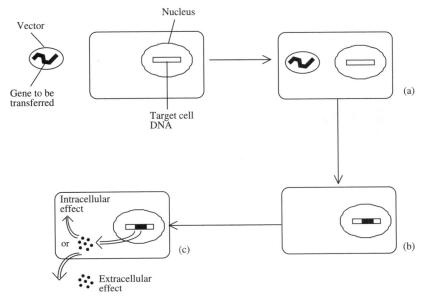

Figure 14.1 Simplified schematic representation of the basis of gene therapy. The genetic material to be transferred is first usually packaged into some form of vector that serves to deliver the nucleic acid to the target cell. (a) Entry of the therapeutic nucleic acid, often still associated with its vector, into the cell cytoplasm. (b) Transfer of the nucleic acid into the nucleus of the recipient cell. This is often, though not always, followed by integration of the foreign genetic material into the cellular DNA. (c) The foreign gene (whether integrated or not) is expressed, resulting in the synthesis of the desired protein product. Regulatory elements of the nucleic acid transferred may be designed to ensure the protein product is retained within the cell, or is exported from the cell, as is necessary. Refer to the text for further details

Once assimilated by the cell, the exogenous nucleic acid must now travel/be delivered to the nucleus. In some cases, the mechanism by which this transfer occurs is understood, at least in part (e.g. in the case of retroviral vectors). In other cases (e.g. use of liposome vectors or naked DNA), this process is less well understood. At a practical level, gene therapy protocols may entail one of three different strategies (Figure 14.3).

Table 14.2 Vector systems used to deliver genes into mammalian cells[a]

Viral-based vector systems	Non-viral-based vector systems
Retroviruses	Nucleic-acid-containing liposomes
Adenoviruses	Molecular conjugates
Adeno-associated virus	Direct injection of naked DNA
Herpes virus	$CaPO_4$ precipitation
Polio virus	Electroporation
Vaccinia virus	Particle acceleration

[a]Prior to 2003 the majority of clinical trials undertaken utilized retroviral vector systems, although adenoviral-based systems have now come to the fore. Non-viral systems have generally been employed least often, although some, e.g. nucleic-acid-containing liposomes, may be used more extensively in the future. Some of the methods tested, e.g. calcium phosphate precipitation, electroporation and particle acceleration, are unlikely to be employed to any great extent in gene therapy protocols.

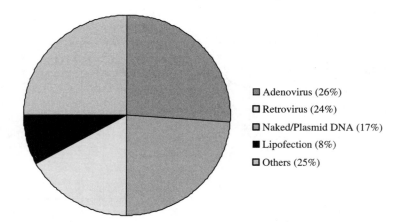

- Adenovirus (26%)
- Retrovirus (24%)
- Naked/Plasmid DNA (17%)
- Lipofection (8%)
- Others (25%)

Figure 14.2   Vectors used thus far in gene therapy trials. 'Others' are mainly viral-based and include the use of pox, vaccinia and adeno-associated viruses, as well as herpes simplex virus. Data adapted from www.wiley. co.uk/genemed/clinical

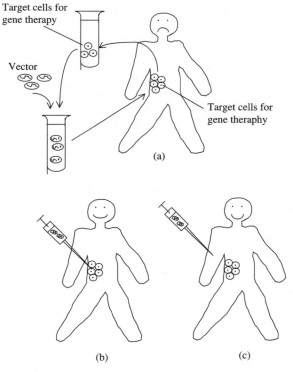

Figure 14.3   The various practical approaches that may be pursued when undertaking gene therapy. (a) *In vitro* gene therapy entails removal of target cells from the body followed by their incubation with nucleic acid-containing vector. After the vector delivers the nucleic acid into the human cells, they are placed back in the body. (b) *In situ* gene therapy entails direct injection of the vector immediately adjacent to the body target cells. (c) *In vivo* gene therapy involves intravenous administration of the vector. The vector has been designed such that it will only recognize and bind the intended target cells. In this way, the nucleic acid is delivered exclusively to those cells. Refer to text for further details

The *in vitro* approach entails initial removal of the target cells from the body. These are then cultured *in vitro* and incubated with vector containing the nucleic acid to be delivered. The genetically altered cells are then reintroduced into the patient's body. This approach represents the most commonly adopted protocol to date. In order to be successful, however, the target cells must be relatively easy to remove from the body, and reintroduce into the body. Such *in vitro* approaches have successfully been undertaken utilizing various body cell types, including blood cells, stem cells, epithelial cells, muscle cells and hepatocytes.

A second approach involves direct injection/administration of the nucleic-acid-containing vector to the target cell, *in situ* in the body. Examples of this approach have included the direct injection of vectors into a tumour mass, as well as aerosol administration of vectors (e.g. containing the cystic fibrosis gene) to respiratory tract epithelial cells.

Although less complicated than the *in vitro* approach, direct *in situ* injection of vector into the immediate vicinity of target cells is not always feasible. This would be true, for example, if the target cells are not localized to one specific area of the body (e.g. blood cells). An alternative (*in vivo*) approach entails the development of vectors capable of recognizing and binding only to specific, predefined cell types. Such vectors could then be administered easily by, for example, i.v. injection. Through appropriate biospecific interactions, they would only deliver their nucleic acid payload to the specified target cells. The simplicity and specificity of this approach renders it the method of choice. However, thus far, no such vector systems have been developed for routine therapeutic use. Intensive efforts to develop these are underway, and a number of different strategies are being pursued. For example, the inclusion of an antibody on the vector surface, which specifically binds a surface-antigen uniquely associated with the target cell, would allow selective delivery. Another approach entails engineering the vector to display a specific hormone that would bind only to cells displaying the hormone receptor. The feasibility of this approach has been demonstrated using retroviral vectors engineered to display EPO on their surface.

14.2.2 Some additional questions

The choice of vector, target cell and protocol used will depend upon a number of considerations. The major consideration is obviously what the ultimate goal of the gene therapy treatment is in any given case. For example, in some instances it may be to correct an inherited genetic defect, whereas in other instances it may be to confer a novel function upon the recipient cell. An example of the former would be the introduction of the cystic fibrosis transmembrane conductance regulator (CFTR) gene (the cystic fibrosis gene) into the airway epithelial cells of cystic fibrosis sufferers. An example of the latter would be the introduction of a novel gene into white blood cells whose protein product is capable of in some way interfering with HIV replication. Such an approach might prove an effective therapeutic strategy for the treatment of AIDS.

An additional consideration that may influence the protocol used is the desired duration of subsequent expression of the gene product. In most cases of genetic disease, long-term expression of the inserted gene would be required. In other instances (e.g. some forms of cancer therapy or the use of gene therapy to deliver a DNA-based vaccine), short-term expression of the gene introduced would be sufficient/desirable.

For most applications of gene therapy, straightforward expression of the gene product itself will suffice. However, in some instances, regulation of expression of the transferred gene would be required (e.g. if gene therapy combating insulin-dependent diabetes mellitus was to be considered). Achieving such expressional control over transferred genes is a pursuit that is only in the early stages of development.

The choice of target cells is another point worthy of discussion. In some instances, this choice is predetermined, e.g. treatment of the genetic condition, familial hypercholesterolaemia, would require insertion of the gene coding for the low-density lipoprotein receptor specifically in hepatocytes.

In other cases, however, some scope may be available to choose a target cell population. Even in the case of redressing some genetic diseases, it may not be necessary to correct genetically the exact population of cells affected. For example, a hallmark of several of the best characterized genetic diseases is the exceedingly low production of a circulatory gene product. Examples include clotting factors VIII and IX, a lack of which leads to haemophilia. It may be possible to correct such defects by introducing the appropriate gene into any recipient cell capable of exporting the gene product into the blood. Is such cases, choosing a target cell could be made upon practical considerations, such as their ease of isolation and culture, their capacity to express (and excrete) the protein product, and their half lives *in vivo*.

Several cell types, including keratinocytes, myoblasts and fibroblasts, have been studied in this regard. It has been shown, for example, that myoblasts, into which the factor IX gene and the growth hormone gene have been introduced, could express their protein products and secrete them into the circulation.

14.3   Vectors used in gene therapy

A list of the various vectors capable of introducing genes into recipient cells has been provided in Table 14.2. These vectors are conveniently categorized as being viral-based or non-viral-based systems. The main vector systems developed, thus far, are discussed in somewhat more detail below.

14.3.1   Retroviral vectors

Some 24 per cent of all gene therapy clinical trials undertaken to date have employed retroviral vectors as gene delivery systems. Retroviruses are enveloped viruses. Their genome consists of ssRNA of approximately 5–8 kb. Upon entry into sensitive cells, the viral RNA is reverse transcribed and eventually yields double-stranded DNA. This subsequently integrates into the host cell genome (Box 14.1). The basic retroviral genome contains a minimum of three structural genes:

Box 14.1

The retroviral life cycle

The retroviral life cycle (Figure 14.B1) begins with the entry of the enveloped virus into the cell. The viral reverse transcriptase enzyme then copies the viral RNA genome into a single (minus) DNA strand and, using this as a template, generates double-stranded DNA. The double-stranded DNA is then randomly integrated into the host cell genome (the proviral DNA). Transcription of the proviral genes host cell's transcription machinery yields mRNA that directs synthesis of mature virion particles. The viral particles bud out from the cell's plasma membrane, picking up a membrane-derived outer coat as they do so.

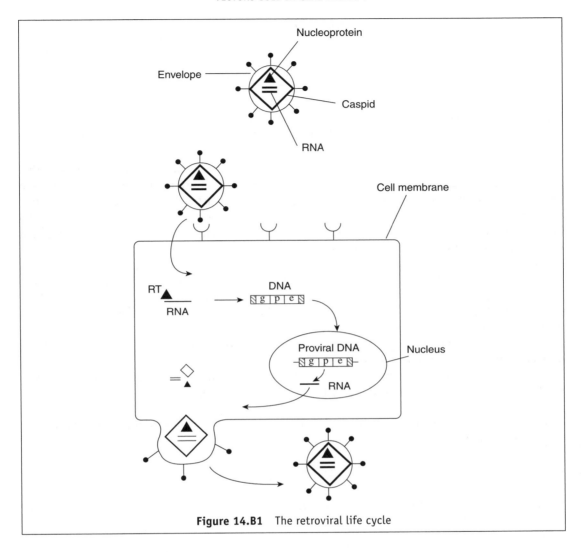

Figure 14.B1 The retroviral life cycle

gag (codes for core viral protein), *pol* (codes for reverse transcriptase) and *env* (codes for the viral envelope proteins). At either end of the viral genome are the long terminal repeats (LTRs), which harbour powerful promoter and enhancer regions and sequences required to promote integration into the host DNA. Also present, immediately adjacent to the 5′ LTR, is the packing sequence (ψ). This is required to promote viral RNA packaging.

The ability of such retroviruses to (a) effectively enter various cell types and (b) integrate their genome into the host cell genome in a stable, long-term fashion, made them obvious potential vectors for gene therapy.

The construction of retroviruses to function as gene vectors entails replacing the endogenous viral genes, required for normal viral replication, with the exogenous gene of interest (Figure 14.4a). Removal of the viral structural genes means that the resulting vector cannot itself replicate. In

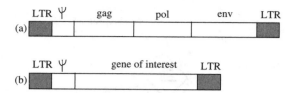

Figure 14.4   Schematic representation of (a) the proviral genome of a basic retrovirus and (b) the genome of a basic engineered retroviral vector carrying the gene of interest. Refer to text for further details

order to generate mature virion particles harbouring the vector nucleic acid (Figure 14.4b), this genetic material must be introduced into a 'packing cell'. These are recombinant cells that have previously been engineered to contain the *gag*, *pol* and *env* structural genes (Figure 14.5). In this way, packing cells are capable of producing mature, but replication-deficient, viral particles, harbouring the gene to be transferred (see Section 14.3.3). These viral particles function as so-called one-time, single-hit gene transfer systems.

More recently, various modifications have been introduced to this basic retroviral system. The inclusion of the 5′ end of the *gag* gene is shown to enhance levels of vector production by up to 200-fold. Additionally, specific promoters have been introduced in order to attempt to control expression of the inserted gene. Most work has focused upon the use of tissue-specific promoters in an effort to limit expression of the desired gene to a specific tissue type.

The most commonly employed (recombinant deficient) retrovirus used in this regard has been derived from the Maloney murine leukaemia virus (MoMuLV).

Retroviruses display a number of properties/characteristics that influence their potential as vectors in gene therapy protocols. These may be summarized as follows:

- retroviruses as a group have been studied in detail and their biochemistry and molecular biology are well understood;

- most retroviruses can integrate their proviral DNA only into actively replicating cells;

- the efficiency of gene transfer to most sensitive cell types is very high, often approaching 100 per cent;

- integrated DNA can be subject to long-term, relatively high-level expression;

- proviral DNA integrates randomly into the host chromosomes;

- retroviruses are promiscuous, in that they infect a variety of dividing cell types;

- complete copies of the proviral DNA are passed on to daughter cells if the original recipient cell divides;

- good, high-level, titre stocks of replication-incompetent retroviral particles can be produced;

- safety studies using retroviral vectors have already been carried out on various animal species.

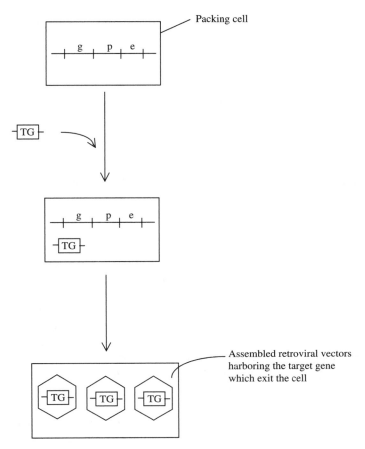

Figure 14.5 The use of packing cells to generate replication-deficient retroviral vectors. The packaging cell is an engineered animal cell into which the retroviral *gag* (g), *pol* (p) and *env* (e) genes have been introduced. The cell line chosen must be one which the (replication-deficient) virus can infect. The engineered retroviral vector genome (which is carrying the target gene; TG) is then incubated with the packing cell. This results in the generation and assembly of mature replication-deficient retroviral vector particles. These exit the cell and will replicate by entering other packaging cells. By completing a number of such replication cycles, large quantities of the desired retroviral vectors are produced

The fact that they have been well studied, display almost 100 per cent transduction efficacy in sensitive cells and that the transferred genes are usually subject to long-term, fairly high-level expression renders retroviruses powerful potential vectors. These advantages form the basis of their widespread use in this regard.

However, many of their other characteristics serve to curtail the application of retroviruses as gene therapy vectors. From a practical standpoint, retroviral vectors are relatively labile. Thus, although retroviruses are relatively easy to propagate, they are often damaged by subsequent purification and concentration, which are steps essential for their clinical use. In most instances, their ability to infect only dividing cells clearly restricts their use. Their lack of selectivity in terms of the dividing cell types they infect is also a disadvantage. They will not infect all dividing cell types: the entry of any specific retrovirus is dependent upon the existence of an appropriate viral receptor on the surface of a target cell. As the identity of most retroviral receptors remains unknown, it remains difficult to predict the entire range of cell types any retrovirus is likely to infect during a gene therapy protocol. Integration and expression of the exogenous gene in cells other than target cells could result in physiological complications.

An additional drawback with regard to retroviral-based vectors is the propensity of the transferred gene to integrate randomly into the chromosomes of the recipient cells. Integration of the transferred DNA in the middle of a gene whose product plays a critical role in the cell could irrevocably damage cellular function. For example, disruption of a central metabolic enzyme could cause cell death, and disruption of a tumour suppresser gene could give rise to cellular transformation. In addition, integration of the proviral nucleic acid to sites adjacent to quiescent cellular proto-oncogenes could result in their activation.

The theoretical complications posed by random chromosomal integration became a medical reality in 2002, when two children who had received retroviral-based gene therapy 2 years previously developed a leukaemic-like condition. The initial clinical trial aimed to treat X-linked severe combined immunodeficiency (SCID-X1), a hereditary disorder in which T-lymphocytes and NK cells in particular do not develop, due to a mutation in the gene coding for the γc cytokine receptor subunit. The clinical consequence is near abolition of a functional immune system.

The trial entailed retroviral-mediated *ex vivo* transduction of haematopoietic stem cells from 10 young SCID-X1 sufferers, with subsequent re-infusion of the treated cells. A marked and prolonged clinical response in which the condition was essentially reversed was observed in 9 out of the 10 patients. The prolonged response was likely due to the transduction of pluripotent progenitor cells with self-renewal capacity (Chapter 10). However, the two youngest patients (1 and 3 months old at the time of treatment) developed uncontrolled proliferation of mature T-lymphocytes 30 months and 34 months after gene therapy respectively.

It has subsequently been shown that this leukaemia-like condition was triggered by proviral integration at a site near the LM02 proto-oncogene promoter, leading to gene activation. This development resulted in an initial ban on further retroviral-based gene therapy trials in some world regions, and the proportion of trials undertaken subsequently using retroviral-based systems has dropped significantly.

14.3.2   Adenoviral and additional viral-based vectors

A number of additional viral types may also prove useful as vectors in the practice of gene therapy. Chief amongst these are the adenoviruses. Adeno-associated virus, the herpes virus, and a number of other viruses, are also being considered (Table 14.2).

Adenoviruses are relatively large, non-enveloped structures, housing double-stranded DNA as their genetic material. Their genome is much larger (approximately 35 kb) and more complex than those of retroviruses. In most instances, only a small fraction of this genome is removed when constructing an adenovirus-based vector. Upon cellular infection, adenoviral DNA becomes localized in the nucleus, but does not integrate into the host cell DNA. Usually, infection by wild-type adenoviruses is associated with, at worst, mild clinical symptoms in humans.

As potential vectors for gene therapy, adenoviruses display a number of both advantages and disadvantages (Table 14.3), and they have been used in over 300 gene therapy trials to date. Their major advantage relates to their ability to infect non-dividing cells efficiently and the usually

Table 14.3 Some characteristic advantages and disadvantages of adenoviruses as potential vectors for gene therapy. Refer to text for further details

Advantages	Disadvantages
Adenoviruses are capable of gene transfer to non-dividing cells	Adenoviruses are highly immunogenic in man
They are easy to propagate in large quantities	The duration of expression of transferred genes can vary, and is usually transient
High levels of gene expression are usually recorded	Infection of permissive cells with wild-type adenovirus usually results in cell lysis
They are relatively stable viruses	Adenoviruses display a broad selectivity in the cell types they can infect

observed expression of large quantities of the desired gene products. However, the failure of the adenoviral-based DNA to integrate into the host cell generally means that its survival and, hence, the duration of gene expression, is limited. Adenovirus-based vectors, carrying various marker genes (i.e. a gene whose expression product is easily detected), have been administered to animals. Marker gene expression has been subsequently noted in various tissues, including heart, liver, muscle, bone marrow, central nervous system and endothelial cells. Duration of marker gene expression ranged from 2–3 weeks to several months.

Whereas short-term, high-level gene expression may be appropriate for some gene therapy applications, it would be of less use for the treatment of, for example, genetic diseases, where long-term gene expression would be required. This could be achieved, in theory, by repeat administration of the adenoviral vector. However, adenoviruses prompt a strong immune response, which limits the efficacy of repeat administration. Indeed, the gene therapy trial death in 1999, as mentioned previously, was apparently caused by a severe and unexpected inflammatory reaction to the adenoviral vector used.

Additional viruses that may prove of some use as future viral vectors include adeno-associated virus and herpes virus. Adeno-associated virus is a very small, single-stranded DNA virus: its genome consists of only two genes. It does not have the ability to replicate autonomously and can do so only in the presence of a co-infecting adenovirus (or other selected viruses).

Although it is found in the human population, it does not appear to be associated with any known diseases. Not surprisingly, only relatively small genes can be introduced into adeno-associated viral vector systems. Such systems, however, do provide a mechanism of gene transfer into non-dividing cells. It also seems to facilitate long-term expression of the transferred genetic material. In contrast to adenoviruses, nucleic acid transferred by adeno-associated viruses appears to be integrated into the recipient cell genome.

The herpes simplex virus represents another potential vector system that is receiving increased attention. Because herpes simplex virus is a neurotrophic virus, it may prove to be particularly useful in delivering genes to neurons of the peripheral and central nervous system. Upon infection, herpes simplex virus usually remains latent in non-dividing neurons, with its genome remaining in an unintegrated form. Thus far, it has proven difficult to generate a replication-incompetent, but yet viable, herpes simplex particle. Moreover, some of the replication-incompetent viruses generated still retain an ability to damage/destroy the cells they infect. Although herpes-based vector systems one day may prove useful in gene therapy, suitable and safe vector variants of herpes simplex virus must first be generated and tested.

An additional virus that has more recently gained some attention as a possible vector is that of the sindbis virus. A member of the alphavirus family, this ssRNA virus can infect a broad range of both insect and vertebrate cells. The mature virion particles consist of the RNA genome complexed with a capsid protein C. This, in turn, is enveloped by a lipid bilayer in which two additional viral proteins (E1 and E2) are embedded. The E2 polypeptide appears to mediate viral binding to the surface receptors of susceptible cells. The major mammalian cell surface receptor it targets appears to be the highly conserved, widely distributed laminin receptor.

The sindbis virus is simple, robust, capable of infecting non-dividing cells and generally supports high levels of gene expression. However, it does display a broad host range and, hence, lacks the inherent targeting specificity characteristic of an idealized viral vector.

Recently, a novel recombinant sindbis virus, displaying altered host cell specificity, has been generated. Scientists inserted a nucleotide sequence coding for the IgG binding domain of *Staphyloccus aureus* into the E2 viral gene. Disruption of the E2 gene renders its protein product incapable of binding laminin (hence, destroying the natural viral tropism). However, the protein A domain allows the chimaeric E2 product to bind monoclonal antibodies. This altered virus may prove to be a useful generic or 'null' vector, potentially capable of being specifically targeted to any desired cell type. This would simply necessitate pre-incubation of the virus with monoclonal antibodies raised against a surface antigen unique to the proposed target cell population (Figure 14.6).

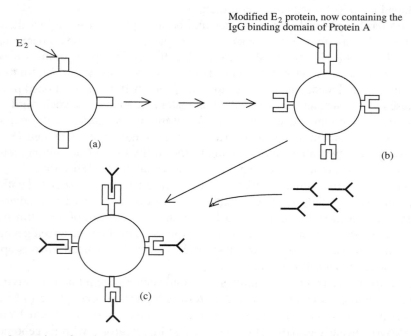

Figure 14.6   Generation of engineered sindbis virus capable of being targeted to bind specific cell types. (a) A simplified depiction of the virus, displaying the surface E2 protein. (b) Genetic engineering facilitates disruption of the E2 gene by incorporation of the IgG binding domain of protein A. (c) Incubation of such engineered viral particles with most monoclonal antibody types results in effective immobilization of the antibody on the viral surface. Thus, the engineered viral vector should be targetable to any specific cell type simply by its pre-incubation with monoclonal antibodies that selectively bind a surface antigen uniquely associated with the target cell

Binding of the monoclonal antibody to the protein A domain would ensue and the immobilized monoclonal antibody would dictate the cell type targeted.

Initial studies using this system have proved encouraging. The altered virus (without associated monoclonal antibody) failed to infect a wide variety of human cell lines. By initially incubating with monoclonal antibody of the appropriate specificity, however, the viral particles were capable of efficiently transducing cells expressing surface receptors such as CD4, CD33 and human leukocyte antigen.

A number of other issues must now be addressed including determining if the IgG–protein A affinity is sufficiently high to keep the antibody associated with the virus *in vivo*. The full potential of this approach will also require more detailed characterization of surface markers uniquely associated with different cell types. However, the approach exemplifies the types of technical innovation now being introduced that will make second-generation vectors more suited to their role in gene therapy.

14.3.3 Manufacture of viral vectors

Viral vector manufacture for therapeutic purposes involves initial viral propagation in appropriate animal cell lines, viral recovery, concentration, purification and formulation. A generalized manufacturing scenario for adenoviral-based vectors is outlined in Figure 14.7. The manufacture of alternative viral vectors likely follows a substantially similar approach.

Master and working banks of both the viral vector and the animal cell line will have been constructed during the drug development process (see Chapter 4). Manufacture of a batch of vector, therefore, will be initiated by the culture of packing cells in suitable animal cell bioreactors. The

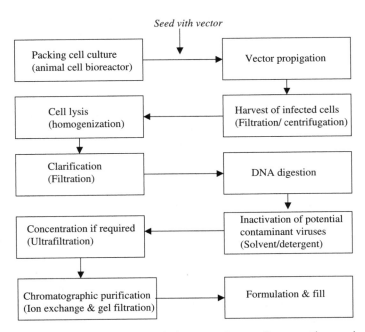

Figure 14.7 Large-scale manufacture of adenoviral vectors for use for gene-therapy-based clinical protocols. Refer to text for details

principles and practice of animal cell culture have been overviewed in Chapter 5. To date, bioreactor size of 100 l or less have been used, which are sufficient to satisfy clinical trial demand. The packing cells are then seeded with the replication-deficient virus, allowing vector propagation (see also Figure 14.5). After a fixed time the viral-infected cells are collected (harvested) by microfiltration or centrifugation and the cells are then homogenized in order to release the viral vector. Traditionally, adenoviral vectors (and indeed many other animal cell viruses) were purified from such a crude mixture by caesium chloride density-gradient centrifugation. While appropriate to laboratory-scale operations, this method is unsuitable for large-scale viral recovery due to scale-up issues and cost.

Alternative purification methods based upon column chromatography are thus employed on an industrial scale. Major 'contaminants' present in this crude viral vector preparation include some intact animal cells, cellular debris, and intracellular molecules, most notably animal cell protein and nucleic acid. Intact cells/cellular debris is removed by filtration. The release of large amounts of cellular DNA increases the solution viscosity and complicates downstream processing. The purification protocol, therefore, usually entails the physical degradation of DNA by addition of a nuclease enzyme. A solvent/detergent treatment step is then undertaken as a safety step in order to inactivate any enveloped contaminant viruses that might also be present. High-resolution purification is usually achieved by a combination of ion-exchange and gel-filtration chromatography, with product concentrations steps being undertaken by ultrafiltration if necessary. The final product is then filter sterilized and filled into glass vials. The purified vectors generally may be stored either refrigerated or frozen, and they display useful shelf lives of 2 years or more.

14.3.4 Non-viral vectors

Although viral-mediated gene delivery systems currently predominate, a substantial number of current clinical trials use non-viral-based methods of gene delivery. General advantages quoted with respect to non-viral delivery systems include:

- their low/non-immunogenicity;

- non-occurrence of integration of the therapeutic gene into the host chromosome (this eliminates the potential to disrupt essential host genes or to activate host oncogenes).

The initial approach adopted entailed administration of 'naked' plasmid DNA housing the gene of interest. This avenue of research was first opened in 1990, when it was shown that naked plasmid DNA was expressed in mice muscle cells subsequent to its i.m. injection. The plasmid DNA concerned housed the β-galactosidase gene as a reporter. Subsequent expression of β-galactosidase activity could persist for anything from a few months to the remainder of the animal's life. The transfection rate recorded was low (1–2 per cent of muscle fibres assimilated the DNA), and the DNA was not integrated into the host cell's chromosomes.

Up until this point, it was assumed that naked DNA injected into animals would not be spontaneously taken up and expressed in host cells. This finding vindicated the cautious approach taken by the FDA and other regulatory authorities with regard to the presence of free DNA in biopharmaceutical products (Chapter 7).

Scientists have also since demonstrated that DNA (coated on microscopic gold beads) propelled into the epidermis of test animals with a 'gene gun', is expressed in the animal's skin cells. Furthermore, the introduction in this fashion of DNA coding for human influenza viral antigens resulted in effective immunization of the animal against influenza. Similar results, using other pathogen models, have also now been generated. It is assumed that expressed antigen is secreted by the cell and, in this way, is exposed to immune surveillance.

Further research has illustrated that systematic administration via i.v. injection rarely achieves meaningful cell transfection. This is most likely due to the high nuclease levels present in serum. In contrast, free nuclease activity in muscle tissue is extremely low.

Modern non-viral-based systems generally entail complexing/packaging the gene of interest (present, along with appropriate promoters, etc., in a circular plasmid) with additional molecules, particularly various lipids or some polypeptides. These generally display a positive charge and, hence, interact with the negatively charged DNA molecules. The function of such carrier molecules is to stabilize the DNA, protect it from, for example, serum nucleases and ideally to modulate interaction with the biological system (e.g. help target the DNA to particular cell types, or away from other cell types).

The most commonly used polymers are the cationic lipids and polylysine chains (Figure 14.8). Cationic lipids can aggregate in aqueous-based systems to form vesicles/liposomes, which in turn

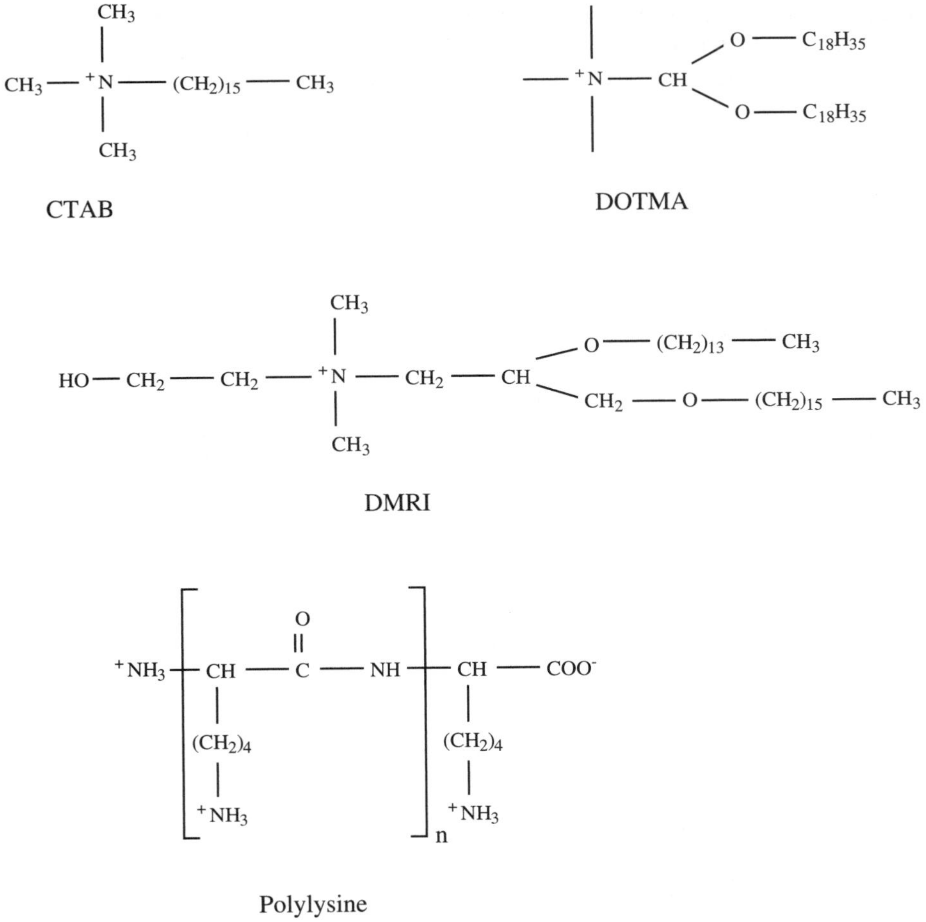

Figure 14.8 Structure of some cationic lipids and polylysine

will interact spontaneously with DNA (Figure 14.9). Initially, the negatively charged plasmid DNA probably acts as a bridge between adjacent vesicles. Further DNA/vesicle interactions quickly generate a complex three-dimensional lattice-like system composed of flattened vesicles (some of which probably rupture) interspersed with plasmid DNA. The lipid component of such 'lipoplexes' should, therefore, provide a measure of physical protection to the therapeutic gene.

Gene therapy results to date using this approach have been mixed. The process of lipoplex formation is not easily controlled; hence, different batches made under seemingly identical conditions may not be structurally identical. Furthermore, *in vitro* test results using such lipoplexes can correlate very poorly with subsequent *in vivo* performance. Clearly, more research is required to underpin the rational use of lipoplexes for gene therapy purposes. The same is true for other polymer-based synthetic gene delivery systems, the most significant of which is the polylysine-based system. Polylysine molecules, due to their positive charge (Figure 14.8), can also form electrostatic complexes with DNA. However, the stability of such 'polyplexes' in biological fluids can be problematic. Furthermore, polyplexes tend to be rapidly removed from circulation, prompting a low plasma half-life. These difficulties can be alleviated in part by the attachment of PEG molecules. PEG attachment is also used to increase the serum half-life of various therapeutic proteins, such as some interferons (Chapter 8).

No matter what their composition, such synthetic gene delivery systems also meet various biological barriers to efficient cellular gene delivery. Viral vector-based systems are far less prone to such problems, as the viral carrier has evolved in nature to overcome such obstacles. Obstacles relate to:

- blood-related issues;

- biodistribution profile;

- cellular targeting;

- cellular entry and nuclear delivery.

Whereas lipoplexes/polyplexes generally protect the plasmid from serum nucleases, the overall positive charge characteristic of these structures leads to their non-specific interactions with cells

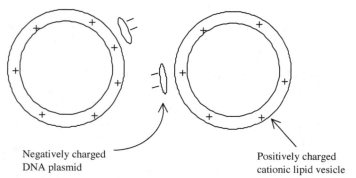

Negatively charged
DNA plasmid

Positively charged
cationic lipid vesicle

Figure 14.9   Initial interaction of plasmid DNA with cationic (positively charged) vesicles. Refer to text for further details (both blood cells and vascular endothelial cells) and serum proteins. Also, following i.v. injection, such DNA complexes in practice tend to accumulate in the lung and liver. Targeting of DNA complexes to specific cell types also poses a considerable (largely unmet) technical challenge. Approaches, such as the incorporation of antibodies directed against specific cell surface antigens may provide a future avenue of achieving such cell-selective targeting. However, it is currently believed that ionic interactions constitute a predominant binding force between the positively charged lipoplexes/polyplexes and the negatively charged eukaryotic cell surface. Such electrostatic interactions may even override more biospecific interactions characteristic of antibody- or receptor-based systems. Currently, probably the most effective means of delivering such vectors to target tissue/cells is to inject them into/beside the target area.

However targeted to the appropriate cell surface, if it is to be clinically effective, the therapeutic plasmid must enter the cell and reach the nucleus intact. Cellular entry is generally achieved via endocytosis (Figure 14.10). A proportion of endocytosed plasmid DNA escapes from the endosome by entering the cytoplasm, thereby escaping liposomal destruction (Figure 14.10). The molecular mechanism by which escape is accomplished is, at best, only partially understood. Anionic lipid constituents of lipoplexes, for example, may fuse directly with the endosomal membranes, facilitating direct expulsion of at least a portion of the plasmid DNA into the cytoplasm. Generally, the DNA is released in free form (i.e. uncomplexed to any lipid).

Some attempts have been made to rationally increase the efficiency of endosomal escape. One such avenue entails the incorporation of selected hydrophobic (viral) peptides into the gene delivery systems. Many viruses naturally enter animal cells via receptor-mediated endocytosis. These viruses have evolved efficient means of endosomal escape, usually relying upon membrane-disrupting peptides derived from the viral coat proteins.

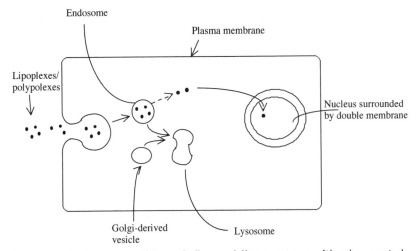

Figure 14.10 Overview of cellular entry of (non-viral) gene delivery systems, with subsequent plasmid relocation to the nucleus. The delivery systems (e.g. lipoplexes and polyplexes) initially enter the cell via endocytosis (the invagination of a small section of plasma membrane to form small membrane-bound vesicles termed endosomes). Endosomes subsequently fuse with golgi-derived vesicles, forming lysosomes. Golgi-derived hydrolytic lysosomal enzymes then degrade the lysosomal contents. A proportion of the plasmid DNA must escape lysosomal destruction via entry into the cytoplasm. Some plasmids subsequently enter the nucleus. Refer to text for further details

Once in the cytoplasm, a proportion of plasmid molecules are likely degraded by cytoplasmic nucleases, effectively further reducing transfection efficiencies. There are two potential routes by which plasmid DNA could reach the nucleus:

- direct nuclear entry as a consequence of nuclear membrane breakdown associated with mitosis;

- transport through nuclear pores, which may occur via passive diffusion or specific energy-requiring transport processes.

Overall, it is estimated that only one in 10^4–10^5 plasmids taken up by endocytosis will enter the nucleus intact and be successfully expressed.

14.3.5   Manufacture of plasmid DNA

The overall generalized approach used to produce plasmid DNA for the purposes of gene therapy trials is presented in Figure 14.11. Prior to its manufacture, researchers would have constructed an appropriate vector housing the therapeutic gene and introduced it into a producer microorganism, usually *E. coli*. Routine large-scale plasmid manufacture then entails culture of a batch of producer microorganisms by fermentation, followed by plasmid extraction and purification. In this regard, the overall approach used resembles the approaches taken in the large-scale manufacture of recombinant therapeutic proteins, as described in Chapters 5 and 6.

Industrial-scale microbial fermentation (upstream processing) has also been described in Chapter 5, to which the reader is referred. Fermentation promotes microbial cell replication and, thus, the biosynthesis of large quantities of plasmid. Subsequent to fermentation, the microbial cells are harvested (collected) by either centrifugation or microfiltration. Following resuspension in a low volume of buffer, the cells must be disrupted in order to release the plasmids therein. This appears to be most commonly achieved by the addition of a lysis reagent consisting of NaOH and SDS. The combination of high pH and detergent action disrupts the microbial cell wall/membranes with consequent release of the intracellular contents. In addition to the desired plasmid DNA, this crude mixture will also contain various impurities, which must be removed by subsequent downstream processing steps. Notable impurities include:

- cell wall debris and some intact cells;

- proteins;

- genomic DNA;

- RNA;

- low molecular mass metabolites;

- endotoxin.

After lysis is complete, the next step can entail the addition of a high-salt neutralization solution, such as potassium acetate. This promotes formation of aggregates of genomic DNA and

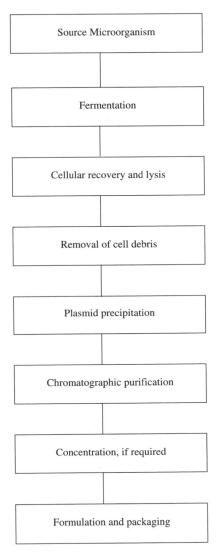

Figure 14.11 Overview of the manufacturing process for the large-scale production of plasmid DNA. Refer to the text for further details

SDS–protein complexes, which can subsequently be removed by centrifugation or filtration. The plasmids can then themselves be precipitated from the resultant solution by the addition of appropriate solvent (usually either isopropanol or ethanol). Upon resuspension, the plasmid preparation can then be subjected to chromatographic purification. The major contaminants likely still present include RNA, genomic DNA fragments, nicked or other plasmid variants, and endotoxins. Gel-filtration chromatography can effectively remove contaminants that differ substantially in shape/size from the desired plasmid. These can include most genomic DNA fragments, RNA and (most) endotoxins. It can also achieve partial removal of plasmid variants, such as open circular plasmids from the main (supercoiled) plasmid preparation. Ion exchange can remove many protein contaminants, as well as RNA. However, genomic DNA and endotoxins generally co-purify with the plasmid DNA. Additional chromatographic approaches based upon reverse-phase and affinity systems have been developed at laboratory scale at least.

A significant feature of plasmid purification employing capture chromatography (i.e. involving plasmids binding to the chromatographic beads) is the low plasmid-binding capacities observed. The pore size of commercially available capture chromatographic media is insufficiently large to allow entry of plasmids, restricting binding to the bead surface. Binding capacities can, therefore, be 100-fold or more lower than those observed when the same media are used to purify (much smaller) therapeutic proteins (Chapter 6).

Purified plasmids may then be analysed using various analytical techniques. Freedom from contaminating nucleic acid/proteins can be assessed electrophoretically. Endotoxin and sterility tests would also be routinely undertaken. The purified plasmid DNA must next be formulated to yield the final non-viral delivery system. Formulation studies relating to such systems remain an area requiring further investigation. Most work reported to date relates to formulating/stabilizing lipoplex-based gene delivery systems. Aqueous suspensions of these (and other) non-viral-based systems tend to aggregate quickly (in a matter of minutes to hours). In order to circumvent this problem, the final delivery systems were often actually formulated at the patient bedside in earlier clinical trials.

Research aimed at identifying appropriate stabilizing excipients/formulation formats is ongoing. Simple freezing is an option, particularly as frozen formulations would be immune to agitation-induced aggregation. However, the process of freezing, particularly slow freezing, in itself induces aggregation. This can be minimized by flash freezing (e.g. by immersion in liquid nitrogen), although this approach may not prove practicable at an industrial scale. The addition of cryoprotectants may help minimize this problem, and initial studies indicate that various sugars (e.g. glucose, sucrose and trehalose) show some potential in this regard. Another avenue under investigation relates to the generation of a final freeze-dried product. Again, issues such as the (relatively) slow freezing process characteristic of industrial-scale freeze-driers complicate attaining this goal in practice.

14.4   Gene therapy and genetic disease

Well over 4000 genetic diseases have been characterized to date. Many of these are caused by lack of production of a single gene product or are due to the production of a mutated gene product incapable of carrying out its natural function. Gene therapy represents a seemingly straightforward therapeutic option that could correct such genetic-based diseases. This would be achieved simply by facilitating insertion of a 'healthy' copy of the gene in question into appropriate cells of the sufferer.

Although simple in concept, the application of gene therapy to treat/cure genetic diseases has, thus far, made little impact in practice. The slow progress in this regard is likely due to a number of factors. These include:

- The number of genetic diseases for which the actual gene responsible has been identified and studied are relatively modest, although completion of the human genome project should rapidly accelerate identification of such genes.

- As discussed previously, none of the first-generation gene-delivering vectors have proven fully satisfactory.

- Some genetic diseases are quite complex, with several organs/cell types being affected. In most instances, it has proven difficult in practice to introduce the required gene into all the affected cell types.

- Regulation of expression levels of the genes transferred has proven problematic.

- Drug companies often display greater interest in applying gene therapy to more prevalent diseases, such as cancer. The patient population suffering from many genetic diseases is relatively modest. In some instances, a limited patient population may not be sufficient to allow the developing company to recoup the cost of drug development.

Some of the genetic conditions for which the defective gene has been pinpointed are summarized in Table 14.4. Many of the initial attempts to utilize gene therapy in practice focused upon haemoglobinopathies (e.g. sickle cell anaemia and thalassaemias). These conditions were amongst the first genetic disorders to be characterized at a molecular level, with the defect centring around the haemoglobin α- or β-chain genes. Furthermore, the target cells in the bone marrow could be removed and subsequently replaced with relative ease. However, these conditions proved to be a difficult initial choice for the gene therapist. The production of the appropriate quantities of functional haemoglobin is dependent not only upon the presence of α- and β-globin genes of the correct

Table 14.4 Some examples of genetic diseases for which the defective gene responsible has been identified

Disease	Defective genes protein product
Haemophilia A	Factor VIII
Haemophilia B	Factor IX
Thalassemia	β-globin
Sickle cell anaemia	β-globin
Familial hypercholesterolaemia	Low-density protein receptor
Severe combined immunodeficiency	Adenosine deaminase
Severe combined immunodeficiency	Purine nucleoside phosphorylase
Niemann–Pick disease	Sphingomylinase
Gaucher's disease	Glucocerebrosidase
Cystic fibrosis	Cystic fibrosis transmembrane regulator
Emphysema	α_1-Antitrypsin
Leukocyte adhesion deficiency	CD18
Hyperammonaemia	Ornithine transcarbamylase
Citrullinaemia	Arginosuccinate synthetase
Phenylketonuria	Phenylalanine hydroxylase
Maple syrup disease	Branched chain α-ketoacid dehydrogenase
Tyrosinaemia type 1	Fumarylacetoacetate hydrolase
Glycogen storage deficiency type 1A	Glucose-6-phosphatase
Fucosidosis	α-L-Fucosidase
Mucopolysaccharidosis type VII	β-Glucuronidase
Mucopolysaccharidosis type I	α-L-Iduronidase
Galactosaemia	Galactose-1-phosphate uridyl transferase

sequence, but also upon detailed regulation of gene expression. Such tight regulation of expression of transferred genes is beyond the capability of gene therapy technology as it currently stands.

Another early genetic disease for correction by gene therapy was SCID. One form of this disease is caused by a lack of adenosine deaminase (ADA) activity. ADA is an enzyme that plays a central role in the degradation of purine nucleosides (it catalyses the removal of ammonia from adenosine, forming inosine, which, in turn, is usually eventually converted to uric acid). This leads to T- and B-lymphocyte dysfunction. Lack of an effective immune system means that SCID sufferers must be kept in an essentially sterile environment.

When compared with treating diseases such as thalassaemia, regulation of the level of expression of a corrected ADA gene was believed to be less important for a successful therapeutic outcome. (In most, though not all, metabolic diseases caused by an enzyme deficiency, it appears that expression of even a fraction of normal enzyme levels is sufficient to ameliorate the disease symptoms.)

Gene therapy trials aimed at counteracting ADA deficiency were initiated in 1990. The first recipient was a 4-year-old SCID sufferer. The protocol used entailed the isolation of the child's peripheral lymphocytes, followed by the *in vitro* introduction of the human ADA gene into these cells, using a retroviral vector. After a period of expansion (by culture *in vitro*), these treated cells were re-injected into the patient. As the lymphocytes (and, by extension, the corrective gene) had a finite life span, the therapy was repeated every 6–8 weeks. This approach appeared successful, in that it has resulted in a marked and sustained improvement in the recipient's immune function. Critically, however, interpretation of this outcome was made more difficult owing to the later revelation that the patient also initiated more conventional SCID therapy just prior to the gene therapy treatment. A second retroviral-based trial aiming to treat a different form of SCID has also been discussed earlier in this chapter.

Haematopoietic (and indeed other) stem cells are attractive potential gene therapy recipient cells because they are immortal. Successful introduction of the target gene into these cells should facilitate ongoing production of the gene product in mature blood cells, which are continually derived from the stem cell population. This would likely remove the requirement for repeat gene transfers to the affected individual.

The routine transduction of haematopoietic stem cells has, thus far, proven technically difficult. They are found only in low quantities in the bone marrow, and there is a lack of a suitable assay for stem cells. However, recent progress has been made in this regard, and routine transduction of such cells will likely be achievable within the next few years.

Additional genetic diseases for which a gene therapy approach is currently being evaluated include familial hypercholesterolaemia and cystic fibrosis. Familial hypercholesterolaemia is caused by the absence (or presence of a defective form of) low-density lipoprotein receptors on the surface of liver cells. This results in highly elevated serum cholesterol levels, normally accompanied by early onset of serious vascular disease. The gene therapy approaches that have been attempted thus far to counteract this condition have entailed the initial removal of a relatively large portion of the liver. Hepatocytes derived from the liver are then cultured *in vitro*, with gene transfer being undertaken using retroviral vectors. The corrected hepatocytes are then usually infused back into the liver via a catheter. Although studies in animals have been partially successful, transduction of only a small proportion of the hepatocytes is normally observed. Subsequent expression of the corrective gene can also be variable. *In vivo* approaches to hepatic gene correction, using both viral and non-viral approaches, are also currently being assessed.

The cystic fibrosis (*cf*) gene was first identified in 1989. It codes for CFTR, a 170 kDa protein that serves as a chloride channel in epithelial cells. Inheritance of a mutant *cftr* gene from both parents results in the cystic fibrosis phenotype. While various organs are affected, the most severely affected are the respiratory epithelial cells. These cells have, unsurprisingly, become the focus of attempts at corrective gene therapy. Cystic fibrosis is the most common inherited mono-genetic disease in Europe and the USA, and sufferers have a typical life expectancy of less than 40 years. Over a third of the 100 or so gene therapy trials thus far undertaken to treat inherited disorders have specifically targeted cystic fibrosis.

Several vectors have been used in an attempt to deliver the cystic fibrosis gene to the airway epithelial cells of sufferers. The most notable systems include adenoviruses and cationic liposomes. Vector delivery to the target cells can be achieved directly by aerosol technology. Delivery of CFTR cDNA to airway epithelial cells (and subsequent gene expression) has been demonstrated with the use of both vector types. However, in order to be of therapeutic benefit, it is essential that 5–10 per cent of the target cell population receive and express the CFTR gene. This level of integration has not been achieved so far; and, furthermore, gene expression has often been transient.

14.5 Gene therapy and cancer

To date, the majority of gene therapy trials undertaken aim to cure not inherited genetic defects, but cancer. The average annual incidence of cancer reported in the USA alone stands at approximately 1.4 million cases. Survival rates attained by pursuit of conventional therapeutic strategies (surgery, chemo/radiotherapy) stand at about 50 per cent.

Initial gene therapy trials aimed at treating/curing cancer began in 1991. Various strategic approaches have since been developed in this regard (Table 14.5). Numerous trials aimed at assessing the application of gene therapy for the treatment of a wide variety of cancer types are now underway (Table 14.6).

Although many of the results generated to date provide hope for the future, gene therapy thus far has failed to provide a definitive cure for cancer. The lack of success is likely due to a number of factors, including:

- A requirement for improved, more target-specific vector systems.

- A requirement for a better understanding of how cancer cells evade the normal immune response.

- For ethical reasons, most patients treated to date were suffering from advanced and widespread terminal cancer (i.e. little/no hope of survival if treated using conventional therapies). Cancers at earlier stages of development will probably prove to be more responsive to gene therapy.

Table 14.5 Some therapeutic strategies being pursued in an attempt to treat cancer using a gene therapy approach. Refer to text for details

Modifying lymphocytes in order to enhance their anti-tumour activity
Modifying tumour cells to enhance their immunogenicity
Inserting tumour suppressor genes into tumour cells
Inserting toxin genes in tumour cells in order to promote tumour cell destruction
Inserting suicide genes into tumour cells
Inserting genes, such as a multiple drug resistance (*mrd*) gene, into stem cells to protect them from chemotherapy-induced damaged
Counteracting the expression of oncogenes in tumour cells by inserting an appropriate antisense gene

Table 14.6 Some specific cancer types for which human gene therapy trials have been initiated[a]

Breast cancer	Colorectal cancer
Malignant melanoma	Tumours of the central nervous system
Ovarian cancer	Renal cell carcinoma
Small-cell lung cancer	Non-small-cell lung cancer

[a]Although several of the strategies listed in Table 14.5 are being employed in these trials, many focus upon the introduction of various cytokines into the tumour cells themselves in order to attract and enhance a tumour-specific immune response.

One of the earliest cancer gene therapy trials attempted involved the introduction of the TNF gene into TILs. The rationale was that if, as expected, TIL cells reintroduced into the body could infiltrate the tumour, TNF synthesis would occur at the tumour site (where it is required). This approach has since been broadened, by introducing genes coding for a range of immunostimulatory cytokines (e.g. IL-2, IL-4, IFN-γ and GM-CSF) into TILs. A variation of this approach involves the introduction of such cytokine genes directly into tumour cells themselves. It is hoped that reintroduction of such cytokine-producing cells into the body will result in a swift and effective immune response, i.e. killing the tumour cells and vaccinating the patient against recurrent

---

Box 14.2

Product case study: Gendicine

Gendicine is the tradename given to the first gene-therapy-based medicine approved anywhere in the world. It gained approval for use in the treatment of head and neck squamous cell carcinoma from China's State Food and Drug Administration in 2003. This is one of the most common cancers in China. The company that developed, manufactures and markets the product is Shenzhen SiBono GeneTech, (Shenzhen, China). Gendicine is a replication-incompetent human serotype 5 adenovirus engineered to contain the native human p53 tumour suppressor gene. The product is administered by direct intratumoural injection and the standard treatment entails Gendicine administration concurrently with the application of radiotherapy.

Product manufacture entails viral vector propagation in a suitable animal packing cell line (known as HEK 293). After cell recovery and lysis, the crude product is clarified by filtration and concentrated by ultrafiltration. The product is then treated with a nuclease preparation in order to degrade contaminant DNA and further downstream processing entails multi-step high-resolution column chromatography (see also Figure 14.7).

Intratumoural injection is believed to facilitate vector uptake and expression of p53 in the adjacent tumour cell, leading to cell cycle arrest and apoptosis. Company data showed complete regression of tumours in 64 per cent of patients treated with Gendicine in combination with radiation therapy, with few associated side effects. By 2006 the product was believed to have been administered to some 50 000 patients in China, and is in late-stage clinical trials for various other cancers.

A broadly similar approach to that of Gendicine is being adopted by some Western companies, including Introgen Therapeutics (USA), whose p53 adenoviral-based drug Advexin has entered phase III clinical trials for sqamous cell carcinoma in 2006.

episodes. In most instances so far, this strategy has been carried out in practice by removal of target cells from the body, culture *in vitro*, introduction of the desired gene (mainly using retroviral vectors), followed by reintroduction of the altered cells into the body.

An alternative anti-cancer strategy entails insertion of a copy of a tumour suppresser gene into cancer cells. For example, a deficiency in one such gene product, p53, has been directly implicated in the development of various human cancers. It has been shown *in vitro* that insertion of a p53 gene in some p53-deficient tumour cell lines induces the death of such cells. A potential weakness of such an approach, however, is that 100 per cent of the transformed cells would have to be successfully treated to fully cure the cancer. Tumour suppressor-based gene therapy in combination with conventional approaches (chemotherapy or radiotherapy) may, therefore, prove most efficacious, and the sole gene-therapy-based medicine approved to date (in China only) is based upon this approach (Box 14.2).

Yet another strategy that may prove useful is the introduction into tumour cells of a 'sensitivity' gene. This concept dictates that the gene product should harbour the ability to convert a non-toxic pro-drug into a toxic substance within the cells - thus leading to their selective destruction. The model system most used to appraise such an approach entails the use of the thymidine kinase gene of the herpes simplex virus (Figure 14.12).

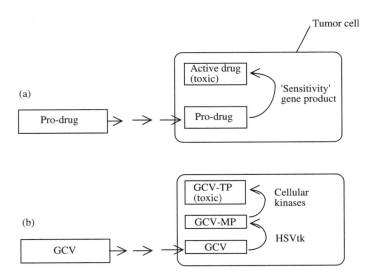

Figure 14.12 Schematic representation of the therapeutic rationale underpinning the introduction of a 'sensitivity' gene into tumour cells in order to promote their selective destruction. As depicted in (a), the gene product should be capable of converting an inactive pro-drug into a toxic drug capable of killing the cell. A specific example of this approach is presented in (b): introduction of the herpes simplex thymidine kinase (HSVtk) gene confers sensitivity to the anti-herpes drug, Ganciclovir (GCV) on the cell. GCV is converted by HSVtk into a monophosphorylated form (GCV-MP). This, in turn, is phosphorylated by endogenous kinases, yielding ganciclovir triphosphate (GCV-TP). GCV-TP induces cell death by inhibiting DNA polymerase. A potential advantage of this system is that some adjacent tumour cells (which themselves lack the HSVtk gene) are also destroyed. This is most likely due to diffusion of the GCV-MP or GCV-TP (perhaps via gap junctions) into such adjacent cells. This so-called 'bystander effect' means that all the transformed cells in a tumour would not necessarily need to be transduced for the therapy to be successful

A different gene therapy-based approach to cancer entails introduction of a gene into haematopoietic stem cells in order to protect these cells from the toxic effects of chemotherapy. Most cancer drugs display toxic side effects which usually limits the upper dosage levels that can be safely administered. One common toxic side-effect is the destruction of stem cells. If these cells could be protected or made resistant to the chemotherapeutic agent, it might be possible to administer higher concentrations of the drug to the patient. In practice, such a protective effect could be conferred by the multiple drug resistance (type 1; MDR-1) gene product. This is often expressed by cancer cells resistant to chemotherapy. It functions to pump a range of chemotherapeutic drugs (e.g. daunorubicin, taxol, vinblastine, vincristine, etc.) out of the cell. Animal studies have confirmed that introduction of the MDR-1 gene into stem cells protects these cells subsequently from large doses of taxol. This approach is now being appraised in patients receiving high-dose chemotherapy for a range of cancer types, including breast and ovarian cancer and brain tumours.

14.6   Gene therapy and AIDS

It is likely that gene therapy will prove useful in treating a far broader range of medical conditions than simply those of inherited genetic disease and cancer. A prominent additional disease target are those diseases caused by infectious agents, particularly intracellular pathogens such as HIV. The main strategic approach adopted entails introducing a gene into pathogen-susceptible cells whose product will interfere with pathogen survival/replication within that cell. Such a strategy is sometimes termed 'intracellular immunization'.

One such anti-AIDS strategy being pursued is the introduction into viral-sensitive cells of a gene coding for an altered (dysfunctional) HIV protein, such as *gag*, *tat* or *env*. The presence of such mutant forms of *gag*, in particular, was shown to be capable of inhibiting viral replication. This is probably due to interference by the mutated *gag* product with correct assembly of the viral core. An additional approach entails the transfer to sensitive cells of a gene coding for antibody fragments capable of binding to the HIV envelope proteins. This may also interfere with viral assembly in infected cells.

Scientists have also generated recombinant cells capable of synthesizing and secreting soluble forms of the HIV cell surface receptor, i.e. the CD4 antigen. It has been suggested that release of such soluble viral receptors into the blood would bind circulating virions, hence blocking their ability to 'dock' at sensitive cells. Although this proved to be the case *in vitro*, early *in vivo* studies have not proved as encouraging. Yet additional therapeutic approaches to AIDS, based upon antisense technology, will be discussed later in this chapter.

14.6.1   Gene-based vaccines

Conventional vaccine technology, including the generation of modern recombinant subunit vaccines, has been discussed in Chapter 13. An additional gene-therapy-based approach to vaccination is also now under investigation. The approach entails the administration of a DNA vector housing the gene coding for a surface antigen protein from the target pathogen. In this way, the body itself would produce the pathogen-associated protein. Theoretically, virtually any body cell could be targeted, the only requirement being that target cells export the resultant antigenic protein such that it is encountered by the immune system. Additionally, gene expression need only be transient, i.e. just sufficiently long to facilitate the induction of an immune response. Target conditions for gene-based vaccines thus far having entered clinical trials include malaria, hepatitis B and AIDS.

14.6.2   Gene therapy: some additional considerations

In addition to some technical difficulties outlined earlier, a number of non-technical issues must be satisfactorily addressed before its practice becomes widespread. Chief amongst these issues are the questions of public perception, ethics and costs.

Gene therapy is not, and will not be, an inexpensive therapeutic tool. The cost of such treatments will likely be broadly similar to the cost of present-day biopharmaceuticals. However, if proven successful in treating many currently incurable conditions, the cost:benefit ratio will almost certainly greatly favour its medical use.

Public perception and ethical considerations are, in some ways, interlinked. The ability to so readily modify our genetic complement holds great therapeutic promise. However, strict regulations overseeing the use of this technology are required (and are currently being enforced). Without proper controls, the danger exists that gene therapy could eventually be used to 'improve' human characteristics. The technical know-how to underpin a new era of eugenics is now almost a reality. The most important safeguard aimed at preventing eugenic-type developments is already in place. Currently, gene therapy is restricted to somatic cells; the genetic manipulation of human germ cells is banned. Any genetic alterations achieved thus will not be transmitted to future generations. Like many other technologies, there is no 'going back' in relation to gene therapy. The challenge is to ensure that human genetic manipulation is used only for purposes that clearly represent the 'common good'.

14.7   Antisense technology

Various disease states are associated with the inappropriate production/overproduction of gene products. Examples include:

- the expression of oncogenes, leading to the transformed state;

- the overexpression of cytokines during some disease states with associated worsening of disease symptoms;

- The overproduction of angiotensinogen, - which ultimately results in hypertension.

An additional example includes the intracellular transcription and translation of virally encoded genes during intracellular viral replication. In all such instances, the medical consequences of such inappropriate gene (over)expression could be ameliorated/prevented if this expression could be downregulated. A nucleic-acid-based approach to achieve just this is termed 'antisense technology'.

The antisense approach is based upon the generation of short, single-stranded stretches of nucleic acids (which can be DNA- or RNA-based) displaying a specific nucleotide sequence. These are generally termed 'antisense oligonucleotides'. These oligonucleotides are capable of binding to DNA (at specific gene sites) or, more commonly, to mRNA derived from specific genes. This

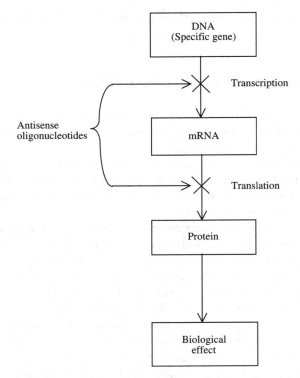

Figure 14.13 Overview of the concept of the antisense approach: the end goal is the prevention of expression of a particular gene product (invariably a protein) by either blocking the transcription or translation of that gene

binding, in most cases, occurs via Watson–Crick-based nucleotide base pair complementarity. Binding prevents expression of the gene product by preventing either the transcription or translation process (Figure 14.13).

14.7.1 Antisense oligonucleotides and their mode of action

The nucleotide sequence of an mRNA molecule contains the encoded blueprint that dictates the amino acid sequence of a protein. Because of this, the mRNA sequence is said to make 'sense'. (This mRNA, therefore, is complementary to an 'antisense' DNA strand, i.e. it is the antisense strand of DNA in a given gene that serves as template for the mRNA synthesis.) As long as at least part of the nucleotide sequences of any mRNA is known, it becomes potentially possible to synthesize chemically an oligonucleotide, either a ribo- or deoxyribo-nucleotide, whose base sequence is complementary to at least a section of the mRNA sequence. As long as such an 'antisense' oligonucleotide can enter the cell, the complementarity of sequences can promote hybridization between the mRNA and the antisense oligonucleotide (Figure 14.14).

Successful binding, however, does not depend alone upon Watson–Crick base complementary. It is also influenced by higher-order secondary and tertiary structures adopted by the RNA.

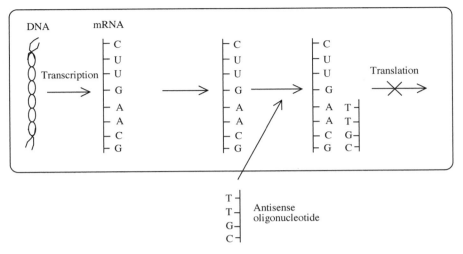

Figure 14.14 Outline of how an antisense oligonucleotide can prevent synthesis of a gene product by blocking translation. In practice, antisense oligos are 12–18 nucleotides in length. In many instances, antisense binding is believed to occur in the nucleus

Intramolecular complementary base pairing can occur (particularly within transfer and ribosomal RNA, but also messenger RNA), resulting in the formation of short duplex sequences, separated by stems and loops. Such higher-order structure seems to be functionally important, conferring recognition motifs for proteins and additional nucleic acids, as well as helping to stabilize the RNA. Regions engaged in intramolecular base pairing are obviously poor targets for antisense oligos. It is thus desirable to synthesize a nucleotide whose sequence is complementary to an accessible sequence along the mRNA backbone. Various approaches are taken to identify such suitable sequences (remember, the entire sequence of the mRNA will be known). The 'blind' or 'shotgun' approach entails synthesizing large numbers of oligos targeted to various (often overlapping) regions of the mRNA. The ability of each oligo to block translation of the mRNA is then directly assessed in an *in vitro* assay system using cell-free extracts. The second design approach entails the use of various computer programs to interrogate the mRNA sequence in an attempt to predict its higher-order structure (and hence identify accessible sequences). This approach remains to be optimized. The translation initiation sites of mRNAs are often popular targets because they are essential to translation and they are generally free from secondary structure. However, sequence homologies can exist within these sequences in unrelated genes. This reduces the specificity of the blocking effect and could lead to clinically significant side effects.

Binding results in the blocking of translation of the mRNA and, hence, prevents synthesis of the mature gene's protein product. The prevention of mRNA translation by duplex formation with antisense oligonucleotides appears to be underpinned by various mechanisms, including: (a) the oligonucleotides act as steric blockers, i.e. prevent proteins involved in translation, or other aspects of mRNA processing, from binding to appropriate sequences in the mRNA; (b) the generation of duplexes also likely allows targeting by intracellular RNases such as RNaseH. This enzyme is capable of binding to RNA–DNA duplexes and degrading the RNA portion of the duplex (most synthetic antisense oligonucleotides are DNA based).

14.7.2   Uses, advantages and disadvantages of 'oligos'

Antisense oligonucleotides (oligos) are being assessed in preclinical and clinical studies as therapeutic agents in the treatment of cancer, as well as a variety of viral diseases (e.g. HIV, hepatitis B, herpes and papillomavirus infections). They also have potential application in treating other disease states for which blocking of gene expression would likely have a beneficial effect. Such medical conditions include restenosis, rheumatoid arthritis and allergic disorders. Cancer, however, remains the most common target indication. The favoured approach is to target genes whose expression/up-regulation triggers or fuels tumorigenesis. These include products of the BCL-2, survivin and clusterin genes. The BCL-2 oncogene products drive neoplastic progression by enhancing cell survival via inhibition of apoptosis. Survivin is generally not expressed in healthy tissue, but expressed at high levels in a range of common cancer types, including lung, colon, breast and prostate cancers. It plays an important role in both promoting cell division and inhibiting apoptosis. The clusterin gene codes for a cytoprotective 'chaperone' protein whose up-regulation is associated with various human cancers.

As potential drugs, antisense oligos display a number of desirable characteristics, the most significant of which is their likely specificity. Statistical analysis reveals that any specific base sequence of 17 or more bases is extremely unlikely to occur more than once in a human cell's nucleic acid complement. It thus follows that an oligonucleotide of 17 or more nucleotide units in length, which is designed to duplex successfully with a specific mRNA species, is unlikely to form a duplex with any other (unintended) mRNA species. Most synthetic oligos, therefore, are in the region of 17 nucleotide units long. These will display virtually an absolute specificity for the target sequence. Additional advantages of the oligonucleotide antisense approach include:

- Relatively low toxicity: thus far, most trials report relatively few significant side effects. This is likely due to the highly specific nature of oligo duplexing, and the fact that they are 'natural' biomolecules. Some toxicity may, however, by triggered by non-specific binding to proteins, and most antisense agents appear to promote pro-inflammatory effects at high dosage levels.

- The requirement for only low levels of the oligo to be present inside the cell, as target mRNA is, itself, usually present only in nanomolar concentrations.

- The ability to manufacture oligos of specified nucleotide sequence is relatively straightforward using automated synthesizers.

However, native antisense oligonucleotides also suffer from a number of disadvantages, which are ultimately responsible for numerous disappointing trial results, and the fact that, after almost two decades of clinical investigation, only a single product has gained approval to date. Disadvantages include:

- sensitivity to nucleases;

- very low serum half lives;

- poor rate of cellular uptake;

- orally inactive.

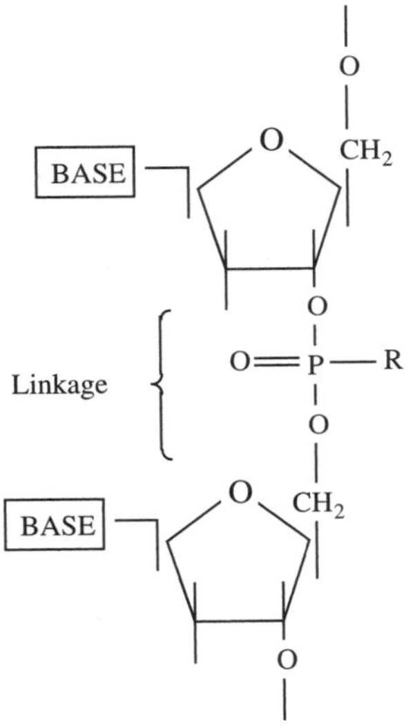

Linkage name	Substituent (R)
Phosphorothioate	-S⁻
Methylphosphonate	-CH₃
Methylphosphotriester	-O-CH₃
Ethylphosphotriester	-O-CH₂-CH₃
Alkylphosphoramidate	-NH-CH₃

Figure 14.15 Major types of modification potentially made to an oligo's phosphodiester linkage in order to increase their stability or enhance some other functional characteristic. The native phosphodiester link is shown to the left

Some progress has been made to overcome such difficulties, and continued progress in the area is expected to render the next generation of oligos more therapeutically effective.

Native oligonucleotides display a 3′–5′ phosphodiester linkage in their backbone (Figure 14.15). These are sensitive to a range of nucleases naturally present in most extracellular fluids and intracellular compartments. The half-life of native oligonucleotides in serum is only of the order of 15 min, and oligoribonucleotides are less stable than oligodeoxynucleotides. Selective modification of the native phosphodiester bond can render the product resistant to nuclease degradation.

Modification usually entails replacement of one of the free (non-bridging) oxygen atoms of the phosphodiester linkage with an alternative atom or chemical group (Figure 14.15). Most commonly, the oxygen has been replaced with a sulfur atom and the resultant phosphorothioates display greatest clinical promise. Phosphorothioate-based oligos, 'S-oligos', display increased resistance to nuclease attack while remaining water-soluble. They are also easy to synthesize chemically and they display a biological half-life of several hours. Most antisense oligos currently assessed in clinical trials are S-oligos, as is the sole antisense agent approved for general medical use to date (Vitravene, Box 14.3). Further modified oligos may well improve product pharmacokinetic and pharmacodynamic properties, as alluded to towards the end of the next section.

Box 14.3

Product case study: Vitravene

In August 1998, Vitravene (tradename) became the first (and thus far the only) antisense product to be approved for general medical use by the FDA. It gained approval within the EU the following year, although it has since been withdrawn from the EU market due to commercial rather than technical reasons. The product is a 21-nucleotide phosphorothioate based product of the following base sequence:

5´-G-C-G-T-T-T-G-C-T-C-T-T-C-T-T-C-T-T-G-C-G-3´

   Developed by the US company, Isis, Vitravene is used to treat cytomegalovirus (CMV) retinitis in AIDS patients. It is synthesized chemically and formulated as a sterile solution (6.6 mg active drug/ml) in WFI using a bicarbonate buffer to maintain a final product pH of 8.7.

   The product inhibits replication of human CMV (HCMV) via an antisense mechanism. Its nucleotide sequence is complementary to a sequence in mRNA transcripts of the major immediate early region (IE2 region) of HCMV. These mRNAs code for several essential viral proteins and blocking their synthesis effectively inhibits viral replication.

   Administration is by direct injection of 0.05 ml product into the eye (intravitreal injection), initially once every 2 weeks and subsequently once every 4 weeks. Animal studies (rabbits) indicated that the product is cleared from the eye over the course of 7–10 days, with direct nuclease-mediated metabolism representing the primary route of elimination. The most commonly observed side effect is ocular inflammation, which typically occurs in one in every four patients.

14.8 Oligonucleotide pharmacokinetics and delivery

Oligo administration during many clinical trials entails direct i.v. infusion, often over a course of several hours. S.c. and, in particular, intradermal administration is usually also associated with high bioavailability.

   Oligos bind various serum proteins, including serum albumin (as well as a range of heparin-binding and other proteins that commonly occur on many cell surfaces). Targeting of naked oligos to specific cell types, therefore, is not possible. Following administration, these oligos tend to distribute rapidly to many tissues, with the highest proportion accumulating in the liver, kidney, bone marrow, skeletal muscle and skin. They do not appear to cross the blood–brain barrier. Binding to serum proteins provides a repository for these drugs and prevents rapid renal excretion.

   The precise mechanism(s) by which oligos enter cells is not fully understood. Most are charged molecules, sometimes displaying a molecular mass of up to 10–12 kDa. Receptor-mediated endocytosis appears to be the most common mechanism by which charged oligos, such as phosphorothioates, enter most cells. One putative phosphorothioate receptor appears to consist of an 80 kDa surface protein, associated with a smaller 34 kDa membrane protein. However, this in itself seems to be an inefficient process, with only a small proportion of the administered drug eventually being transferred across the plasma membrane.

   Uncharged oligos appear to enter the cell by passive diffusion, as well as possibly by endocytosis. However, elimination of the charges renders the resultant oligos relatively hydrophobic, thus generating additional difficulties with their synthesis and delivery.

Attempts to increase delivery of oligos into the cell mainly centre on the use of suitable carrier systems. Liposomes, as well as polymeric carriers (e.g. polylysine-based carriers), are gaining most attention in this regard. Details of such carriers have already been discussed earlier in this chapter.

An alternative system, which effectively results in the introduction of antisense oligonucleotides into the cell, entails application of gene therapy. In this case, a gene, which when transcribed yields (antisense) mRNA of appropriate nucleotide sequence, is introduced into the cell by a retroviral or other appropriate vector. This approach, as applied to the treatment of cancer and AIDS, is being appraised in a number of trials.

Oligos, including modified oligos, appear to be ultimately metabolized within the cell by the action of nucleases, particularly 3′-exonucleases. Breakdown metabolic products are then mainly excreted via the urinary route.

Even phosphorothioate oligos display serum and tissue half-lives of less than a day. As a consequence, continuous or frequent i.v. infusions are required for product administration. Some progress has been reported in the development of second-generation phosphorothioate oligos with improved pharmacokinetic characteristics. The most promising development entails the modification of the ribose sugar found in the repeat nucleotide structure. Attachment (at the 2′ position) of methyl ($-CH_3$) or methoxy ethyl ($-CH_2CH_2OCH_3$) groups increases product stability, as well as product potency (by enhancing binding affinity for RNA). However, these changes also abrogate the product's ability to activate RNaseH, a primary mechanism of inducing its antisense effect. This, in turn, may be overcome by the more recent development of chimaeric phosphorothioate oligos, in which 2′-modified sugar nucleotides are placed only at the ends of the molecule, leaving a nuclease-compatible gap in the middle.

14.8.1 Manufacture of oligos

In contrast to the biopharmaceuticals discussed thus far (recombinant proteins and gene therapy products), antisense oligonucleotides are manufactured by direct chemical synthesis. Organic synthetic pathways have been developed, optimized and commercialized for some time, as oligonucleotides are widely used reagents in molecular biology. They are required as primers, probes and for the purposes of site-directed mutagenesis.

The nucleotides required (themselves either modified or unmodified as desired) are first reacted with a protecting chemical group. Each protected nucleotide is then coupled in turn to the growing end of the nucleotide chain, itself attached to a solid phase. After coupling, the original protecting group is removed and, when chain synthesis is complete, the bond anchoring the chemical to the solid phase is hydrolysed, releasing the free oligo. This may then be purified by HPLC. The most common synthetic method used is known as the phosphoramidite method, which uses a dimethoxytrityl protecting group and tetrazole as the coupling agent. Automated synthesizers that can quickly and inexpensively synthesize oligos of over 100 nucleotides are commercially available.

14.8.2 Additional antigene agents: RNA interference and ribozymes

RNAi and ribozymes represent two additional approaches to gene silencing/down-regulation with therapeutic potential. RNAi is an innate cellular process that achieves silencing of selected genes via an antisense mechanism. It shares many characteristics with the antisense-based approach described above, but also some important differences, e.g. in the exact mechanism by which the antisense effect is achieved.

RNAi probably evolved initially in primitive organisms in order to protect their genomes from viruses, transposons and additional insertable genetic elements, and to regulate gene expression. The RNAi pathway was first discovered in plants, but it is now known to function in most, if not all, eukaryotes.

RNAi represents a sequence-specific post-translational inhibition mechanism of gene expression, induced ultimately by dsRNA, be it produced naturally or synthesised *in vitro* and introduced into a cell. Entry of dsRNA triggers its cleavage into short (21–23 nucleotide long) sequences called short interfering RNAs (siRNAs). This cleavage is catalysed by a cellular nuclease enzyme called 'Dicer'. The siRNA is incorporated into a multi-subunit effector complex known as an RNA-induced silencing complex (RISC), which also contains several nucleic acid processing enzymes (a helicase, an endonuclease and an exonuclease). The double-stranded siRNA then unwinds (a process promoted by the helicase activity), and the 'sense' strand of the dsRNA is discarded. The remaining 'antisense' siRNA strand then facilitates RISC binding to a specific mRNA via Watson–Crick base complementarity, which is then degraded by RISC nuclease activity.

RNAi technology has obvious therapeutic potential as an antisense agent, and initial therapeutic targets of RNAi include viral infection, neurological diseases and cancer therapy. The synthesis of dsRNA displaying the desired nucleotide sequence is straightforward. However, as in the case of additional nucleic-acid-based therapeutic approaches, major technical hurdles remain to be overcome before RNAi becomes a therapeutic reality. Naked unmodified siRNAs for example display a serum half-life of less than 1 min, due to serum nuclease degradation. Approaches to improve the RNAi pharmacokinetic profile include chemical modification of the nucleotide backbone, to render it nuclease resistant, and the use of viral or non-viral vectors, to achieve safe product delivery to cells. As such, the jury remains out in terms of the development and approval of RNAi-based medicines, in the short to medium term at least.

Certain RNA sequences can function as catalysts. These so-called ribozymes function to catalyse cleavage at specific sequences in a specific mRNA substrate. Many ribozymes will cleave their target mRNA where there exists a particular triplet nucleotide sequence G–U–C. Statistically, it is likely that this triplet will occur at least once in most mRNAs.

Ribozymes can be directed to a specific mRNA by introducing short flanking oligonucleotides that are complementary to the target mRNA (Figure 14.16). The resultant cleavage of the target

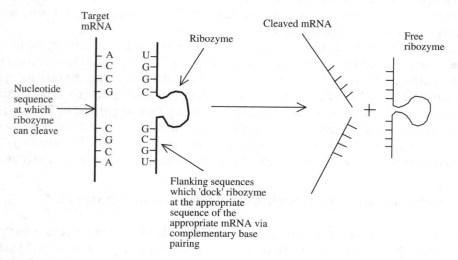

Figure 14.16 Outline of how ribozyme technology could prevent translation of specific mRNA, thus preventing synthesis of a specific target protein

obviously prevents translation. One potential advantage of ribozymes is that, as catalytic agents, a single molecule could likely destroy thousands of copies of the target mRNA. Such a drug should, therefore, be very potent. Again, however, ribozymes suffer from similar complications to antisense-based products in terms of their development as biopharmaceuticals, and no such product is likely to gain approval for some time to come.

14.9   Aptamers

Aptamers are single-stranded DNA or RNA-based sequences that fold up to adopt a unique three-dimensional structure, allowing them to bind a specific target molecule.

Binding displays high specificity, and aptamers capable of distinguishing between closely related isoforms or different conformational states of the same protein have been generated. Binding affinity is also high. It is in the low nanomolar to picomolar range, which is comparable to the binding affinity of an antibody for the antigen against which it was raised.

Aptamer technology was first developed in 1990. It entails the initial generation of a large aptamer library, with subsequent identification of individual aptamers binding a target ligand via an appropriate selection strategy. DNA aptamer libraries are usually generated via direct chemical synthesis and amplified by PCR (Chapter 3). RNA libraries are usually generated by *in vitro* transcription of synthetic DNA libraries. Identification of specific aptamers binding the target molecule is most easily undertaken by an automated *in vitro* selection approach known as systematic evolution of ligands by exponential enrichment (SELEX). Most libraries contain up to 10^{15} species.

Because of their high binding specificity and affinity, aptamers (like antibodies) are/may prove useful for affinity-based purification, target validation and drug discovery, diagnostics and therapeutics. One such product (Macugen, Box 14.4) has been approved for general medical use to date. A modest number of additional aptamers are in clinical trials, aimed at treating conditions including infectious diseases, cancer and haemophilia.

Aptamers appear to display low immunogenicity; but, when administered systemically, they are quickly excreted via size-mediated renal clearance. In order to prevent renal removal, such aptamers are usually conjugated to PEG. PEG may also help further protect the aptamers from degradation by serum nucleases; native aptamers are prone to nuclease attack, but their half-lives can most effectively be extended via chemical modification, as discussed earlier in the context of antisense agents.

14.10   Cell- and tissue-based therapies

Recent progress relating to the identification, isolation and manipulation of stem cells has made this cell type a focus of enormous attention, both within the scientific and general communities. Although stem cells harbour great medical potential, their routine application to the treatment of medical conditions (so-called regenerative medicine) remains a distant prospect, as discussed below. In contrast, fully differentiated cells or groups of cells (organs and tissues) are currently in routine medical use. Such products include cells or tissues used for the purposes of transplantation, as well as a small number of engineered cell-based products.

Box 14.4

Product case study: Macugen

Macugen (tradename) is the first and thus far only aptamer approved for general medical use. It was approved by the FDA in December 2004. The product is a synthetic PEGylated oligonucleotide with binding specificity for vascular endothelial growth factor (VEGF). It is indicated for the treatment of neovascular ('wet') age-related macular degeneration (AMD).

The RNA-based aptamer consists of 28 nucleontides of predefined sequence, chemically modified in order to render it resistant to nuclease degradation. Two 20 kDa PEG molecules are also covalently attached at one end of the nucleotide. The overall product molecular mass is 50 kDa. Final product is presented as a sterile solution containing sodium chloride and sodium phosphate as excipients.

The target indication (age-related neovascular or 'wet' macular degeneration) results from the proliferation of abnormal blood vessels in the eye, which leads to retinal damage and loss of vision. The process of vascularization (angiogenesis) is driven by VEGF. Macugen adopts a three-dimensional shape that allows it to interact specifically with VEGF, thereby inhibiting its activity (the monoclonal antibody-based product Avastin, Chapter 13, achieves a similar effect, but in the context of cancer). The product is administered directly to the site of action by intravitreous injection. Absorption from the eye into the general bloodstream occurs only very slowly, allowing a frequency of administration of once every 6 weeks. Subsequent product metabolism is via nuclease degradation.

Initial clinical assessment involved some 1200 wet AMD patients. Although both control and product groups continued to experience vision loss, the rate of vision decline experienced by macugen-treated patients was significantly slower than in the case of control patients. The most frequent/potentially serious side effects noted during these trials were endophthalmitis, retinal detachment, eye inflammation/irritation and blurred vision, although rare cases of anaphylaxis have also been reported. Macugen is marketed by Eyetech Pharmaceuticals and Pfizer.

Transplantation entails the transfer of living cells/tissue/organs from a donor to a recipient. In some cases (e.g. many skin grafting procedures) the donor and recipient are actually the same individual, and this is termed autologous transplantation. More usually, however, the donor and recipient are different individuals, and this is termed allogeneic transplantation.

Common forms of transplantation include whole blood transfusions (Box 14.5), bone marrow transplantations, skin grafting and transplantation of a wide range of organs, including kidneys, liver, pancreas, lungs and heart. Improvements in surgical transplant techniques, along with the availability of effective immunosuppressive drugs (including some antibody-based drugs; Chapter 13), render 1-year success rates for most organ transplants in the 75–95 per cent range.

Tissue/organs destined for transplant are rarely considered to be pharmaceutical products. The material for transplant is usually harvested directly by clinicians via surgical or other appropriate techniques, followed by direct transplantation without significant *in vitro* processing.

'Tissue- or cell-engineered' products represent a small but significant subgroup of cell-based products. Such products also consist of/contain fully differentiated cells but do undergo some

Box 14.5

Whole blood

Whole blood is blood that has been aseptically withdrawn from humans. A suitable anticoagulant is added (often heparin or a citrate–dextrose-based substance), although no preservative is present. The blood is usually stored at temperatures ranging from 1–8 °C, and has a short shelf life (48 h after collection if heparin is used as the anticoagulant, or up to 35 days if citrate–phosphate–dextrose with adenine is employed).

The blood is generally warmed to 37 °C immediately prior to transfusion. Whole blood is often used to replace blood lost due to injury or surgery. The number of units (one unit equals approximately 510 ml) administered depends upon the health and age of the recipient, along with the therapeutic indication. Administration of whole blood may also be undertaken to supply a recipient with a particular blood constituent (e.g. a clotting factor, immunoglobulin, platelets or red blood cells). However, this practice is minimized, in favour of direct administration of the specific blood constituent needed.

Associated with the administration of blood or blood products is the risk of accidental transmission of infectious agents such as hepatitis viruses or HIV. The prevention of accidental pathogen transmission relies upon:

- careful screening of all blood donors/donations;

- introduction of methods of pathogen removal/inactivation during the processing steps;

- careful screening of all finished products.

The identity of each blood donor should be recorded, and all donor blood bags must be labelled carefully. Traceability of individual blood donors/donations is essential, in case the donor or product is subsequently found to harbour blood-borne pathogens. The risk of contamination of blood during collection/processing is minimized by using closed systems and strict aseptic technique.

Before any blood donation is released for issue/processing, it must be tested for the presence of various pathogens particularly likely to be present in blood. In most countries, these tests include immunoassays capable of detecting:

- hepatitis B surface antigen (HBsAg);

- antibodies to HIV;

- antibodies to hepatitis C virus;

- syphilis antibodies.

However, no immunoassay is 100 per cent accurate and all will report a low number of false negatives (and false positives). It is believed that in the order of 1 in every 42 000 blood units reported to be HIV antibody-negative actually harbours the virus.

In general, processes capable of inactivating viral or other pathogens (e.g. heat or chemical treatment) may not be applied to whole blood or most blood-derived products. Thus, for whole blood at least, effective screening of donations is relied exclusively upon to prevent pathogen transmission. Many of the processing techniques used to derive blood products from whole blood (e.g. precipitation, but especially chromatographic purification) can be effective in separating viral or other pathogens from the final product (see also Chapters 6 and 7). Fractionated products, therefore, are less likely to harbour undetected pathogens.

In addition to being screened for likely pathogens, the ABO blood group and the Rh group is also determined. In the USA alone, in the region of 35 transfusion-related deaths occur annually due to errors in blood group typing or the presence of bacteria in the product.

Box 14.6

Product case study: Carticel

Carticel is a preparation of autologous cultured chondrocytes used in the treatment of symptomatic cartilage defects of the femoral condyle (the rounded protuberance at the end of the hip bone), caused by acute or repetitive trauma, in patients for whom surgical repair has proven inadequate. It is produced by Genzyme Biosurgery and gained approval for medical use in 1997. Each single-use product container contains approximately 12 million cells devoid of microbial contamination, in a final volume of 0.4 ml sterile, buffered cell culture media (Dulbecco's modified Eagle's medium; Chapter 5).

Chondrocytes are found embedded in the cartilage matrix that they originally produced. Cartilage itself is a dense connective tissue consisting mainly of proteoglycans (biomolecules composed of a protein backbone to which extensive carbohydrate side-chains are attached, such that the molecule is predominantly carbohydrate based). It is capable of withstanding considerable pressure and effectively acts as a shock absorber in joints. There are three types of cartilage: hyaline cartilage is found at bone joints, larynx, trachea, bronchi and nose; elastic cartilage is found in the external ear; and fibrocartilage is found in the intervertebral discs and in tendons.

Chondrocytes are initially harvested from the patient's own body and undergo *in vitro* expansion via cell culture (Chapter 5) in a medium containing foetal bovine serum. The cells are aseptically filled into vials and are tested for viability and microbial sterility. They are implanted by trained surgeons into the affected area via injection, using a catheter. Typically, between 1 million and 2 million cells are introduced back into the damaged joint area in this way. Both preclinical studies in rabbits, goats, dogs and horses and clinical studies in humans have illustrated the ability of the cultured chondrocytes to promote cartilage repair. However, a significant number of patients treated with Carticel require later surgical intervention, due to locking, clicking or painful joints that is believed to occur due to overgrowth of grafted tissue.

modification or formulation *in vitro* prior to their medical use. Examples include Carticel (Box 14.6) and Apligraf, a skin substitute used in the treatment of certain ulcers, which is composed of keratinocytes and fibroblasts derived from human neonatal foreskin tissue and bovine collagen.

14.10.1 Stem cells

The therapeutic application of stem cells has long been a dream of medical sciences, but recent discoveries and technical advances have brought this dream somewhat closer to being a reality. Stem cells are usually defined as undifferentiated cells capable of self-renewal that can differentiate into more than one specialized cell type.

Such cells are often classified on the basis of their original source as either embryonic or adult stem cells. As the name suggests, embryonic stem cells are derived from the early embryo, whereas adult stem cells are present in various tissues of the adult species. Much of the earlier work on embryonic stem cells was conducted using mouse embryos. Human embryonic stem cells were first isolated and cultured in the laboratory in 1998. Research on adult stem cells spans some four decades, with the discovery during the 1960s of haematopoietic stem cells in the bone marrow (Chapter 10). However, the exact distribution profile, role and ability to manipulate adult stem cells (particularly those outside of the bone marrow) are subjects of intense current research, and for which more questions remain than are answered.

Embryonic stem cells are derived from pre-implant-stage human embryos, usually at the blastocyst stage (the blastocyst is a thin-walled hollow structure containing a cluster of cells, known as the inner cell mass, from which the embryo arises). These embryos are invariably ones initially generated as part of *in vitro* fertilization procedures but which are destined to be discarded, either due to poor quality or because they are in excess to requirement. There are an estimated 400 000 *in vitro* fertilization-produced embryos in frozen storage in the USA alone, of which some 2.8 per cent are likely to be discarded.

Culture of human embryonic stem cells starts with the recovery of the blastocyst's inner cell mass (Figure 14.17). One common recovery procedure is termed 'immunosurgery'. The process

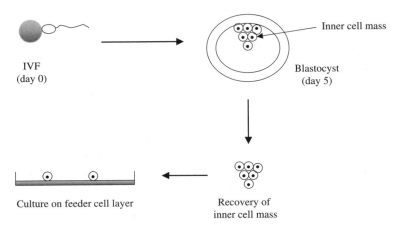

Figure 14.17 Overview of the generation and culture of human embryonic stem cells. IVF: *in vitro* fertilization. Refer to text for further details

entails the initial treatment of blastocysts with pronase (a cocktail of proteolytic enzymes), which effectively degrades the outer protective membrane known as the 'zona pellucida'. The blastocysts are next treated with anti-human whole serum antibody and guinea pig complement, which triggers complement-mediated lysis of the blastocyst outer cell layers (the trophoblast), allowing recovery of the inner cell mass. The latter cells are then cultured under defined conditions in order to allow them to multiply while remaining undifferentiated.

In addition to cell culture media, the culture vessels often contain a layer of 'feeder' cells (e.g. mouse fibroblasts), irradiated in order to prevent their growth and division. These feeder cells can serve two functions: (a) to provide a suitable substratum with which the embryonic stem cells can interact, aiding in their growth and division; (b) feeder cells can release often ill-defined nutrients into the medium, which can again support stem cell growth. The presence of a feeder cell layer would represent a complication in the downstream processing of stem cells for therapeutic use, and could represent a potential source of pathogenic contaminants. More recently, culture systems have been developed in which the feeder cell layer is replaced by fibronectin (a glycoprotein found on the cell surface) or matrigel (a protein-rich membrane extract from a mouse sarcoma cell line). Substantial research work remains ongoing in order to identify an optimal cell culture medium composition that will facilitate strong cell growth while remaining in an undifferentiated state. Basic animal cell culture media are often supplemented with serum as a nutrient source (Chapter 5). It is known that the addition of the cytokine LIF can sustain mouse embryonic stem cells in the undifferentiated state, but LIF alone cannot achieve this in the context of human embryonic stem cells. Research, therefore, continues with a view to optimize culture media composition for such human cell lines.

Whereas the culture of human embryonic stem cell lines requires the maintenance of cells in an undifferentiated state, the application of such cells in regenerative medicine requires the subsequent controlled differentiation of such cells to generate a specific desired cell type (e.g. a specific neuron type to treat a specific neurodegenerative disease, etc.). The process by which any stem cell differentiates naturally to form a specific cell is hugely complex and understood only in outline and only for a few cell types. Differentiation is dependent upon several concerted signals from effector molecules such as cytokines. A major challenge, therefore, is to gain a more complete understanding of how differentiation into specific cell types is driven and controlled. Only with such knowledge will come the ability to grow specific cells (and ultimately tissue/organ types) from stem cells for the purposes of regenerative medicine.

Although only in its infancy, some progress has been reported in elucidating details of selected directed differentiation pathways, initially in the context of mouse embryonic stem cells, but latterly also in the context of human embryonic stem cells (Figure 14.18). This progress has largely been the result of empirical studies and is largely achieved in one or more of three ways: (a) manipulation of culture media composition; (b) alteration of the surface characteristics of the matrix on which the cells are grown (e.g. adhesive feeder cells or specific protein-based matrices); (c) via introduction of specific regulatory genes into the stem cells themselves.

One example of a relatively recently elucidated pathway that directs differentiation of dopaminergic neurons is outlined in Figure 14.19. The ability to generate dopaminergic-like neurons represents a significant milestone in the attempt to apply regenerative medicine to the treatment of Parkinson's disease. This neurodegenerative condition, which effects some 2 per cent of adults over the age of 65, is triggered by the death of this cell type in the brain. Parkinson's disease, therefore, is likely to be one of the first clinical targets in the development of regenerative medicines.

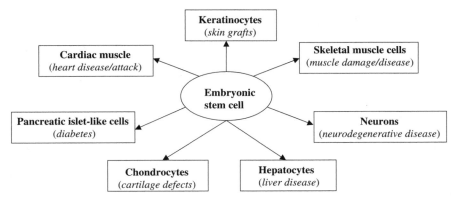

Figure 14.18 Some cell types reported to have been produced via *in vitro* directed differentiation from either mouse or human embryonic stem cells. Potential uses for such cell types in regenerative medicines are listed in italics

14.10.2 Adult stem cells

The main focus of stem cell research over the last few decades has been directed to embryonic stem cells. However, more recently, research upon and an understanding of various populations of adult stem cells has gathered pace. Adult stem cells are undifferentiated cells found amongst differentiated cells in a tissue or organ. These cells can renew themselves and can differentiate to yield the major cell types characteristic of the tissue in which they reside. The main physiological role of adult stem cells, therefore, appears to be to maintain and to repair (to a certain extent at least) the tissue in which they reside.

For many years it was believed that adult stem cell populations were present in a very limited number of tissue types, and that they could only differentiate into cells characteristic of the tissue in which they reside. Recent research challenges both of these assertions. Adult stem cells are being discovered in a growing number of tissues, including bone marrow, peripheral blood and

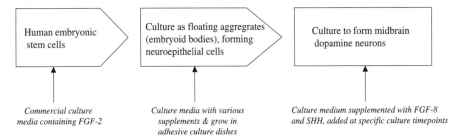

Figure 14.19 Simplified schematic overview of the directed differentiation of human embryonic stem cells to form differentiated dopamine like neurons. The full pathway details are available in Yan, Y., Yang, D., Zarnowska, E.D., Du, Z., Werbel, B., Valliere, C., *et al.* 2005. Directed differentiation of dopaminergic neuronal subtypes from human embryonic stem cells. *Stem Cells* **23**, 781–790. FGF: fibroblast growth factor; SHH: sonic hedgehog (a regulatory protein)

blood vessels, the brain and spinal cord, skeletal muscle, skin, liver, pancreas, digestive system, cornea and retina. Identification and study of such cells can be made difficult by the low levels in which they are normally found, the presence of many additional cell types and the difficulties in culturing them. Much basic research is required to answer fundamental questions regarding adult stem cells, including: How many types exist and where are they located? What was their ultimate source? What level of plasticity (see below) do they exhibit? What factors stimulate their relocation and differentiation at a site of tissue damage?

The use of adult, as opposed to embryonic, stem cells in regenerative medicine would have a number of significant advantages. It would overcome moral/ethical difficulties associated with blastocyst destruction. It would also allow for autologous transplantation of cells, i.e. adult stem cells could be harvested from a patient, cultured and differentiated *in vitro* and then reintroduced back into the patient. This would overcome potential immunological complications and a requirement to use immunosuppressive drugs. Hurdles to this approach not only include the difficulties in isolating and successfully culturing these cells, but also ascertaining the level of plasticity exhibited by adult stem cells. This refers to the range of potential fully differentiated cell types that could be produced from adult stem cell populations, and such investigations represent a very active area of current stem cell research.

14.11   Conclusion

Every few decades a medical innovation is perfected that profoundly influences the practice of medicine. Widespread vaccination against common infectious agents and the discovery of antibiotics serve as two such examples. Many scientists now believe that the potential of nucleic-acid- and cell-based technologies rivals even the most significant medical advances achieved to date.

It is now just over a decade since the first nucleic-acid-based drugs began initial tests. Several such drugs will likely be in routine medical use in less than a decade more. The application of gene technology could also change utterly the profile of biopharmaceutical drugs currently on the market. Virtually all such products are proteins, currently administered to patients for short or prolonged periods, as appropriate. Gene therapy offers the possibility of equipping the patient's own body with the ability to synthesize these drugs itself, and over whatever time-scale is appropriate. Taken to its logical conclusion, gene therapy thus offers the potential to render obsolete most of the biopharmaceutical products currently on the market. Regenerative medicine, too, although still in its infancy, harbours enormous future medical potential. Of all the biopharmaceuticals discussed throughout this text, nucleic-acid- and cell-based drugs may well turn out to have the most profound influence on the future practice of molecular medicine.

Further reading

Books

Anonymous. 2001. *Stem Cells: Scientific Progress and Future Research Directions*. Department of Health and Human Services. Available at http://www.stemcells.nih.gov. (Note: this website also provides a number of additional reference materials on the subject of stem cells.)

Kresina, T. (ed.). 2001. *An Introduction to Molecular Medicine and Gene Therapy, Parts I and II*. Wiley-Liss

Blankenstein, T. (Ed.) 1999. *Gene Therapy, Principles and Applications.* Birkhauser Verlag.

Lowrie, D. 1999. *DNA Vaccines.* Humana Press.

Phillips, M. 2000. *Antisense Technology.* Methods in Enzymology, volume 313. Academic Press.

Crooke, S. (Ed.) 2001. *Antisense Drug Technology.* Marcel Dekker.

Stein, C. and Krieg, A.1998. *Applied Antisense Oligonucleotide Technology.* Wiley.

Articles

Gene therapy

Buchschacher, G. and Wong-Staal, F. 2001. Approaches to gene therapy for human immunodeficiency virus infection. *Human Gene Therapy* **12**(9), 1013–1019.

Davies, J.C., Geddes, D.M., and Alton, E.W. 2001. Gene therapy for cystic fibrosis. *Journal of Gene Medicine* **3**(5), 409–417.

Demeterco, C. and Levine, F. 2001. Gene therapy for diabetes. *Frontiers in Bioscience* **6**, D175–D191.

Edelstein, M.L., Abedi, M.R., Wixon, J., and Edelstein, R.M. 2004. Gene therapy clinical trials worldwide 1989–2004 – an overview. *Journal of Gene Medicine* **6**(6), 597–602.

Eliyahu, H., Barenholz, Y., and Domb, A.J. 2005. Polymers for DNA delivery. *Molecules* **10**(1), 34–64.

Ferguson, C., Larochelle, A., and Dunbar, C.E. 2005. Hematopoietic stem cell gene therapy: dead or alive? *Trends in Biotechnology* **23**(12), 589–597.

Ferreira, G.N., Monteiro, G.A., Prazeres, D.M., and Cabral, J.M. 2000. Downstream processing of plasmid DNA for gene therapy and DNA vaccine applications. *Trends in Biotechnology* **18**(9), 380–388.

Furlan, R., Butti, E., Pluchino, S., and Martino, G. 2004. Gene therapy for autoimmune diseases. *Current Opinion in Molecular Therapeutics* **6**(5), 525–536.

Gardlik, R., Palffy, R., Hodosy, J., Lukacs, J., Turna, J., and Celec, P. 2005. Vectors and delivery systems in gene therapy. *Medical Science Monitor* **11**(4), RA110–RA121.

Lewin, A. and Hauswirth, W. 2001. Ribozyme gene therapy: applications for molecular medicine. *Trends in Molecular Medicine* **7**(5), 221–228.

Mhashilkar A, Chada S, Roth JA, and Ramesh R. 2001. Gene therapy – therapeutic approaches and implications. *Biotechnology Advances* **19**(4), 279–297.

Ohlfest, J.R., Freese, A.B., and Largaespada, D.A. 2005. Nonviral vectors for cancer gene therapy: prospects for integrating vectors and combination therapies. *Current Gene Therapy* **5**(6), 629–641.

Palmer, D.H., Young, L.S., and Mautner, V. 2006. Cancer gene therapy: clinical trials. *Trends in Biotechnology* **24**(2), 76–82.

Phillips, A. 2001. The challenge of gene therapy and DNA delivery. *Journal of Pharmacy and pharmacology.* **53**(9), 1169–1174.

Pfeifer, A. and Verma, I. 2001. Gene therapy: promises and problems. *Annual review of genomics and Human Genetics* **2**, 177–211.

Robertson, J. and Griffiths, E. 2001. Assuring the quality, safety and efficacy of DNA vaccines. *Molecular Biotechnology* **17**(2), 143–149.

Schatzlein, A. 2001. Non-viral vectors in cancer gene therapy: principles and progress. *Anti-Cancer Drugs* **12**(4), 275–304.

Smith, H. and Klinman, D. 2001. The regulation of DNA vaccines. *Current Opinion in Biotechnology* **12**(3), 299–303.

Verma, I. and Weitzman, M. 2005. Gene therapy: twenty first century medicine. *Annual Review of Biochemistry* **74**, 711–738.

Wu, N. and Ataai, M. 2000. Production of viral vectors for gene therapy applications. *Current Opinion in Biotechnology* **11**(2), 205–208.

Yechoor, V and Chan, L. 2005. Gene therapy progress and prospects: gene therapy for diabetes mellitus. *Gene Therapy* **12**(2), 101–107.

Young, L.S., Searle, P.F., Onion, D., and Mautner, V. 2006. Viral gene therapy strategies: from basic science to clinical application. *Journal of Pathology* **208**(2), 299–318.

Antisense technology

Aboul-Fadt, T. 2005. Antisense oligonucleotides: the state of the art. *Current Medicinal Chemistry* **12**(19), 2193–2214.

Adah, S.A., Bayly, S.F., Cramer, H., Silverman, R.H., and Torrence, P.F. 2001. Chemistry and biochemistry of 2′, 5′-oligoadenylate-based antisense strategy. *Current Medicinal Chemistry* **8**(10), 1189–1212.

Akhtar, S., Hughes, M.D., Khan, A., Bibby, M., Hussain, M., Nawaz, Q., Double, J., and Sayyed, P. 2000. The delivery of antisense therapeutics. *Advanced Drug Delivery Reviews* **44**(1), 3–21.

Galderisi, U., Cipollaro, M., and Cascino, A. 2001. Antisense oligonucleotides as drugs for HIV treatment. *Expert Opinion on Therapeutic Patents* **11**(10), 1605–1611.

Hughes, M.D., Hussain, M., Nawaz, Q., Sayyed, P., and Akhtar, S. 2001. The cellular delivery of antisense oligonucleotides and ribozymes. *Drug Discovery Today* **6**(6), 303–315.

Jason, T.L.H., Koropatnick, J., and Berg, R.W. 2004. Toxicology of antisense therapeutics. *Toxicology and Applied Pharmacology* **201**(1), 66–83.

Lebedeva, I. and Stein, C. 2001. Antisense oligonucleotides: promise and reality. *Annual Review of Pharmacology and Toxicology* **41**, 403–419.

Rubenstein, M., Tsui, P., and Guinan, P. 2004. A review of antisense oligonucleotides in the treatment of human disease. *Drugs of the Future* **29**(9), 893–909.

Stein, C.A., Benimetskaya, L., and Mani, S. 2005. Antisense strategies for oncogene inactivation. *Seminars in Oncology* **32**(6), 563–572.

Vidal L., Blagden, S., Attard, G., and de Bono, J. 2005. Making sense of antisense. *European Journal of Cancer* **41**(18), 2812–2818.

RNA interference

Amarzguioui, M., Rossi, J.J., and Kim, D. 2005. Approaches for chemically synthesized siRNA and vector-mediated RNAi. *FEBS Letters* **579**(26), 5974–5981.

Bagasra, O. 2005. RNAi as an antiviral therapy. *Expert Opinion on Biological Therapy* **5**(11), 1463–1474.

Campbell, T. and Choy, F. 2005. RNA interference: past, present and future. *Current Issues in Molecular Biology* **7**, 1–6.

Caplen, N. 2003. RNAi as a gene therapy approach. *Expert Opinion on Biological Therapy* **3**(4), 575–568.

Chowdhury, D. and Novina, C. 2005. RNAi and RNA-based regulation of immune system function. *Advances in Immunology* **88**(88), 267–292.

Pai, S.I., Lin, Y.Y., Macaes, B., Meneshian, A., Hung, C.F., and Wu, T.C. 2006. Prospect of RNA interference therapy for cancer. *Gene Therapy* **13**(6), 464–477.

Uprichard, S. 2005. The therapeutic potential of RNA interference. *FEBS Letters* **579**(26), 5996–6007.

Ribozymes

Alvarez-Salas, L.M., Benitez-Hess, M.L., and DiPaolo, J.A. 2003. Advances in the development of ribozymes and antisense oligodeoxynucleotides as antiviral agents for human papillomaviruses. *Antiviral Therapy* **8**(4), 265–278.

Bagheri, S and Kashani-Sabet, M. 2004. Ribozymes in the age of molecular therapeutics. *Current Molecular Medicine* **4**(5), 489–506.

Lilley, D. 2005. Structure, folding and mechanisms of ribozymes. *Current Opinion in Structural Biology* **15**(3), 313–323.

Kahn, A. and Lai, S. 2003. Ribozymes: a modern tool in medicine. *Journal of Biomedical Science* **10**(5), 457–467.

Kashani-Sabet, M. 2004. Non-viral delivery of ribozymes for cancer gene therapy. *Expert Opinion on Biological Therapy* **4**(11), 1749–1755.

Sioud, M. and Iversen, P. 2005. Ribozymes, DNAzymes and small interfering RNAs as therapeutics. *Current Drug Targets* **6**(6), 647–653.

Aptamers

Hoppe-Seyler, F., Crnkovic-Mertens, I., Tomai, E., and Butz, K. 2004. Peptide aptamers: specific inhibitors of protein function. *Current Molecular Medicine* **4**(5), 529–583.

Nimjee, S.M., Rusconi, CP., and Sullenger, B.A. 2005. Aptamers: an emerging class of therapeutics. *Annual Review of Medicine* **56**, 555–583.

Proske, D., Blank, M., Buhmann, R., and Resch, A. 2005. Aptamers – basic research, drug development, and clinical applications. *Applied Microbiology and Biotechnology* **69**(4), 367–374.

Yan, A.C., Bell, K.M., Breeden, M.M., and Ellington, A.D. 2005. Aptamers: prospects in therapeutics and biomedicine. *Frontiers in Bioscience* **10**, 1802–1827.

You, KM., Lee, S.H., Im, A., and Lee, S.B. 2003. Aptamers as functional nucleic acids: *in vitro* selection and biotechnological applications. *Biotechnology and Bioprocess Engineering* **8**(2), 64–75.

Stem cells

Hoffman, L. and Carpenter, M. 2005. Characterization and culture of human embryonic stem cells. *Nature Biotechnology* **23**(6), 699–708.

Mayhall, E., Paffett-Lugassy, N., and Zon, L. 2004. The clinical potential of stem cells. *Current Opinion in Cell Biology* **16**, 713–720.

Wagers, A. and Weissman, I. 2004. Plasticity of adult stem cells. *Cell* **116**, 639–648.

Index

Note: Figures and Tables are indicated by *italic page numbers*, Boxes by **emboldened numbers**

Pharmaceutical biotechnology: concepts and applications   Gary Walsh
© 2007 John Wiley & Sons, Ltd   ISBN 978 0 470 01244 4 (HB)   978 0 470 01245 1 (PB)